PLANTS AND ANIMALS IN THE YOEME WORLD

ETHNOECOLOGY OF THE YAQUIS OF SONORA AND ARIZONA

Richard Stephen Felger
Felipe Silvestre Molina

Foreword by Exequiel Ezcurra

2024

Library of Congress cataloging-in-publication data

Felger, Richard Stephen.
Plants & Animals in the Yoeme World: Ethnoecology of the Yaquis of Sonora and Arizona.
Richard Stephen Felger and Felipe Silvestre Molina; foreword by Exequiel Ezcurra.
Includes bibliographic references and appendices.
ISBN: 979-8-218-37111-1

This book is dedicated to the people who worked with us, to Richard's wife Silke Schneider, and to Rosario Vakame'eri Castillo and Anselma Tonopuame'a Castillo who inspired us.

-Richard and Felipe

CONTENTS

FOREWARD

On the 4th of October of 1533, eleven years after the fall of the Aztec Empire to the hands of the Spanish conquistador Hernán Cortés, Captain Diego de Guzmán with a company of Spanish soldiers arrived at the banks of the Yaqui River. The purpose of their incursion was taking natives prisoner to force them into slave labor in the rapidly-growing mining endeavors in the *Altiplano*, the highlands of Zacatecas, Durango, and San Luis Potosí. The military incursion was forced to retreat by the *Yoeme* (Yaqui) warriors. The skirmish was the first clash of the legendary Yaqui Wars that went on for the next four centuries.

Two years later, in 1535, the Spanish explorer Álvar Núñez Cabeza de Vaca wandered into central and southern Sonora, and was amazed by the fertility of the soil, the sheer beauty of the land, the abundance of water, and the vitality of the rivers. His written narrative describes the rich harvests of maize, beans, squash, and cotton, the quality of the permanent houses and the cotton clothing worn by the local inhabitants, and the immense respect with which women were treated in this desert-river society. Alas, Cabeza de Vaca soon realized that the Spaniards had been organizing incursions into this generous land, pillaging villages and taking slaves. His narrative described, with deep sadness, how parts of the land had become "unpopulated because its inhabitants were fleeing into the mountains, not daring to have houses, nor to farm, from fear of the Christians." Indeed, the people of the rich coastal valleys, the farmers of the Yaqui and Mayo rivers, were finding refuge hidden between mountain ridges and organizing themselves in a fierce defense against Spanish incursions. Apart from slave labor, the sheer richness of the land was a powerful magnet for Spanish greed. "This land," Cabeza de Vaca wrote, "is without doubt the best of all that there are in the Indies and most fertile and abounding in supplies, and they plant three times a year. They have many fruits and very beautiful rivers and many other very fine waters."

Over the course of 400 years, the Spanish and Mexicans repeatedly launched military campaigns into the Yoeme territory, which resulted in several serious battles and massacres, and forced a significant proportion of the Yoeme nation into exile. Some level of peaceful coexistence was achieved with the arrival of the Jesuit priests Andrés Pérez de Ribas and Tomás Basilio in 1617, without military escort. They brought with them new agricultural practices, European cereals, and domestic animals that blended with the Yoeme traditional use of the land in a sort of ecological syncretism. The Jesuit Arcadia lasted 120 years but began unraveling in the later years. As Felger and Molina describe in detail, growing tensions put the Jesuits in an increasingly difficult position, and in 1767 a decree by King Charles III of Spain ordered the expulsion of all Jesuits from the Americas.

First the Spanish colonial government, and later the Mexican federal authorities, drove thousands of Yoemem into exile. Some went to work in the mines; many others were brutally enslaved in the *henequén* plantations of the Yucatán Peninsula. Many of the bountiful lands around the Yaqui River valley were appropriated by *yorim*, the non-Yoeme people, the foreign invaders. A system of modern industrial agriculture replaced the ancient riverside fields of maize, beans, squash, and cotton with industrialized, high-input monocultures of wheat, rice, feed-plants, and other water-squandering crops that leach immense amounts of pesticides and nitrogen pollutants into the local rivers and coastal lagoons. Many historians consider the Yaqui Wars to have formally ended in 1937, when President Lázaro Cárdenas reserved for the Yoeme people 500,000 hectares of ancestral lands on the north bank of the river, ordered the construction of a dam to provide irrigation water for them, agricultural equipment, and water pumps. Social conflicts have continued to this day, however, around the access to their ancestral lands and the appropriation of the Yaqui River's water to supply Sonora's growing urban population.

It is remarkable —in the context of the Yoeme history of hardships, war, and plunder— that the Yoeme culture has survived and thrived through these long centuries of attrition. Although the deep knowledge that the Yoemem have about their environment, their plants and their animals has been fragmented, it has persisted nonetheless. The cultural survival of the Yoeme nation is extraordinary, and this is what this book is all about. Felipe Silvestre Molina, one of the two authors and a Yoeme himself, describes in his preface how the Yoeme nation as a whole, and especially the elders, have been tenaciously determined to carry on the Yoeme culture from one generation to the next that even war and exile did not stop them from teaching or learning the Yoeme way of life.

This book shows that, despite vicissitudes and challenges, the history of the Yoemem, the Yaqui people, is not one of decline and hopelessness, but rather a hopeful effort of cultural survival. Even after centuries of aggression, the Yaquis of Sonora and Arizona maintain an exceptionally sharp and vivid knowledge of their natural world. In this book, Richard Felger and Felipe Molina show us how the Yoeme vision of the world, their extraordinary sense of place, is vibrantly alive and plays a vital role in their cultural survival. From generation to generation, surrounded by the growth of unsustainable monocultures supported by pesticides and polluting nitrogen fertilizers, they have maintained their ancient, rich lore of native plants and wild animals, which they still celebrate with poetry and songs, and still gather from the surrounding wild ecosystems.

This book compiles in a masterful way the biological ethnography of the Yaqui people in both their ancestral lands in Sonora and their diaspora communities in Arizona. The traditional Yaqui authorities are still demanding, with a firm and growing voice, the restitution of the lands and waters that were taken away from them. And here lies the importance of this book: As long the Yoeme people keep singing to their wild plants and animals, keep gathering fruits from their traditional grounds, keep celebrating their ancestral biodiversity, there are still reasons for hope in environmental justice and in a more sustainable world.

In this extraordinary volume, Richard Felger and Felipe Molina document brilliantly the endurance of an ancient way of life and the survival of a unique vision of the world, not only as a celebration of the past but also as a living tribute to all the Yoemem that built and maintain this legacy. The Yoem Lutu'uria —Yoeme Truth— is a celebration of the present and the keystone for a hopeful future for this nation and their extraordinary land.

Exequiel Ezcurra

PREFACE

Richard Stephen Felger

My first trip to the place I eventually learned is the Yoem Bwiara was with my high school biology teacher and her colleagues. It was year-end vacation and we were driving from Los Angeles to Álamos, Sonora. It was thrilling to see so many new cactus and other plants. Giant, leafless morning glory trees crowned in white flowers north of Hermosillo were just one of many wonders. The main west coast highway was paved only as far south as the vicinity of Empalme. Farther south we passed through a dusty town of low adobes and a crumbling fort. I learned it was a Yaqui town and the military presence was left over from years of warfare. At a long lunch break beneath giant cottonwoods on the right bank of the Yaqui River, I noticed new leaves already unfolding as the old ones were only beginning to fall away. I was waking to being at the edge of the tropics.

I had my first plant press and was making collections for later identifications. I had Howard Scott Gentry's Rio Mayo Plants, already well worn. Gentry's book held intriguing native uses of plants—an entry into the world of ethnobotany.

I went to the University of Arizona because it was in the desert and the closest I could get to the tropical deciduous forest of southern Sonora. On various field trips, especially long summertime ones, I started learning indigenous names and lore of plants and animals. Some years later, while working on the botany of Gulf of California islands, I would ask Seris questions such as where a certain plant or animal could be found. I soon became aware of their immense knowledge of the natural world. I met Ed and Becky Moser in El Desemboque, and Becky had a list of medicinal plants. I suggested we write it up, maybe three weeks of effort. Twenty-five years later we finalized the People of the Desert and Sea, Ethnobotany of the Seri Indians.

After meeting Felipe Molina, it seemed only natural to embark on a similar work with Yaquis. I already had the privilege of some learning from traditional Yaquis. I was surprised to witness the differences in working with Yaquis and Seris, agriculturalists with rich farmland in a vast riverine valley neighboring sea-faring desert dwellers in a land too dry for agriculture.

Felipe and I made several extensive fieldtrips to the Yaqui (Yoeme) homeland in the late 1980s, together with photographer Bill Steen. We stayed in Pótam with Felipe's relatives. The results became the foundation of the present book. We made numerous additional fieldtrips, together and separately, as well as working with Yoemem in Arizona. We also mined the treasure trove of recorded information, especially the early seventeenth century Jesuit literature. Due to circumstances we put aside our Yoeme collaborations. Eventually we returned to spend a number of years working on our mutual interest of

Richard and Felipe, having breakfast at the home of Antonia and Nacho Amarillas in Pótam. 15 Dec 1988 (WS).

the Yoeme world. I am pleased that we finalized our work.

I emphasize that after all these years, there remains considerable Yoeme knowledge of the natural world that is not included in this book, both among community members and the archival and published records. I hope others will follow us.

Felipe Sylvester Molina

Our ancestors were from Wiivisim and Veenem and were farmers and fishermen. For that reason my maternal grandparents were so knowledgeable about the life in the Yoem Bwiara. Many of our elders were so determined to carry on the Yoeme culture that even war and other obstacles did not stop them from teaching or learning the Yoeme way of life. I am glad that I met non-Yoeme people that were interested in our culture and history. I have collaborated with the following people who will help carry on our Yoeme understanding of life on this planet: Dr. Larry Evers; Dr. Richard Felger; Dr. David Shorter; and Dr. David Shaul. I would say that many of our elders valued the tradition of sharing knowledge. In this way, I was raised in my extended family and community members in different Yoeme communities of Arizona and Sonora. I am forever grateful to each of these people. What our elders taught us is called the Yoem Lutu'uria (Yoeme Truth). Today, this truth continues among some traditional Yoemem.

Takalaim (center background) near San Carlos. 1996 (FM).

ACKNOWLEDGEMENTS

Richard extends deep appreciation to Silke Schneider for continued encouragement on this project.

We extend gratitude to the following individuals who helped in many ways, including but not limited to fieldwork, research and expertise, illustrations and photographs, review, technical support, and general advice and encouragement. We thank the many Yoemem acknowledged in Consultants and Teachers in Part 1.

Kelly W. Allred. Botanical help including grasses, etymology, literature, and more.

James Aronson. Fellow traveler and friend in fieldwork who shared information on plants, economic botany, and ecology.

Daniel F. Austin. Contributed information on the morning glory family, bibliographic references, and ethnobotany.

Chris Baisin. Assistance with Sonora fieldwork during early years of our research.

Marc André Baker. Botanical information and reviews, including cacti and especially *Cylindropuntia*.

James Anthony Bauml. Computer assistance in the early years of the research, and comparisons with other cultures in northwestern Mexico.

David Bruce Berry. Provided essential computer help, making it possible to finalize our work.

Susan M. Berry, Ann Lane Hedlund, Bonnie Buckley Maldonado, and **Sharman Apt Russell**: Silver City writers' group friends gave encouragement and feedback for writings.

Hanna Blood. Logistics, botanical specimens, and help with presentations for the Yoeme project at professional conferences.

Peter Blystone. Interviews, filming, photos, helped with presentations, and essential fieldwork in Sonora.

Michael Thomas Bogan. Photos, information for sites in Sonora, aquatic invertebrates, and wetlands.

Adrienne Booth. Copy editing, archiving text and data. Adrienne helped write and administered our 2018–2019 grant from the Christensen Fund.

Tomas Bowen. Friend, mentor, and advisor on numerous projects. His expertise and patience in responding to questions, editing, and references is gratefully acknowledged.

David Earl Brown. Ecologist at Arizona State University, he provided data for wildlife, conservation, and Sonoran ecology.

Bill Broyles and **Joan Scott**. Hospitality and transportation while working at University of Arizona and in Ajo, Arizona; as well as fieldwork, literature, and more.

Richard C. Brusca. Authority for Gulf of California invertebrates, Rick provided identifications and reviews.

Alberto Búrquez-Montijo. Friendship and ecological, botanical, and geographic information, Spanish translations, reviews, and fieldwork. Thanks to Alberto and Angelina Martínez for hospitality in Hermosillo

Susan Davis Carnahan. Extensive help across the scope of our Yoeme project including plant life, avifauna, editing, and literature. Co-author with Richard and Jesús Sánchez for *The Desert Edge: Flora of the Guaymas–Yaqui Region of Sonora,* Mexico.

Aaron Cooper. Director of International Sonoran Desert Alliance, he administered the Christensen Fund grant 2018–2019.

Thomas F. Daniel. Information for northwestern Mexico plants, especially Acanthaceae.

Mark Allan Dimmitt. Fieldwork and extensive information on plants and animals.

Larry Evers. We thank Larry for encouragement with our project.

Exequiel Ezcurra. Geographic, ecologic, and historical information, reviews, and Spanish translations, and hospitality.

George McNeil Ferguson. Support with collections, herbarium data, identifications, and research at the University of Arizona herbarium.

Walter Frank Fertig. Generous copy editing and reviews including plant life and avifauna.

Lloyd T. Findley. Information and review of Gulf of California fishes, and hospitality in San Carlos.

Mark Fishbein. Assistance with plant-related data, especially Apocynaceae.

Aaron David Flesch. Contributed substantial ecological and bird life information and reviews. Our modest coverage of bird life would not have been possible without Aaron's generous assistance.

Juan-Pablo Gallo-Reynoso. Fieldwork, Sonoran vertebrates, local names, geography, photos, and literature.

Edward Erik Gilbert. Sonoran fieldwork and substantial help relating to SEINet.

Powell (Gil) B. Gillenwater III. Friendship and fieldwork, including travel in Sonora and hospitality in Phoenix.

Jesús Armando Haro. Assistance with cultural and geographic information for northwestern Mexico.

Philip Alan Hastings. Identifications, information, and reviews for Gulf of California fishes.

Trica Oshant Hawkins. Generous reviewing and checking the text, and constructive suggestions.

Ann Lane Hedlund. Helped with organizing the text, cultural methodology, literature, and myriad questions relating to manuscript preparation.

James Henrickson. Long term botanical help, sharing unpublished work, and hospitality when visiting University of Texas, Austin.

Osvel Hinojosa-Huerta. Review and information on Sonoran avifauna.

Wendy Caye Hodgson. Reviews and information including agaves, food plants, and herbarium records at the Desert Botanic Garden.

Sky Jacobs. Material for Sonoran plants, animals, and places, and mapping data.

Andrea Jacques. Librarian at Western New Mexico University, she obtained interlibrary book loans and located references.

Matthew Brian Johnson. Shared data and illustrations, especially desert legumes.

Richard (Rick) Alan Johnson. Computer help over a number of years.Tomas Michael Kolaz. Archival literature and sources of photos, and generous encouragement and sharing of published and unpublished information on Yoeme and Yoreme culture. Tom participated our initial grant from the Smithsonian Institute.

Cathy Moser Marlett. Identifications, data, and review for Gulf of California molluscs, and Sonoran history and geography.

Stephen Marlett. Linguistic information, translations, and reviews.

Angelina Martínez-Yrízar. Vegetation and ecology of the Sonoran Desert, and hospitality in Hermosillo.

Kathryn Mauz. Provided extensive copy editing, information, and text organization. Her far-reaching scholarship greatly improved the book.

Michelle Mary McMahon. Support with collections, herbarium data, identifications, and research at the University of Arizona herbarium.

Raquel Padilla-Ramos. Yoeme historic information; faculty member Instituto Nacional de Antropología e Historia, Centro INAH Sonora. Her murder left us in sorrow.

Ronald J. Parry. Identifications, data, and references for insects.

William R. Radke. Bill generously provided photos and information of freshwater fish.

Amadeo M. Rea. Amadeo provided reviews and shared knowledge of traditional regional cultures, ethnobiology, linguistics, and animal life, especially avifauna.

Jon P. Rebman. Over the years Jon has provided botanical information, especially relating to the Baja California Peninsula.

Frank W. Reichenbacher. Our summertime fieldwork contributing to documenting Yoeme region flora.

Ana Lilia Reina-Guerrero. Sonoran plant collections, information, and Spanish translations.

Arlene Grace Ripley. Arlene contributed a significant number of photos and information of animal life.

James Douglas Ripley. Doug hosted Felipe and Richard for Arizona Native Plant Society conferences and helped with production of published articles.

Iris Erika Rodden. Expert computer and other technical help, friendship and encouragement.

James C. Rorabaugh. Jim contributed numerous photos, especially reptiles and amphibians, and information on herpetofauna, as well as editing and reviews.

Phillip Clark Rosen. Shared information and provided reviews for Sonoran Desert herpetofauna, as well as photos of animal and plant life.

Alexander "Ike" Russell. Richard's friend and mentor. We went just about anywhere in Sonora in his airplanes and to the Gulf of California in his panga-style boat. Bowen (2002) and Felger (2002).

Jean Straub Russell. Jean often accompanied Richard in fieldwork in Sonora. We would camp just about anywhere in Sonora. One day found Condominios Pilar rising at our favorite campsite next to Estero Soldado at San Carlos. Jean and Ike purchased a studio to store the camping gear. Times changed and we stayed inside the condominium.

Susan Rutman. Finalized this manuscript for publication; provided photos and hospitality; conducted fieldwork; and co-authored publications including the Flora of Southwestern Arizona.

Francisco Saavedra. He shared knowledge of plant uses, especially for medicine, with Richard in 2001. Francisco was living on an *ejido* north of Guaymas and worked part-time as a cowboy. Although not a Yoeme, he had extensive knowledge of plant uses.

Andrew Michael Salywon. Records from the Desert Botanical Gardens herbarium, information on agaves and more, and host when working at the garden and Phoenix area conferences.

José Jesús Sánchez-Escalante. Fieldwork, floristics, photos, Sonoran geography, plant collections including critical records from the Yoem Bwiara, and co-author with Richard and Sue Carnahan of *The Desert Edge: Flora of the Guaymas–Yaqui Region of Sonora*, Mexico.

Andrew C. Sanders. Botanical expertise, identifications, and reviews, especially for northwestern Mexico.

Justin Schmidt. Biology, toxicity, and identifications of terrestrial invertebrates.

Anna Seiferle-Valencia. For comprehensive reviewing and editing of the final drafts, and guiding our book to publication.

Andrew Semotiuk. Reviews and information for Yoreme and Yoeme ethnobotany, geography, literature, and linguistic and cultural data.

David Delgado Shorter. Shared photos, literature sources and review of Yoeme culture from his fieldwork and archives.

Curtis Latham Smith. Contributed photos and enduring patience with the accompanying botanical work by his wife Sue Carnahan and Richard.

Richard W. Spellenberg. Botanical information and reviews, especially for Nyctaginaceae.

Edward Holland Spicer. Richard writes: Early in my career Professor Spicer encouraged my research and shared information on Sonora and Yaqui culture.

Edward ("Ned") Holland Spicer (1906–1983). Felipe writes: I remember the time when my grandfather introduced me to Dr. Spicer. My grandfather said that Dr. Spicer could speak *Yoem noki* better than some Yoemem. I thought that was amazing. I met Dr. Spicer in New Pascua in 1976. I remember he was wearing a straw hat and was quite happy to meet me. I guess grandfather really wanted me to meet him. Many years later he invited me to his house and gave me some photographs of Marana and some of my family. Mrs. Spicer is the one that took those photographs in 1936 and 1937 in *Ili Huu'upa* (Little Mesquite) settlement and the *Kampo Wiilo* (Skinny Camp) settlement. Kampo Wiilo was named after Dolores Garcia because he was tall and thin. He played the flute and drum for the pahko'ola dancers.

Richard writes: University of Arizona Professor Ned Spicer was an important and generous mentor early in my career. He encouraged me to pursue my interests. I remember our meaningful conversations about Sonoran geography, people, and places.

William B. Steen. Accompanied us with fieldwork in the Yoem Bwiara and supplied photos of our work. Bill contributed details for agave food preparation and Sonoran geography.

Victor Werner Steinmann. Plant collections and floristic data, in particular Euphorbiaceae.

Bryan Stevens. Generous sharing of information including Yoeme masks, ritual, and history, literature, and text editing.

Barbara Straub. Fieldwork, hospitality in Tucson and Condiminios Pilar in San Carlos.

Gordon C. Tucker. Plant identification, taxonomic expertise, and sharing unpublished works.

Raymond Marriner Turner. Admired mentor, he encouraged Richard's early research, and shared information on desert plants and ecology.

Thomas R. Van Devender. Over the years Tom has generously shared a wide range of biological and ecological information, especially pertaining to Sonora.

Sula E. Vanderplank. Helped with herbarium work, fieldwork, hospitality, and friendship.

James Thomas Verrier. Jim contributed extensive information and reviews of the flora, plant uses, identifications of insect life, and assisted with herbarium work. Jim and Iris Rodden extended hospitality for Richard in Tucson.

Robert Anthony Villa. Information on Sonoran herpetofauna, plants, Yoeme and Yoreme culture, photos, literature, and edited text for animal life. He also translated Spanish language works and provided extensive and critical reviews.

Benjamin Theodore Wilder. Fieldwork, editor, colleague, friend and co-author on related botanical and ecological works. We thank Ben as editor and publisher of this book.

Michael Francis Wilson. Fieldwork and extended assistance with early phases of the Yoeme project. Information on insect life including saturnid moths and plant collections.

George Alfred Yatskievych. Reviews and significant botanical information including related botanical publications and herbarium data.

We thank the many organizations who contributed to this publication and have provided significant, public access information. These include the following:

Arizona State Museum, Tucson, for access to archives, collections, and library.

Christensen Fund, 2018–2019, grant for completion of the project.

International Sonoran Desert Alliance administered the Christensen grant.

SEINet Portal Network, for extensive information on plants specimens and records.

Smithsonian Institution, grant for initial research for the Yoeme project.

Tropicos.org. Missouri Botanical Garden. 2020 [continuously updated].

University of Arizona 24/7 IT Support Center for technical assistance.

The University of Arizona Herbarium and **Herbario de la Universidad de Sonora** have been major resources.

Wallace Research Foundation provided generous support for a number of years of Richard's earlier research.

We thank the staff of these and other herbaria for loans, specimen images, and information. See Part 3 for further herbarium-related acknowledgments.

PART 1
THE PEOPLE, SETTING, AND HISTORY

INTRODUCTION

This is a book about plants and animals in the Yoeme world, including the Yoem *Bwiara* in Sonora, Mexico, and that region of south-central Arizona where Yoeme communities formed during the diasporas of the nineteenth and twentieth centuries (Figure 1). We document more than 415 plant species and over 600 kinds (taxa) of animal life, and describe from historical and first-person accounts many of their relationships of the *Yoemem* and the ecology of the region. More extensive coverage of the plant life and the region is presented in a companion work, *The Desert Edge: Flora of the Guaymas-Yaqui Region of Sonora, Mexico* (Felger et al. 2020). We present an overview of plants and animals in the Yoeme world, and especially those in the Yoem Bwiara, including ones not known to have to have a specific use.

Yoeme (*Yoemem*, plural) is the people's own name for themselves, meaning "the People," while *Yaqui* is the name used by the general public as well as the official designation by the Mexican and United States governments for the people and their language. We mostly use Yoeme or Yoemem, rather than Yaqui or Yaquis, although the terms are interchangeable. We also choose to use Yoem Bwiara instead of Yaqui homeland, or Yoeme lands or territory in Sonora.

Yoem Bwiara is the term for the sacred, traditional land of the Yoemem centered along the lower Río Yaqui, or *Hiak Vatwe*, in Sonora, Mexico. This introduction provides a brief overview of the Yoem Bwiara and Yoeme history and culture, which helps to understand Yoeme knowledge of plants and animals. The cultural history is hugely complicated, and we offer only selected highlights, emphasizing some older, traditional insights. References to plants and animals from literature and archival material on the Yoemem are incorporated throughout this book. Although documentation of Yoeme culture is the most extensive of any cultural group in northern Mexico, those works have not been written or recorded by botanists or other biologists, so there is a substantial lack of information on plants and animals in the Yoeme world. There are often errors and amusing naiveté in available identifications, translations, and information for the biological world. For example, the works of Carlos Castañeda, which have been associated with the Yaquis, are fiction (Spicer 1969; Kelley 1978; Fikes 1993).

Yoem Lutu'uria, Yoeme Truth, is a core belief that includes respect for all life, animals, plants, and sea life, as well as water, sea, rocks, clouds, rain, and wind. Yoem Lutu'uria includes the *Aniam*, the spiritual worlds or realms; they are all connected. Evers and Molina (1987:18) wrote, "Yaquis have always believed that a close communication exists among *all* the inhabitants of the Sonoran desert world in which they live: plants, animals, birds, fishes, even rocks and springs. All of these come together as a part of one living community which Yaquis call *Huya Ania*, the wilderness world."

Yoem Lutu'uria, is almost the same as *Yo'ora Lutu'uria* (elders' truth). Felipe says, "One of the biggest responsibilities of a human being on this earth is to pass on the traditional knowledge to the younger generation. All of the communities should guide and instruct the children on all the important aspects of Yoeme culture, so that children will grow up and learn and understand how it is to move about on this earth." That philosophy underlies the spirit of this book. Our coverage of Yoeme knowledge of plants and especially animal life is far from complete. Much information has slipped away with time and there is further knowledge we were not able to record. Yoeme culture is vibrant and dynamic, leading to diverse views that include new interpretations of traditional knowledge. We do not expect complete agreement on what we are able to present. We look forward to future contributions about Yoeme plant and animal knowledge by others.

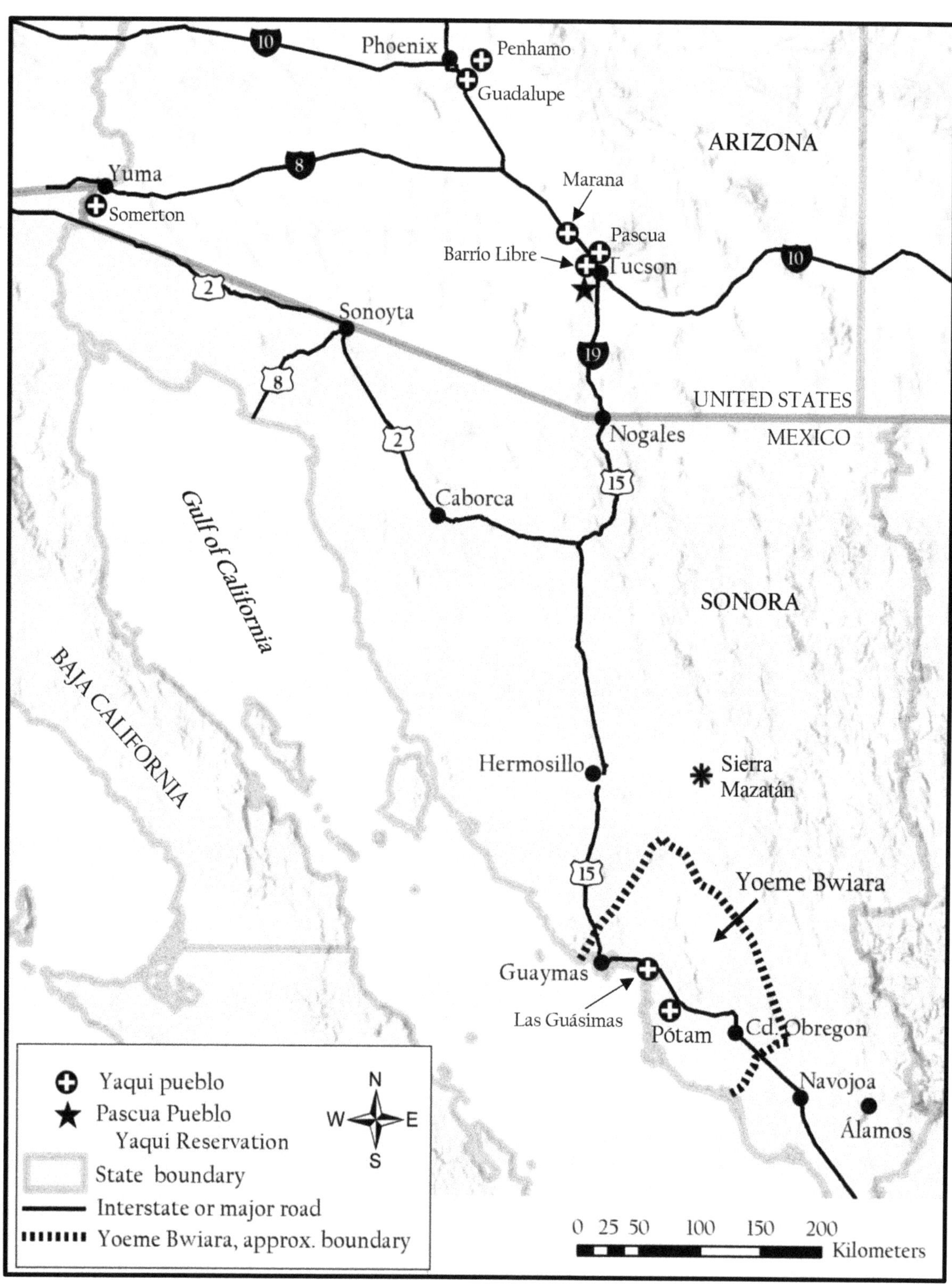

Map 1. Location of Yoem Bwiara and distribution of some Yoeme villages or settlements in Sonora, Mexico, and southern Arizona, United States.

SELECTED BACKGROUND REFERENCES

There is a vast literature addressing Yoeme history and culture. Some of the major works consulted are listed below. For the sake of easy reading, we do not reference them every time they are used. Also see Literature Cited for a more extensive bibliography.

Arte de la Lengua Cahita, por un padre de la Compañia de Jesús (con una introducción, notas y un pequeño diccionario). Tomás Basilio (1890 [1634]).

Historia de los Triumphos de Nvestra Santa Fee Entre Gentes las Mas Barbaras, y Fieras del Nuevo Orbe. Andrés Pérez de Ribas (1645), and in a translation by Daniel T. Reff et al. (1999).

Studies of the Yaqui Indians of Sonora, Mexico. William Curry Holden et al. (1936).

The Contemporary Culture of the Cáhita Indians. Ralph L. Beals (1945).

El Idioma Yaqui. Jean B. Johnson (1962).

The Yaqui Deer Dance: A study in Culture Change. Carlton S. Wilder (1963).

A Yaqui Life: The Personal Chronicle of a Yaqui Indian. Rosalio Moisés et al. (1971).

The Yaquis: A Cultural History. Edward H. Spicer (1980).

Missionaries, Miners, and Indians, 1533–1820. Evelyn Hu-DeHart (1981).

Yaqui Resistance and Survival: The Struggle for Land and Autonomy, 1821–1910. Evelyn Hu-DeHart (1984).

With Good Heart: Yaqui Beliefs and Ceremonies in Pascua Village. Muriel Thayer Painter (1986).

Yaqui Deer Songs/Maso Bwikam. Larry Evers and Felipe S. Molina (1987).

People of Pascua. Edward H. Spicer (1988).

Hiakim: The Yaqui Homeland. Larry Evers and Felipe S. Molina, editors (1992).

Yaqui Homeland and Homeplace: The Everyday Production of Ethnic Identity. Kirstin C. Erickson (2008).

We Will Dance Our Truth: Yaqui History in Yoeme Performances. David Delgado Shorter (2009).

Norma de escritura de la lengua jiak noki (yaqui). INAL (Instituto Nacional de Lenguas Indigenas) (2019).

YOEM NOKI—THE LANGUAGE

In Yoeme narratives the first people created were the *Surem*, the ancestors of the Yoemem. *Yoem Lutu'uria* (Yoeme Truth) tells that the Yoemem received or inherited their language, *Yoem noki*, from the Surem.

In this book we use the orthography of the official alphabet adopted by the Pascua Yaqui Tribe of Arizona. Pronunciation and orthography are essentially those used in North American English. In Sonora, the Yoeme language is generally written with a modified Spanish orthography. Many words in Yoem noki include a glottal stop, represented by an apostrophe ('), which is a pause between vowels (the English approximation is in words or expressions such as "uh oh!"). A final "m" in nouns almost always represents a plural, for example Yoemem. A noun drops one of the letters (a vowel) when used as an adjective, for example: *maaso* (deer) becomes *maso aawam* (deer antlers).

By the year 2015, in Sonora about 90 percent of the Yoemem spoke Yoem noki fluently, and all spoke Spanish. In Arizona about 10 percent of the Yoemem spoke Yoem noki fluently, and all spoke English, and many were also fluent in Spanish. As in so many other traditional cultures, the elders are concerned that the younger generation is not speaking the language. In the Yoem Bwiara where children used to listen to their elders discuss Yoem Lutu'uria, they now tend to be occupied with electronic devices (Shorter 2009). However, many families and schools in both Arizona and Sonora are making an effort to maintain the Yoeme language. (We use the term "elder" for the oldest person, or an elderly person.)

Yoeme traditions and language are closely related to those of the Mayo (*Yoreme*) people of Sonora and Sinaloa, whose homelands are south of the Yoem Bwiara along the Río Mayo in southern Sonora and the river valleys of northern Sinaloa. Felipe writes: "The original name the Mayos of Sonora and Sinaloa call themselves is Yoreme (*Yoleme* is an alternate spelling). We speak the same language and only a small percentage of words are quite different. Some words are pronounced the same but have different meaning. In Yoem noki, for example, a centipede is called *masiwe* and in *Yoremnokki* it is *masiaka*, which means Centipede Mountain. There is a Yoreme village named after the centipede."

The languages of the Yoemem and Yoremem are Uto-Aztecan, in the grouping known as Cahita in academic literature and in Yoeme publications and communications in Mexico. The term *cahita*, as a cultural term, was pointed out as being inappropriate as early as 1911 by the eminent anthropologist Alfred Kroeber (Thomas and Swanton 1911:12). Richard writes: "I remember one of my early trips to the Yoem Bwiara when my background included publications on 'Cahita culture'. I was puzzled by frequently hearing the word '*kahita*,' but was soon informed *es nada*, that it means nothing." There is, however, no other convenient term to describe this language and cultural grouping, which includes people traditionally living in regions associated with the great southern Sonora and northern Sinaloa rivers. Spicer (1983:250) pointed out that this language grouping "once contained 18 or more closely related dialects spoken by culturally similar natives of the region now included in southern Sonora...and Sinaloa." Cahita, however, continues to be used for this language and cultural grouping, especially in Mexico and without a demeaning connotation.

There is a long list of published works on the Yaqui/Yoeme language, including bilingual dictionaries beginning with Basilio (1890 [1634]). These publications include works by Spicer (1943), Kurath and Spicer (1947), Taub (1950), Johnson (1962), Barber (1973), Brewer (1976), Zavala (1989), Molina and Shaul (1993), Silva Encinas et al. (1998), Dedrick and Cassad (1999), Molina et al. (1999), Estrada et al. (2004), and INALI (2009).

We generally have not attempted linguistic analyses of Yoem noki words, although translations are given when possible. Some words show obvious derivation from Spanish, especially names for non-native plants and animals. Other names, including ones recorded during the Jesuit era, are often from neighboring, mutually intelligible dialects or languages (Shaul 2014), most notably from Basilio (1890 [1634]) and Peréz de Ribas (1645; Reff et al. 1999).

SOME COMMON YOEME TERMS

Hiak Vatwe. The Río Yaqui.

Hitevi. Traditional healer; curandera, curandero.

Huya. Huya has many meanings, all related to plants; the whole plant, any live plant, or tree, or wilderness.

Maaso. Deer (white-tailed deer), deer dancer.

Maso bwikam. Deer songs.

Pahko. A Yoeme religious ceremony, for the general public or at a home.

Pahko'ola (singular) **Pahko'olam**, the old men of the pahko (*o'ola* is an affectionate term for an old man). "Pascola"dancer.

Sea. Flower.

Tampaleo. Sitting against a vertical board during a pahko, he simultaneously plays a drum and flute for the maaso and pahko'olam.

Yoem Bwiara. The sacred homeland of the Yoemem, centered along the lower Río Yaqui.

Yoeme. The Yaqui people; Yoemem (plural).

Yoem noki. The Yoeme language.

Yoreme, Yoleme. The Mayo people of Sonora and Sinaloa; Yoremem (plural).

Yori. Yorim, plural. Outsider(s), especially non-Yoeme Mexicans; from Espanyoolim *espanoles*). Variants: Yoi, yoori.

YOEM BWIARA—THE HOMELAND

According to Yoeme tradition, the Yoem Bwiara, or the Yoeme homeland, was defined by the Singing of the Boundary, creating the Holy Dividing Line (Spicer 1980; Evers and Molina 1992). The Singing of the Boundary is included in the various narratives of the Talking (or Singing) Tree, which will be described in The Talking Tree in Way of Life. The region has been described as follows:

> The supernaturally sanctioned land comprised about 650 square miles. It

> extended for 200 miles along the coast from a few miles west of Guaymas, including the land on which that city was built during the 19th century, to a point between the mouths of the Yaqui and Mayo rivers. It included the whole of the Bacatete Mountains in which some Yaquis had begun to graze goats, sheep, and cattle during the 1600s; it also embraced more than half of what the Spaniards came to call the Ostimuri mining district, rich in silver until mined out during the 18th century. It encompassed the fertile south of the lower Yaqui River, which Mexicans began to take over for large-scale irrigation during the 1890s as well as the river lands on which the Eight Towns were situated. The sacred boundary line was about 375 miles in length. Yaquis, through the sanctions of the myth of the Singing of the Boundary, regarded themselves as the proprietors, in the service of God, of this large area of land. (Spicer 1980:170)

The general area of coverage for the Yoem Bwiara in Sonora is centered around 27.6°N latitude, 110.3°W longitude. It includes the Guaymas region and the vicinity of San Carlos on the Gulf of California, southward to the Río Yaqui valley on both sides of the river and inland about 100 km (about 62 mi), including the Sierra Bacatete.

CLIMATE AND SEASONS OF THE YOEME BWIARA

Pérez de Ribas (1645; Reff et al. 1999:84) provided the following observations that well represent the two extremes in the climate of the Yoem Bwiara:

> The climate is extremely hot. This is especially true near the South Sea [Pacific Ocean], as it is hot all along the coast of Mexico. Nevertheless, it is usually very cold during the months of December and January. The rest of the time is usually excessively hot, so much so that even the animals suffer greatly. Many times livestock become so tired while walking that the heat melts the fat in their bodies and they drop dead; other times the heat makes them so stiff that they cannot be used for a long time, and then only if they are bled immediately. The rains are few, especially along the coast.
>
> Because of the dryness, this land of Sinaloa [Sonora was then considered part of Sinaloa] would be completely uninhabitable for human beings, and even for animals, if it were not traversed and replenished by the rivers that run through it and into the Sea of California...during the rainy season...they flow with such great force that the rising waters entirely flood the fields, spreading over one or two leagues at the mouth of the river. Due to the loose soil of the floodplain, the rivers sometimes even change course...During these floods, which do not ordinarily subside for four, six, or eight days the Indians remain safe by using a peculiar means adapted to their way of living. Because there are no hills or knolls where one can seek safety, they erect platforms made of poles in the low crowns of scrub trees. They cover these with branches and then dirt so they can build a fire, and there they live while the fields are flooded.

Meecham is the term for months or moons. The seasons as they are known in this arid and subtropical climate (the Yoem Bwiara is centered around 27.6°N latitude, 110.3°W longitude) especially recognize changes in temperature and moisture.

Tataria/Tasaria meecham (hot months). March to July. Since the introduction of Christianity, May is called Itom Ae Meecha (Our Mother's Month). Late spring and early summer comprise a long and very hot dry season. By the end of April many of the trees and shrubs in the open desert and thornscrub are leafless, and by May and June even the chollas and prickly-pear cacti are withered and drooping. Summer, called tasaria or tataria, is likewise long and hot. Life-giving monsoon rains begin after San Juan's day (June 24). Renewal of the summer monsoon turns the leafless, parched landscape into lush jungle-like greenness.

Tevuhlia meecham (rainy months). Late July (when the rains start) to September—the monsoon months. The full force of summer rain usually arrives about a month after summer solstice and continues through early September. The time of the summer rains is called *tevuhlia*.

Pérez de Ribas (1645; Reff et al. 1999:328) wrote:

> When the river rises and overflows, which ordinarily happens almost every year, the fields are irrigated so that summer planting is possible. Rainfall is therefore not needed for their crops to ripen and for them to enjoy abundant harvests. The Indians have already harvested one crop [by the time] the river overflows, which is usually at the beginning of July. This crop is their main harvest, but nevertheless, during the rainy season some of them plant again, although this harvest is of less importance, their main sustenance comes from the summer crop. This harvest produces an abundance of maize, beans, squash, cotton, and other things that they cultivate.

Flooding of Yoem Bwiara towns and settlements from excessive cool-season and sometimes summer–fall rains was a common occurrence before construction of upriver dams. Exceptionally heavy rains, however, can still bring on destructive flooding.

Vali hupte meecham (coolness-is-coming months). October and November. Since the introduction of Christianity, November is called *Animam Meecha* (month of the departed souls). Fall season is called *severia hupte* (coolness coming). This is often a dry season, although sporadic hurricane-fringe or tropical depression storms, and occasional hurricanes, may bring huge amounts of rainfall in a relatively short time, and extend the summer-monsoon plant growth.

Severia meecham (winter months). December, January, February, and can include part of March. Winters are mild. Light frost is common; more severe freezing temperatures are rare but can be economically devastating for some cultivated plants. Winter is called *severia* and winter rain is called *severia yuku* (coolness rain). Heavy maritime dew is common along the coast during the cooler months, enough so that people living along the coast used to depend on dew to water family gardens.

VEGETATION OF THE YOEME BWIARA

Both climatically and geographically, the Yoem Bwiara is transitional between the Sonoran Desert to the north and subtropical dry forest to the south and southeast. As a consequence, the vegetation is generally comprised of desertscrub and subtropical foothills thornscrub in the drier areas and tropical deciduous forest in the less harsh areas, especially along drainages. The Yoem Bwiara lies south of the Plains of Sonora and includes the southern reaches of the Central Gulf Coast subdivisions of the Sonoran Desert. Most of the lowland vegetation consists of Sinaloan thornscrub. (Shreve 1951; Felger and Lowe 1976; Turner et al. 1995; Friedman 1996; Búrquez et al. 1999.)

Mangroves, *Rhizophora mangle* (Rhizophoraceae), Bahía Concepción, Baja California Sur. 25 Feb 2016 (SC).

The flora of all vascular plant species in the Yoem Bwiara includes an estimated 700 plant species (Appendix B; Felger et al. 2020). These are mostly of tropical or subtropical origin—their closest relatives are generally in more tropical parts of Mexico. Although some of the same animal and plant species occur in the Yoem Bwiara and in southern Arizona, many of the more subtropical

Río Yaqui. A. At low water, Tórim, including Yaqui cottonwoods (*Populus mexicana*, Salicaceae) and mesquites (*Prosopis glandulosa*, Fabaceae), with *Washingtonia robusta* (Arecaceae) in background. 1 Jun 2019. (PB). B. At Tórim, including mesquite trees (*Prosopis glandulosa*, Fabaceae) and Yaqui cottonwood (*Populus mexicana*, Salicaceae). 1 Jun 2019. (PB).

species of southern Sonora are not found in southern Arizona, where freezing temperatures and other factors are limiting.

Expansive sandy beaches face onto the sea along much of the coast. These beaches often rise to low dunes and in some places higher, shifting dunes. Elsewhere there are extensive coastal inlets or *esteros*. *Vai komia* (water arm) is an estero, such as the tidal inlets with mangroves near the coastal fishing camps of Chiinim (Los Algodones) and Las Cruces. Three prominent plants form the mangrove forests: red mangrove (*Rhizophora mangle*, Rhizophoraceae) in deepest water, black mangrove (*Avicennia germinans*, Acanthaceae) nearest and bordering the shore, and white mangrove (*Laguncularia racemosa*, Combretaceae) reaching maximum density in between. Teresa Amarillas recounted that during the Mexican wars (probably early twentieth century), instead of going to the mountains to escape soldiers, she and her family and neighbors hid in the mangroves.

Gallery forests of cottonwoods (*Populus mexicana*) and willows (*Salix gooddingii*, Salicaceae), as well as other trees, once lined the banks of the Río Yaqui and portions of a vast floodplain across its delta. These great riparian trees shaded riverside pathways and roads, places of no spines. The river coursed through this lowland floodplain, sinuously changing course over time, fostering rich soil and an ideal setting for people and agriculture. Spring floods brought by melting snows in the high Sierra Madre of Chihuahua and an extensive upriver drainage system deposited fresh alluvium on river terraces along the floodplain and fostered the development of huge bosques of mesquite (*Prosopis glandulosa*, Fabaceae) flanking the galleries cottonwoods and willows. The riverine forests and associated plant and animal life have been retreating since the middle of the twentieth century, when upriver dams and diversions reduced river flow and seasonal flooding. Remnants of these forests remain along portions of the river, irrigation ditches, and other wetlands, and in protected pockets around homes, but the changes have been vast and ecologically and culturally devastating.

Desertscrub in dry season. Aguaje Los Pilares, Sierra Bacatete, including *Carnegiea gigantea*, *Prosopis glandulosa*, and *Parkinsonia praecox*. May 2008 (MB).

The lower 100 km of valley-bottom spreading out from the Río Yaqui embrace the agricultural lands where alluvial soil seems endlessly deep. The

Desertscrub in wet season. Rancho Aguajito, Sierra Santa Ursula, including *Carnegiea gigantea*, *Ipomoea arborescens*, and *Mariosousa heterophylla* on highest rock slope in upper right, and *Stenocereus thurberi*. 28 Dec 2019 (MB).

Riparian canyon, Aguaje Los Pilares, Sierra Bacatete, including *Ficus pertusa*, *Guazuma ulmifolia*, *Prosopis glandulosa*, and *Typha domingensis*. May 2008 (MB).

Yoemem̦ have relied on floodwaters to nourish crops since pre-contact times. In modern times the river has provided irrigation water for large-scale agriculture. However, the majority of the water has been diverted to non-Yoeme agriculture on the left (south) bank of the river, lands originally belonging to the Yoemem.

Most of the original vegetation away from the river is desertscrub and dense to sometimes sparse thornscrub. Many of the common trees and shrubs shed their leaves in the late spring–early summer dry season. Some trees, such as the *Bursera* (Burseraceae) and *Jatropha* (Euphorbiaceae), shed their leaves soon after the summer rains cease. Others, such as the boat-spine acacia (*Vachellia campechiana*) and *mauto* (*Lysiloma divaricatum*, Fabaceae), gradually shed leaves. Many of these plants are leafless during the dry spring months, and burst into green with the first summer rains. Others, such as mesquite (*Prosopis glandulosa*), may shed most or all of their leaves in the winter–early spring dry seasons but leaf out in spring in response to warm weather. The bright green of newly leafed mesquite groves or *bosques* stands in sharp contrast to the parched surrounding vegetation. Columnar cacti, *echo* (*Pachycereus pecten-aboriginum*) and organpipe (*Stenocereus thurberi*), poke flower-bearing stems above the surrounding spinescent trees and shrubs.

Basilio (1890: [1634]:139) provided indigenous names for 27 "*arboles chaparros*" ("short trees"), in addition to naming various other kinds of trees. Much of the historical coastal plain vegetation has been cleared and replaced with irrigated agricultural fields. The remaining trees and shrubs have been heavily exploited for firewood and many of the mesquite bosques or forests (*hu'upa soyo'oria*, mesquite lush-grove) have been decimated for charcoal production. Centuries of livestock grazing has taken a toll on the original vegetation.

In some areas near the coast, such as at Ráhum, there are expanses of salt scrub dominated by native iodine bush (*Allenrolfea occidentalis*, Amaranthaceae), and sea blite (*Suaeda nigra*, Amaranthaceae), and other halophytes (salt-tolerant plants) such as the non-native tamarisk (*Tamarix chinensis*, Tamaricaceae). Places on hillsides or mountains of large lava rocks where few or no plants grow are called *tuukum* (black volcanic rocks, from *tuukuk*, turned off, such as fire or burning).

Riparian canyons, supporting a rich flora of tropical affinity, are found in the Sierra Bacatete—the range east of Guaymas and north of the coastal plain of the Río Yaqui and other mountains in and around the Yoem Bwiara. One such canyon system is at Rancho Bacatetito near the abandoned Mexican military fort or *cuartel*. Groves of buttresstrunked *chalates* (*Ficus insipida*, Moraceae) grow along the broad canyon stream below the old fort, providing deep shade along a trickling stream. This was the water source for the fort. One of the tactics of Yoeme rebels was to cut off or divert water away from the fort. Steep slopes inaccessible to cattle surround the canyons where vegetation remains relatively intact. Another riparian canyon, the well-known Cañón del Nacapule near San Carlos, likewise supports a rich flora (Felger et al. 2017).

The higher elevations, the mountaintops of the Sierra Bacatete—with several summits exceeding 600 m (about 2,000 ft)—remain botanically unexplored. From time to time during the prolonged and recurring Mexican-Yaqui wars, many people fled to the rugged Sierra Bacatete. In the mountains the people relied on their knowledge of food plants and wild game. *Saawa* (*saiya*, *Amoreuxia palmatifida*, Cochlospermaceae) is often mentioned as one of the resources; it yields edible tuberous roots, leaves, fruits, and seeds. Many other plants sustained the people in their mountain fastness, including agaves, fruits of three species of wild figs (*Ficus*), *uvalama* (*Vitex mollis*, Lamiaceae), cactus fruits such as echo (*Pachycereus pecten-aboriginum*) and *pitaya dulce* or organ pipe (*Stenocereus thurberi*), and the large, succulent roots of *pochote* (*Ceiba aesculifolia*, Malvaceae). We will never know the full extent and details of wild food resources during those harsh times.

Bahía Las Guásimas, with *Tamarix aphylla* (Tamaricaceae). 1 Jun 2019 (PB).

WAY OF LIFE

YOEM LUTU'URIA—YOEME TRUTH

Yoem Lutu'uria includes respect for all life, animals, plants, and sea life, as well as water, sea, rocks, clouds, rain, and wind. Yoem Lutu'uria includes the *Aniam*, the spiritual worlds or realms, and they are all connected. The concepts can be flexible and overlapping. Some of the primary Aniam and related concepts are listed here (excerpted largely from Evers and Molina [1987]).

Yo Ania, the Enchanted World. "The *Yo Ania* is an ancient world, a mythic place, outside historic time and place, yet it can be present in the most immediate way. Felipe's grandfather, Rosario Castillo [Rosario Vakame'eri-Castillo] and Don Lupe [Guadalupe Molina] translated *yo* as '*encanto*'; we translate it as 'enchanted'" (Evers and Molina 1987:62). It is said that Yo Ania exists in certain small mountains or hills such as *Takalaim* (The Forked One at San Carlos), *Sikili Kawi* (Red Mountain, near Empalme), and a small hill near Picacho Peak in Arizona. These are well-known ones, although there are others.

Huya Ania, the Wilderness World. Huya Ania refers to places where there is abundant plant life, and can be physical or spiritual. One can obtain spiritual powers by entering and walking in the Huya Ania. It is always a wild, natural place, away from towns and human settlements. Huya Ania also can be described as the Plant World or the Desert World. Huya Ania is said to be the physical manifestation of the Yo Ania. When entering the Huya Ania one should give a blessing and ask for permission to enter, and after the visit one should leave an offering.

Sea Ania, the Flower World. Yoem Lutu'uria narratives tell that Sea Ania was created after Yo Ania and after people. Sea Ania is in all life, in all creatures and nature, including the rivers, the wind, the clouds, ocean, rain, sun, moon, sky, and stars. Sea Ania is Huya Ania with flowers and is where the deer lives. The Flower World is the living beauty of the natural world and the final resting place of the Yoemem.

Don Lupe (Guadalupe Molina), a deer singer from Vicam Suichi, described the physical Sea Ania to Felipe. This conversation took place at Felipe's house in Marana near Tucson, in the late 1970s. Don Lupe talked about the Sea Ania after a deer singing session in the night:

> One can see the Flower World during the spring or fall flower season. That place can offer *utte'ewa* [power or strength]. Walking among plants with flowers allows one to have a peaceful experience and well-being in mind, spirit, and body. Those flowers are gathered as offerings at house altars. [Utte'ewa, power or strength, can be both physical and mental].

Anselma Tonopuame'a Castillo, Felipe's maternal grandmother, said that a person should get up early in the morning, before sunrise, and walk to the east. In that hour the wind will blow from the east and that wind is called *machiwa heeka* (dawn wind), *taewa heeka* (day wind), or *sewa heeka* (flower wind). Anselma informed her children and grandchildren that one must walk during this time because the breeze is a purifying and blessing wind.

Tuka Ania, the Night World. This realm includes the moon, called *mala mecha* (mother moon), which is a modern term, because the old Yoem noki term for mother is *ae*. When singing deer songs and ceremonial prayers or speeches, mother is called *aye*. It is interesting to note that in the Yoreme (Mayo) language mother is also *aye*.

Seatakaa, Flower Body. Seatakaa is special or spiritual power. It can be power from Huya Ania, available to certain people. "Seatakaa is a protective energy of the *hiapsi*, the heart or spirit of a person" (Felipe *in* Evers and Molina 1987:53). This scarcely translatable concept is also discussed by Shutler (1977), Painter (1986), and Spicer (1980:88, 327). Seatakaa "is the innate and mysterious power possessed by an individual which enables him [or her] to cure. *Seatakaa*, however, is not limited to curing power. It is in greater concentration, as it were, in certain individuals, animals, plants, and places...Animals possess *seatakaa*, but most especially the deer [maaso]" (Shutler 1977:186, 188).

HARVESTING AND HONORING

Felipe tells us that before harvesting, gathering, or taking a part of a plant, one should petition the natural world, talk to the plant, out loud or to one's self, and ask for permission to use it. An offering should be left at the plant. For example, people talk to the saguaro, orally or mentally, before picking the fruit. They ask permission to pick and thank the saguaros for a good year of fruit and for one's health or that of one's family, hoping to be there again next year. This request is done for the first fruit of any major plant harvest before gathering or picking, and before eating beans, corn, organpipe cactus fruit, squash, watermelon, etc.

Only the east side of the plant should be used for medicinal purposes, because the rising sun strikes it first on the east side; it has more power there. Plants are best collected during the time of the full moon. The three days around the time of full moon would be best for medicinal plants or anything for the house, such as household posts, beams, and wall materials. This does not mean going out at night, but rather during daytime. Traditional healers (*hitevim*) or medicinal plant collectors (*hittoaream*) gather plants during the full moon, unless it is an emergency.

Huya kimuwa (wilderness being-entered) means going into the wilderness or desert world to gather or hunt. Special preparations should be made; you should purify your mind, body, and soul. To purify yourself is *emo bwasawame* (cooking/smudging yourself) with smoke of green mesquite or other leaves, or with *kopal* (copal-like resin from brittlebush, *Encelia farinosa*, Asteraceae, or more likely from *torote, Bursera*, Burseraceae). In modern times people have used orange peelings or leaves and commercial incense. Emo bwasawame is done especially for deer hunters so that the deer will not smell the humans. Creosotebush (*Larrea tridentata*, Zygophyllaceae) branches also are used for purification.

Similar tradition or reverence also applies for ceremonial regalia, and other activities such as making a clay pot or a basket. This is what Felipe tells us:

> If you decide to make a clay pot, you also should ready yourself by purifying yourself, mind, body, and soul. A red cloth would be left at the clay pit, the *Vavu Ania* (Clay World); it is alive so you talk to it and leave an offering. The red cloth represents an offering; you are leaving a possession from your life, a reverence or respect as an exchange for success to produce useful and beautiful objects.

Yoeme pottery was not decorated but was perceived as beautiful sitting anywhere around the house. When making a clay pot, or any artistic object, such as a weaving, you leave a little of your spirit inside it for all times, therefore, such objects are created with reverence and treated with respect. The color red is a healing color. Since the Clay World will lose part of itself a red cloth is offered, so it will heal from the separation. It is also a way to

Water olla on a 3-forked mesquite stand, Pótam. 15 Dec 1988 (WS).

thank the Clay World for allowing a portion of its world to be taken.

THE PAHKO, DEER SONGS, AND THE FLOWER WORLD

Maso bwikam (deer songs) are extensive portrayals of the natural world. These poetry-songs represent the spirit of *saila maaso*, little brother deer (Evers and Molina 1987:7). The songs are the voice of *saila maaso*. "All that he should talk about, that is what we sing."

> Deer songs [are]...usually sung by three men to accompany the performance of the deer dancer. Yaquis regard deer songs as the most ancient of their verbal art forms. Highly conventionalized in their structure, their diction, their themes, and their mode of performance, deer songs describe a double world, both "here" and "over there," a world in which all the actions of the deer dancer have a parallel in the mythic, primeval place called by Yaquis *sea ania*, flower world. Deer songs describe equivalences between these two real parts of the Yaqui universe. They are verbal equations, developed richly with phonological, syntactic, and rhetorical parallelisms and repetitions...Readers with a vantage from the Yaquis' ecological homeland will recognize common things from the natural world of the Sonoran desert imaged in the songs: insects and plants, birds and animals; and those rarer ones, more powerful in combination: rain, flowers, the deer.

A pahko at Marana, Arizona; early 1960s. The pahko'ola is Felipe's grandfather, Rosario Vakame'eri-Castillo and the maaso (deer dancer) is Juan Cruz from Vikam Suichi (Painter and Sayles 1962).

Flowers pervade Yoeme narratives. "One of the most cherished stories...tells of the flowers which poured from the side of Christ [on the cross] ...a symbol of divine grace" (Evers and Molina 1987:54). Without apparent conflict, flowers are regarded as all that is good and beautiful, including the deer and other beings in the magical pre-Christian world.

When *Wok Vake'o*, said to be the first man to compose deer songs, encountered a fawn, he sang something like this (Evers and Molina 1987:50):

Little one, born in the night,
Caressed by the fresh wind.
Little one, born in the night,
Caressed by the fresh wind.
Where are you going then,
Flower fawn among the flowers?
Dressed in flowers, I am going.

Pahko is the term for a religious ceremonial occasion, for the general public or at a home. *Pahko'olam* (pascola dancers, the "old men of the pahko") are present at every pahko and serve as hosts for the celebration (Griffith and Molina 1980). A typical pahko includes the pahko'olam, a deer dancer (*maaso*), deer singers (*masobwikame*), and a *tampaleo*, as well as harp and violin musicians. However, the deer dancer, deer singers and tampaleo are not always present, especially at a smaller pahko.

Pahko heka (Ceremonial Arbor) at the home of Ignacio Amarillas Sombra, Pótam. *Pahko heka* is a ceremonial structure for the church group and the ceremonial dancers. 1987 (FM).

The pahko'olam wear specific regalia and mime, dance, and joke with each other and the public observers. These actions of the pahko'olam are called *yeuwame*, translated as the act of playing, or games or skits. However, the translations are too simplistic (Painter 1986: 264–266; Evers and Molina 1987:134–137). Although *yeuwamem* can provide amusing satire about Yoeme society as well as society as a whole, there are established sequences and intricate although varied relationships known to traditional Yoemem. There are *yeuwamem* involving farmers pestered by a raccoon, the pursuit of the deer (maaso) by mountain lions, and the drama of rain. Most *yeuwame* are accompanied by deer songs (Evers and Molina 1987:137).

One *yeuwame* is the *Maso Me'ewa* (Killing the Deer), a ritual drama of death and renewal. The Killing the Deer ceremony takes place during a *Lutu Pahko* (lutu is from the Spanish word *luto*, mourning). The Lutu Pahko marks the first anniversary of a relative's death and the end of a year of mourning. The culmination of the Maso Me'ewa ritual is the stalking of the deer. One pahko'ola becomes a hunting dog, while the other three may pretend to be hunters. Their burlesque stands in contrast to the hunt, which is communicated by the deer dancer and the deer singers. Like so many other deer songs, the songs of the Maso Me'ewa ceremony are filled with images of flowers.

The maaso, deer dancer, is the best known of all Yoeme traditions. Felipe wrote:

> Everything the deer dancer uses in his dance has held life. The cocoon rattles around his legs were once homes of the butterflies [actually saturniid moths]. As we dance we want the butterfly to know that, even if he is dead, his spirit is alive and his house is occupied. The gourd rattles in the dancer's hands give life to the plant world. The rattles around the dancer's waist are deer hooves. They represent the millions of deer who have died so that men might live. (Evers and Molina 1987:129)

THE TALKING TREE

Huya Nokame (or *Kuta Nokame*) the Talking Tree, also called the Singing Tree or the Talking Stick, is a prominent Yoeme narrative. There are many versions. Shorter (2009) discusses twelve versions. Felipe recorded the following version when Luciano Velasquez told it in Yoem noki at Kompuertam (a village near Vícam) in August 1982. Here is Felipe's translation (Evers and

Molina 1987:37–38):

> In the time before the Spanish conquest the Surem lived in the area that is now west of Ciudad Obregón. Their river was called *Yo Vatwe*, Enchanted River. In this region Surem had their homes, and they lived on both banks. Their houses were called *hukim*. They were built of sticks and mud and were about four or five feet high. They hunted, fished, and farmed to stay alive.
>
> Well, anyway, during those early times a tree was heard talking on a small hill called *Tosai Bwia*, White Earth. Some say it was heard at *Omteme Kawi*, but we say it was at Tosai Bwia. This tree was an old dead mesquite tree, and it made strange humming sounds. Nobody could understand the sounds. That bothered the Surem. All the intelligent men in the Surem land were notified and told to visit the tree. None of them could figure it out. They all had to admit failure. They could not decide what the meaning of the tree was.
>
> These wise men knew of one other wise man who lived near a little mountain called *Asum Kawi*, Grandmother's Mountain. This man's name was *Yomumuli*, and he had twin daughters. The wise men visited him and requested that his two daughters interpret the Talking Tree. Yomumuli told them that the girls didn't have a good vocabulary, so they would be incapable of doing such a task. The men insisted. Finally they convinced Yomumuli to take the girls to the Talking Tree. But Yomumuli did one thing first. He took his twin daughters to the ocean. There they talked to a fish, so that they might better understand the talking tree.
>
> At Tosai Bwia the girls stood on either side of the tree and they began to interpret. The tree predicted Christianity and baptism, wars, famine, floods, drought, new inventions, even drug problems, and so on. After the tree had given all the information it stopped making the sounds.
>
> The Surem were happy, but they didn't really like some of the things they heard, so they planned a big meeting. The meeting was held near a water hole called *Yo Va'am*, Enchanted Waters. This is in the region between Vicam and the modern town of Colonia Militar. There the Surem held both a meeting and a dance of enchantment. At this meeting some of the Surem decided to leave the Yaqui region, while others decided to stay and to see these new things. At this dance of enchantment, they say a real live deer came to dance for the Surem. After the dance the Surem who were leaving cut up a portion of the *Yo Vatwe*, wrapped it up in a *carrizo* mat, and took it north to a land of many islands. Other Surem stayed around and went into the ocean and underground into the mountains. There in those places the Surem now exist as an enchanted people. Those who stayed behind are now the modern Yaquis and they are called the Baptized Ones.

Yoemem in Sonora sometimes say the Talking Tree predicted the coming of an enormous black snake, an allegory for the Mexican Highway 15 (the major north–south highway of western Mexico). The highway brings fatal accidents and assaults traditional Yoeme culture and the Yoem Bwiara (Sheridan 1996).

ARC OF HISTORY

EUROPEAN ARRIVAL

Prior to the arrival of the first Europeans, there probably were more than 30,000 Yoemem living along the lower 100 kilometers of the Hiak Vatwe (Río Yaqui). Some estimates say 60,000. However, as much as half or more of the population may have died from epidemics of Old World origin including smallpox, influenza, measles, and typhus, which devastated native populations in Sinaloa and Sonora between 1593 and 1617 (Reff 1991). Episodic epidemics of Old World diseases continued to devastate Yoemem and other native peoples into modern times.

There is a wealth of Spanish colonial documents as well as published accounts involving discussions, interpretations, speculations, and translations of the early Spanish-Yoeme history. Sheridan (1988) reminds us that these accounts are based on European perceptions. We include a few paragraphs and brief dialogs of the early recorded history.

The first major European intrusion was Captain Diego de Guzmán's expedition of 1533, "ranging up the west coast of New Spain on slave raids" (Spicer 1962:46). The Yoemem were militarily well-organized and met the Spaniards in a force of thousands of fighters, which resulted in the intruders defeat and retreat. One would expect that the Yoemem knew what was in store for them if they were conquered. The expedition was chronicled by the "Anonymous Reporter" (Hu-DeHart 1981) and identified by Folsom (2014) and others as Jorge Robledo.

> Leaving Culiacán in July, the Spaniards reached the Mayo River by September, having crossed the Sinaloa and Fuerte (Zuaque or Suaque) rivers a month earlier...On October 4, 1533, the Guzmán expedition arrived at the Yaqui River, or, as the Anonymous Reporter spelled it, the Yaquimí...The Indian warriors...were assembled in a large field [awaiting the Spaniards] (Hu-DeHart 1981:14–15).

An abridged excerpt of the translation of Jorge Robledo's account indicates what subsequently happened (Folsom 2014:25):

> From among them there emerged an Indian distinguished from the others by a black *sambenito* like a scapulary, which was sown [*sembrado*] with meticulously crafted conchs of pearls, [representing] many little dogs, deer and many other things, and since it was the morning, and the sun shone on him, the garment glittered in the manner of silver, and he had his bow and quiver of arrows, and a well-crafted staff in his hand, and he came ruling over the people. We came together to the distance of two stone's throws; and since we were so close, this Indian who governed the others stepped out in front of all and with his bow he made a very long line in the ground, and he knelt on it and kissed the ground, and after doing this he rose up, and, standing up he and his people began to speak, telling us to stop, that we would not pass by that line he had made, and if we did pass, they would kill us all...
>
> Through an interpreter Guzmán tried to tell the Yaquis that he did not want a fight. The Yaquis replied that they would bring out something to eat. They went on to tell the Spanish to remain where they were and they would come tie up the Spanish horses. They took out cords for the purpose, and they seemed 'to have great arrogance among them.'

Ultimately the Spaniards and their impressed native auxiliaries were driven away with heavy losses. It was the first time Spanish marauders were defeated by a native force, and set the stage for how outsiders regarded Yoeme culture and history.

In 1536, Álvar Núñez Cabeza de Vaca and his three companions, along with a large contingent of indigenous supporters, arrived near the Yoeme region. This was toward the end of their amazing journey across the continent after their 1527 disaster in Florida (Nuñez Cabeza de Vaca 1542, 1555; Adorno and Pautz 1999). The four travelers were well received by a large number of native people, which presumably included Yaquis.

In the following decades there were few recorded contacts. Near the beginning of the summer rains in 1565, there was an extended Yoeme encounter with Francisco de Ibarra's party. Traveling with them, Balthazar de Obregón kept a meticulous chronicle of the expedition (Hu-DeHart 1981:18–19):

> At this welcome were five hundred handsome and brilliantly-dressed Indians. They wore their typical costumes, decorated with bright feathers, conches, beads, and sea shells.
>
> This river Yaquimi is the most thickly populated of all the regions traversed by the general. It must contain fifteen thousand men in the ten leagues [about 42 km, or 26 mi] from the sea to the mountain. The town is situated amid a luxurious grove, a fourth of a league [about 1 km, or more than half a mile] in extent. This river is large, cool, and contains quantities of good fish. On its margins are many fields of maize, beans, and squashes. The people are kind. The women are beautiful...The general explored the river as far as the sea. There we found clusters of coral and quantities of pearl-bearing shells. The natives presented the Christians with gifts of fish, game and other foods which they had in their land.

This enthusiastic description followed the hardships suffered earlier by Ibarra's eight years of exploration and failed ambitions. The ensuing decades of the sixteenth century were marked with violence and tragedy among the regional indigenous people and Spanish colonials and military (Folsom 2014).

A major contact between the Spanish and the Yoemem involved the military conflicts led by Capitán Diego Martínez de Hurdaide in 1610–1611. Hurdaide was defeated in multiple assaults. Unlike so many other places in Mexico, the Yoeme lands had not been lost (Spicer 1954:25):

> There was no burning or sacking of Yaqui villages. In fact, in all these encounters the Yaquis were victorious, driving both Guzmán and Hurdaide out of their territory. It was as victors, rather than as defeated people, that they asked to have missionaries sent to them. The Jesuits, Andrés Pérez de Ribas and Tomás Basilio came into the Yaqui country in 1617 without military escort.

Four Zuaque assistants accompanied the two Jesuits. Thus missionaries, rather than soldiers and settlers, came to live among the Yoemem and for the most part peace reigned for 120 years, although Old World diseases and subterfuge persisted.

Tomás Basilio (1580–1645) remained in the region for the rest of his life. He authored an impressive indigenous–Spanish dictionary (Basilio 1890 [1634]). Andrés Pérez de Ribas (1576–1655) left the Yaqui–Mayo region in 1619 after being recalled to Mexico City, supposedly due to failing health, and continued working on his *Historia* (Pérez de Ribas 1645). We are able to identify most of the plant and animal names in these seventeenth-century works. Many of these names are still used nowadays.

Pérez de Ribas' account was "published in Spain in 1645 and ostensibly is a history of the Jesuit missions of northern Mexico during the period from 1591 to 1643" (Reff et al. 1999:3). This book is a treasure trove of natural history information, including observations such as, "In the rivers there are wild ducks and an abundance and variety of fish that enter from the sea, especially when they lay their eggs" (Reff et al. 1999:84). However, like Basilio, Pérez de Ribas generally did not distinguish between Yaqui and Mayo uses, names, and information, and that of neighboring Cahitan groups. Reff et al. (1999) provided an excellent translation, although some identifications of animals and plants are not accurate. Pérez de Ribas (1645; Reff et al. 1999:329) also reported that, "These Indians are generally taller and more robust than those of other nations." Their healthy appearance was likely due at least in part to favorable nutrition from a diversity of food resources.

During the Jesuit missionary times the Yoemem generally accepted Christianity, combining it with elements of their indigenous culture. The Jesuits introduced new agricultural practices and new crops—most significantly, wheat, barley, and oats—and domestic animals. These agricultural successes eventually resulted in surpluses that the Jesuits sometimes sold, but mostly sent to their Californian missions in need of supplies. Following the mutually beneficial relationships of the early Jesuit times, the later missionary era was often marked deteriorating conditions. Yoemem generally resented harsh treatments and not sharing in their hard-earned surpluses. A rebellion in 1740 was fueled in part by poor harvests in September 1739 and the refusal of local Jesuit authorities to readily share food surpluses designated for the Californian missions and Upper Pimería.

The Spanish empire continued to decline while the Jesuit Order, The Society of Jesus, was on the upswing. Development in northwestern Mexico was waking up. The rich agricultural lands of the Yoem Bwiara attracted continued outside interest, fueling terrible conflicts.

CHRISTIANITY AND YOEME CEREMONIES

Jesuit missionaries introduced Catholic customs and celebrations practiced in seventeenth- and eighteenth-century Europe, including the public ceremonies of Lent and Holy Week (*Waehma* in Yoem noki, from *La Cuaresma*, the Spanish for Lent). After the Jesuit expulsion of 1767, Yoemem continued on their own to carry on Catholic-based practices.

Spicer (1980:62) described Yoeme religion as a complex tapestry, "a conjunction of world views." In his opinion, it is a new religion, one that draws from both European and Native American traditions. However, "After more than three and one-half centuries of dialogue with Catholicism, not to mention other versions of Christianity in this century, Yaquis continue to hold the figure of Christ and the *maaso* [deer dancer] explicitly apart in their ceremonies" (Evers and Molina 1987:130). For example, the maaso, pahko'olam, and deer singers are seen alongside Catholic-derived rituals, including the *Kantooram* (women singing and chanting hymns).

In the late nineteenth and early twentieth centuries, warfare and terror during the Porfiriato (times of the dictatorship of Mexican president Porfirio Díaz) compelled halting of public religious events. Traditional outlets for public expression were gradually regained in later decades.

Yoeme public rituals are open to outsiders and are said to offer blessings for all. The intricate Waehma ceremonies and events culminate in the *Looria Tenniwa* (Running of the Gloria) on Holy Saturday, the day before Easter. Detailed descriptions of the Waehma are provided by Painter and Sayles (1962), Spicer and Crumrine (1997), Spicer (1980), and Moctezuma Zamarrón et al. (2016).

Authorities in charge of maintaining the Waehma ceremonies include the *Kavayeom* (horsemen; *caballeros* in Spanish). The long, symbolic drama includes the *Fariseros* (Pharisees), who try to invade the church and capture the image of Jesus. The forces of good, who defend Jesus, include the "Baptized Ones," especially the Kavayeom and the *Anhelesim* (the Angel Guards, or Little Angels, represented by little girls and boys). The forces of good also include the maaso and pahko'olam. The invaders charge the church three times, and are repelled each time by the forces of good throwing "flowers" of confetti, cottonwood leaves, and flower petals. Church bells ring and the forces of good prevail. All the while the deer singers chant *Looria Bwikam* (Gloria Songs), bringing Huya Ania (the Wilderness World) into the enchanted drama. Here is one of the Looria Bwikam translated by Felipe (Evers and Molina 1987:58):

Wilderness world,
flower freely, is blowing,
wilderness world.
Wilderness world,
flower freely, is blowing,
wilderness world.
[repeated 3 more times]

Over there, I, in Yevuku Yoleme's
flower-covered, Enchanted,
flower patio,
flower freely, is blowing,
wilderness world.

Wilderness world,
flower freely, is blowing,
wilderness world.

YOEME SETTLEMENTS

The Yoemem have farmed and lived in the rich bottomlands along the Hiak Vatwe (Río Yaqui) since ancient times. In pre-Jesuit times, homes of extended families were clustered along the river near agricultural fields. The homes were often placed close to the river among the sheltering shade of giant cottonwoods, willows, mesquites, and other trees such as hooso (*Albizia sinaloensis*, Fabaceae). Pérez de Ribas told of 80 rancherias along the river. Because of the biannual floods, the people were often forced to temporarily move to higher ground or onto raised platforms (Pérez de Ribas 1645; Reff et al. 1999:84).

House front yard, Pótam; cane poles (*Arundo donax*, Poaceae) beneath a willow (*Salix gooddingii*), and *Ricinus communis* (Euphorbiaceae). 20 Nov 1985 (RF).

Yoeme homes are greatly respected, and each has its own personality and spirit. They are alive. Traditional housebuilding was a communal affair, with men, women, and children partaking in gathering and processing materials, as well as actual construction. At Yoem Pueblo in Marana in the mid-twentieth century, the finishing off, usually mud-packing the walls, was done by women: "mud parties" were entertaining, and boys helped.

Freshly cut *vaaka* (*Arundo donax*, Poaceae), Pótam. Fence and wall also of cane, and a willow house cross (*Salix gooddingii*, Salicaceae). 1 Jun 2019 (PB).

Kompuertam, home garden, including canna lilies (*Canna indica*, Cannaceae) along fence. 14 Mar 1989 (WS).

Following the arrival of the Jesuits, many of the people gradually settled into eight towns along the Río Yaqui, although there may have already been movement into villages (Beals 1932; Hu-DeHart 1981:18–19). These settlements, established around the Jesuit mission churches, are the *Wohnaiki Pweplom*, the *Ocho Pueblos*. The eight holy pueblos are *Ko'oko'im* (Cócorit), *Potam* (Pótam), *Raahum* (Ráhum), *Torim* (Tórim), *Vahkom* (Bácum), *Veenem* (Belén, or Belém), *Vikam* (Vícam), and *Wiivisim* (Huírivis).

From early Jesuit times to the late nineteenth century Yaqui men were a major source of regional labor in northwestern Mexico well beyond the sometimes as coerced or forced hard labor. They worked in mines and on haciendas, as pearl divers, fishers, and on Gulf of California islands mining guano and salt. Their families sometimes went

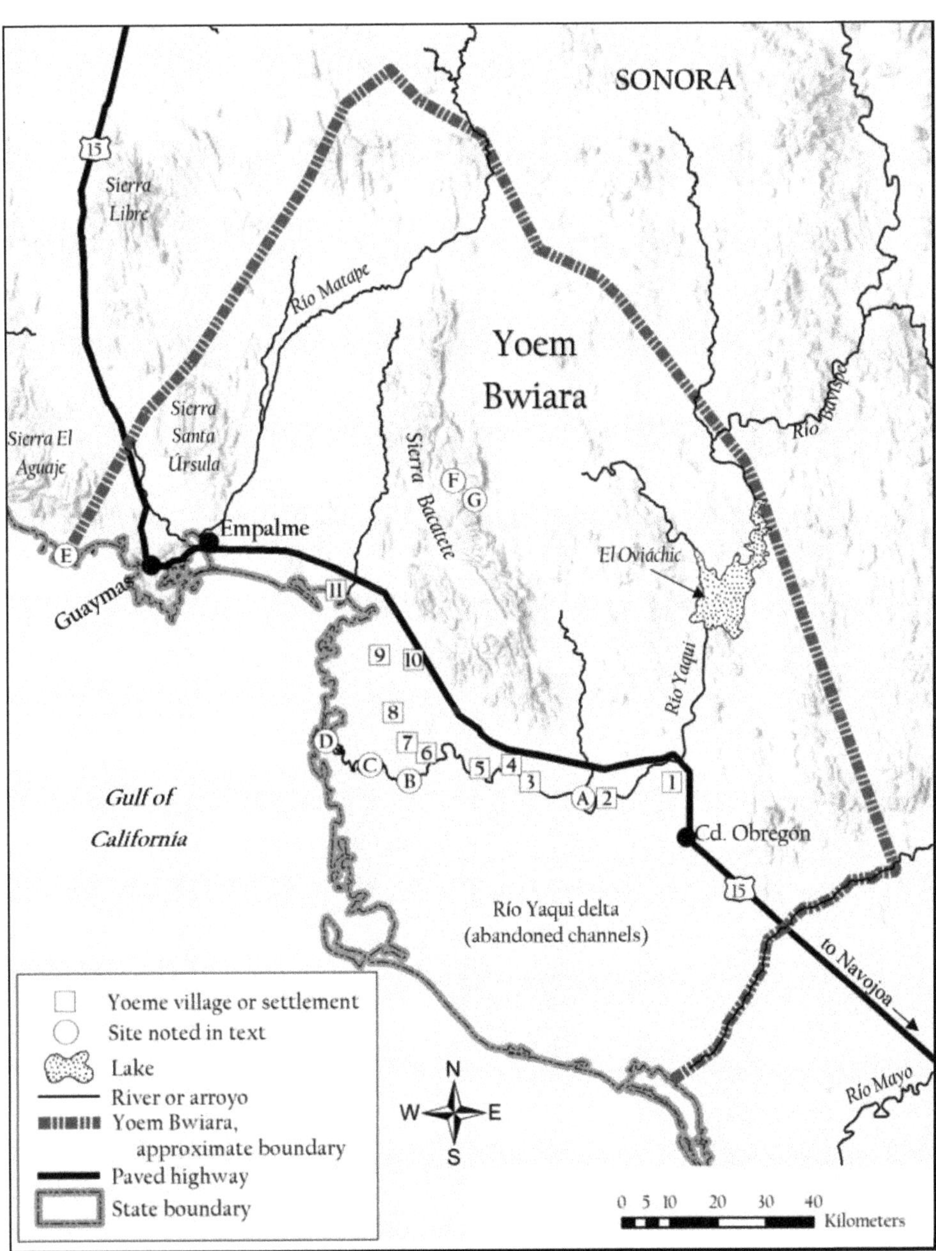

Map 2. Distribution of Yoeme villages or settlements and some other sites in Sonora, Mexico. See Table 1 for the definition of Yoeme villages or settlements, represented as circles enclosing numbers, and of other sites mentioned in the text, represented as squares enclosing letters.

Table 1. Definition of Yoeme villages or settlements and other sites on map 2.

Number on map	Village or Settlement	Type of Site
1	Ko'oko'im, Cócorit, Cócorim	Yoeme settlement at Jesuit mission
2	Vahkom, Bácum	Yoeme settlement at Jesuit mission
3	Tórim	Yoeme settlement at Jesuit mission
4	Compuertas, Kompertam, Kompuertam	Yoeme settlement
5	Vícam, Bícam	Yoeme settlement at Jesuit mission
6	Pótam	Yoeme settlement at Jesuit mission
7	Ráhum	Former Yoeme settlement at Jesuit mission
8	Wiivisim, Huírivis, Huiribis	Yoeme settlement at Jesuit mission
9	Veenem, Belém, Belén	Abandoned Yoeme settlement at Jesuit mission
10	Pitahaya	Yoeme settlement
11	Las Guásimas	Yoeme fishing village
Letter on map	Site Noted in Text	Type of Site
A	Loma de Bácum	Landmark hill near Bácum
B	Cuesta Alta	Plant collection site along the Río Yaqui
C	Kopas, Copas	Plant collection site along the Río Yaqui
D	Chiinim, Los Algodones	Yoeme fishing camp
E	Takalaim	Prominent peak on the coast at San Carlos
F	El Cuartel	Ruins of a Mexican fort & plant collection site
G	Cerro El Gallo	Landmark hill

with them, forming small communities. Women and children were also often employed at mines and haciendas.Yoem Bwiara—sometimes as paid workers and

During the nineteenth and early twentieth centuries many people fled persecution and warfare in their homeland and found refuge in Arizona. There are four major Yoeme communities in the Tucson area (Barrio Libre, New Pascua, Pascua, and Yoem Pueblo), and two in the Phoenix area (Guadalupe and Penjamo). A number of others existed from time to time in places such as at Coolidge, Eloy, Mesa, Nogales, and Yuma, as well as in California and other western states (Spicer 1940,1988; Spicer and Hill 1947). Yoemem in Arizona have kept close cultural relationships with people in Sonora despite more than 500 km (310 miles) distance and the international border. However, militarization of the United States border after September 11, 2001 has made border crossing ever more difficult.

MEXICAN-YOEME WARS

Following Mexico's independence from Spain in 1821, relations with the *Yorim* (outsiders, non-Yoeme people) became increasingly hostile, and from time to time escalated into warfare. Disputes generally centered around the usurpation of Yoeme lands and cultural independence. Troncoso (1905), Dabdoub (1964), Spicer (1980), Hu-DeHart (1984), Zatarain Tumbaga (2018), and others provide details of the complicated history of violence inflicted by the Mexican military and racist government policies. Many Yoeme insurgents resorted to raiding, but by the end of the nineteenth century the Yoemem were losing the struggle.

Three events are especially poignant. In February 1868 during the Bácum Massacre, 120 Yoemem, mostly women and children, were gunned down by Mexican military while trying to escape the burning church in which they were incarcerated. On January 18, 1900, the hard-fought battle of *Maso Kova* (*Masokova*, *Mazocoba*) ended with more than 400 Yoemem killed on the battlefield. "Over 1,000 persons, many of them women, children, and wounded men were taken prisoners; only 834 managed to survive the long trek to the campaign headquarters at Las Guásimas" (Hu-DeHart 1984:143). Over 120 Yoeme men, women, and children were killed by Mexican federal troops at the Battle of Sierra Mazatán in June 1902. Hrdlička (1904) made a gruesome collection of human remains and artifacts at the Mazatán site, which were sent to the American Museum of Natural History. In 2009, the remains were returned to Yoemem in Sonora for proper burial (Padilla and Moctezuma 2015).

By the early twentieth century, the Mexican and Sonoran governments escalated warfare to genocide. Many Yoemem fled to remote places in the Sierra Bacatete, along the northeastern reaches of the Yoeme territory. They lived there for extended times, making use of their knowledge of edible plants, hunting, and wilderness survival. They also received clandestine supplies from their brethren working on haciendas. Many fled to Arizona.

In the mid-1980s, Miguel Romero provided the following narrative documented in a tape-recorded interview. A copy of the recording was given to Felipe by Ángel Romero, Miguel's nephew living in Pótam. Miguel is referring to the early part of the twentieth century when the people were in the Sierra Bacatete fleeing from the Mexican military. Some of the words are archaic terms no longer in use. *Saawa* is the edible plant *Amoreuxia palmatifida* and *peronim* for Mexican soldiers is a derogatory word taken from the Spanish *pelones*.

...ne hiva haha'e ala
hak yaiwak haivu haene
huname ili kuu choam kia inian ili
kokoela chuktim huname nu'une
hakun yaiwaka'apo vea haivu am bwa'ane
bwe si hiokot machi omme si ka tu'i
kia si rohiktiachi
ko'oko hiokotmachi
ili uusim hi'ibwavaeka bwanne
haisa mamachi ume ili uusim
ta hitasa am bwatuane
waate hak ili kuu'uta
puannamame ili am mimikne
ta kuu'uta hiva ili pinne
bwa'ame into kaita
saawam hitasa ket aayuk o'ovek
ta ume peronim into kia
um itot cha'aka na katne
haisa a hita bwa'amta hariune

So I always cut agave
when we arrived at a place, I cut agave
Those little agave crowns just like these little
round cuts, those were used for food later
So when they arrived at a place, they could
already be eating them. Well, it is so painful,
it is, man, so bad Just so sad. Painful, it was pitiful.
The little children wanted to eat and they would cry
That is how little children are.
But what did we have to give to them to eat?
Some people somewhere there carried a little agave
They gave a little bit of that to the children,
but they had only little pieces of agave to suck.
There was no other food.
There are other foods there in the mountains,
saawam and others,
but we didn't have time to get them because
of the pelones. They were after us there,
how could we look for food?

DEPORTATIONS AND DIASPORA

Dr. Manuel Balbás, a surgeon from 1899 to 1901 with the Mexican army that invaded Yoeme territory, recognized how deeply the Yoemem were attached to their land. In his words (Balbás 1927):

> The Yaqui alleges property rights in a belt of land, which includes the eight primitive towns of the river...and in addition the mountains of Bacatete which are on the North...Outside this zone, which is very large and rich as a result of its excellent irrigable land, its woodlands and pastures, the Yaqui pretends absolutely no dominion, recognizes without argument private property rights and is submissive to all constituted authorities. But inside the zone...he recognizes no owners other than themselves, no authorities other than those named by themselves. That is to say that they desire to establish within this land an isolated fatherland from the rest of the world, absolutely free, independent and sovereign. (Spicer 1980:141)

Despite his awareness of Yoeme desires, Balbás was typical of the *cienetíficos* who advised Mexican president Porfirio Díaz. Balbás believed "progress" was inevitable and just. Expressing the widespread sentiment (propaganda) of the time, he wrote:

> The margins of the Yaqui River can constitute one of the greatest agricultural riches of the country. This river, like the Nile, inundates each year a great extension of land, leaving as it recedes to its bed, a magnificent accumulation of organic substances...
>
> It cannot be permitted that such riches remain indefinitely in the power of men who do not know how or care to exploit them usefully...
>
> Progress has great requirements and people who do not adjust to them must succumb, because the irresistible force of the universal action must push them ahead and absorb them in the course of its constant and powerful development. (Spicer 1980:141)

Along with the subjugation of Yoeme lands, one brutal aspect of "progress" was the literal fragmentation of the Yoeme population during the late 1800s and especially the early 1900s. Until the Mexican Revolution of 1910, the government mounted a program of:

> ...massive and systematic deportation of all Yaquis of Sonora. Most of the unfortunate Yaquis ended up in the notoriously brutal henequen plantations of Yucatán, where few survived the rigors of virtual slave labor. More than any military campaigns, deportation eventually broke down the spirit of the Yaquis, who could not sustain the resistance when they were physically dispersed and their families wantonly torn apart. (Hu-DeHart 1984:155)

Like so many others, Anselma Tonopuame'a Castillo's aunt, Joaquina Galaviz-Tonopuame'a, was deported to Yucatán in the first decade of the twentieth century, where she worked for a family as a slave. She silently wept when she had to prepare food for the family. Her tears went into the food and she said, "The Yoim ate my tears." She was weeping, thinking of her family back in Wiivisim. She escaped and walked back to Sonora. She told Anselma that many people got sick on the ships and were thrown overboard when they died. They were taken to Manzanillo and made to walk to Yucatán (over 2,000 km, or 1,200 mi, by road). Many more died. Although the precise number of deported Yoemem has not been determined, Governor Rafael Izábal of Sonora bragged he had shipped out 2,000 by 1907. The following year the deportation program peaked, so the final figure must have been considerably higher. To escape the net, most Yoemem fled their homeland and fanned out across northwestern Mexico. They worked in mines and haciendas, changing their names and masking their identities. Public religious ceremonies were abandoned. During the last decade of the Díaz regime, in the time of Sonoran governor Rafael Izábal, being Yoeme or indigenous meant brutal peonage or death.

It was during these times that many Yoemem crossed the border into the United States, especially Arizona. They were railroad workers or labored in the fields around cities such as Phoenix, Tucson, and Yuma. According to Spicer (1980:158), "the Yaquis had become the most widely scattered native people in North America."

The Yoeme population within the tribal lands was reduced to a low point of fewer than 3,000 as a result of wars, deportation of thousands to elsewhere in Mexico, and emigration of thousands more to the United States. They established communities in Arizona and others sought employment in places such as California, New Mexico, Oregon, Texas, and Washington.

"BOUND IN TWINE"

Slave labor and plantations provided raw material for the henequen-wheat complex, a geopolitical economic triangle: Yaquis-Yucatán-North American wheat ("Bound in Twine," Turner 1910; Evans 2007; see *Agave fourcroydes*, Asparagaceae). The 1910 revolution ended slave labor in Mexico, but exploited labor continued to support the Yucatán plantations.

The Great Plains farms of the United States and Canada produced a wealth of wheat that changed global food production and politics. Henequen from Yucatán was made into commercial twine in the United States factories, commonly in prison factories. Henequen fiber was needed for manufacturing binder twine used for tying bundles of cut wheat. Binders, essential harvesting equipment, cut grain stalks and tied them into bundles that were gathered into shocks before threshing. The henequen-wheat complex spanned the 1880s to the 1930s and persisted into the 1950s. The innovation of a combined harvester-thresher machine (known as a combine) ended twine-binding and the henequen–wheat complex.

MEXICAN REVOLUTION

The onset of the Mexican Revolution in 1910 ended the deportation program and *de facto* slavery in Mexico. During the conflict, many Yoemem participated in that struggle, joining the armies of different factions. Some formed all-Yaqui battalions, fighting for leaders like Álvaro

Yoemem prisoners, ca. 1908. The photo shows women and children, although men were also in this group. They are being marched from Hermosillo to Guaymas to be deported. Univ. of Southern CA Digital Library, CA Historical Society Collection, accession #1520. Photographer unknown.

Totoitakuse'epo (Hill of the Rooster). 13 Mar 1989 (WS).

Obregón, who was also from Sonora. Thousands of Yoemem returned to the Yoem Bwiara to rebuild their lives and culture. However, encroachment and expropriation of Yoeme land by Yorim (non-Yoeme people) continued. Sporadic violence flared between the Yoemem and the Mexican military. Dispossessed Yoemem engaged in guerilla warfare.

Francisco Madero, the first new president, reportedly told a group of Yoeme survivors that they would receive compensation for their losses and their lands would be restored. Madero was assassinated in 1913 and promises were forgotten. Former general Álvaro Obregón established a vast estate and commercial enterprises on Yoeme and Yoreme lands. Warfare flared in 1916–1917 and led to renewed deportations to distant prisons, including the Islas Marías Federal Prison established in 1905.

At Kompuertam (near Vícam) in March 1989, Luciano Velasquez told us that he was a young boy when he and his family came back to their homes near Pótam from the mountains (Sierra Bacatete). There they found soldiers had strung up Yoeme men with wire and hung them in mesquite trees. So they cut them down and buried them.

Mexican soldiers would kill boys because boys could become rebel fighters. People dressed boys as girls, but the soldiers would pull up their skirts and kill them. Soldiers killed boys in front of their parents. These events are described in *A Yaqui Life* (Moisés et al. 1971) and other accounts.

The last military campaign began in 1926 and was the most destructive one. Álvaro Obregón apparently forgot that Yoemem helped bring him to power during the revolution and led assaults against Yoeme forces. Thousands of Yoemem

Field cross of mesquite, Kopas (between Pótam and Vícam). 13 Dec 1988 (WS).

noncombatants fled again to the Sierra Bacatete, but the army dropped bombs from airplanes to ferret the people out of their hiding places. The Yoemem fought their last major battle in 1927 at *Totoitakuse'epo* (Hill of the Rooster, *Cerro del Gallo*). They were defeated and Mexican garrisons were established in Yoeme towns and remained until the 1970s.

Rosario Vakame'eri-Castillo said that, after the end of the wars, men from the *Wohnaiki Pweplom* (eight pueblos) came to Arizona asking Yoemem to return and rebuild the villages. The men from Sonora said there is peace and that the holy crosses of mesquite were once again standing at the households. Some returned, most did not. Rosario and his family were happy in Marana and did not want to return to Mexico.

YOEMEM IN THE MODERN WORLD

In 1937, Mexican president Lázaro Cárdenas issued a decree that stated:

> To the Yaqui tribe is ceded the whole extension of the workable land located on the right bank of the Yaqui River, with the necessary water from the dam in construction at Angostura, similarly all the mountain lands known as 'Sierra del Yaqui' [the Bacatete Mountains], for which areas shall be provided the resources and elements necessary for the better utilization of their lands. (Spicer 1980:263)

The Yaqui Indigenous Zone included only one-third of the historical Yoem Bwiara, all of it on the north side of the Río Yaqui. Yet, for the first time, the Mexican government recognized Yoeme land rights. Promises were made and, as usual, were mostly forgotten. The government promised and in 1952 built a giant river dam, forming the lake known as "El Oviáchic," and engineered canals for irrigation. Eventually some of the water reached Yoeme farmers, and by the 1970s many were cultivating their rich farming heritage. Most of the water, however, went south of the river to serve Yorim, American, and Canadian interests.

Half a million acres of irrigated land south of the Río Yaqui brought wealth to Yorim and North Americans. That part of the Yaqui Valley became the cutting edge of industrial agriculture. It was there, on the south side of the Río Yaqui, that the Rockefeller Foundation funded the wheat-breeding program of Nobel laureate Norman Ernest Borlaug that triggered the Green Revolution of the 1940s and later decades. Ciudad Obregón (more properly known to the Yoemem as Cajemé) on the southern margin of the historic Yoem region, became a city of 120,000 in 1950 and had grown to 430,000 by 2015. The burgeoning population in Sonora craved and obtained more and more water from the Río Yaqui, leading to less water for the Yoemem. By the early twenty-first century, the river in the Yoem Bwiara was reduced to a trickle (Padilla Ramos and Moctezuma Zamarrón 2017).

Felipe quotes a letter written by Yoemem in Sonora around 1990:

> We belong to a people of indomitable fighting spirit. Our race never gave in during the dark days of the Spanish and we set an example as a people who never allowed ourselves to be conquered, nor for our lands to be taken away during the colonial period. [The writer then talks about how President Cárdenas recognized Yoemem title to their land.] With the strength which has been recognized as ours, and the great fighting and working spirit of our people, the Yaqui community began to be a prosperous and united people. However, our richness was coveted; the renters came, the profiteers, the great landholders, who with the help of evil officials grew rich at the expense of our people. (Evers and Molina 1992:19)

In Arizona, the Pascua Yaqui Association obtained 200 acres of land southwest of Tucson in 1964, which became New Pascua. The Association gained recognition as a United States Indian Tribe in 1978. Older Yoeme communities including Pascua and Barrio Libre in Tucson, Yoem Pueblo in Marana, and Guadalupe and Penjamo near Phoenix remain vibrant cultural enclaves. Yet poverty reigns in the Yoem Bwiara, driven along by banks, mafia, racism, politics, and diverting water from the Río Yaqui.

Around the year 2015, about 30,000 Yoemem lived in Sonora and over 10,000 in the United States, primarily in Arizona.

YOEME CONSULTANTS AND TEACHERS

It is our intent to acknowledge people who have been instrumental in our efforts to put together this publication. We sit on their shoulders and pass the information in this book to the next generations of Yoemem and scholars. Any omissions are not for lack of appreciation or debt, and sincerely regretted.

We are reminded that in earlier traditional Yoeme culture, peoples' names were often fluid, not firmly fixed as in Anglo-Hispanic culture. Minor differences in spelling were often not considered important or resulted in publications of an author imposing American English or Mexican Spanish spelling. Furthermore, Spicer (1980:160) tells us that in efforts to escape persecution and deportations of the late nineteenth and early twentieth centuries, "Names were changed and rechanged constantly; Yaqui surnames were abandoned for Spanish; so that Husacameneas overnight became Valenzuelas; but Valenzuelas of yesterday became Molinas of today." Also see Jane Holden Kelley's introduction in "A Yaqui Life" (Moisés et al. 1971) and Rosalio Moisés (below). People in Sonora have used accents with their names, while people in Arizona generally have not used accents with their names.

We have not attempted to provide comprehensive or even similar narratives for the different people mentioned here. We include a few Yoemem we did know or work with, but whose legacy is significant. In using the present tense here, we are often referring to a cutoff date of about 2015.

Jesús Alvarez-Vasquez, see **Jesús Yoilo'i**

Guillermo Alvarez-Valenzuela

He resides in Tucson and is originally from Pótam. He has been a pahko'ola dancer since the age of eight.

Pedro Alvarez

Deceased. He lived in the Eloy–Casa Grande area in Arizona, and in about 1978 he moved to *Waalupe* (Guadalupe), Arizona.

Steve Armadillo

He lives in Tucson and is Felipe's younger brother. At various times he has assisted with our project. He helped with the English to Yoeme translations for the Las Guásimas interview sessions in May 2019.

Antonia Amarillas-Flores

Deceased in 2000. She lived in Pótam, but was originally from *Kobwabwa'im* (Great Blue Heron), the old Pótam village. She was very knowledgeable about plants in the mountains and the valley of the Yoem Bwiara, and had extensive knowledge of traditional ways. Her husband was Ignacio ("Nacho") Amarillas-Sombra.

She said that once she and Nacho went to the mountains to gather *huvahe taakam* (*Vitex mollis*) fruits and it started to rain and continued for days. They were trapped there in the mountains. When the rain stopped they went back to Pótam by canoe because the river was so high, and they saw live chickens on top of logs drifting out to sea. They felt sorry for the chickens but could not do anything to save them. The whole village was flooded.

On various occasions in the late 1980s, Felipe, Richard, and photographer Bill Steen stayed with Antonia and her family in Pótam and shared meals with them.

Ignacio ("Nacho") Amarillas-Sombra

Deceased in 2005. He was a very knowledgeable man from Pótam. He provided us with a wealth of traditional information. He was well-known for his mat- and basket-making skills using *vaaka* (*Arundo donax*). His wife was Antonia Amarillas-Flores. Nacho was Felipe Molina's cousin. We worked with him in December 1988 and March 1989, mostly at Pótam, and also at Chiinim and Rahum.

Guillermo ("Checho") Amarillas-Flores

He was Ignacio Amarillas' son. He obtained permission from the traditional authority in

Pótam for us to travel through the Yoeme lands in Sonora. He also introduced us to many traditional families and provided information on animals and plants, including modern crop cultivation. Guillermo was killed in an assault and robbery in February 2017.

Teresa Amarillas

Deceased in late 1990s. She was a *kantoora* (hymn singer) and skilled in Yoeme traditions. She lived most of her life in Pótam but was born in Arizona. An Anglo farmer provided her family and Alfonso Leyva-Flores' family with 10 acres of land near Rillito, Arizona, to live and garden. The families lived there for a few years and then returned to Sonora when Teresa was an infant. She was witness to warfare and atrocities by Mexican soldiers. During the wars, instead of going to the mountains, her family and neighbors hid in the mangroves along the Yoem Bwiara coast. In her later years she would visit her uncle Tomas Martinez in Marana. She was a cousin of Ignacio Amarillas and Felipe Molina.

Estanilausio Bacasegua

He lives in Tucson and is originally from Loma de Guamúchil in Sonora. He has vast knowledge of plant and animal life in the Yoem Bwiara. He told Felipe how Loma de Guamúchil was settled. His grandmother's family was one of the many families who were forced to leave their home in Vahkom (Lagoons, now named Bácum) by the encroaching Yoorim (Mexicans).

Clementina Bainori

She lives in Las Guásimas and is originally from the Yorem Bwiara (Mayo Land). She is married to Hipolito Flores-Romero.

Vicente Baltazar

Deceased 1994. He fought with the Mexican federal troops against the Villistas and the traditional *Kau Homem* (Mountain Dwellers). Vicente lived in Pótam, was a farmer, and a member of an agricultural *socio* (*ejido*). He shared some of his considerable knowledge of plants and history.

Roman Borbón

A pahko'ola dancer and mask maker. Originally from Hermosillo, he later moved to Tucson where he lived with his wife Frances Flores. They passed away in the 1990s. He provided Richard with considerable traditional plant knowledge on shared trips to Hermosillo and also in Tucson.

Anselma Castillo, see **Anselma Tonopuame'a-Castillo**

Rosario Castillo-Bacaneri, see **Rosario Vakame'eri-Castillo**

Armando Felix

Armando (deceased) was a farm worker in the Rillito-Marana area. He had extensive knowledge of plants in the Santa Cruz River area. His wife was Gabriela Felix.

Gabriela Felix

Deceased 1993. She lived in Rillito with her husband Armando Felix. Gabriela was originally from the *Bwe'u Hu'upa* settlement near Marana. The family moved to Pascua and then to *Kampo Wiilo* in Marana and finally to Rillito.

Alejandra Flores-Romero

She lives in Las Guásimas and is the daughter of Juana Romero and Alfonso Leyva-Flores. She is married to Leoncio Valenzuela. She learned some of her knowledge of the sea and fishing from her brothers and mother.

Ángel Flores-Romero

He lives in Las Guásimas and is a fisherman. Felipe and Richard worked with him on sea life and land plants in the late 1980s and Felipe interviewed him again in May 2019. (His older brother, from Pótam, is also named Ángel Flores-Romero.)

Ángel Flores-Romero

He lives in Pótam and is the oldest son of Alfonso Leyva-Flores. Juana Lugo-Osuna is his wife. Ángel gave Felipe a cassette tape of his uncle talking about the suffering during the wars with the Mexican Federal Army. Felipe interviewed him in June 2019 in Pótam.

Claudio Flores-Romero

He is the youngest son of Alfonso Leyva-

Flores. He lives in Las Guásimas and is a *kuchureo* (fiherman) with his brothers. In the early 2000s Claudio participated in inter-tribal runs from Vícam to Tucson and on to Mount Graham in Arizona. Felipe interviewed him in Las Guásimas in 2019.

Esteban Flores-Romero

He is a fisherman and lives in Las Guásimas, where Felipe interviewed him in 2019. He is a son of Alfonso Leyva-Flores.

José Guadalupe ("Lupe") Flores

He was born in Ko'oko'im (Cocorit) in 1925. He moved to Arizona and passed away about 2000. He was a deer singer from Barrio Libre and moved to New Pascua in the mid-1980s. He made pahko'ola masks, rosaries, beadworks, musical instruments, and other traditional crafts and ceremonial items that he sold to community members as well as to art dealers.

Hipolito Flores-Romero

Hipolito was originally from Pótam and later moved to Las Guásimas with his family. He became a fisherman in his younger years. He is married to Clementina Bainori. Richard worked with him in March 2007 and Felipe interview him in May 2019.

Martha Flores-Gonzalez and her husband **Mateo ("Saaro") González-O.**

They live in Vícam. When they were visiting in Tucson in 1994, we worked with them at the Arizona-Sonora Desert Museum and elsewhere. They identified plants and animals at the Museum and provided traditional knowledge. They have a fine vegetable garden and travel extensively in Sonora and bring seeds for their home garden. When we saw them in 1994 they had just returned from Yécora with vegetable seeds.

Mateo is a harp player and used to come to Tucson in the 1990s for the Lenten ceremonies. He and Felipe participated in the Folklife Festival at the Smithsonian Institution in Washington, D.C. in 1988.

Joaquina Galaviz-Tonopuame'a

She was Anselma Tonopuame'a-Castillo's aunt. Joaquina was also known as *Ne'esa Wachina* ("Aunt Joaquina," *ne'esa* is the older sister of a father and *wachina* is the Yoeme transliteration of Joaquina).

In the years before 1910, Ne'esa Wachina was captured by the Mexican Federal Army for deportation. She was taken to Guaymas to be transported by ship, and fated for Yucatán. She said many people died on the way and were thrown overboard. She was a house slave and said she cried every day thinking about her home in Wiivisim. When she cried her tears fell into the dough she kneaded for the family's tortillas. The people ate my tears she would say to her family including Felipe's grandmother. Joaquina escaped and walked all the way back home by herself, and related the things she saw and experienced in Yucatán. She passed away in the mid-1920s.

Estefana Garcia

She lived in New Pascua and previously lived in Marana. Estefana was Felipe's maternal aunt. She was born in South Tucson in 1918 and died in 1996. Her husband was Juan Luis Garcia. Before marriage she traveled with her father Rosario Vakame'eri-Castillo and mother Anselma Tonopuame'a-Castillo to western Arizona as her father worked for the Southern Pacific Railroad Company. She was a family historian.

Juan Luis Garcia

Deceased 1988. He was born in Sells and lived most of his life in Marana, and in his later years he moved to New Pascua. His mother, Josefa Garcia was a *hitevi* (traditional healer). He knew of a treatment for the recluse spider bite, and once healed his daughter of such a bite. Felipe Molina is his nephew.

Ignacio ("Nacho") González

He lives in Las Guásimas. Richard collected plant specimens with him in 1985.

Mateo ("Saaro") González-O., see **Martha Flores-Gonzalez**

José María ("Chema") Jaimez

We worked with him in March 1989 at his

home in Kompuertam. José María (deceased) had extensive knowledgeable of plant life in the Yoem Bwiara valley and the nearby mountains. His wife is María Valenzuela. They had a fine garden at their home compound surrounded by a beautifully woven carrizo fences (*Arundo donax*).

Alfonso Leyva-Flores

He was born in Yuma and when he was a young boy the family moved back to Pótam. An Anglo rancher in Arizona had provided his family and Teresa Amarillas' family some land, about 10 acres, north of the Mortero Hohokam site near Rillito. The families decided to return to Pótam and left their Arizona land. Alfonso grew up in Pótam and in the early 1950s he moved to Las Guásimas where we worked with him in the 1980s. He passed away in 1991. He used to come to Marana to visit his uncle, Tomas Martinez, and other relatives. He is identified as Don Alfonso Florez Leyva in "The Elders' Truth" (Maaso et al. 1993).

Fernaldo Leyva-Flores

He worked with us on sea life and land plants at Las Guásimas and again with Felipe in 2019. Fernaldo was originally from Pótam. The family moved to Guásimas when Fernaldo was young, and he became a fisherman. He is also known as Fernando Flores-Romero and is Alfonso Leyva-Flores' son.

Carmen Liogue-Garcia

Deceased 1971. She lived in New Pascua and was originally from Vícam. She grew medicinal plants for a neighbor *hitevi* (traditional healer). In some publications she is known as Carmen Liowe. The Liogue Senior Center of the Pascua Yaqui Tribe is named in her honor. She was married to Rosalio Valencia-Vega (see Rosalio Moisés).

Juana Lugo-Osuna

She lives in Pótam and is the wife of Ángel Flores-Romero (the oldest Flores-Romero brother). She has vast knowledge of plants and animals in the Yoem Bwiara. She also has knowledge of the supernatural world and the Huya Ania around Pótam. Felipe interviewed her in 2019.

Miki Maaso

Miki Maaso (also known as Miguel Cinfuego) passed away during Holy Week 2018. He lived in Sonora and had been to Tucson to participate in the Easter ceremonies as a tampaleo. He was a deer singer and deer dancer in Sonora for ceremonies at the large pueblos. He contributed to The Elders' Truth (Maaso et al. 1993) and Yaqui Deer Songs (Evers and Molina 1987).

Eddie Martinez

He lives in the Tucson region and is the son of Frank Martinez. Eddie began making pahko'ola masks in 2000.

Francisca Martinez

Deceased 1985. She lived in Marana and was originally from Tórim. Her husband was Tomas Martinez. Francisca was a great gardener. She planted *hiak viva* (*Nicotiana rustica*, Solanaceae) in containers, for use as medicine. She had many edible and flowering plants in her yard.

Frank ("Chico") Martinez

Deceased summer 2006. During the 1980s, his house at Old Pascua was a scene for Yoeme craft production including pahko'ola masks, flutes, and other items sold to dancers and musicians for Yoeme ceremonies as well as to art dealers. He was a pahko'ola dancer and used his own masks.

Frankie Martinez, Jr.

Deceased October 2004. He lived in the Tucson region and was Frank Martinez's son. Frankie was a promising young artist, a deer singer and mask maker. He was killed in an accident at the age of 21.

Tomas Martinez

Deceased ca. 1991. He lived in Marana and was originally from the Yoem Bwiara in Sonora. He was married to Francisca Martinez. Tomas was knowledgeable about plant and animal life both in Arizona and Sonora.

Cruz Matus

In 1985 he was about 80 years old and had lived in the Guaymas region since childhood. He accompanied Richard to the surrounding desert and shared traditional knowledge of plants and life in the Guaymas region, including from the early twentieth century.

Jose Matus

Jose Matus (deceased) was also known as Jose Yoeta. He lived in *Waalupe* (Guadalupe), Arizona. Yoeta is an enchanted and spiritual sound heard in the Huya Ania. Matus is a common Yoeme surname, related to or derived from *maatu*, the term for charcoal. Jose was a good historian and was concerned about Yoeme political and social life both in Arizona and Sonora.

Rosalio Moisés (Rosalio Valencia-Vega)

Deceased 1969. He was born 1896 at the Colorado Mine in Sonora and at various time he lived in Arizona, Sonora, and Texas. Rosalio's autobiography is recorded in *A Yaqui Life: The Personal Chronicle of a Yaqui Indian* (Moisés et al. 1971). In the extensive introduction by Jane Holden Kelley (page xiv) we learn, "He was given the name Rosalio Moisés Valenzuela at birth," and later in life used several other names. He served as interpreter and guide for William Holden because he was trilingual (Holden et al. 1936; Rosalio's name is altered as Ramón Torrey). At various times Rosalio worked with Professor Holden in Texas. In his later years he lived in Arizona where he was known as Rosalio Valencia-Vega, and was married to Carmen Liogue-Garcia. Although we did not know Rosalio, we find his firsthand information on plants and animals to be credible and significant.

Guadalupe ("Lupe") Molina

Deceased in the mid-1980s. He is also known as Don Lupe or Don Lupe Masobwikame. He was a deer singer from Vikam Suichi and sometimes visited relatives and friends in Arizona, especially as a deer singer. As a young man he was drafted into the Mexican Federal Army. He fought against the revolutionaries in the Mexico City region, especially at Tlaxcala. He said that in addition to taking their weapons and provisions, he and other Yoeme soldiers also took musical raspers and gourds. In between fighting they would sit together and sing deer songs for the deer dancer. He said the people of the Tlaxcala region would come and listen.

Maria Moreno

Maria lived in Rillito (near Tucson) and moved to Yoem Pueblo (Marana) in the early 1970s. She had extensive knowledge about medicinal plants, especially in the vicinity of Marana. Maria passed away in the mid-1990s.

Julian ("Julio") Morillo

Deceased 1995. The Mark Rossi Studio in Tucson had an exhibit featuring the work of Julian Morillo in April 1995. The artist was ill and could not attend and died in October. The show featured 35 years of Julian's work.

Julian Morillo was a prolific carver of sculptures related to Yoeme traditions, including pahko'ola masks. He was a groundskeeper at the University of Arizona during the 1980s. He often would take exotic woods that were cut down or trimmed from the University of Arizona campus in Tucson—any that caught his eye, and try his hand at carving the wood.

His masks and small sculptures were highly traditional but because of his access to non-traditional woods, his work was different from others in the Tucson area. The sculptures were made from cedar, cottonwood, fir, ironwood, oak, maple (from flooring removed from the old Bear Down gym on campus), mesquite, olive, and redwood. His masks were mainly made from cottonwood root and desert willow, although cedar, redwood, and some unidentified woods were also used.

Fernando Murillo-Rendón

Originally from Pótam, he moved to Tucson after marrying a young woman from Tucson. Richard met him in the 1980s when he came to Tucson as a pahko'ola dancer. Together they made a number of trips from Tucson to Pótam and elsewhere in the Yoem Bwiara. Fernando taught Richard traditional knowledge of plants, animals, and Yoeme culture.

Maria ("Chira") Murrieta

Maria Murrieta (deceased) was a *kiyohtei* (altar woman) at Pascua and took care of the church altar. She helped verify plant names and also helped with words for the Yaqui Bilingual Dictionary Project in the Tucson Unified School District. She was knowledgeable of plant life in the Tucson and Marana area.

Luis Ochoa

He lived in Marana and was married to Meregilda Ochoa. Luis (deceased) was originally from Guadalupe, Arizona, where he was a Pilato in the Pharisee Society.

Meregilda Ochoa

Meregilda (deceased) lived in Marana. She learned a great deal of traditional knowledge from her mother, Romana Sanchez who was one of the original settlers of *Ili Huu'upa* village in Marana in the early 1900s. Meregilda had a beautiful garden where she grew fruits, vegetables, medicinal plants, and flowers. Felipe and Richard worked with her on various occasions.

Miguel Romero

When Felipe met with him in the 1980s he was an elderly blind man living in Pótam; and later deceased. He was very knowledgeable about Yoeme history. He said that every Yoeme person had to register with the Mexican government and get an identification card and have it at all times; because if they caught you without this card you would be deported to Yucatán or killed. (That would have been in the early twentieth century.)

Romana Sánchez

Deceased in the mid-1980s. Doña Romana was a traditional person and lived most of her life in Marana. She had witnessed the wars against the Mexican federal troops, because her parents were in Pancho Villa's army. She credited Pancho Villa and Francisco Madero for the survival of the Yoemem. Felipe learned from her in the 1970s. She was the mother of Meregilda Ochoa.

Anselma Tonopuame'a-Castillo

She was Rosario Vakame'eri-Castillo's wife and Felipe's grandmother. She passed away in 1975. Originally from Guaymas, she lived most of her life in Marana. She said that her sister, Fernalda, knew and used ancient Yoeme terms no longer in use. In 1967, she was very interested in giving Felipe the names of sea life, birds, and other animals from the Yoeme lands. She came from a seafaring family.

Rosario Vakame'eri-Castillo

Deceased January 1979. His mother's family name was Paz Vakame'eri (Bacaneri in Spanish), and he was also known as Rosario Castillo-Bacaneri. Vaka is cane (*Arundo donax*) and *me'eri* means killed, "Cane Killed" (for some reason the cane is killed).

He was Felipe's grandfather and mentor. A substantial part of the information reported in this book derives from his teachings, including notes that Felipe wrote at the time.

Rosario was born in Guaymas on October 4, 1888. He came from a fishing family in Guaymas, at a Yoeme community there called *Matanza*. Soldiers in Guaymas took his father away and he was never heard from again. Soldiers shot his older brother, Geronimo, in the knee and then they took him away. Geronimo was a pahko'ola dancer.

On January 18, 1900, the Maso Kova massacre occurred and Rosario witnessed the transporting of men, women, and children to Tórim. His mother told him not to play with the children brought down from the mountain because she was afraid he would be locked up with them and be deported. The people brought down from the mountain were kept at Tórim with federal soldiers guarding them in a big corral. Some were executed and the rest deported to work as slaves in henequen fields in Yucatán or perhaps sugar plantations in Oaxaca.

Rosario's mother died in 1901 and then Rosario was on his own. When he was about 12 years old he went to Santa Rosalia, Baja California, with a Mexican (non-Yoeme) fishing crew. He became a fisherman and regularly traveled between Guaymas and Santa Rosalia on large fishing boats. The fishermen would hide him below deck when soldiers came aboard to inspect the boat looking for Indians to deport.

Rosario was taken into Pancho Villa's army and traveled around fighting, mostly in the Guaymas region. He was a trumpeter for one of the battalions. He sounded the trumpet for battles and remembered once being in Hermosillo in front of the government palace. At the same time the Yoeme coyote drum (*wiko'i ya'ura*) would be sounded by another boy walking in the form of a cross as a blessing to the four directions. Romana Sanchez in Marana said that the grandmother of the coyote drummer was afraid of the boy's fate, so she walked with him as he drummed.

Eventually Rosario was told, "the best thing was to go north," because the battle was being lost. So he went north and lived in Nogales, at *Nogaliitom*. From there he went to Continental, Arizona, and lived in "box car" or section car (*sesionnim*) with his future wife Anselma Tonopuame'a, who he met in Nogaliitom.

Rosario moved to Marana in 1921 and worked for the Anway family. Rosario and Anselma married in Tucson in 1922. They first lived in a tent. The family settled at Kampo Wiilo in the 1930s and moved to Kampo Uuno in 1942 because there were abandoned houses there from people who had returned to Sonora. (Kampo Uuno was named Yoem Pueblo in 1980.) He also worked for the Southern Pacific Railroad Company.

Rosario had been a chapayeka and became a pahko'ola dancer in the 1930s. He carved his own masks. He continued dancing until 80 years old, when he had an accident in 1967 (Molina et al. 2003).

Bernaldo Valencia

About 1920 to 2005. He lived in Tucson and was originally from Pótam. As a boy he lived with his elders in the Vakatetteve (Bacatete Mountains) when they were hiding from Mexican soldiers.

He was a noted tampaleo and people say that he was also a deer dancer. He also performed in the Wiko'i Ya'ura (Bow Leaders' or coyote dance). He enjoyed sharing his considerable knowledge and collecting medicinal plants from the Tucson Mountains. He shared knowledge with Felipe and Richard, especially in the 1990s. Bernaldo was traveling with Felipe in 1977, and at a rest stop about 10 miles east of Globe, Arizona, he saw plants he called *yerba pahmo*. Bernaldo was excited to see the plants and gathered bundles of yerba pahmo to take home to Tucson for medicinal purposes.

Rosalio Valencia-Vega, see **Rosalio Moisés**

Angelina Valenzuela-Rodriguez

Deceased 2001. She lived in Tucson and was originally from Eloy. She grew medicinal plants for various *hitevim* to use. Rosalio Valencia-Vega (see Rosalio Moisés) was her maternal grandfather.

Celina Valenzuela-Flores

She lives in Las Guásimas and was interviewed by Felipe in May 2019.

José María ("Chema") Valenzuela-Matus

José lives in Tucson and is originally from Vikam Suichi. He first came to Arizona to play the harp for the pahko'olam. He continues be a harp player at ceremonies in the Arizona Yoeme communities. He was interview by Felipe in August 2018.

Leoncio Valenzuela

He lives in Las Guásimas and is married to Alejandra Flores-Romero. Leoncio was interviewed by Felipe in May 2019.

Louis David Valenzuela

David lives in Tucson, where he is well-known for his pahko'ola masks and other Yoeme arts. His maternal great-grandfather was Rosalio Valencia-Vega.

Luis ("Veeteme") Valenzuela, Sr.

Luis (deceased) lived most of his life in Guadalupe, and more recently at New Pascua. He was originally from Pótam. Veeteme can be translated as "burning."

María Valenzuela

She lives in Kompuertam. Her husband was José María Jaimez.

Erasmo Valle

Deceased September 1996. He lived in New Pascua. Originally from Texas, he married Paula Valle. He has contributed much southwest Hispanic information. He was Felipe Molina's stepfather.

Juanita Paula Valle

Deceased in 2000. She lived in Marana most of her life. She was Felipe Molina's mother. She was noted for knowledge of the history of Old Pascua and Marana, and collected historic photos from the region. She learned traditional knowledge from her great aunt, Fernalda Anguis Tonopuame'a, who was from Guaymas. Juanita was a *kiyohtei* (altar woman) in Pascua.

Luciano Velasquez

When he was a young boy he witnessed atrocities of the Mexican-Yaqui wars. Felipe recorded a version of the *Talking Tree* narrative when Luciano told it at Kompuertam in 1982 (Evers and Molina 1987:37–38). Felipe and Richard worked with Luciano in Kompuertam in March 1989. He is deceased.

Jesús Yolo's

Deceased 1982 in Pótam. Don Jesús was a primary consultant for Larry Evers and Felipe Molina's (1987) *Yaqui Deer Songs* book.

[He] was also known by his Mexican name, Jesús Alvarez Vasquez. Born in Pótam, Sonora, around the turn of the century, Don Jesús grew up during the turbulent times surrounding the Mexican Revolution of 1910. While still a boy, he served in the *Yoem Vatayonnimmake*, the Yaqui Battalion, as a soldier and was wounded. When the fighting stopped, he came north across the border to work on the railroad along borderland tracks from Yuma into Texas. For a time he lived in one of the Yaqui camps near Scottsdale, Arizona, and worked in the cotton fields there. Very early in his life, Don Jesús lived with hardship, suffering, war and death. (Evers and Molina 1987:13)

Porfirio Yokiwa

Deceased mid-1990s. He was originally from Vícam and lived in the Marana area for the last 20 years of his life. Porfirio was a deer singer and later became a *Tampaleo*. In the 1970s and 1980s he taught Felipe about plants, animals, including the making of *tesguin* from agaves, and coyote and *nahi* songs and dances.

PART 2
BIOLOGICAL ETHNOGRAPHY

In this summary, a listing is provided for plants and animals in Yoeme culture, including ones from Sonora, primarily the Yoem Bwiara, and southern Arizona in the areas of the Yoeme communities. The individual species and taxonomic accounts in Parts 3 and 4 provide more detailed information. An asterisk (*) indicates plants and animals not native to the Yoem Bwiara and nearby regions in Sonora, as well as ones not native to southern Arizona.

PLANTS AS FOOD

In pre-contact times, Yoemem relied on farming and wild harvesting of food plants. Crops were watered by the twice-yearly floodwaters of the Río Yaqui. Cultivated food plants included maize, beans (teparies), squash, and amaranths. Other produce included cotton, gourds, and tobacco. These are warm-season crops.

Beginning with the arrival of Jesuit missionaries around 1617, new food plants and domestic animals were introduced, along with farming implements and tools, and associated agricultural practices. The Yoemem successfully adopted new crops and animal husbandry, forever changing their patterns of subsistence. Significant introductions from the Old World to the Yoem Bwiara include the following:

Cucurbitaceae – Gourd Family
- *Citrullus lanatus*
- *Cucumis melo*

Fabaceae – Legume Family
- *Cicer arietinum*
- *Pisum sativum*

Lythraceae – Loosestrife Family
- *Punica granatum*

Moraceae – Mulberry Family
- *Ficus carica*

Poaceae – Grass Family
- *Avena sativa*
- *Hordeum vulgare*
- *Saccharum officinarum*
- *Triticum aestivum*

Rosaceae – Rose Family
- *Prunus armeniaca*
- *Prunus persica*

Rutaceae – Citrus Family
- *Citrus sinensis*

Wheat (*Triticum aestivum*) was the single most important introduction and served as an economic base for the Jesuit missions. Planted in fall, it ripened with spring floods and was harvested in late spring with the advent of the pre-summer drought. Wheat provided a supplement for reliance on wild harvests such as mesquite and cactus fruits.

At least 20 other food plants of Old World origin became part of Yoeme cuisine. These include plants that have been cultivated in the Yoem Bwiara lands, as well as ones purchased in modern times at stores:

Amaranthaceae – Amaranth Family
- *Beta vulgaris*

Amaryllidaceae – Amaryllis Family
- *Allium cepa*
- *Allium sativum*

Anacardiaceae – Sumac Family
- *Mangifera indica*

Arecaceae – Palm Family
- *Phoenix dactylifera*

Asteraceae – Daisy Family
- *Carthamus tinctorius*

Brassicaceae – Mustard Family
- *Brassica oleracea*
- *Raphanus sativus*

Apiaceae – Celery Family
- *Coriandrum sativum*
- *Daucus carota*

Cucurbitaceae – Gourd Family
- *Luffa aegyptiaca*

Fabaceae – Legume Family
- *Glycine max*
- *Sesamum indicum*
- *Tamarindus indica*
- *Vigna unguiculata*

Moraceae – Mulberry Family
- *Morus alba*

Musaceae – Banana Family
- *Musa ×paradisiaca*

Solanaceae – Nightshade Family
- *Solanum lycopersicum*

Rosaceae – Rose Family
- *Cydonia oblonga*

Vitaceae – Grape Family
- *Vitis vinifera*

Non-native cultivated food plants originally native elsewhere in Mexico or the Caribbean region include the following:

Anacardiaceae – Sumac Family
- *Spondias purpurea*

Asparagaceae – Asparagus Family
- *Agave americana*

Bromeliaceae – Bromeliad Family
- *Ananas comosus*

Cactaceae – Cactus Family
- *Opuntia ficus-indica*

Cannaceae – Canna Family
- *Canna indica*

Fabaceae – Legume Family
- *Phaseolus vulgaris*

Solanaceae – Nightshade Family
- *Capsicum annuum*
- *Physalis philadelphica*

Myrtaceae – Myrtle Family
- *Psidium guajava*

In addition to cultivated crops, wild plant foods such as mesquite pods (*Prosopis glandulosa*), cactus fruits (especially *Pachycereus pecten-aboriginum* and *Stenocereus thurberi*), certain agaves (*Agave*), amaranth grain (Amaranthus), saiya roots (*Amoreuxia palmatifida*), and others were important parts of the diet. Basilio (1890 [1634]:198) listed nine herbs that were eaten, as well as 17 herbs that were not eaten. He also listed 16 trees from which edible fruits were gathered (Basilio 1890 [1634]:138), in addition to several other fruits and foods derived from plants. More than 65 wild plants, both native and non-native, have featured in Yoeme indigenous cuisine.

Amaranthaceae – Amaranth Family
- *Blitum nuttallianum*
- *Chenopodium neomexicanum*
- **Dysphania ambrosioides*

Apocynaceae – Dogbane Family
- *Marsdenia edulis*
- *Vallesia glabra*

Arecaceae – Palm Family
- *Brahea brandegeei*
- *Sabal uresana*
- *Washingtonia robusta*

Asparagaceae – Asparagus Family
- *Agave angustifolia*
- *Agave colorata*

Asteraceae – Daisy Family
- **Sonchus oleraceus*

Brassicaceae – Mustard Family
- **Sisymbrium irio*

Cactaceae – Cactus Family
- *Carnegiea gigantea*
- *Cylindropuntia arbuscula*
- *Cylindropuntia fulgida*
- *Cylindropuntia thurberi*
- *Ferocactus herrerae*
- *Lophocereus schottii*
- *Mammillaria grahamii*
- *Opuntia engelmannii*
- *Opuntia gosseliniana*
- *Pachycereus pecten-aboriginum*
- *Pachycereus pringlei*
- *Peniocereus marianus*

Peniocereus striatus
Stenocereus alamosensis
Stenocereus thurberi
Cannabaceae – Hemp Family
Celtis pallida
Cochlospermaceae – Cochlospermum Family
Amoreuxia palmatifida
Cucurbitaceae – Gourd Family
**Cucumis melo* var. *dudaim*
Cyperaceae – Sedge Family
Cyperus esculentus
Fabaceae – Legume Family
**Melilotus albus*
Phaseolus acutifolius
**Pithecellobium dulce*
Prosopis spp.
Sesbania herbacea
Lamiaceae – Mint Family
Vitex mollis
Malpighiaceae – Acerola Family
Malpighia emarginata
Malvaceae – Mallow Family
Ceiba aesculifolia
Guazuma ulmifolia
Martyniaceae – Devil's Claw Family
Proboscidea altheifolia
Proboscidea parviflora
Moraceae – Mulberry Family
Ficus insipida
Ficus pertusa
Nyctaginaceae – Four-o'clock Family
Boerhavia coccinea
Passifloraceae – Passion-vine Family
Passiflora arida
Polygonaceae – Buckwheat Family
Antigonon leptopus
Portulacaceae – Purslane Family
**Portulaca oleracea*
Resedaceae – Mignonette Family
Forchhammeria watsonii
Rhamnaceae – Buckthorn Family
Sarcomphalus obtusifolius
Rubiaceae – Madder Family
Randia echinocarpa
Randia thurberi
Sapotaceae – Sapote Family
Sideroxylon occidentale
Solanaceae – Nightshade Family
Capsicum annuum var. *glabriusculum*
Lycium andersonii
Lycium fremontii
Physalis crassifolia
**Physalis pubescens*
**Solanum americanum*
Solanum erianthum
Verbenaceae – Verbena Family
Lippia palmeri
Viburnaceae – Viburnum Family
**Sambucus cerulea*

In addition to food resources, plants used as sources of emergency water include the stem of the Arizona barrel cactus (*Ferocactus wislizeni*) and the roots of the kapok tree (*Ceiba aesculifolia*).

ANIMALS AS FOOD

Wildlife was abundant in the Yoem Bwiara until modern times. Native turkeys (*Meleagris gallopavo mexicana*) were the only domesticated food animals in pre-contact times. Domestic animals of Old World origin were introduced by Jesuit missionaries in the early seventeenth century, and some had been obtained in the sixteenth century. Introduced domestic animals include burros, cattle, chickens, goats, horses, pigs, and sheep.

Some of the animal species commonly utilized for food are listed here. These and others are discussed in the species accounts.

Crustaceans
- *Callinectes bellicosus*
- *Panulirus*, various species
- *Penaeus*, various species
- *Sicyonia*, various species

Insects
- **Apis mellifera*
- *Brachygastra mellifica*
- *Melipona*, unidentified species
- *Trigona*, unidentified species

Mollusks
- *Anadara tuberculosa*
- *Chione*, various species
- *Crassostrea*
- *Dosidicus gigas*
- *Lolliguncula panamensis*
- *Octopus*, various species
- *Ostrea*, various species

Fishes

The diverse marine and freshwater fish fauna of the Yoem Bwiara is discussed in the species accounts. Fish were variously important for subsistence in traditional times and more recently for commercial fisheries, such as at Las Guásimas.

Reptiles
- *Chelonia mydas*
- *Ctenosaura macrolopha*
- *Gopherus*, various species
- *Kinsperson*, various species
- *Trachemys yaquia*

Birds
- *Callipepla gambelii*
- **Gallus gallus*
- *Geococcyx californianus*
- *Meleagris gallopavo mexicana*
- *Pelecanus occidentalis*
- *Zenaida asiatica mearnsi*

Mammals

Domestic mammals:
- **Bos taurus*
- **Capra hircus*
- **Ovis aries*
- **Sus scrofa domesticus*

Wild mammals:
- *Lepus californicus, L. alleni*
- *Neotoma phenax*
- *Odocoileus virginianus*
- *Sylvilagus audubonii*
- *Tayassu tajacu*

PLANTS AND ANIMALS IN MEDICINE AND FOR HEALING

Traditional healers, hitevim, either women or men, would have an extensive inventory of medicinal plant species. Working with an hitevi in 1958–59 in Tucson, Shutler (1977) reported that this man had a repertoire of more than 70 healing herbs. Much has been written about traditional Yoeme medical practices. Primary references include works by Wagner (1936), Shutler (1967, 1977), Painter (1986), and Olavarría (1992). These and others mostly provide English or Spanish names and sometimes Yoeme names, few or no scientific names, and no specimen documentation; some are incorrectly identified. We identify approximately 140 plant taxa used in medicine and curing, of which 118 are native plants and 24 are non-native plants. Basilio (1890 [1634]:138) described medicinal or household uses for several kinds of trees. Indigenous concepts, including seatakaa (healing power) and other philosophies, are often associated with the use of medicinal plants. In a summary of medical practices for the Yoemem in Sonora in1934, Wagner (1936:79) observed that "Yaquis are a fairly healthy people," yet found people succumbing to tuberculosis and other maladies:

> Diseases prevalent are whooping cough, pneumonia, diphtheria, malaria and typhoid. Epidemics of smallpox occur and are widespread.
>
> Headache is treated with curative moistened leaves held to the head by moist cloths or leaves applied to face and wrists. Cathartics—herbs eaten or taken as tea—are used, while dysentery is controlled by other herbs. Small wounds, abrasions and ulcers are covered with curative wood-scrapings or leaves, and bandaged.
>
> The mortality is not at childbirth but within the first six or eight years when whooping cough, diphtheria and smallpox take their terrible toll. (Wagner 1936:79–80).

While most medicinal plants were collected locally, some were sought from distant places. Wagner (1936:86) reported that the hitevi he interviewed "at Torim said that it took a lot of work to keep up his collection of herbs and medicines. They had to be collected from over a wide area. He had to go as far as three hundred miles for some of them. With the utmost frankness he said that none of his remedies were sure."

Medicinal plants are often prepared as a tea or used as a poultice. The east side of the plant should be used for medicinal purposes, because the rising sun hits it first on the east side, it has more power there. Among others, Wagner (1936:84) confirmed this practice, "a brew is made from mesquite bark. The bark must come from the tree on the side of the rising sun." It is said that plants should be collected during the time of the full moon, unless it is an emergency.

On September 24, 1988, Felipe asked Teresa Amarillas, from Pótam, "Which are the plants you like the most for medicine?" She replied:

> A white cloth, or white color signifies purity and healing.
>
> Kosawe [*Krameria*], use only the stem, for cough.
>
> Si'iya, harave ya'ane [**Matricaria chamomilla*], for fevers, used for high fevers.
>
> Kau vattai [*Rumex hymenosepalus*], boil roots for sores and wash the sores with it; in Spanish it is called yerba colorado; also used for vokiam (fever blisters).
>
> Wiroa, any vine, ko'okoemta koktuane; when someone is sick, tie a vine around the neck; viivai a hittone [a vine cured with cigarettes, *Nicotiana*]

Plants in Medicine and for Healing

Acanthaceae – Acanthus Family
- *Avicennia germinans*
- *Elytraria imbricata*
- *Tetramerium nervosum*

Achatocarpaceae – Achatocarpus Family
- *Phaulothamnus spinescens*

Aizoaceae – Aizoon Family
- *Sesuvium portulacastrum*

Amaryllidaceae – Amaryllis Family
**Allium sativum*
Apiaceae – Celery Family
**Ligusticum porteri*
Apocynaceae – Dogbane Family
Sarcostemma clausum
Vallesia glabra
Arecaceae – Palm Family
Various species
Aristolochiaceae – Birthwort Family
Aristolochia watsonii
Asparagaceae – Asparagus Family
Agave spp.
Asphodelaceae – Asphodel Family
**Aloe vera*
Asteraceae – Daisy Family
Ambrosia ambrosioides
Ambrosia confertiflora
Ambrosia monogyra
Ambrosia salsola
Baccharis spp.
Bebbia juncea
Encelia farinosa
Encelia halimifolia
Isocoma tenuisecta
**Matrica chamomilla*
Pectis spp.
Perityle californica
Perityle microglossa
Porophyllum gracile
Porophyllum pausodynum
Verbesina encelioides
**Xanthium strumarium*
Xylothamia diffusa
Bataceae – Saltwort Family
Batis maritima
Bignoniaceae – Bignonia Family
**Tecoma stans* var. *stans*
Brassicaceae – Mustard Family
Descurainia pinnata
Bromeliaceae – Bromeliad Family
Tillandsia exserta
Tillandsia recurvata
Burseraceae – Frankincense Family
Bursera laxiflora
Bursera microphylla
Cactaceae – Cactus Family
Cylindropuntia spp.
Echinocereus leucanthus
Lophocereus schottii
Mammillaria grahamii
Pachycereus pringlei
Peniocereus greggii
Peniocereus striatus
Capparaceae – Caper Family
Atamisquea emarginata
Convolvulaceae – Morning Glory Family
**Ipomoea carnea*
Cucurbitaceae – Gourd Family
**Citrullus lanatus*
**Cucumis melo* var. *dudaim*
Ibervillea sonorae
**Luffa aegyptiaca*
Cyperaceae – Sedge Family
**Cyperus rotundus*
Ephedraceae – Ephedra Family
Ephedra aspera
Ephedra trifurca
Euphorbiaceae – Spurge Family
Cnidoscolus palmeri
Euphorbia spp.
Jatropha cardiophylla
Jatropha cinerea
**Ricinus communis*
Fabaceae – Legume Family
Haematoxylum brasiletto
Piscidia mollis
Prosopis glandulosa
Prosopis velutina
Senna covesii
Vachellia farnesiana
Fouquieriaceae – Ocotillo Family
Fouquieria spp.
Krameriaceae – Ratany Family
Krameria sonorae
Lamiaceae – Mint Family
**Mentha* spp.
**Ocimum basilicum*
**Salvia hispanica*
Vitex mollis

Lauraceae – Bay Family
- **Cinnamomum verum*

Malvaceae – Mallow Family
- *Guazuma ulmifolia*
- *Malva parviflora*
- *Sphaeralcea emoryi*

Moraceae – Mulberry Family
- *Dorstenia drakena*

Myristicaceae – Nutmeg Family
- **Myristica fragrans*

Nyctaginaceae – Four-O'Clock Family
- *Boerhavia* spp.

Oleaceae – Olive Family
- *Fraxinus velutina*

Poaceae – Grass Family
- **Arundo donax*
- **Zea mays*

Polygonaceae – Buckwheat Family
- *Rumex* spp.

Primulaceae – Primrose Family
- *Bonellia macrocarpa*

Ranunculaceae – Ranunculus Family
- *Clematis drummondii*

Resedaceae – Mignonette Family
- *Forchhammeria watsonii*

Rhamnaceae – Buckthorn Family
- *Sarcomphalus amole*
- *Sarcomphalus obtusifolius*

Rubiaceae – Madder Family
- **Coffea arabica*
- *Hintonia latiflora*
- *Randia echinocarpa*

Rutaceae – Citrus Family
- **Ruta graveolens*

Salicaceae – Willow Family
- *Populus spp.*
- *Salix gooddingii*

Santalaceae – Sandalwood Family
- *Phoradendron californicum*

Sapotaceae – Sapote Family
- *Sideroxylon occidentale*

Saururaceae – Lizard-Tail Family
- *Anemopsis californica*

Selaginellaceae – Spike-moss Family
- **Selaginella* spp.

Simmondsiaceae – Jojoba Family
- *Simmondsia chinensis*

Solanaceae – Nightshade Family
- *Capsicum annuum* var. *glabriusculum*
- *Datura discolor*
- **Nicotiana glauca*
- **Nicotiana rustica*
- **Nicotiana tabacum*
- **Solanum americanum*
- **Solanum tuberosum*

Stegnospermataceae – Stegnosperma Family
- *Stegnosperma halimifolium*

Verbenaceae – Verbena Family
- *Lippia palmeri*

Viburnaceae – Viburnum Family
- **Sambucus cerulea*

Zygophyllaceae – Caltrop Family
- *Guaiacum coulteri*
- *Larrea tridentata*

Animals in Medicine and for Healing

Spiders
- Cobweb of a certain spider

Insects
- **Apis mellifera*
- *Eleodes* sp.

Reptiles
- *Crotalus* spp.
- *Chelonia mydas*

Mammals
- *Odocoileus virginianus*
- *Canis latrans*
- *Neotoma phenax*

PLANTS AND ANIMALS IN BELIEFS, RELIGION, AND RITUAL

Animals and plants pervade Yoeme spiritual life. These topics are described in nearly every major work involving Yoeme culture and in many places in this book.

Plants Associated with Beliefs, Religion, and Ritual

Asteraceae – Daisy Family
Pluchea sericea
Burseraceae – Frankincense Family
Bursera fagaroides
Cactaceae – Cactus Family
Carnegiea gigantea
Cordiaceae – Cordia Family
Cordia parvifolia
Cucurbitaceae – Gourd Family
Cucurbita argyrosperma
**Lagenaria siceraria*
Euphorbiaceae – Spurge Family
Croton texensis
Fabaceae – Legume Family
Haematoxylum brasiletto
**Mucuna pruriens*
Olneya tesota
Phaseolus vulgaris
Rhynchosia precatoria
Prosopis glandulosa
P. velutina
Lamiaceae – Mint Family
**Ocimum basilicum*
Martyniaceae – Devil's Claw Family
**Martynia annua*
Poaceae – Grass Family
**Arundo donax*
Primulaceae – Primrose Family
Bonellia macrocarpa
Rosaceae – Rose Family
**Prunus persica*
Rubiaceae – Madder Family
Randia obcordata
Randia thurberi
Rutaceae – Citrus Family
**Citrus sinensis*
Salicaceae – Willow Family
Populus deltoides
Populus mexicana
Salix gooddingii
Solanaceae – Nightshade Family
Capsicum annuum var. *glabriusculum*
**Nicotiana rustica*
**Nicotiana tabacum*
Zygophyllaceae – Caltrop Family
Larrea tridentata

Animals Associated with Beliefs, Religion, and Ritual

Insects
Gryllidae (crickets)
Formicidae (ants)
Lepidoptera, a black butterfly or perhaps a moth
Tabanus spp.
Amphibians
Toads
Reptiles
Gopherus spp.
Heloderma suspectum
Lampropeltis spp.
Masticophis flagellum
Phrynosoma solare
Snakes
Birds
Bubo virginianus
Corvus sinaloae
**Gallus gallus*
Geococcyx californianus
Pelecanus erythrorhynchos
Trochilidae (hummingbirds)
Mammals
Canis familiaris
**Felis catus*
Lynx rufus
Lepus spp.
Mephitidae (skunks)
Odocoileus virginianus

PLANTS AND ANIMALS IN MATERIAL CULTURE

In this summary a listing is provided for the many uses and cultural relationships of plants and animals in ways other than for food and medicine. Detailed information is provided in the individual species accounts.

Many traditional materials have become scarce in modern times and Yoemem often make do with creative substitutions.

Adornment

The maaso (deer dancer) and pahko'olam dancers wear an intricate, specific necklace, called hopo'oroosim that incorporates abalone shell (*Haliotis*, Gastropoda). Seeds from ho'opo (*Piscidia mollis*), as well as beads made from the wood of kuh kuta (*Lycium andersonii* or *Phaulothamnus spinescens*), are used to make rosary necklaces. The bright red seeds of chirikoote (*Erythrina flabelliformis*) are sometimes strung as necklaces. Edward Palmer in 1887 (Watson 1889:59) reported that flowers of the tahsi'o tree (*Bonellia macrocarpa*) were strung as necklaces.

Feathers, especially from macaws and parrots (*Ara militaris* and *Amazona albifrons*), were widely used for decoration and ritual. Perfume was prepared from kuka flowers (*Vachellia farnesiana*).

Basketry

Baskets, waarim, were primarily made from split canes of vaaka (*Arundo donax*). Baskets made of vaaka have made by men, some of whom had an artisanal business making baskets. Traditionally every household had at least several waarim. Baskets were sometimes also made from nawi'o (*Mariosousa heterophylla*) or wata (*Salix gooddingii*). Women wove palm-fiber baskets and hats (see Arecaceae, palm family).

Dye

Blue dye has been produced from chiihu (*Indigofera suffruticosa*), an important dye plant in the Yoem Bwiara. Other plants in Sonora are used to produce yellow dyes (*Argemone gracilenta, Bonellia macrocarpa*) and red dyes (*Haematoxylum brasiletto, Rivina humilis*). Arizona plants used to produce reddish dyes include *Calliandra eriophylla, Jatropha cardiophylla*, and *Krameria bicolor*.

Fencing

Home compound yards are enclosed with fences that provide a degree of privacy and exclude unwanted animals, and also help reduce dust from unpaved streets. Fencing made with the following plants, some used as living stems, is described in the species accounts:

- Cupressaceae – Cypress Family
  - **Juniperus* spp.
- Cactaceae – Cactus Family
  - *Carnegiea gigantea*
  - *Pachycereus pecten-aboriginum*
  - *Stenocereus thurberi*
- Fabaceae – Legume Family
  - *Prosopis* spp.
- Fouquieriaceae – Ocotillo Family
  - *Fouquieria splendens*
- Poaceae – Grass Family
  - **Arundo donax*
- Rutaceae – Citrus Family
  - **Ruta graveolens*
- Salicaceae – Willow Family
  - *Salix gooddingii*

Fiber, Cordage, Paper, and Weaving

Agave leaves yield fibers that are made into rope and twine, called ku'u sawa suma. Agave fiber in Mexico is generally known as ixtle or pita. Fiber from *Agave fourcroydes* was the basis for the henequen plantations in Yucatán during the late 1800s and early 1900s, where many Yoemem were sent as slave laborers (see Bound in Twine in Part 1).

Cordage and fiber for weaving was also made from palm leaves (see Arecaceae, palms), hemp stems (*Cannabis sativa*), and mesquite bark or roots (*Prosopis*).

Cotton (*Gossypium hirsutum*) has been grown in the Yoem Bwiara since pre-contact times. Women made cotton clothing and blankets. Yoeme looms, in use until the end of the nineteenth century, were used to weave cotton and wool. The wool was from locally raised sheep (*Ovis aries*). Old World flax (*Linum usitatissimum*) was cultivated to produce fiber for weaving on introduced Spanish looms.

Paper was made from the bark of *Bursera fagaroides* or *Mariosousa heterophylla*.

Rawhide and leather straps and cordage were commonplace, made from animal hides, including deer and cattle. However, specific information is scarce. It is interesting to note that translations of the Spanish chronicle of Guzmán's 1533 failed expedition relate: "Through an interpreter Guzmán tried to tell the Yaquis that he did not want a fight. The Yaquis replied that they would bring out something to eat. They went on to tell the Spanish to remain where they were and they would come tie up the Spanish horses" (Folsom 2014:25; also see Hu-DeHart 1981:14–15). If the narrative is correct, the Yoemem obviously had strong cords.

Fuel

Mesquite (*Prosopis*) in Sonora and Arizona has been the most important cooking and heating fuel. Writing about Yaquis and Mayos, Beals (1945:4) noted, "Wood is the principal fuel of all classes, and Indians supply the towns, transporting wood on burros." Harvesting mesquite wood for fuel in Arizona and Sonora, and in Sonora for making charcoal, has provided an important income for Yoeme men.

Other plants commonly used for fuel include: *Allenrolfea occidentalis, Atriplex polycarpa, Juniperus, Populus*, and *Tamarix aphylla*.

Games, Play, and Toys

Children in Marana played with shelf fungus (Polyporaceae) and green algae that grew in canals and irrigation ditches. Mesquite (*Prosopis*) thorns were sometimes used as darts, using tree tobacco (*Nicotiana glauca*) as a target.

Boys hunted and played with slingshots and small bows and arrows, often made from canes of carrizo (*Arundo donax*). Canes were also used to make toys including bows, kite frames, and miniature furniture.

Children sometimes played with insects, including green fig beetles (*Cotinus mutabilis*), Sonoran Desert cicadas (*Diceroprocta apache*), and the nests of leafcutter bees (*Megachile*).

Pérez de Ribas (1645; Reff et al. 1999) described gambling games that employed cane stems (*Arundo donax*).

Household and Utilitarian

Plant materials are used for various practical purposes in the home.

Brooms are made from *Abutilon incanum, Ambrosia monogyra*, and *Chloracantha spinosa*.

Various utensils and containers are made from the wood of *Albizia sinaloensis*, the stems of *Arundo donax*, and the fruit (gourds) of *Crescentia alata* or *Lagenaria siceraria*.

Cane (*Arundo donax*) may be used to make a variety of household items.

Adhesives are made from lac produced by *Tachardiella fulgens* scale insects on *Coursetia glandulosa*, and from a resin exuded by *Encelia farinosa*.

Jatropha cuneata roots were used "to cure leather."

Furniture is made from the wood of *Salix gooddingii* and *Stenocereus thurberi*. A three-branched post from mesquite (*Prosopis*) is used to make a support for household water ollas.

Hunting, Fishing, and Weapons

Several plants were used in hunting, fishing, and warfare. The bark of *Bonellia macrocarpa* was the source of a fish poison. An arrow poison was derived from the sap of *Pleradenophora bilocularis*.

Bows were made from the wood of *Celtis pallida*, and presumably other hardwoods. Arrow shafts were made with stalks of *Arundo donax*. Arrow points were made of hardwoods including *Condalia globosa, Haematoxylum brasiletto, Senegalia greggii* (and probably *S. occidentalis*), and *Vachellia campechiana*. Bows and arrows found at the Sierra Mazatán massacre site of June 1902 were described as "plain, nearly 5 feet in length, flat, but slightly arched, and occasionally are strengthened with sinew...The long [arrow] point of more or less prismatic shape, is made, often crudely, of hard, sometimes knotty, white or reddish wood" (Hrdlička 1904:66–67). The hardwood of *Albizia sinaloensis* has been used to make rifle stocks.

Music

Materials from the natural world are used in making musical instruments and in producing sounds during ceremonies. Many plants and animals are the subjects of songs or the inspiration for sounds produced in ceremonies that are described in the species accounts.

Plant Materials Used in Music

Pinaceae – Pine Family

- *Pinus* spp. Commercial lumber sometimes for sena'asom frame, the disk-rattle instrument.

Apocynaceae – Dogbane Family

- *Vallesia glabra*. Drumsticks for tampaleo drum; sena'asom (disk-rattle) frame.

Cucurbitaceae – Gourd Family

- **Lagenaria siceraria*. Half gourd, resonator for musical rasper and water-drum; gourd rattles.

Fabaceae – Legume Family

- *Haematoxylum brasiletto*. Musical rasper and rasping stick (hirukia aso'ola).
- *Mariosousa heterophylla*. Drumstick for tampaleo drum; rasping stick for musical rasper.
- *Parkinsonia florida*. Resin for violin bow.
- *Parkinsonia praecox*. Resin for violin bow.
- *Prosopis glandulosa* and *P. velutina*. Deer singers' rasping sticks.
- *Rhynchosia precatoria*. Seeds for sounding in ayam gourd rattles.
- *Senegalia greggii*. Deer singers' rasping sticks.
- *Vachellia constricta*. Rasping sticks for the musical rasper.

Malvaceae – Mallow Family

- *Guazuma ulmifolia*. Frame for the tampaleo drum.

Poaceae – Grass Family

- **Arundo donax*. Stalks (stems) for the tampaleo flute; water-drum stick; musical instruments, and split cane instrument.
- **Zea mays*. Drumstick for water drum is wrapped in cornhusks.

Rhamnaceae – Buckthorn Family

- *Condalia globosa*. Musical rasper.

Animal Materials Used in Music

Silk moths (*Rothschildia* and *Eupackardia*). Cocoons used to make leg rattles for ceremonial use.

Deer (*Odocoileus*) or pig (**Sus scrofa*). Hooves used to make rattles for ceremonial use.

Shampoo and Washing

Several plants are used to produce shampoo and soap as described in the species accounts:

Asphodelaceae – Asphodel Family

- **Aloe vera*

Cucurbitaceae – Gourd Family

- **Cucumis melo* var. *dudaim*
- *Cucurbita digitata*

Fabaceae – Legume Family

- *Parkinsonia florida*
- *Prosopis glandulosa*
- *Prosopis velutina*

Fouquieriaceae – Ocotillo Family

- *Fouquieria macdougalii*

Poaceae – Grass Family

- **Arundo donax*

Sapindaceae – Soapberry Family

- *Sapindus saponaria*

Simmondsiaceae – Jojoba Family

- *Simmondsia chinensis*

Zygophyllaceae – Caltrop Family

- *Guaiacum coulteri*

Shelter

Yoeme homes are greatly respected, each has its own personality and spirit—it is alive. Traditional housebuilding has been a communal affair, with men, women and children partaking in gathering and processing materials, as well as actual construction. The finishing off, often mud-packing the walls, was done by women; "mud parties" were entertaining, and boys helped.

House walls and posts

House posts for the traditional houses have most often been made from huu'upa, mesquite (*Prosopis*). Walls of the traditional houses incorporated materials from the following plants:

Cactaceae – Cactus Family
Carnegiea gigantea
Pachycereus pecten-aboriginum
Stenocereus thurberi
Fouquieriaceae – Ocotillo Family
Fouquieria spp.
Poaceae – Grass Family
**Arundo donax*
Salicaceae – Willow Family
Populus mexicana

Ramadas

Plants used for ramadas (including pahko heka, or pascola ramada) and temporary shelters include the following:

Asteraceae – Daisy Family
Baccharis salicifolia
Baccharis sarothroides
Pluchea sericea
Fouquierieaceae – Ocotillo Family
Fouquieria spp.
Fabaceae – Legume Family
Lysiloma divaricatum
Prosopis glandulosa
Prosopis velutina
Poaceae – Grass Family
**Arundo donax*
Salicaceae – Willow Family
Populus deltoides
Populus mexicana
Salix goodingii
Tamaricaceae – Tamarisk Family
**Tamarix aphylla*

On May 17, 1980, the Yoem Pueblo community in Arizona celebrated the purchase of their land from the Cortaro Water Users Association, and the people set up a pahko heka for a blessing ceremony for the celebration. The posts were abandoned railroad ties. The beams were commercial lumber, and in place of vaaka (*Arundo donax*) latticework they used commercial boards (1 × 8 inch) to support the weaving of fresh, leafy willow branches (*Salix gooddingii*). The roofing was also of fresh, leafy willow branches. Part of the side-wall for protection from the elements was a hipetam (vaaka mat) from Pótam and another was a commercial woven mat. A photo of this pahko heka by Dorothy Fanin shows the Arizona, American, and Yoeme flags, the deer-dance area, the altar, and the dining area.

Roofing

The traditional house roofs in Sonora were made with thatching, more recently often covered with corrugated tarpaper, and then covered with earth. (After about 2005 traditional roofing was being covered or replaced with plastic sheeting.) Wildflowers, grasses and various other plants grew on the roofs. Plants used in roofing included the following:

Arecaceae – Palm Family
Brahea brandegeei
Asparagaceae – Asparagus Family
Agave spp.
Asteraceae – Daisy Family
Baccharis salicifolia
Cactaceae – Cactus Family
Carnegiea gigantea
Fabaceae – Legume Family
Prosopis glandulosa
Sesbania herbacea
Fouquieriaceae – Ocotillo Family
Fouquieria spp.
Poaceae – Grass Family
**Arundo donax*
**Sorghum halepense*
Sporobolus airoides
Sporobolus cryptandrus
Tamaricaceae – Tamarisk Family
**Tamarix aphylla*
Typhaceae – Cattail Family
Typha domingensis

Smoking

Home-grown hiak viva (*Nicotiana rustica*) was used for cigarettes, smoked for pleasure, rituals, and for medicinal purposes. Cigarettes were rolled in corn husks or the papery bark of saa tooro (*Bursera fagaroides*). The supernatural "flying cigarette" is discussed with *Nicotiana rustica*. Commercial tobacco (*Nicotiana tabacum*) was grown for sale or trade. Siari viiva (*Cannabis sativa*) has been used recreationally and in illicit trade.

Tools

Hibwisia is the word for any stick, or wood, to be used for a certain purpose, for hand-held objects, such as a tool handle. For example, the saguaro fruit-picking pole, made with *Arundo donax* or the wood of *Carnegiea gigantea*, is called hiabwai. A planting stick made with the wood of *Haematoxylum brasiletto* is called wi'ika. Plants with hardwood used for making handles for axes and other tools include *Celtis pallida*, *Piscidia mollis*, *Cordia sonorae*, and *Solanum erianthum*. Hardwoods, including *Lycium* (various species) and *Cordia parvifolia*, are used to make the cross-pieces of cactus fruit-picking sticks.

Transport and Carrying

Women carried all manner of items on their head, as shown in many early photographs. The carrying yoke, shoulder yoke, or palanka (palanca in Spanish) was used by men to transport water containers and other heavy items. The palanka was often made from wood of *Celtis pallida*, *Haematoxylum brasiletto*, *Havardia sonorae*, or *Piscidia mollis*. Prior to acquiring horses, mules, and burros, all burdens were carried by people.

Wood of *Populus mexicana* was sometimes used for wagon and wheelbarrow wheels in Sonora. Wagons, made from various woods, were apparently not common until the 1940s (Beals 1945; Spicer 1980), and continued to be used into the 1990s (Burckhalter 1992).

Dugout canoes, kuta bwakti (wood scooped-out), were used along the coast and the lower Río Yaqui for fishing and travel to other villages or communities. Dugouts were made from river trees, mostly avaso (*Populus mexicana*) and sometimes hooso (*Albizia sinaloensis*), or anything with a large trunk (Beals 1945). Felipe's grandfather, Rosario Vakame'eri-Castillo, carved model kuta bwaktim from cottonwood roots.

PAHKO'OLA MASKS

Pahko'ola dancers, the "old men of the pahko," wear a mask called kuta puhva (wood face) or mahka (from the Spanish mascara) and specific regalia and attire. They perform at every pahko with musicians (harp and violin) and usually with the deer dancer, tampaleo, and deer singers. The mask is worn on the front of the face when pahko'olam dance to the tampaleo flute and drum, deer dancer, and deer singers, and on the side or back of the head when the harp and violin are playing and when not performing. Griffith (1972), Griffith and Molina (1980), Lutes (1983), Felger and Molina (1988), and Kolaz (2007) provide insight, descriptions, and the cultural importance of pascola masks.

Mask Making Materials and Techniques

Yoeme masks used in pahko'ola dances and ceremonies have been made from the wood of the following trees:

Bignoniaceae – Bignonia Family
- *Chilopsis linearis* (Arizona)

Burseraceae – Frankincense Family
- *Bursera fagaroides* (Sonora)
- *Bursera laxiflora* (Sonora)

Euphorbiaceae – Spurge Family
- *Jatropha cordata* (Sonora)

Fabaceae – Legume Family
- *Albizia sinaloensis* (Sonora)
- *Erythrina flabelliformis* (Sonora)
- *Parkinsonia* sp. (Arizona)
- *Senegalia occidentalis* (Sonora)

Malvaceae – Mallow Family
- *Guazuma ulmifolia* (Sonora)

Salicaceae – Willow Family
- *Populus deltoides* var. *fremontii* (Arizona)
- *Populus mexicana* subsp. *dimorpha* (Sonora)
- *Salix gooddingii* (Arizona & Sonora)

In Sonora, avah naawa (cottonwood root)

from *Populus mexicana* is the most commonly used and preferred wood. Some say that chirikoote (*Erythrina flabelliformis*) is excellent for mask making, but it is found too far away in the mountains, so they use cottonwood because it is more readily accessible. Many mask makers prefer a lightweight wood because it is easy to carve, and the hardwoods sometimes crack. Arizona mask makers have used many kinds of wood. In the Tucson area, they occasionally carved masks from commercial pine and "cedar" (juniper), desert willow, as well as the more common and preferred cottonwood roots and limbs (Kolaz 1985). A few masks in Tucson were made from the bases of large palm leaves, or chunks of redwood sawed off a highway rest stop near Marana.

Human face types comprise most of the masks until about 1930. Goat face styles seem to be the first departure from the human face style and the earliest examples are from the 1930s in Sonora. Goat faces were the only animal masks until about 1960 when the monkey (chango) face styles began appearing in Yoeme ceremonies. In the 1970s, dog- or coyote-style masks seem to have made their debut, when "recently made ones with canine snouts and bared fangs" began to be worn at ceremonies (Griffith 1972; Griffith and Molina 1980:26). Pig-style masks appeared about the same time as the dog/coyote style (Kolaz 1997). Many different animal face styles were being carved beginning about 1980. This change was probably a result of giving buyers a wider range of masks. New masks with faces of badgers, eagles, lizards, parrots, roosters, and other desert animals flooded the market. Some Yoemem did not regard these masks as an accepted part of the Yoeme mask making traditions. Animal style masks retain the traditional mask colors, designs, and beard and brows.

A mask is first painted black, which provides the background color. The designs are of various forms and colors. According to some elders, the more traditional colors are black and white, and other colors, red, blue, yellow, and pink, came in later. Nowadays commercial, oil-based paints are used. Various animals such as centipedes, lizards, scorpions, snakes, butterflies, dragonflies, and other insects, as well as flowers, stars, moons, and crosses may be painted on the sides of the mask. In the 1950 and 1960s lizards were the most common, and in Pótam in the 1970s and 1980s snakes were commonplace. Felipe tells us that these represent the natural or animal world.

Masks are embellished with small bundles of cow or horsehair, tied with string, to form the eyebrows and beard. A few older masks have ixtle (agave fiber). Rosario Vakame'eri-Castillo used horsetail hair brought to him by a Mexican rancher. Horsehair was cut in clumps from the middle of the tail, below where it was hard, or sometimes from the mane. The eyebrows and beard represent wisdom, or yo'owe, a wise elder, a person of great wisdom. According to many elders, yo'owe is a person with special talents and spiritual powers who talks to the sun, the moon, the stars, the clouds, and the rain. Many Yoreme (Mayo) masks have goat hair, but goat hair is not common on Yoeme masks. None use human hair.

Tucson mask carvers have primarily used cow-tail hair. This mainly has to do with communities being close to local slaughterhouses. The carvers would go to the back door and knock and ask for the blonde cow tails. Prior to the mid-1980s the butchers used to give the tails to the carvers, but then started charging $1.00 per tail. By the 1990s, the local slaughterhouses had closed, and carvers had to find other places to procure hair for their masks. Some carvers knew people who owned horses and they would use hair from the tail or mane. At that time various colors, other than white (blonde), began to be used in Tucson. In addition, when animal hair was not available, a last resort was hemp-type rope, which would be unwound and brushed out to use for beard and brow hair (Kolaz 1985, 2007).

Head straps for the mask utilized various materials. In modern times the most common strap is a shoestring, usually white or black in color. Other straps have been made from tanned cowhide, store-bought cotton string, and, less often, twisted agave fiber.

Felipe tells how his grandfather, Rosario Vakame'eri-Castillo, crafted masks:

> Cottonwood root [*Populus*] was the only wood we knew. Once Rosario tried chinaberry wood [*Melia azederach*]. It

cracked and he put in some sort of putty and it worked. It was just an experiment to see how it worked.

Rosario took a one-foot cylindrical section of cottonwood root and trimmed it with a small hatchet, and when he got the right size, he would continue with a pocketknife. This describes the making of a mask working with my grandfather:

Stage 1. A log from cottonwood root is made into a cylinder about one foot long.

Stage 2. The log, if large enough it is cut in half, lengthwise.

Stage 3. Rosario drew lines, with a thick pencil in a cross pattern, on the outside or convex surface, lengthwise and across the width.

Stage 4. He then started to carve out the eye area, in a triangle shape. The nose is indicated next.

State 5. In this step the eyes appear, carved deep into the wood, and then holes are made for the eyes. The wood is still a half cylinder. The mouth is placed and initially carved, and at this time the nose is also carved.

Stage 6. A small hatchet is used to take off excess wood from the front.

Stage 7. The eyes, nose, and mouth are essentially completed, and the teeth are carved.

Stage 8. The face is turned around and the inside carved or scooped out with a sharp chisel.

Stage 9. Pencil markings indicate placement of holes for hair making up the eyebrows and the hair at the sides of the mask. Thick wires are placed in a fire and when red hot they are used to burn holes into the wood. About five wires would be used, a wire put back in the fire when it cooled. This stage almost finishes the form. Then sandpaper is used to make the mask nice and smooth.

Stage 10. The painting is started. The whole front of the mask is painted black, using store-bought paint. Black signifies ka nuklak, infinity, or pala ania, which is the universe. Rosario said that after Christianity the black would be the dark side of life, the devil. The black backdrop is where the stars, moon, and sun are painted. At this stage, one would start painting a series of triangles, commonly red, black, and white, around the rim of the mask; these are taa himsim (sun's moustache, actually sun flares) or tam chuktim (broken teeth design). When I was about 15 years old, Rosario would draw a design on paper and I would draw and paint it on the mask, because Rosario's hands were no longer steady. Then white triangles are painted under the eyes, which signify rainwater or tears.

The triangles and other features of the mask are sometimes highlighted by incised carving. Sometimes images of an animal such as a lizard, snake, or butterfly are painted on the side of the mask, and these too might be painted on incised carving. The snake and other animals represent Huya Ania, the wilderness world. It could be a spiritual or a physical snake, and needs to be respected because the snake helps a person who is on a spiritual path through dreams and visions.

Stage 11. The hair, using horsehair, is woven and stitched into the holes, using an intricate pattern. The hair is cut to the desired length and formed into small bundles and inserted into each hole. Each bundle is tied and woven in place with strong string. When Rosario did not have thick enough string, he would spin thread to make a thick enough string. The eyebrows are clipped off short. In contrast, the Sonoran Yoremem (Mayos) leave the eyebrow hair long, but the Sinaloan mask makers clip the eyebrows short. The length of the rest of the hair on the mask is also distinctive for the people of each region. Another grandson also helped Rosario with the hair. The three of us worked on a mask. It was common for kids to help elder men making a mask, so they could learn.

Stage 12. The cross on top (forehead) and bottom (chin) is always done last, after everything else is done. The crosses are centered. The reason the cross is there is to protect the dancer (sometimes people may wish you evil). You often see the ancient cross (like a Maltese cross) as well as the usual Christian cross. The design of the Maltese cross is the ancient sun symbol, representing Itom Achai Taa'a (Our Father the Sun). It is four triangles pointing inward. Also used is the star symbol, which is also an older traditional design.

Rosario took his time. He might make only one mask a year. He made them mostly for himself. Sometimes he would be dancing and someone would want to buy one, mostly Anglos. He didn't really make them for sale, but was always willing to sell one and make another. My grandmother, Anselma Tonapuame'a would say, "Poor things, do you have to sell them?" She became attached to them. Grandfather would work on one after work and on weekends.

A mask was always put away in a safe place, such as in a closet or shed. The mask is always kept hanging outside of the bag containing the other regalia, you don't want to wrinkle up the hair or beard.

The painting was fun, the drilling of the holes fascinating because I liked the smell of the smoke. You can have a lot of wires in the fire at once; use one, and then pick up another, not waiting for the last one to heat up. The wire would be glowing red-orange. Carving was done both inside the house and outdoors. That's what I remember. When grandfather was

Pahko'ola masks. A. Goat mask made by Viviano Valenzuela in Pótam, about 1965 (RF). B. Mask made by Antonio Bacasewa in Vicam Pueblo, early 1980s (BS).

finished, he hangs it up on the wall in the house, in the kitchen, central place of family gatherings. People would come to see the new addition to the family, *uhyoi* (beautiful).

History of mask use

Yoeme use of dance masks has an ancient history. Raphael Folsom's translation of a Spanish colonial document is the earliest reference we know for Yoeme pahko'ola masks:

> On July 10, 1744, The Jesuit Lorenzo José García wrote...that Yaquis were performing dances and rituals that Governor Agustín de Vildósola had expressly forbidden...In the town of Potam...'almost the whole river' [people from the Yaqui towns] [came] to watch the Paccola ceremony...'The lord governor has correctly forbidden you to dance with masks, drums, and flutes.'...the next day when he went to Huiveris to say Mass. 'There...the same speech of exhortation was repeated, and the same dance, with different masks, drums, and flutes' [was performed].
>
> Yaquis wanted and obtained the freedom to practice rituals such as the Pascola, while at the same time obtaining Catholic spiritual care and paid work within the structure of empire. (Folsom 2014:182–183)

The earliest documentations of actual pahko'ola masks date to the 1880s and there are very few. Human face types comprise most of the masks until about 1930. The earliest example known to us is a mask collected in 1883 by Dutch anthropologist Herman ten Kate (Kolaz 2015). This is not only the earliest but the best preserved and documented nineteenth century mask as it retains original paint, head strap, and most of the beard and brow hair. It is in the collections of the Rijksmuseum voor Kolkenkunde (National Museum of Ethnology) in Leiden, The Netherlands. Ten Kate wrote about his travels in the La Paz region of Baja California Sur, where he observed Easter rituals at the Yoeme community where he obtained the mask (Hovens et al. 2004).

A pahko'ola mask at the Arizona State Museum that came from the San Xavier Mission near Tucson is probably from the nineteenth century. The wood is tentatively identified as palo verde (*Parkinsonia*). The mask no longer has visible paint but under a microscope it shows traces of paint; it does not have hair, just the holes. A few other masks collected in the early 1900s, at the Heye Foundation, now at the National Museum of the American Indian, Smithsonian Institution, are also human face types. These masks have a black base coat of paint with designs painted primarily

Pahko'ola mask collected by Herman ten Kate in 1883 at La Paz, Baja California Sur (from Kolaz 2015).

in white and red with small amounts of other

colors. All the masks have forehead cross and most have a cross on the chin as well. The beard and brow hairs are light in color and appear to be horse or cow tail. This traditional face type, color pattern, and hair position are essentially unchanged in 140 years (Kolaz 2015). The earliest masks are highly developed without clear relationship to other mask traditions (except among the related Yoremem), indicating an ancient indigenous origin.

INTRODUCTION TO PARTS 3 & 4

CONVENTIONS AND NOTATIONS FOR PLANT AND ANIMAL LIFE

In the following accounts for plant and animal life, the accepted scientific names for genera and species are in bold, and selected synonyms are italicized within brackets [—]. Non-native plants and animals are indicated with an asterisk (*); these include plants animals not native to the Yoem Bwiara and nearby regions as well as ones not native to southern Arizona. Yoeme names follow the scientific names in small caps, and translations are provided, except for core or unanalyzable names (or at least when translations are not known). Common names, when known or worthwhile, follow in Spanish (italics) and English (plain text). Yoeme names are not always available, and the same holds for English- and Spanish-language common names.

The brief descriptions of plants and animal life pertain to the Yoem Bwiara, and southern Arizona, mostly the Tucson–Phoenix region where the majority of Yoemem live outside of Sonora. Geographic and ecologic distributions likewise focus on southern Sonora and southern Arizona.

For measurements and sizes, the qualifications about and approximately are generally omitted, with the obvious understanding that such quantitative values are seldom exact. The abbreviation "sp." in a scientific name is used for any one species not identified (e.g., Prosopis sp.).

Months of the year in specimen citations are abbreviated as the first three letters (e.g., Jan, Feb, Mar).

We present an overview of plants and animals in the Yoeme world, and especially those in the Yoem Bwiara including ones not known to have to have a specific use.

PART 3
PLANT LIFE IN THE YOEME WORLD

INTRODUCTION

The term *huya* has many meanings, all related to plants. For example, huya can refer to the whole plant, including any live grass, vine, shrub, annual, perennial, herb, etc. *Huya* also refers to "tree," and *Huya Nokame* is the Talking Tree. The whole Wilderness World is *Huya Ania* and the Flower World is *Sea Ania* (these are discussed earlier, in the Introduction).

The following Yoeme terms are used to name the parts of plants and stages of plant life:

huya	plant
naawa	root
nahsakaria	branch
ota	stem
sawa	leaf
seewa	flower
siwe	a growing plant
taaka	fruit
vasiwe	sprouting
waake	drying plant

Ouwo is a general term for tree, apparently an archaic term, and may be used for a fruit tree or any tree. Nowadays many people use *huya* to describe a tree. *Ouwo* is used in some plant names, such as *na'aso ouwo* (orange tree) and *liimon ouwo* (lemon tree).

In the following accounts we first list the fungi (although biologically not plants) and algae, followed by the vascular plants: ferns and spike mosses, cone-bearing plants (gymnosperms), and flowering plants (angiosperms). The arrangements within these sections are alphabetical by family, genus, and species. Scientific names of families follow generally accepted botanical conventions (Stevens 2001 onward). Scientific names are not static and are subject to change with new information—science marches on.

We include about 415 species or taxa (kinds) of plants. Our documented, specimen-based listing of the flora of the Yoem Bwiara includes over 485 taxa. The total flora, including the botanically unexplored reaches of the Sierra Bacatete, is probably at least 650 species.

We cite selected herbarium (botanical) specimens for plants included in this publication for the purpose of verifiability and bio-cultural historical context. Richard has seen most specimens or images of all specimens cited. When no collection number is provided, the specimen is identified by the date of collection. Generally only the first collector's name is given. All specimens cited are at the University of Arizona Herbarium unless otherwise indicated by the standard abbreviations for herbaria (Index Herbariorum, Thiers 2020) as follows:

ARIZ, University of Arizona (default location of specimens, if not otherwise marked)
ASU, Arizona State University
BH, Bailey Hortorium Herbarium
CAS, California Academy of Sciences
DES, Desert Botanical Garden
GH, Gray Herbarium, Harvard University
HNT, Huntington Botanical Gardens Herbarium
K, Royal Botanic Gardens, Kew
MEXU, Universidad Nacional Autónoma de México
MO, Missouri Botanical Garden
NMC, New Mexico State University Herbaria, Las Cruces
NY, New York Botanical Garden
P, Muséum national d'histoire naturelle
RSA, Rancho Santa Ana Botanic Garden
SD, San Diego Natural History Museum
UC, University of California, Berkeley
UCR, University of California, Riverside
UNM, University of New Mexico Herbarium, Albuquerque
US, United States National Herbarium, Smithsonian Institution
USON, Universidad de Sonora

FUNGI

Fungi belong to their own taxonomic kingdom, separate from those of plants and animals, but are included here because there are so few for which we have Yoeme information.

"MOLDS"

Common names: VOAK; *molde de alimentos*; food mold

Arizona and Sonora. *Voak* is the greenish mold that grows on food such as old bread and tortillas. Food mold may include various unrelated fungi.

MACROFUNGI or "MUSHROOMS"

Battarrea digueti
Common name: sandy stiltball
and
Battarrea phalloides
Common name: sandy stiltball, scaly stalked puffball
and
Podaxis pistillaris
Common name: VURU SISI, VUU SISI (burro piss); *punta perdido*; desert shaggy-mane

Arizona and Sonora. These common Sonoran Desert mushrooms are found around homes and in the desert. These stalked puffballs are described as "looking like regular mushrooms." Until about the 1960s, children in Marana called them *kavai sisi* (horse piss).

NAAKA
Common name: shelf fungus

Arizona and Sonora. Shelf fungi (polypore fungi) when growing on cottonwood trees, are called *avah naakam* (cottonwood *naakam*).

Play: In the 1950s, children in Marana liked to play with these shelf fungi and poke them, saying they were poisonous.

ALGAE

NAMU ROKOA (*namu*, cloud + *rokoa*, an untranslatable core word)
Common names: *algas marinas/lamas*; algae, seaweed

Arizona. At Marana, *namu rokoa* was the name used for the large green algae that grew in canals and irrigation ditches. Children played with it when swimming.

Sonora. At Las Guásimas, Fernaldo Leyva-Flores told us *namu rokoa* is a kind of seaweed (marine algae) and is not edible.

FERNS AND SPIKE-MOSSES (LYCOPODS)

MARSILEACEAE – WATER-CLOVER FAMILY

Marsilea vestita [*Marsilea mucronata*]
Common names: WOKOVAVASE'ELA ; *trébol de agua*; hairy water-clover

Arizona and Sonora. These ferns are small perennials with leaves resembling a "four-leaf clover." The plants are found in wetland habitats including arroyos, roadside ditches, and slow-moving river water. We saw *Marsilea* in shallow water along the Río Yaqui.

Wokovavase'ela is also the general name for a swallow (Passeriformes: Hirundinidae).

Arizona: East of Tucson, Rillito River, in reservoir, 24 Aug 1913, *Thornber 7502.*

Sonora: Cuesta Alta, Río Yaqui, *Felger 85-1370.*

SELAGINELLACEAE – SPIKE-MOSS FAMILY

Selaginella
Common names: *flor de Piedra*; resurrection plant, spike-moss

Sonora. Some members of the genus, known as resurrection plants, curl up into a tight ball when dry and are brown on the outside. When placed in water, or after a rainfall, the branches open out (unfold) to reveal the green interior of the plant.

Some "resurrection" species of spike-moss occur in the mountains in Sonora to the east of the Yoem Bwiara. *Selaginella lepidiphylla* or similar species from mountains in other parts of Mexico are often sold in herbal stores.

Medicine: Wagner (1936:84) reported, "for a blow on the chest, as when one is kicked by a horse, the resurrection plant is used. The plant is placed in water, and when it begins to grow the patient drinks the water. If the plant fails to open up, it means the person will die."

CONE-BEARING PLANTS (GYMNOSPERMS)

CUPRESSACEAE – CYPRESS FAMILY

Several members of the cypress family occur in Sonora and southern Arizona but are found in mountains and canyons above the desert. Although none occur within the Yoem Bwiara, the wood is put to use. Basilio (1890 [1634]:207) cited *cauábaso* as the word for *cedro*, a "cedar" indicating the Montezuma cypress (*Taxodium mucronatum*), a tree known from riparian habitats in east-central and southeastern Sonora and perhaps formerly in the Yoem Bwiara.

Juniperus arizonica
Common name: Arizona juniper
and
Juniperus coahuilensis
Common names: TAHKALI; *taskale*, *taskate*; redberry juniper

Arizona. Large shrubs to small trees. Common in northern Arizona and above the desert in southern Arizona.

Commercial cedar boxes, with their characteristic aromatic wood, are also called *tahkali*, and are made from various juniper species.

Fencing and fuel: This aromatic wood was brought into Marana from northern Arizona for firewood, and for fence posts because the wood is long lasting.

Juniperus arizonica: Arizona: Pima County, 5 mi east of Vail Junction, 12 Jun 1939, *Benson 9461*.

EPHEDRACEAE – EPHEDRA FAMILY

Ephedra
Common names: KANUTIO; *canutillo*, *tepopote*; joint-fir

Arizona. Desert shrubs with slender, green stems and small scale-leaves; male and female cones are on separate plants. *Ephedra aspera* (boundary ephedra) has two scale-leaves at each node; *E. trifurca* (rough joint-fir) has three scale-leaves at each node.

Beverage and medicine: Tea made from the stems is taken as a blood purifier and a tonic for well-being. *Kanutio* is used in the same way as for *kosawe* (*Krameria*, Krameriaceae) and as a coffee substitute. The tea is kept in a large jar with pieces of stems floating in it.

Ephedra aspera: Arizona: Tortolita Mountains, Tortolita Mountain Park, *Lindley 192*.

Ephedra trifurca: Arizona: 2.4 mi by road south of Sasco Road on east side Cerro Prieto, Samaniego Hills, *Felger 88-500*.

PINACEAE – PINE FAMILY

Pinus
Common names: WOKO; *pino*; pine

A number of pine tree species occur in the higher mountains in Arizona and Sonora. A pinecone is called *woko* taaka (pine fruit).

Musical instrument: Commercial pine wood, 2 × 4 inches, has sometimes been used in Tucson to make *sena'asom*, disk-rattle instruments.

FLOWERING PLANTS (ANGIOSPERMS)

ACANTHACEAE – ACANTHUS FAMILY (includes Avicenniaceae)

Avicennia germinans

Common names: PASEO; *mangle blanco*; black mangrove

Sonora. Large shrubs to small trees forming thickets in shallow tidal waters of mangrove esteros or inlets. The many slender, upright breathing-roots (pneumatophores) emerging from the mud beneath the branches are called *vai kanteelam* (water candles). The leaves are alternate, dull, grayish green, and encrusted with salt from salt-excreting glands. The flowers are small and white, and on calm, warm nights the moist air in the esteros is filled with their sweet fragrance.

Paseo is a general term for a mangrove. Basilio (1890 [1634]:138) recorded *paseo* as one of four names for *árboles de marisma* (marsh trees).

Avicennia germinans is one of the three common mangrove species along the Sonora coast; the others are *Conocarpus racemosa* (Combretaceae) and *Rhizophora mangle* (Rhizophoraceae).

Beverage: The leaves have been used as a coffee substitute. A tea is made from a small branch with about a dozen leaves.

Medicine: The tea, made with about a cup of leaves from a small branch tip, is taken as a remedy for fever or cough.

Sonora: Guaymas Bay, 29 Dec 1951, *Blakley B-822*

Acanthaceae, *Avicennia germinans*, San Carlos; note salt crystals on leaves. 28 Apr 2015 (SC).

(DES). Between Empalme and Guaymas, 27 Feb 1933, *Shreve 6120*.

Elytraria imbricata

Common names: *cordoncillo*; scaly-stem

Arizona and Sonora. Small perennials; the stems tough and wiry, with scale-like leaves. Flowers small and bright blue.

Common in Sonora, often in shaded places, especially washes and canyons. In Arizona mostly at elevations above the desert.

Medicine: Edward Palmer recorded "cordoncillo, used as a remedy for fevers, venereal diseases, etc." in Guaymas in 1887 (Watson 1889:66).

Arizona: Rincon Mountains, near Vail, 10 Nov 1913, *Thornber 7136*.

Sonora: Cañón Bacatete, 17 Dec 1988, *Felger & Molina*, observation. Guaymas, *Palmer 285 in 1887* (GH, NY, US).

Justicia californica [*Beloperone californica*]

Common names: SEWALULUKUT (*sewa*, flower + *semalulukut*, hummingbird; hummingbird flower); *chuparrosa*; desert hummingbird-bush

Arizona and Sonora. Desert shrubs, usually sparsely leaved, with the plants becoming nearly leafless during dry seasons. Flowers (corollas) tubular, red-orange, and attracting hummingbirds. Flowering response non-seasonal, often with massive flowering following rains. Found along dry watercourses, coastal dunes, and desert plains.

Arizona: Saguaro National Park, Tucson Mountains, *Bertelsen 26 Mar 1989*.

Sonora: Guaymas Bay, 12 Aug 1961, *Turner 61-39*.

Ruellia californica

Common names: HUPA CHUMI (skunk ass), JAVELINA; *chuparrosa, rama parda*

Sonora. Small shrubs with sticky, pungently aromatic leaves and showy, lavender flowers. Flowering response non-seasonal; often with mass flowering following rains. The dry fruits pop open audibly when the plants are watered.

Some Yoemem in Sonora call it *javelina* because javelinas eat this plant and hide among the bushes.

Widespread and often abundant in the Yoem Bwiara, along arroyos and canyons, and dry slopes. Cultivated in southern Arizona as a landscape plant.

Sonora: Las Guásimas, *Felger 85-277*. Bacateve, Cañón Bacatete, 13 Mar 1989, *Felger & Molina*, observation. Guaymas, *Palmer 196 in 1887* (NY, US).

Tetramerium nervosum

Common names: *tapa culo*; hairy fournwort

Arizona and Sonora. Delicate perennials. Flowers (corollas) white and lavender, opening in the morning and falling by mid-day or with mid-morning heat. Generally flowering during hot weather.

Common in Sonora, often beneath shrubs in arroyos and canyons, and north- or east-facing slopes. Horses greedily eat the whole plant. Also in southeastern Arizona, mostly above the desert.

Medicine: As indicated by the impolite common name, the plant is used in Sonora as a remedy for "*mucho diarrea*." Fresh, leafy stems are placed in a liter of water and the liquid consumed.

Sonora: San Carlos, *Felger 85-300C*.

ACHATOCARPACEAE – ACHATOCARPUS FAMILY

Phaulothamnus spinescens

Common names: KUH KUTA, KUS KUTA (cross wood), VA'AKO; *mal de ojo*, *putilla*; snake eyes

Sonora. Spinescent, woody shrubs. Leaves narrow, semi-fleshy, and smooth. Flowers inconspicuous, 2.5 mm wide, unisexual, male and female flowers on the same plant. Fruits globose, 5 mm wide, fleshy, translucent white, and 1-seeded.

The fruit is described as "small and white, like *sita'avao* but smaller." *Sita'avao* is *Vallesia glabra* (Apocynaceae). *Va'ako* is also the name for a species of *Lycium*, which is a similar-appearing shrubs.

Achatocarpaceae, *Phaulothamnus spinescens*, north of Hermosillo. 9 Sep 2006 (PR).

Common on slopes and especially along drainageways and the coastal plain.

Medicine: The fruit continues to be used to alleviate acne. For this purpose, the fruit is crushed and applied to the afflicted area. **Religion and Ritual**: The wood is used to make rosary beads (also see *Lycium andersonii*, Solanaceae). The stem is described as having "a soft fibrous center" (the pith). The stem is cut into small sections, the "fibrous center" removed, and the pieces are carved into beads. The beads are fried in lard until blackened and hard, and are then water-resistant.

José ("Lupe") Guadalupe Flores, in New Pascua, had relatives bring him kus kuta wood from Sonora to carve the rosaries he sold to community members during Lent. He carved the beads with a pocket knife and then boiled them in lard to darken them. Tom Kolaz was at his house at Barrio Libre and New Pascua many times when Yoemem dropped by to purchase *kuusim* (rosarios).

Sonora: South of Pótam, *Felger 85-1415*. North-northwest of Torocobampo, Sierra Bacatete, 14 Dec 1988, *Felger & Molina*, observation. Guaymas, *Palmer 68 in 1887*.

AGAVACEAE, see ASPARAGACEAE

AIZOACEAE – AIZOON FAMILY

Sesuvium portulacastrum

Common names: VAI MUUNI (*vai*, water + *muuni*, bean; water bean); *verdolaga de playa*; sea purslane

Sonora. Low, spreading, herbaceous perennials, often forming expansive mats. The stems and leaves are thick and succulent, and the flowers small and pink.

Sea purslane grows along the coast on upper beaches and at the inland margins of mangrove swamps, often with *Batis maritima* (Bataceae). The name *vai muuni* is also given to *Batis maritima*.

Medicine: This plant is used to alleviate the pain of a stingray wound in the same way as *Batis maritima*. It is applied to the wound as a poultice.

Sonora: Miramar, saline-alkaline soil at edge of cattails, *Felger 85-509*. Las Guásimas, 13 Mar 1993,

Fishbein 925.

Sesuvium verrucosum

Common names: BWAAROM; western sea-purslane

Arizona and Sonora. Warm-season annuals. Flowers small and pink, the anthers often bright magenta.

Bwaarom is also the name for *Portulaca oleracea* (Portulacaceae), a common edible plant.

Common near the mouth of the Río Yaqui on saline mud flats, often in open, nearly barren areas with iodine bush (*Allenrolfea occidentalis*, Amaranthaceae). The stems and leaves are not as thick and succulent as those of *Sesuvium portulacastrum*.

Arizona: Pinal County, Picacho Reservoir, 25 Oct 1985, *Rea 878*.

Sonora: West of Pótam on road to Chiinim, *Felger 88-578*.

Trianthema portulacastrum

Common names: KOWI BWAAROM (*kowi*, pig + *bwaarom*, the word for *Portulaca oleracea*); *verdolaga de cochi* (pig greens); horse purslane

Arizona and Sonora. Common annuals growing with the summer rains. The plants are often semi-succulent; the leaves rounded and opposite with one leaf in the pair usually smaller. The flowers are small and pink. It is sometimes confused with the edible *bwaarom* (*Portulaca oleracea*, Portulacaceae).

Often growing in disturbed, weedy areas including agricultural areas and in gardens, and also in natural habitats.

Animal food: *Kowi bwaarom* is said to provide good food for pigs.

Food: It is generally considered not edible, although a few people said it can be eaten as greens, but it is too bitter to be worthwhile.

Arizona: Tucson, 8 Sep 1916, *Shreve 4906*.

Sonora: Miramar, *Felger 85-896*. Rincón del León, *Felger 85-913*.

ALISMATACEAE – WATER-PLANTAIN FAMILY

Sagittaria longiloba

Common names: PATO PUUSI (*pato*, duck in Spanish + *puusi*, eye; duck eye), VAA VIKAM (water arrows); *flechas de agua*; arrowhead, broadleaf arrowhead

Arizona and Sonora. Herbaceous aquatic perennials with the leaves of mature plants and flower stalks raised above the water. The leaves have arrow-shaped blades with lateral lobes larger than the middle one. The flowers are white. The plants have an underground corm or "bulb" and die back during dry seasons.

In the Yoem Bwiara this arrowhead grows in roadside ditches and borrow pits, and temporary summer ponds in low, swampy savanna areas, and probably was once common along the Río Yaqui.

Gardens: In 1939, Howard Scott Gentry recorded this arrowhead grown in flower pots at Vícam Pueblo.

Arizona: 40 mi south of Tucson, Sasabe road, 10 Sep 1932, *Harrison 9057*.

Sonora: Mex Hwy 15, 1.7 mi southeast of Pitahaya, *Felger 85-1247*. "Near Mori, Yaqui country, adobe mud of water standways on heavily brushed coastal plain, observed also in pots in Bicam pueblo," 26 Oct 1939, *Gentry 4741*.

ALLIACEAE, see AMARYLLIDACEAE

ALOACEAE, see ASPHODELACEAE

AMARANTHACEAE – AMARANTH FAMILY (includes Chenopodiaceae)

The goosefoot family, Chenopodiaceae, has long been considered a separate family from the amaranth family. Many studies verify that they form a single, natural plant family.

Allenrolfea occidentalis

Common names: VAI MO'OKO (*vai*, water + *mo'oko*, the word for *Suaeda nigra*); *chamiso*; iodine bush

Sonora. Small succulent shrubs, appearing leafless. The branches are alternate (one at a node) and the flowers are minute.

Common in saline habitats near the shore, often at the inland margins of the mangroves. *Vai mo'oko* is also used for the perennial pickleweed *Arthrocnemum subterminale*. *Allenrolfea* is distinguished in part by its alternate branching.

Fuel: This plant is used for kindling and said to burn even when green.

Sonora: Playa del Sol, *Felger 85-1129*. Estero Soldado, *Felger 85-1440*. Chiinim, 15 Mar 1989, *Felger & Molina*, observation.

Amaranthus fimbriatus

Common names: WEE'E; *bledo*, *quelitillo*; fringed amaranth

Arizona and Sonora. Warm-season annuals, sometimes persisting until December or even spring. The stems and leaves are smooth (glabrous), without glands, and the leaves narrowly lanceolate. Male and female flowers are on the same plant; flowers green and white; the female flowers have incised or fringed bracts and sepal margins. The seeds are small, blackish, and like those of *Amaranthus palmeri* and *A. watsonii*.

This species is recognized as being different from the usual *wee'e* (*Amaranthus palmeri* or *A. watsonii*). For example, when asked if this *wee'e* was different from the usual *wee'e*, several people said they thought this one (*A. fimbriatus*) was male and the other one female, *wee hamuchia*. Similar answers were given to questions about various other plants of different species with the same name.

Arizona: [Tucson], Tumamoc Hill, 25 Aug 2006, *Wilder 06-234*.

Sonora: Near Chiinim, sand hummocks near shore, *Felger 88-563*. Guaymas, *Palmer 154 (in part) in 1887* (UC).

Amaranthaceae, *Amaranthus palmeri*, Santa Cruz County, Arizona. 16 Aug 2014 (SC).

Amaranthus palmeri
and
Amaranthus watsonii

Common names: WEE'E; *bledo*, *quelite*; careless weed, pigweed

Arizona and Sonora. These are generally upright plants with terminal flowering branches. The plants are highly variable in size according to temperature and soil moisture, from less than 20 cm to more than 1.5 m tall. As in other members of this family, the flowers are small and have sepals but lack petals. In these two species, male (pollen-producing) and female (seed-producing) flowers occur on separate plants. The bracts and sepals of the female flowers are spinescent. The seeds are small, blackish, and lens-shaped. The young, tender plants are sometimes called *aho*, and the older, taller, seed-bearing plants are called *wee'e*.

Amaranthus palmeri occurs in Arizona and Sonora, while *A. watsonii* is almost entirely restricted to the Gulf of California region including western Sonora. In the Yoem Bwiara, *A. watsonii* occurs near the coast, and is replaced eastward (inland) by *A. palmeri*. *Amaranthus watsonii* has many small, yellow-brown glands on the leaves, especially along the veins, while *A. palmeri* lacks these glands. They are otherwise essentially indistinguishable. Both are common, weedy annuals growing during hot weather, and *A. watsonii* also grows at other seasons if there is sufficient rainfall.

Widespread and abundant in natural and disturbed habitats including roadsides and as urban and agricultural weeds.

Food: The young plants are prepared as cooked greens and the seeds were an important food resource in the Yoem Bwiara. Spicer (1980:11) wrote, "Amaranth (pigweed) was cultivated by Yaquis, its small black seeds harvested to make flour and the new leaves cooked and eaten as greens." In Arizona, in late summer and fall the seeds were prepared as *saktusi* (*pinole* or flour); at this time of year the plants are already drying up.

Amaranth was a traditional food but is seldom utilized nowadays. The seed-bearing tops of the mature plant were cut off, beaten onto a rock or other hard object over a canvas or large cloth. The seeds were then wind-winnowed from the chaff and placed in containers for future use. The seeds were often thrown into a large *olla* (clay pot), hot coals added with some sand, and stirred with a stick until they popped. The popped seeds were winnowed and eaten as snacks or ground into flour to make *wee vannaim* (amaranth atole). The popped seeds were ground on a *mata* (metate) with a *tutuha* (mano) and the flour could be stored or used right away.

Rosaria Sombra grew a considerable amount of amaranth in her garden in Pótam. This was the wild, native amaranth. Rosaria saved the seeds and planted them in her garden. When the seeds were harvested, Rosaria and others would toast and grind them to make *wee nohim* (amaranth tamales).

Rosalio Moisés provided the following narrative (Moisés et al. 1971:158):

> Through all these hard years, when there was no money to buy food, we went out into the brush and collected things to eat. From the *hue-e* plant, we ate the leaves of the young plants, and when the plants matured we gathered the seeds. In the leafy stage, we called it *ajo*. The tiny brown seeds can be parched and made into a sort of pinole, or boiled, ground and made into *masa* for tortillas and tamales. Tortillas made of *hue-e* seeds look and taste like dirt. The first time I ate them, at the house of Panocha and Juana Valencia, I asked, 'Where did you get this dirt? I want to make tortillas too.' They thought it very funny that I believed I was eating dirt.

Amaranthus palmeri: Arizona: Tucson, Santa Cruz Valley, 27 Oct 1927, *Thornber 2543*.

Amaranthus watsonii: Sonora: Miramar, *Felger 85-522*. Rancho Bacatetito, 13 Mar 1989, *Felger & Molina*, observation. Guaymas, *Palmer 312 in 1887* (US).

Amaranthus – Grain Amaranth

Domesticated amaranths, known as grain amaranths, were widely grown in southwestern North America in precontact and historic times. These are warm season crops with requirements similar to that of maize. Wild amaranths have dark-colored, blackish seeds, whereas grain amaranths typically have light-colored seeds (Sauer 1967; Thapa and Blair 2018). Guarijíos grew amaranths, probably grain amaraths (Gentry 1963; Yetman and Felger 2002) and presumably Yoemem, Yoremen and their neighbors did likewise.

Spicer (1980:11) illustrated "*Amaranthus hybridus*" as one of the "Plants important to Yaquis." But we have no indication of seed color or an herbarium specimen. However, the Yoemem in Sonora almost certainly once grew one or more of the grain amaranths.

Evidence for grain amaranths in the region comes from Basilio (1890 [1634]:144–145). He cited *hue* as the term for *bledo*, which is similar to the present-day Yoeme term *we'e* for *Amaranthus*. *Bledo* is the Spanish name for young, green amaranths. Basilio recorded names for different colors:

> *Bledos negros*. HUECHUCULI (black amaranths)
> *Bledos amarillos*. HUESAHUALI (yellow amaranth)
> *Bledos blancos*. HUETOSALI (white amaranth)
> *Bledos otras tadíos*. TAHUE (another late bledo; perhaps indicating late harvesting?)
> *Bledos colorados*. SIQUILIHUE (red amaranths)

Black may refer to *Amaranthus palmeri*, the common black seeded, wild amaranth. Yellow and white seem to refer to seeds of grain amaranths. Red amaranths might refer to color of the foliage and inflorescence. These colors, apart from black, imply domesticated, grain amaranths. Grain amaranths in northwestern Mexico were probably *Amaranthus hypochondriacus* (Das 2016).

Arthrocnemum subterminale [*Salicornia subterminalis*]

Common names: VAI MO'OKO (*vai*, water + *mo'oko*, the word for *Suaeda nigra*); pickleweed

Sonora. Low, spreading shrubs or subshrubs with succulent stems that appear leafless. The branches are opposite.

Common names: Common in coastal salt scrub, often in saline, tidally-wet mud at the edge of mangroves. *Vai mo'oko* is also used for iodine bush (*Allenrolfea occidentalis*), which grows in similar habitats.

Sonora: Guaymas, Isla Pajaros, 7 Nov 2006, *Búrquez-Montijo 2006-51*. Chiinim, *Felger 88-559, Molina, & Steen*.

Atriplex barclayana

Common names: VAI MAMYA (*vai*, water + *mamya*, the word for *Solanum americanum*); *saladillo*; coast saltbush

Sonora. Herbaceous perennials or low shrubs. The leaves are somewhat fleshy, grayish-green to whitish. The flowers are inconspicuous; male and female flowers are usually on separate plants. Fruiting bracts (enclosing the seed) are small and hard with short, blunt spines.

Occurs in sandy or sometimes rocky sites along the coast.

Medicine: This plant is said to be a good remedy for cough.

Sonora: Chiinim, *Felger 88-565*. Guaymas, 26 Jan 1927, *Jones 3292* (UCR).

Atriplex elegans

Common names: TOOKO HUYA, TOROKO HUYA (gray or light blue plant); *chamiso cenizo*; wheel-scale orache

Arizona and Sonora. Annuals with grayish leaves and small, disk- or wheel-shaped fruiting bracts. Generally found in disturbed, weedy places.

Animal food: The plants are eaten by goats.

Arizona: Tucson, 27 Jul 1913, *Thornber 4970*.

Sonora: Miramar, *Felger 85-892*. San Carlos, overgrazed field, *Felger 85-1226*.

Atriplex polycarpa

Common names: TOH OUWO (white plant); *chamiso cenizo*; desert saltbush

Arizona and Sonora. Shrubs with grayish-green leaves. Male and female flowers on separate plants. The fruiting bracts have many small, blunt teeth or ridges.

The term *ouwo* for a plant or tree is an uncommon usage, and used by people who grew up in places away from the villages in Sonora and in Arizona.

This saltbush occurs in many desert habitats from Arizona to the Yoem Bwiara at least as far south as Las Guásimas.

Fuel: It is considered good for firewood, but burns quickly.

Arizona: Tucson, West Branch of the Santa Cruz River, 21 Jul 2001, *Mauz 21-61*.

Sonora: North-northeast of Las Guásimas, 16 Dec 1988, *Felger & Molina*, observation. Bahía San Carlos, *Russell 28 Nov 1964*.

***Beta vulgaris** subsp. **vulgaris** convariety **altissima**

Common names: VETAVEEL; *betabel, remolacha azucarena*; sugar beet

Arizona. During the mid-twentieth century, some Yoemem from Marana and elsewhere in southern Arizona worked on sugar beet farms, especially with irrigation and harvesting.

Blitum nuttallianum [*Monolepis nuttalliana*]

Common names: PETOOTAM; poverty weed

Arizona and Sonora. Small winter–spring annuals with semi-succulent stems and leaves, and minute, inconspicuous flowers.

Widespread and often common in valley bottoms in southern Arizona and northern Sonora, and apparently rare as far south as Guaymas.

Food: In Arizona the leafy shoots have been eaten as greens, often cooked with onions and cheese.

Arizona: Roadside, Interstate Hwy 10 near Cortaro, 22 Feb 1968, *Barr 68-40*.

Sonora: Guaymas, disturbed habitat, *Felger 85-416A*.

***Chenopodiastrum murale** [*Chenopodium murale*]

Common names: CHUUHI, KOOWI CHOALI (*koowi*, pig + *choali*); *choali, chual*; netleaf goosefoot

Arizona and Sonora. Annuals with rather smooth, broad leaves.

Widespread and common, often in weedy places, in many parts of the Sonoran Desert. Reported to be native to the Old World.

Yoeme people do not consider it to be edible, although the Seris made use of the young herbage and especially the seeds for food (Felger and Moser 1985).

Animal food: In Las Guásimas we were told that when cows eat it they may become bloated and die.

Arizona: Tucson, Santa Cruz Valley, 27 Oct 1902, *Thornber 2548*.

Sonora: Southeast of Pótam, weed at edge of fields, *Felger 88-589*. Las Guásimas, *Felger 88-615*.

Chenopodium neomexicanum

Common names: CHOALI, KAPA; *choali, chual*; lamb's quarters

Arizona and Sonora. The young, tender plants are not strongly scented. Mature plants, however, are notably stinky. Choali often reaches 0.5 to 1+ m tall in moist habitats in the Yoem Bwiara.

Choali is said to be related to *vakoe* (an unidentified plant, perhaps another species of *Chenopodium*).

Food: The very young shoots are eaten as greens (*quelites*). Meregilda Ochoa of Marana told us that people pick only the bottom leaves to eat, in order to allow the plant to continue growing and producing. It is washed thoroughly and boiled with onions and salt. *Choali* is also used by non-Yoemem and often called *chual*.

Arizona: Tucson, West Branch of the Santa Cruz River, 19 Jul 2001, *Mauz 21-41*.

Sonora: South of Pótam, *Felger 85-1416*.

***Dysphania anthelmintica** [*Chenopodium ambrosioides* var. *anthelminticum*]

Common names: EPASOOTE; *epazote*; wormseed

Arizona and Sonora. Annuals, mostly growing during warm weather, sometimes becoming more than 1 m tall. Stems reddish. Leaves are bright green and highly aromatic.

This plant is often grown in gardens in Arizona and Sonora and is naturalized along irrigation ditches and the riverbed of the Río Yaqui.

Food: It is an important culinary herb in Mexico, and is especially savored cooked with beans.

Sonora: Palo Parado, Río Yaqui, 7.3 mi southwest of Potam, gallery forest of *Populus mexicana*, very aromatic, 1.5 m tall, 19 Nov 1985, *Felger 85-1394B*.

Monolepis nuttalliana, see **Blitum nuttallianum**

Suaeda nigra [*Suaeda moquinii*, *S. ramosissima*, *S. torreyana*]

Common names: MO'OKO; *chamiso*; sea blite, seepweed

Arizona and Sonora. Scarcely woody shrubs, forming large, dense mounds. The leaves are succulent and rounded in cross section. The flowers are inconspicuous.

In the Yoem Bwiara it is common on the coastal plain, especially near the shore, and also in weedy places. It also occurs on saline/alkaline soils in southern Arizona, often near farmland and on floodplains.

Arizona: Labeled as *Suaeda torreyana*: Tucson, West Branch of the Santa Cruz River, *Mauz 21-129*. Pinal County, near Picacho Reservoir, *Landrum 10885*.

Sonora: South of Pótam on road to Vícam Pueblo, *Felger 85-1418*. Ráhum, abundant, 13 Dec 1988, *Felger & Molina*, observation. Chiinim, abundant, 13 Dec 1988, *Felger & Molina*, observation. Guaymas, *Palmer 13 & 329 in 1887*.

Tidestromia lanuginosa

Common names: KAU EE OONA (mountain ant salt); *hierba ceniza, hierba lanuda*; honeysweet

Arizona and Sonora. Hot weather annuals, often low and spreading with red stems, and usually with scurfy-white foliage; flowers minute and yellow.

Common and widespread in natural and disturbed habitats. We were told, "You see it in the mountains, it is pretty but has no use."

Arizona: Tucson, West Branch of the Santa Cruz River, 10 Aug 2001, *Mauz 21-97*.

Sonora: 15 km north of Guaymas, *Bartholomew 3654* (ASU). Northeast of Tórim, *Felger 88-597*.

AMARYLLIDACEAE – AMARYLLIS FAMILY
(includes Alliaceae)

***Allium cepa**

Common names: SEVORA; *cebolla*; onion

Cooked onions are much used in Yoeme cuisine. In the early twentieth century Yoemem sometimes worked as laborers in onion fields in the Hermosillo region and in southern Arizona. It was not pleasant work. Rosalio Moisés described working as a child near Hermosillo:

> The worst part of our job was planting onions. When you plant onions for a week, down on your hands and knees, you itch all over in a terrible way. We were paid twenty-five centavos a day, while grown men working in the fields made thirty-three centavos a day. Of course no one really got any money: every Sunday we drew rations instead. Each man got ten kilos of corn, five kilos of beans, one kilo of sugar, one kilo of coffee, and one kilo of lard. (Moisés et al. 1971:23)

***Allium sativum**

Common names: AASOS; *ajo*; garlic

Garlic was cultivated in home gardens in Arizona, but since the 1970s has been mostly store-bought.

Food: It is extensively used as seasoning, for example greens are cooked with garlic. Felipe says every household has garlic and if you don't have it, you go to a neighbor's house and ask to borrow some.

Medicine: Garlic is said to be used as a blood purifier, but *hitevem* warned parents not to eat too much of it. Garlic cloves, inserted into the rectum, have been used as a remedy for *womtila* (*susto* or fright).

Hymenocallis sonorensis

Common names: LIRIO (from the Spanish, *lilio*, for lily); *cebolla de coyote* (coyote onion); Sonoran spider lily

Sonora. Robust, herbaceous perennials from deeply-set bulbs, with strap-shaped leaves, and showy, spidery white flowers.

Occasionally found in the low-lying coastal plain southeast of Guaymas and often locally common along irrigation ditches and the Río Yaqui.

Gardens: It is sometimes cultivated in local gardens in the Yoem Bwiara.

Sonora: 13.5 km (by air) northwest Yavaros, 3 Sep 1994, *Friedman 382-94* (ASU). Bácum, locally abundant along irrigation ditch, 21 Nov 2006, *Phil Rosen*, photo.

Amaryllidaceae, *Hymenocallis sonorensis*, Bácum. 21 Nov 2006 (PR).

ANACARDIACEAE – SUMAC FAMILY

***Mangifera indica**

Common name: mango (Yoem noki, Spanish, and English)

Sonora. Large tropical trees grown in home gardens in the Yoem Bwiara. The trees sometime suffer minor frost damage to new growth. Originally native to southeast Asia.

Food: It is widely grown in southern Sonora for the delicious fruit.

Sonora: Kompuertam, 11 Mar 1989, *Felger & Molina*, observation.

***Spondias purpurea**

Common names: *ciruela*, *yoyoma*; red mombin, Spanish plum

Sonora. Shrubs or trees with soft wood and thick trunks and branches. Leaves drought-deciduous. Fruits 3 to 3.5+ cm long, fleshy and plum-like, ripening in early summer; different trees bear yellow or red-purple fruits. Native to tropical America and widely cultivated.

Food: The fruits are esteemed in Mexico; the fleshy pericarp is slightly acidic and eaten fresh or sun dried. The dried fruits, rolled in salt, are sold as *ciruelas*. In southern Sonora some people use the term *ciruelo* for the red-fruited form and *yoyoma* for the yellow-fruited form.

Sonora: Wiivisim, cultivated, 15 Mar 1989, *Felger & Molina*, observation.

APIACEAE (UMBELLIFERAE) – CELERY FAMILY

***Apium graveolens** var. **dulce**

Common names: AAPIO; *apio*; celery

One of the store-bought vegetables in Arizona and Sonora. In the Yoem Bwiara often purchased at village stores.

***Coriandrum sativum**

Common names: SILAANTRO; *cilantro*; cilantro, coriander

Arizona and Sonora. Aromatic, cool-season annuals, often grown in home gardens. Native to Europe.

Food: The fresh leaves are widely used as seasoning.

Sonora: Kompuertam, garden plant, 11 Mar 1989, *Felger & Molina*, observation.

***Daucus carota**

Common names: SANOORIA; *zanahoria*; carrot

One of the store-bought vegetables in Arizona and Sonora. In the Yoem Bwiara often purchased at village stores and grown as a cool-season vegetable in home gardens. Felipe says, "Some families on both sides of the border will have a small garden of vegetables including carrots. Nowadays, people in Arizona are too busy at their jobs that much gardening is not happening."

***Ligusticum porteri**

Common names: PAWIS; *chuchupate*; osha

This widely esteemed medicinal plant is an aromatic perennial herb that has a thick root and basal stem, and parsley-like leaves. The dried rootstock is widely sold in Mexico and the United States. It grows wild in mountains in the southwestern United States and the Sierra Madre Occidental in Mexico. It does not occur in lowland regions of Arizona and Sonora.

Medicinal: Many Yoeme people in Arizona and Sonora have used *pawis* for a wide variety of medicinal purposes. We were told it "is considered good for cleansing the stomach." Anselma Tonopauame'a as well and most of the elders in Marana always had some on hand. Painter (1986:55) reported that "pawis...is made into a paste for broken bones."

The people in Marana would purchase pawis from a traveling Mexican herbal vendor who was given the nickname "Colima" because he always talked about his homeland in Colima, Mexico. He would come to Marana driving a big truck in the 1950s and 1960s. He was then perhaps in his 70s. He also sold fruits and vegetables. Colima was especially happy when he saw people from Mexico such as Yoeme women dressed in their traditional, long, colorful skirts and matching blouses, and shawls, and men with traditional attire including straw hats.

APOCYNACEAE – DOGBANE FAMILY
(includes Asclepiadaceae)

Perennial herbs, shrubs, and vines. Various milkweed vines are called *talayote* in northern Mexico. The fruits of most native milkweed vines are reported to be edible when young and tender, often roasted in coals, and some are reported to be edible fresh. Some local members of the family, however, are poisonous.

***Asclepias curassavica**

Common names: blood-flower, Mexican butterfly weed, scarlet milkweed

Sonora. Herbaceous plants to about 1 m tall with bright red-orange and yellow flowers; mostly flowering in the warmer times of the year.

This milkweed is common in the lower Río Yaqui region, often along irrigation and drainage ditches. It is not native to the region, and in 1985 Guillermo Moroyoki and others in Pótam said it is a "new" plant in the region. It is native in tropical regions of the Americas, and often weedy.

Sonora: Cuesta Alta, Río Yaqui, *Felger 85-1384*.

Funastrum clausum [*Cynanchum clausum; Sarcostemma clausum*]

Common names: WICHO'E, WICHO'E KAUWA'ARA (vine with much milk); *huiroa*

Sonora. These robust vines often grow to the tops of trees. The plants have milky latex (sap), the

Apocynaceae, *Funastrum clausum*, San Carlos. 3 Nov 2017 (SC).

leaves narrow to broadly oval, the flowers white to greenish and somewhat waxy.

Riverine and riparian habitats, along the Río Yaqui and seasonally flooded lowlands and swamps in the Yaqui Valley.

Medicine: José María Jaimez told us, "When an insect bites, use this plant with salt and apply to the area."

Sonora: Southeast of Tórim, *Felger 88-598*. Flats near standing water, 28 mi southeast of Guaymas, 18 Aug 1956, *Waterfall 12831*.

Funastrum heterophyllum

[*Funastrum hartwegii; Sarcostemma cynanchoides* subsp. *hartwegii, S. heterophyllum*]

Common names: MASO PIPI (deer teat); *güirote*; vining milkweed

Arizona and Sonora. This common vine has copious milky latex (sap), slender leaves, and clusters of white and maroon flowers. The plants have a strong aroma called *eoktiachisi huuva* (foul/offensive smell).

The vines grow on trees and shrubs, telephone poles, and fences, and even on columnar cacti. Flowering can occur almost all year long. Widespread in Arizona and in Sonora in arroyos, canyons, riverbanks, and plains, from the coast to the mountains.

Arizona: Tucson, West Branch of the Santa Cruz River, 19 Jul 2001, *Mauz 21-35*.

Sonora: Strand at Estero Soldado, *Felger 85-681A*. Southeast of Tórim, *Felger 88-599*.

Marsdenia edulis

Common names: *talayote, tonchi*

Sonora. Robust, glabrous vines with copious milky latex (sap). Leaves to 15+ cm long. Flowers small and pale pink. Fruits green, smooth and ovoid, 7.5 to 12 cm long.

Mostly in arroyos, canyons, and brushy slopes.

Food: In 1887, Edward Palmer reported that the green fruit was eaten in Guaymas (Watson 1889). The Seris roast and eat the unripe fruits (Felger and Moser 1985).

Sonora: Guaymas, *Palmer 150 in 1887* (GH). Cerro Onteme, coastal plain 3.5 mi (airline) S of Vícam, 13 Dec 1986, *Sanders 8776* (ARIZ, UCR).

*Nerium oleander

Common names: LAUREEL; *laurel*; oleander

Arizona and Sonora. Large shrubs or small trees with showy, white, pink, or red flowers. Commonly cultivated, often as a hedge or pruned as a small tree. Every part of the plant is poisonous, for people as well as livestock. Just a few leaves can be lethal to a horse.

Plumeria rubra [*Plumeria acutifolia*]

Common names: *cacalosúchil, súchil*; frangipani, plumeria

Sonora. Large shrubs or small trees, the stems thick and succulent with copious milky latex (sap); flowers white, showy, and fragrant. Native to the sierras including Sierra Bacatete. At the onset of summer rains, the high rocky slopes are spotted with the bright white flowers poking through the rapidly-greening, low canopy. This is essentially a tropical species, especially of the tropical deciduous forest. The usual common name in Sonora is *súchil*. The Nahuatl name by which it is known through much of Mexico is *cacalosúchil*.

Cultivation: Cultivated forms are grown in tropical regions around the World, including the Yoem Bwiara and elsewhere in southern Sonora. These horticultural forms can have variously colored flowers including white, pink, red or yellow. *Plumeria* is frost-sensitive.

Sonora: Cañón la Pintada, 31 Dec 1983, *Sánchez-Escalante*, photos.

Vallesia glabra

Common names: SITA'AVAO; *citavaro, huevito*

Some Yoemem use the variant name *sita'avaro*, which would be the Yoreme spelling.

Sonora. Evergreen shrubs with milky latex (sap). Foliage dense with shiny leaves 3 to 8 cm long. The flowers are 3.5 to 4 mm long, white, star-shaped, in small clusters, and not noticeably fragrant. The fruits are opalescent, whitish and about 1 cm long.

Yoemem in Tucson sometimes grow sita'avao, purchasing the plants from Desert Survivors Nursery.

Usually found in low-lying, heavy soils along larger washes, irrigation ditches, canyons and arroyos in mountains and in the dense thornscrub of the coastal plain and along the lower Río Yaqui. It is not palatable to cattle and is often common in overgrazed areas.

Cocoons of the cincta silkmoth (*Eupackardia cincta*) are found on *sita'avao* shrubs. The cocoons are used in the *tenevoim* leg rattles (see Lepidoptera: Saturniidae).

Food: The fruit is eaten but is said to cause a rash in your mouth.

Medicine: The fruit is used as eye drops. In 1887, Edward Palmer recorded that, "the fruit is eaten by children, and its juice is used for inflammation of the eyes" (Watson 1889:59). It is also used with the staghorn cholla (*Cylindropuntia thurberi*, Cactaceae) and *palo jito* (*Forchhammeria watsonii*, Resedaceae), and applied to sores that are slow to heal. The leaves are also gathered together, warmed in hot ashes, and applied to painful areas, tied in place with a cloth.

Music: The wood is the preferred material for drumsticks for the *tampaleo* drum as well as for the *sena'asom* (disk-rattle) frame. Since these woods are not available in Arizona, various other kinds of wood suitable for carving may be substituted.

Sonora: Cuesta Alta, Río Yaqui, *Felger 85-1376*. North-northeast of Las Guásimas, *Felger 88-605*.

Apocynaceae, *Vallesia glabra*, San Carlos. 16 Dec 2013 (SC).

ARECACEAE (PALMAE) - PALM FAMILY

Three genera of palm are native to Sonora, including the Yoem Bwiara region: *Brahea*, *Sabal*, and *Washingtonia*. These palms have sturdy, solitary trunks and sometimes become quite tall. The leaves are large and fan-shaped. Across much of Sonora the native palms have been used for thatching, weaving, building materials, and food (Felger and Moser 1985, Felger and Joyal 1999, Felger et al. 2001).

There are a number of references to Yoeme uses of palms, but these are mostly not specific to the kind of palm. When Yoemem were in the *Vakatetteve Kawim* (Sierra Bacatete) they would have had access to palms for food and other uses.

Basketry and weaving: The new, tender, emerging leaves, primarily of *Sabal* and sometimes *Brahea*, have been highly valued for weaving twilled and coiled baskets, mats (*hipetam*; *petates*), and other items. People often traveled considerable distances to harvest palms. The young, tender *Sabal* leaves probably was the preferred source for hat making (*move'im*, hats). Hrdlička (1904:66–67) described baskets and hats made from palms.

Rosalio Moisés told of Yoeme women in Sonora making "palm fiber hats for sale" (Moisés et al. 1971:22). Spicer (1980:107) showed a 1942 photo of a *moro* (pahko'olam manager) "wearing [a] palm-

Arecaceae, palm-leaf crosses, Las Guásimas. 31 May 2019 (PB).

frond hat." Until the latter decades of the twentieth century, men in Sonora preferred hats woven from palm leaves.

Construction: Coville and MacDougal (1903:19) recorded, "About eight miles west of Torres ...One abandoned house had a ridgepole made of a palm trunk. This was notched in the manner followed by the aborigines in pre-Columbian times in making a ladder, and it is evident that it had been put to such a use before it was employed as a ridgepole."

Food: The blackish, sugary mesocarp of the ripe fruit of *Sabal*, is eaten fresh in northeastern and east-central Sonora. The fruits of *Washingtonia* are sweet and edible, and the fruits of certain *Brahea* were used for food in Sonora, but information is fragmentary. For example, *Brahea* fruits are eaten by the Guarijíos and in the Ures region (Felger and Joyal 1999). In the eighteenth century Nentvig (1977) reported the use of palm hearts (terminal buds) for food in Sonora, and Gentry (1942, 1963) likewise reported that the Guarijíos ate palm hearts. Palm hearts are cut from the center of young plants, which kills the plant.

Medicine: When a person has a bad nightmare, a dry palm frond is burned in a bowl or can and the ashes are applied in the form of a cross to the forehead, temples, palms of the hands, nape, inside the elbow joint, behind the knee, and soles of feet.

Brahea brandegeei [*Brahea elegans*, *B. roezlii* sensu Wiggins 1964; *Erythea brandegeei*, *E. clara*, *E. elegans*]
Common names: OVEI, TOVEI; *babiso*, *palmilla*; Sonoran hesper palm

Sonora. These palms can reach more than 10 m tall but are usually much shorter. The leaves are dull bluish-green and have stout spines along the leaf stalk (petiole). The fruits, larger than those of the other native palms, are rounded, mostly 18 to 20 mm wide, with a thin, fleshy, sweet, date-like pulp (pericarp) surrounding a large seed. The fruits generally ripen in late August and early September.

Hesper palms are abundant in the Sierra El Aguaje, Sierra Libre, and Sierra Bacatete in canyons, often on steep rock slopes, and even cliffs and bedrock where they grow from rock crevices. The mountaintops support many thousands of these palms.

Alfonso Leyva-Flores said this is a palm-like tree (meaning like tako, Sabal uresana) that grows in the Sierra Bacatete. Mateo González identified *tovei* at the Arizona-Sonora Desert Museum in Tucson. He said it is like *tako*, grows about six meters tall, the leaves are closer [together], the Mexicans [in Sonora] use it for roofing because those roofs do not leak, it has small seeds, and has flowers like the date tree and both have spines.

This palm was a potential food resource for Yoemem when they were living in the Sierra Bacatete, but direct evidence is lacking. The seeds are edible when roasted (Felger and Joyal 1999).

Sonora: Nacapule Canyon: 14 Apr 1936, *Bailey 263* (BH); May 1982, *Parfitt 3030* (ASU).

***Phoenix dactylifera**
Common names: LAATIKO; *datillo*; date palm

Arizona and Sonora. The date palm is native to the Middle East. Date palms are grown for ornament and the sweet, edible fruits (dates). This palm branches ("suckers") from the base, forming large, multiple-trunk clumps. Date palms are grown from seed as well as from the large basal branches or offshoots. The leaves are large, pinnate (elongated, in the shape of a feather), and dull, bluish green. Leaflets near the base of the leaves are modified as large, sharp spines.

Dates were introduced into northwestern Mexico in early colonial times. The name *laatiko* is from the

Arecaceae, *Phoenix dactylifera* with Sue Carnahan, San Carlos. 22 Apr 2016 (CS).

Spanish *datillo*. Date palms are cultivated as well as feral in the Yoem Bwiara, and are from seed-grown stock.

Selected cultivated varieties with larger and more desirable dates are clonal, propagated from offshoots, and were not brought into the greater Southwest (e.g., California deserts) until the later part of the nineteenth century when live offshoots were imported from the Middle East.

Sonora: San Carlos, weedy arroyo, 220 m inland from shore, associated species: *Sabal uresana*, *Washingtonia robusta*, 22 Apr 2016, *Carnahan 1719*. Kompuertam, 11 Mar 1989, *Felger & Molina*, observation. Pótam, 11 Mar 1989, *Felger & Molina*, observation.

Sabal uresana

Common names: TAKO; *palma del taco*; Sonoran palmetto

Sonora. This large palm has very large, bluish, fan-shaped leaves; the leafstalk (petiole) is spineless (entire) and extends into the leaf blade. The fruits are blackish when ripe and edible.

This palm is especially common in some foothill areas, mostly along canyon and valley bottoms in the mountains north of San Carlos and mountain areas along the northeast side of the Yoem Bwiara. Orioles often attach their nests, which are woven from *Sabal* leaf fibers, to the leaf blades.

Sonora: Back of San Carlos Bay, 27 Mar 1934, *Bailey 1* (BH). San Carlos, 14 Mar 1988, *Zona 259* (HNT).

*Washingtonia filifera

Common name : Desert fan palm

and

Washingtonia robusta [*Washingtonia sonorae*]

Common names: *abanico*; Mexican fan palm

Arizona and Sonora. The two *Washingtonia* species are large, tall fan palms that are widely cultivated in Arizona and Sonora as well as many subtropical regions worldwide. *Washingtonia robusta* is readily distinguished from *W. filifera* by its more slender and often taller trunks (sometime to 20+ m) and leaf blades that are shiny rather than dull green. Bailey (1936) pointed out that petioles of mature trees have few or no marginal spines, while the leaf stalks of the young palms are armed with stout, sharp teeth.

Washingtonia robusta is native to the Baja California Peninsula and Sonora in riparian canyons in the Sierra El Aguaje north of San Carlos. *Washingtonia filifera* is native to desert canyons in southeastern California and nearby Baja California, and Arizona in mountains north of Yuma.

Rose (1899: 255) found *Washingtonia robusta* growing at Guaymas and reported "most of them have been cut out and used as rafters for houses."

Ceremonial: Leaves collected for Palm Sunday are kept in the *pahko heka* (pascola ramada) during the all-night ceremonies. These leaves are distributed on Palm Sunday and each person makes a small cross with one or two leaf segments.

Left: Arecaceae, *Sabal uresana*, San Carlos. 31 Oct 2025 (SC). **Right:** Arecaceae, *Sabal uresana*, and *Phragmites australis* and *Prosopis glandulosa*, Rancho Santa Ursula, Sierra Santa Ursula. 4–5 Apr 2009 (MB).

The crosses are taken home and hung inside or outside of the house.

Food: Both *Washintonia* species produce prodigious quantities of small edible fruits that have a date-like flavor (Felger and Moser 1985; Hodgson 2001). In 1887, Edward Palmer reported that *Washingtonia sonorae* "seeds are used for food by the Indians" in the Guaymas region (Watson 1889:79). Although the seeds are edible, usually parched and ground, Palmer may have been referring to the whole fruits.

Masks: During the 1980s and occasionally in later decades, when palm leaves were trimmed at the University of Arizona, some artists carved pahko'ola masks from the large petiole (leaf stalk) bases. A few of these masks are in the permanent collection of the Arizona State Museum.

Washingtonia filifera: Arizona: Phoenix, 11 Feb 1950, *Blakley B-18* (DES).

Washingtonia robusta: Sonora: Cañón Nacapule, *Bailey 3 & 262* (BH); *Felger 92-1027*. Near Kompuertam, cultivated, 11 Mar 1989, *Felger & Molina*, observation.

Arecaceae, *Washingtonia robusta*, near San Carlos. 23 Jan 2018 (CS).

Secluded canyons in the mountains about Guaymas, *Palmer 311 in 1887* (CAS, GH).

ARISTOLOCHIACEAE – BIRTHWORT FAMILY

Aristolochia watsonii

Common names: BWASU'UBWILA; *hierba del indio*; desert pipevine, Indian-root

Bwasu'ubwila is not translatable.

Arizona and Sonora. Small vines from a perennial, carrot-shaped root; dying back to the root during drought, and with winter cold in Arizona. The leaves are arrow-shaped. The strange, semi-tubular flowers consist largely of a tooth-shaped calyx, yellow-green with dark maroon. The leaves are eaten by the reddish-orange caterpillars of the pipevine swallowtail butterfly. The caterpillars sequester aristolochic acid, a toxin that renders them unpalatable to birds.

Aristolochiaceae, *Aristolochia watsonii*, Cañón del Nacapule. 11 Mar 2015 (SC).

This species is one of the smallest in the genus. Many other species occur around the world and many are important medicinal plants, largely due to unusual, biologically active compounds; all have bizarre flowers.

In the Yoem Bwiara this plant occurs mostly along arroyos and canyons. We were told it is more common in some areas of southern Arizona than in the Yoem Bwiara. Juan Luis Garcia told Felipe that "it is a medicinal plant, like a vine, and can be found at *Chukui Kawi*" ("Black Mountain"; described as a black volcanic mountain northwest of Marana, which is now Picacho Peak State Park).

Medicine: *Aristolochia watsonii* is an important medicinal plant. In 1887, Edward Palmer reported, "the root used for aches and pains" (Watson 1889:73). Johnson (1962:292) reported it was used as a remedy for "*viento o resfriado*."

The mashed, ground root can be boiled, then strained and the liquid consumed, but one should only drink a small amount at a time. This is taken as a remedy for stomach problems. To treat headaches, the back of the neck and temples are bathed with this liquid. For continued use, the rest of the mixture can be steeped in a large jar, together with the residue of the roots and its fibers. The dried roots are ground and often used in combination with other medicinal plants (for example see *Baccharis salicifolia*, Asteraceae).

People from Sonora often request friends and relatives to bring *bwasu'ubwila* from Arizona, because it is relatively scarce in the Yoem Bwiara in Sonora. Teresa Amarillas saw *Aristolochia* when visiting Marana in October 1987. She identified it as *bwasu'ubwila* and said it is dried, ground into a powder, and used for dizziness. The powder is applied to forehead and temple, "just lie down and somebody, usually a *hitevi* [healer], rubs it on you." "*Polvota va'apo to'one pottiata vetichi'ivo*," pour the powder into some water and drink it for upset stomach after eating too much.

Arizona: Tucson, southwest of Tumamoc Hill, 5 Aug 1983, *Bowers 2698*. Pinal County, Santa Cruz Flats near Eloy, in abandoned field, 3 Apr 1978, *Karpiscak 1978-74*.

Sonora: Cañón Nacapule, *Felger 85-239*.

ASCLEPIADACEAE, see APOCYNACEAE

ASPARAGACEAE – ASPARAGUS FAMILY (includes Agavaceae)

Agave

Common names: KUU'U; *maguey*; agave, century plant

Several native species of *kuu'u* occur in the Yoeme region in Sonora and different native species are found in southern Arizona. The tall flowering stalk is called *kuu varoa* (agave flowering-stalk). The opening of the flowers is called *vikoe*.

Basilio (1890 [1634]:208) recorded *cuu* as the word for "*mezcal, planta*." Cultivated agaves are called *wah kuu'u*. The term *wah* is an adjective that signifies a cultivated plant.

There are two readily recognized subgenera (Gentry 1982):

Subgenus *Littaea* in the Yoem Bwiara is represented by *Agave chrysoglossa*, *A. felgeri*, and *A. vilmoriniana*. Members of this subgenus in northwestern Mexico generally have slender leaves with a sharp spine at the tip and entire margins (without teeth), and relatively narrow inflorescences. These agaves are generally avoided for food and beverage due to bitterness from high saponin content.

Subgenus *Agave* in the Yoem Bwiara includes *Agave angustifolia* and *A. colorata*, and in southern Arizona includes *A. chrysantha*, *A. palmeri*, and *A. simplex*. *Agave americana* and *A. fourcroydes* are also in this subgenus. These species are characterized by leaf margins with sharp teeth as well as a sharp spine at the tip, and prominently branched inflorescences. These agaves are the ones most desirable for food and beverage.

Cultural knowledge and use of agave

Yoemem would have found useful agaves almost everywhere they went during their diaspora, including the Baja California Peninsula, southern Arizona, and the mining camps and haciendas in Sonora.

Pérez de Ribas (1645; Reff et al. 1999:88) wrote:

> The mescal plant, which in the shape and form of its foliage is like a large aloe, is also a source of food and enjoyment. It is thought that there are many varieties of this plant...Wine, honey, and vinegar are made from it. Thread and twine are also extracted from the tender blades of the plant, the sharp points of which are used for needles. Although it is true that the plant provides all these products, these peoples use it mainly for food. When it is in season they cut it at the base of the

> trunk and roast it in a pit into which heated stones have been placed and subsequently covered over with branches and dirt. After the slow heat has softened the trunk and a portion of the blades they are made into a type of preserve, which is very sweet because of the way in which the plant has been roasted. This is the only beneficial plant they are in the habit of raising near their houses.

Perez de Ribas' report of harvesting agaves "in season" supports our knowledge that the plants are harvested seasonally, just as a plant is beginning to produce a flowering stalk. By the time an agave plant produces its elongated flowering stalk, it has spent its store of carbohydrate and is no longer suitable for food or wine making (Gentry 1982; Felger and Moser 1985; Hodgson 2001). It is noteworthy that Pérez de Ribas mentioned that agaves were cultivated in the early seventeenth century.

Beverages: Pérez de Ribas (1645; Reff et al. 1999:90) reported, "Their wine was made from various native plants...such as the mescal and its stalks when it is in season and bears fruit. When these plants are ground or crushed and added to water they soften up in two or three days and take on that pleasing taste that snatches away whatever judgment these rational souls still possessed." The plants would have been roasted before being "ground and crushed" and harvested before the "stalks" enlarged and were bearing fruit.

Fermented agave beverage, called *kuu vino* (agave wine), was probably made mostly from *Agave angustifolia* and *A. colorata*. *Kuu vino* was much used in ceremonial celebrations.

After the mid-twentieth century, *bacanora*, instead of *kuu vino* or *vachi vino* (corn wine), became widely used in Sonora during the big summer *pahkom*. Bacanora generally was bought from ranchers in the mountains in eastern Sonora. This famously powerful and semi-illicit liquor often led to drunkenness. Tequila (a product of distillation) also has been used in place of the traditional fermented beverages.

Rosalio Moisés provided firsthand accounts of the illicit mescal (probably bacanora) trade during the 1930s: "One way a person could make a few pesos was by bootlegging mescal from Bacum. Beer was sold openly in all the Yaqui villages north of the river, but mescal was prohibited. Since Bacum was on the south bank of the river, mescal was legal there. Many Yaquis from every pueblo kept up a steady traffic to Bacum" (Moisés et al. 1971:147).

In his detailed account of Pótam, Spicer (1954:52) wrote: "There were a dozen or more families in both years [1942 and 1947] whose adult members drank mescal wine to excess and so constantly as to interfere with their performance of much work." He observed, as well:

> Young Mexican and Yaqui men frequently get together in the evening and drink the illicit mescal wine. They gather in small groups of a half dozen or more out in the bush surrounding the village. There they drink, often throughout the night, singing to the accompaniment of guitars and talking around fires. Sometimes they ride horses, particularly on fiesta nights, rarely dismounting, swaying in their saddles and passing their bottles from rider to rider. (Spicer 1954:108)

In the early 1970s Guadalupe Compoy and some of her followers, men and women, were trying to stop the sale of liquor, mostly tequila and bacanora, in the Yoeme lands. Among their concerns were men who were farmers not tending to their work. Guadalupe was struck by a car while waiting at a bus stop on Highway 15 in southern Sonora. Some think it was murder. The excess drinking abated for a while. She and her followers were responsible for bringing running water to the various villages by working with the state and federal governments. Lutes (1977) provided detailed accounts of alcoholic drinking in Pótam.

Cordage and weaving: Basilio (1890 [1634]:224, 226) recorded *tasequi* as the word for "*el ixtle*," a kind of agave fiber, and *soom* as the word for "*pita, filamento que se extrae del maguey*." Pérez de Ribas (1645; Reff et al. 1999:92) reported, "The women also practiced the art of spinning and weaving cotton or other fibers such as Castilian hemp or pita." Castilian hemp would be *Cannabis sativa* (Cannabaceae) grown for fiber, while "*pita*" refers to agave fiber (Santamaria 2000). Reff et al. (1999:92) suggested, "This fiber (ixtle) probably derived most often from the narrow leaves of *Agave angustifolia*." However, Pérez de Ribas did not distinguish between Mayo and Yaqui usage, so "pita" could also have come from other long-leaved agave species in the region.

Rope or twine made from agave fiber is called *ku'u sawa suma* (agave leaf tie) (see *Agave angustifolia* and *A. fourcroydes*; also Beals 1945). *Ku'u sawa suma* has been used in Sonora to tie grass bundles or thatch as roofing for traditional homes, twine or rope to carry a gourd canteen (see *Lagenaria*, Cucurbitaceae), and many other practical purposes. Some older pahko'ola masks have a head strap made from agave twine, and sometimes the beard was made from agave fiber.

Food: The large cores, "hearts" or "*cabezas*," of agaves were roasted and eaten as a staple, and used for making alcoholic beverages, especially in Sonora. Plants that are about to flower concentrate carbohydrates to supply the energy to produce the massive flowering stalk, flowers, and seeds. One can recognize these plants by the smaller and narrower leaves produced at the center of the plant, and the heart of the plant becomes noticeably larger. The plants are dug out or severed at about ground level. The leaves (except the broad leaf bases) are trimmed away with a knife or machete, leaving the large, white heart or *cabeza*. These are roasted in a pit in the ground, to be cooked at least one or two days. Thoroughly cooked agave hearts are sweet and delicious. Uncooked or undercooked agave can be highly caustic to the skin (sometimes producing a painful rash) and dangerous to eat. The young, emerging flower stalk is also edible.

The sweet, cooked agave could be dried for future use, probably made into cakes, and as such could be stored for a long time if kept dry (the process is described by Gentry 1982; Felger and Moser 1985; Hodgson 2001; and Yetman and Van Devender 2002). Maguey hearts are sometimes offered for sale in local Sonoran markets, but were likely not harvested by Yoemem.

Agaves were one of the major wild foods harvested in the Yoem Bwiara. In spring, 1949, Rosalio Moisés and his family were in dire straits following a major flood earlier in the year: "I went out into the brush to gather wild foods. I gathered mescal hearts and baked them for twenty-four hours, buried in the ground. I could cook up to two-dozen hearts at once. Lázaro [his little boy] liked eating the cactus [sic], and it kept us alive" (Moisés et al. 1971:216).

In the mid-1980s, Miguel Romero told of harvesting agaves in the early twentieth century when the people were in the Sierra Bacatete fleeing from the Mexican military (his narrative is given in the introduction).

Medicine: Wagner (1936:83) reported, "the juice of the century plant is applied to sores. The plant is mashed in such a way that the juice falls directly on the sore."

***Agave americana**

Common names American agave

Arizona and Sonora. This is the common, large, blue-leaved, cultivated agave. It has been grown in home gardens in Arizona and Sonora, and can persist from cultivation long after being abandoned, but has not become established in Arizona or Sonora. It is not native to the region and has been cultivated in Mexico since ancient times.

Arizona: Tucson, wash at southwest base of Tumamoc Hill, 6 Feb 1985, *Bowers 2981*. Phoenix, Desert Botanical Gardens, 18 Jun 1977, *Gentry 23674*.

Sonora: Hermosillo, Campus de la Universidad de Sonora, *Hernández-Zazueta 22 Feb 2015* (USON).

Agave angustifolia [*Agave owenii*, *A. pacifica*, *A. yaquiana*]

Common names: KUU'U; *bacanora*, *lechugilla*; narrow-leaf agave

Sonora. Medium-sized agaves, often freely producing offsets. Leaves narrow, bluish-green, and with sharp marginal teeth. Flowers green and yellow, commonly flowering in late February. At about dusk the flowers produce a strong, musky odor reminiscent of apricot fruit; the odor attracts bats, which are the major pollinators.

This is the most common agave in the Yoem Bwiara; found on the coastal plain, canyons, and rocky slopes.

Beverage: This species is one of the major sources of bacanora, the famous Sonoran liquor distilled in mountain areas east of Hermosillo.

Left and above: Asparagaceae, *Agave angustifolia*, Bahía San Pedro. 4 Feb 2015 (SC).

Food: *Agave angustifolia* has been extensively utilized for food. A rocky hill on the southwest side of Kompuertam, where many of these plants occur, is called *Kuubwa'e Kawi* (eating the agave).

Because of its importance, people moved these plants around, planting them near their rancherias especially farther south in Mexico. This is one reason why this species has the largest distribution of any agave.

Sonora: Las Guásimas, *Felger 85-273*. North-northwest of Torocobampo, Sierra Bacatete, 14 Decc 1988, *Felger & Molina*, observation. Vicinity of Empalme, *Rose et al. 12260* (US). Hills above Río Yaqui, Tonichí, *Spicer 8 Sep 1941*.

Agave chrysoglossa

Common name: *amole*

Sonora. Medium-sized agaves. Leaves narrow, thick, gray-green, and without marginal teeth. Flower stalks 2 to 3 m tall. As the name *amole* indicates, the crushed leaves can be used as soap, a characteristic of most members of the subgenus *Littaea*.

Rocky canyon slopes, cliffs, and higher elevations in the Sierra El Aguaje.

Sonora: Cerro 4–5 mi north of Bahía San Carlos, 29 Mar 1963, *Gentry 19882*.

Agave colorata

Common names: KUU'U; *maguey*; banded agave

Sonora. Medium-sized to sometimes large agaves, solitary or producing offsets. Leaves often 30 to 60+ cm long and relatively wide, ashy gray and with broad, ashy brown to pinkish horizontal bands, the margins bearing stout, sharp teeth.

Asparagaceae, *Agave colorata*, Bahía San Pedro. 29 Mar 2016 (SC).

Flower stalks often 2 to 3 m tall; inflorescences branched. Flowers yellow, from late spring to June.

Agave colorata is mostly found on rocky slopes, especially in the mountains at the northeast side of the Yoem Bwiara, in the mountains around Guaymas, and in the Sierra El Aguaje. This attractive agave is in the nursery trade in southern Arizona, with a number of horticultural selections.

Food: This agave was harvested in the Yoem Bwiara (Spicer 1980:11). Gentry (1972:119) reported apparent over-harvesting "by Mayo and Yaqui Indians in precolonial times."

Sonora: Sierra Bojihuacame [southeast of Cd. Obregón], 1 Mar 1952, *Gentry 11641*. Cerro, 4–5 mi north of San Carlos, 29 Mar 1963, *Gentry 19881*.

Agave felgeri

Common name: *mescalito*

Sonora. This small agave produces offsets to form colonies. Leaves slender, often about 30 cm long with a sharp terminal spine, and without marginal teeth. The upper leaf surface flattened, generally with a reddish band along each margin of a pale-green midstripe. Flowers yellow, mostly early summer or fall, on slender flowering stalks often 1.5 to 2 m tall.

Rocky slopes, from near sea level at Bahía San Carlos to mountain peaks including Cerro El Vigía and Sierra Bacatete. Dense colonies at San Carlos have mostly been destroyed for resort development.

Asparagaceae, *Agave felgeri*, San Carlos. 12 Feb 2017 (SC).

Agave felgeri is similar to *A. schottii*, the shin-dagger agave of southern Arizona and northern Sonora. The Seris used the slender flowering stalks of *Agave schottii* for arrow shafts (Felger and Moser 1985) and Yoemem may have found a similar use for the stalks of *A. felgeri*.

Sonora: Bahía San Carlos, 13 Dec 1951, *Gentry 11343*.

*Agave fourcroydes

Common names: KUU'U; *henequén*; henequen

This large agave has sword-shaped leaves 1.2 to 1.8 m long growing out of a thick stem that may reach 1.5 m or more in height. The leaves have sharp teeth along the edges, and a large, stout, sharp terminal spine.

Agave fourcroydes has been grown or harvested for fiber in Yucatán since pre-contact times, and in modern times grown as henequen in Tamaulipas, Veracruz, and Yucatán. The leaf fiber is "coarse, strong, excellent for rope and coarse twine, as baling twine, [and] is resistant to seawater" (Gentry 1982:575). Gentry (1982:630) also reported, "Until recently, all fiber plantation *Agave* there [Yucatán] were henequen, *A. fourcroydes* or near relatives with armed leaves." (Also see Colunga-GarcíaMarín 2003).

During the reign of the dictator Porfirio Díaz in the late nineteenth and early twentieth century, Yoemem and many other native people in Mexico were summarily rounded up and sent to work on henequen plantations in Yucatán in horrific slave conditions. *Kuu chuktiwan* (agave were-cutting) suggests the forced labor work in the henequen fields in Yucatán. The slave labor was grueling in the hot, humid climate, and injuries from the sharp spines of the plant added to the pain and suffering.

There were attempts to cultivate henequen in southern Sonora on former Yoeme lands. Gentry (1982:575) reported, "good leaf growth and quality fiber production require a minimum of ca. 20 inches (750 mm) of annual rainfall in frostless climates. The Obregón plantation south of Ciudad Obregón, Sonora, for instance, with less than 20 inches (500 mm) of annual rainfall, never produced fiber of sufficient length and quality to meet trade standards." Evans (2007) wrote of efforts at henequen production in southern Sonora in the 1920s, but by 1927 the results were negligible.

Sonora: Obregón plantation, 12 mi southeast of Cd. Obregón, in plantation on argillaceous plain, 17 Dec 1951, *Gentry 11361* (ARIZ, DES, US; Gentry 1982:592).

Agave vilmoriniana

Common names: *amole*; octopus agave

Sonora. Leaves usually more than one meter long, very succulent, grayish green, and without marginal teeth. Gentry (1982:82–84) called it a "cliff-dweller...when viewed from a distance [they] resemble giant spiders on a wall....This species is distinguished by its unarmed, large, gracefully arching, deeply guttered leaves." Flowering stalks 2–4 m long, upright or arching. Flowers bright yellow.

Michael Bogan's photos of the narrow, riparian canyon at Aguaje Los Pilares show this agave growing on sheer rock walls. It undoubtedly occurs elsewhere in the Sierra Bacatete, especially at higher elevations. Eastern Sonora and Chihuahua to Jalisco.

As the name amole indicates, the crushed leaves can be used as soap, a characteristic of most members of the subgenus Littaea.

Octopus agave is in the nursery trade and commonly grown as an ornamental plant in southern Arizona and elsewhere. It is propagated by bulbils that form in the flowering stalks of some plants.

Sonora: Sierra Bacatete, Aguaje Los Pilares, 28.174590, -110.465898, 18 May 2008, Michael Bogan, photos.

ASPHODELACEAE – ASPHODEL FAMILY (includes Aloaceae)

*Aloe vera

Common names: SAAVILA; *sávila*; aloe vera

Arizona and Sonora. *Aloe vera* produces rosettes of thick, succulent green leaves with short, thick spines along the margins. The flowers are bright yellow and are borne on erect stalks.

Aloes are native to the Old World, and *Aloe vera* is widely cultivated around the world for its medicinal properties and as an ornamental. Jesuit missionaries apparently introduced it into northwestern Mexico.

Aloe vera is cultivated in Arizona and especially in Sonora. It is easily propagated from offsets. Established colonies in the Yoem Bwiara may persist almost indefinitely. In some places on the coastal plain there are extensive, localized colonies, such as near Las Guásimas; the largest ones, covering perhaps a hectare, were seen near the abandoned townsite of Kopas and also between Kopas and Chiinim.

Asphodelaceae, *Aloe vera*, Tórim. 1 Jun 2019 (PB).

Medicine: *Saavila* is one of the more widely used medicinal plants. The short spines along the leaf margin are sliced away and the thick "skin" of the leaf is cut away to reveal the gelatinous, succulent part of the leaf that is used medicinally.

The fresh succulent part of the leaf (with the "skin" removed), applied directly, is used to treat burns, cuts, and other wounds, and to prevent scarring of the skin.

The thick inner part of the leaf is boiled, cooled, and applied to a wound as a dressing, used to clean the skin, or the liquid is consumed to alleviate stomachache: "*Se usa para curar heridas, limpiar el cutis, para dolor de estómago; se hierve y se coloca sobre la herida ó se toma licuada.*" Mateo González told us, "If you are having a hard time breathing, drink it. Also drink the clear aloe vera for bloating. For high blood pressure drink it like coffee. For bad breath, drink it like water. Also for loss of hair apply the inner part of the leaf." *Aloe vera* also has been used to treat mange in dogs.

Shampoo: The gelatinous part of the leaf is soaked in water, mashed, and stirred, and then used as a scalp and shampoo rinse; this makes the hair grow thicker and keeps your hair dark and prevents getting gray hair: "*Se usan para champu también. Se hierve, se saca liquido, se rebana y cuando hervido* *se usa la liquido, como jabón para que le cresca el cabello*" (Nacho González, Las Guásimas, 1985).

Arizona: Pima County, 8 mi west of San Miguel, 18 Mar 1935, *Shreve 7126.*

Sonora: Vicinity of Kopas, 13 Dec 1988, *Felger & Molina*, observation. 7.5 mi by road northeast of Vikam Suichi, *Felger 89-118 & Molina.* Base of Kuubwa'e Kawi, 1 km west of Kompuertam, 12 Mar 1989, *Felger & Molina*, observation. Northeast of Cuartel at Cañón Bacatete, 13 Mar 1989, *Felger & Molina*, observation.

ASTERACEAE (COMPOSITAE) – DAISY OR COMPOSITE FAMILY

The majority of the composites are herbaceous or small shrubs, and a few in the Yoem Bwiara are large shrubs. Flower structures and arrangements are complex and occur in bewildering combinations. Composites and orchids are the largest plant families in the world in terms of number of species.

The usual composites inflorescence is a flower head with a collection of individual florets (small flowers). Daisy-like flower heads have petal-like ray florets (sometimes just called "rays" although technically the ray is just the corolla and not the whole floret) surrounding a central disk of much smaller disk florets. Some have flower heads with only ray florets while others have only disk florets, and even others have female flowers without a corolla and develop into burs. Sepals are modified as highly varied bristles, called the pappus; or the pappus may be absent. The fruit is called an achene (technically a cypsela) and has only one seed. We use the term "fruit" interchangeably for an achene. Each flower head may bear a few (or sometimes only one) to many achenes.

Composites contribute the greatest diversity of medicinal herbs in this and other arid regions of the world. The great diversity of comps, especially in dry regions, is attributed to their complex evolution of the flower head (resembling a single "normal" flower) and secondary compounds, which serve to ward off herbivores from insects to vertebrates and provide protection from viruses and bacteria. It is these secondary compounds, sometimes expressed as aromatic compounds or gummy exudates, that are exploited by people for medicinal purposes. These secondary compounds often have antiviral and antibacterial properties.

Ambrosia ambrosioides

Common names: HIOWE; *chicura*; canyon ragweed

Arizona and Sonora. Multiple-stem bushes often 1 to 1.5 m tall, with slender, leafy stems. The leaves are rather large, narrowly triangular, rough, and sticky (due to glands), and are often studded with insect galls. The male flowers produce hay fever-causing pollen and the female flowers develop inside a bur.

Common along arroyos and canyons, river margins, and roadside depressions and other disturbed habitats.

Medicine: The dried root, *raíz de chicura*, is especially used by women. *Raíz de chicura* is sold in markets and herbal shops, including in Guaymas and Ciudad Obregón. The woody part of the root is used for "cold in the uterus." It is prepared by boiling five pieces in a liter of water for five minutes, and the tea consumed three times per day: "*Frío del útaro. Se prepara cociendo cinco trocitos en un litro de agua durante cinco minutos, y se toma el agua tres veces al día.*" In addition, Mateo González said "the root is really good medicine for rheumatism (*riumam*)."

Arizona: 15 mi south of Tucson, *Spalding 19 Mar 1906.*

Sonora: San José de Guaymas, *Felger 85-489.* 8 mi by road northeast of Vikam Suichi, *Felger 89-93 & Molina.* Near Kompuertam, 11 Mar 1989, *Felger & Molina*, observation.

Ambrosia confertiflora

Common names: CHICHIVO, CHI'ICHIVO (not translatable); *estafiate*; slimleaf ragweed

Arizona and Sonora. Herbaceous perennials with leafy stems. The male flowers produce hay fever-causing pollen. The female flowers develop inside small burs usually with hooked spines.

Often found on clayish soils and other poorly-drained, low-lying places in arroyos and canyons, and especially common in weedy places, including roadsides, around homes, in fields, and along canal banks.

Medicine: The leafy stem tips are mashed and made into a little ball like a pill, and eaten to treat stomach pain.

Arizona: Tucson, Santa Cruz River bottom, Tucson, 24 Sep 1903, *Thornber 41.*

Sonora: Cuesta Alta, Río Yaqui, *Felger 85-1364.*

Southeast of Potam, *Felger 88-593*.

Ambrosia monogyra [*Hymenoclea monogyra*]
Common names: KAU HEEKO (*kau*, mountain + *heeko*, the word for *Baccharis sarothroides*; mountain desert-broom); *hierba de pasmo, jécota*; slender burrobush

Arizona and Sonora. Upright, broom-like shrubs 2 to 2.5 m tall, with gummy, resinous herbage. Male and female flowers occur on different plants; flowering is in fall. The flower heads are small and straw-colored; the female flower heads form winged burs.

Larger gravelly washes of drainageways and dry riverbeds.

Household: The branches were tied together to make brooms; especially in Arizona.

Medicine: The concept of *yerba pahmo* extends to various other shrubby composites with resinous/viscid and aromatic herbage, including *Baccharis pteronioides*, *Isocoma tenuisecta*, and *Xylothamia diffusa*. The medicinal uses are probably interchangeable among these plants.

Arizona: [Tucson], Ft. Lowell, 24 Oct 1903, *Thornber 86*.

Sonora: Cañón Bacatete (Vacateve), Sierra Bacatete, broad arroyo bottom, *Felger 88-625*.

Ambrosia salsola var. **pentalepis** [*Hymenoclea salsola* var. *pentalepis*]
Common names: HEEKO NAWIA (*nawia*, coward, weak or delicate + *heeko*, the word for *Baccharis sarothroides*; cowardly desert broom); white burrobush

Arizona. Globose, resinous shrubs to about 1 m tall, with reduced leaves. The flower heads are somewhat similar to those of *Ambrosia monogyra*, but flowering is in spring.

Open desert plains, floodplains of washes and riverbeds, and often in overgrazed places since it is not palatable to horses or cattle.

Medicine: *Heeko nawia* is used to treat sores. The leafy stems are boiled in water, left to simmer, and the sore then washed with this liquid.

Arizona: Tucson, 5 Apr 1951, *Parker 7410*.

Baccharis pteronioides
Common names: YERBA PAHMO; *hierba de pasmo*

Arizona and northern Sonora. Low, spreading shrubs, glandular-viscid with short, linear leaves mostly falling before flowering. Flower heads small and whitish, with disk florets only.

The plants are generally localized along small washes and upper bajadas, often at elevations above the desert.

Medicine: This plant is well-known in the treatment of *pahmo* (*pasmo*, shock). We were told that when someone goes from one extreme temperature to another (such as hot to cold), one's face may become "twisted" (spasms). For example, a woman in New Pascua was making tortillas indoors over a hot stove in winter. When her little boy yelled she ran outside to him, into the cold weather. Her face went into spasms, pulled to one side. In another incident, a woman from Old Pascua was brought to an *hitevi* (healer) at Barrio Libre with a similar problem. The *hitevi* explained the uses for *yerba pahmo* and began curing the patient. It took several treatments for her face to return to normal, and even then her face remained slightly twisted.

Arizona: Nogales, hills about city, 6 Dec 1914, *Thornber 7061*. Oracle, Santa Catalina Mountains, *Thornber 28 May 1905*.

Baccharis salicifolia [*Baccharis glutinosa*]
Common names: VACHOMO; *batamote*; seepwillow

Arizona and Sonora. Shrubs often more than 1.5 m tall. The leaves and new shoots produce a gummy, viscid exudate. The leaves are bright green, about 5 to 10 cm long, and somewhat willow-like. Male and female flowers are on different plants; the flower heads are small and white, with only disk florets.

Basilio (1890 [1634]:205) recorded *bachomo* as the word for "*batamote, planta*."

This shrub is common in wetlands or at least temporarily wet places. In the Yoem Bwiara it is abundant along irrigation canals, at the edges of agricultural fields, and along the river; also in water-retaining roadside ditches and arroyo beds. It is occasionally seen in home gardens in Sonora. In Arizona it is common in riparian and other wetland habitats.

Ceremonial: The stems are tied together to form a mat, called *tapehtim*, used in tabletop altars in household yards during *Animam Mikwa* (All Souls Day; November 1), for offerings to the departed souls. *Vachomo* was also sometimes used to make heavy burial mats (Moisés et al. 1971:21), especially if *vaaka* (*carrizo*, *Arundo donax*, Poaceae) was not available.

Asteraceae, *Baccharis salicifolia*, Santa Cruz County, Arizona. 21 Aug 2014 (SC).

Construction: The leafy branches have been incorporated into roofing for traditional houses and ramadas. Rosalio Moisés reported that at Old Pascua, "We built a large ramada of cottonwood logs covered with *batamote* branches gathered along the river" (Moisés et al. 1971:53).

Medicine: The young, green shoots are eaten as a remedy for stomachache (presumably only a small amount would be ingested). If one is "bloated up," the leaves or young green shoots can be applied with spit and rolled over the stomach. It is one of several plants reported as excellent to treat earaches. One person said, "Find a few of the freshest leaves on the plant, rinse them and then roll in the fingers into a little ball, apply a little bit of olive oil, and insert it into the ear." Also see tobacco (*Nicotiana rustica*, Solanaceae), basil (*Ocimum basilicum*, Lamiaceae), and rue (*Ruta graveolens*, Rutaceae).

Rosalio Moisés reported that his grandmother cured him of "worms" with "medicine made out of *batamote*, *saus* tree, and *yerba del indio* ground up together and dissolved in water with a little salt. She made me drink a very big glass of this horrible, bitter medicine, following by as much water as I could hold" (Moisés et al. 1971:37). *Saus* is willow (*Salix gooddingii*, Salicaceae) and *yerba del indio* is desert pipevine (*Aristolochia watsonii*, Aristolochiaceae).

Arizona: Rillito Creek, 4 mi north of Tucson on Campbell Ave, 12 Sep 1944, *Gould 2519*.

Sonora: Cuesta Alta, Río Yaqui, *Felger 85-1360*. Las Guásimas, growing in a home garden, 16 Dec 1988, *Felger & Molina*, observation. Yaqui River, *Palmer 16 in 1869* (US). Tórim, 24 Mar 1934, *Studhalter 1493* (US).

Baccharis sarothroides

Common names: HEEKO, HEKO; *escoba amargo*, *romerillo*; desert broom

Arizona and Sonora. Broom-like woody shrubs 2 to 3 m tall. The numerous, many-branched, green stems are often nearly leafless. The flower heads are small and greenish to white. Male and female flowers occur on separate plants. The female plants produce large quantities of fluffy, white "seeds" (the tiny fruits with long, white hairs).

Basilio (1890 [1634]:212) cited *hecco* as the word for *romerillo*, a word also applied to several other broom-like plants.

Common in southern Arizona and rather scarce in the Yoem Bwiara where it generally occurs on disturbed sites such roadsides. The cultural information generally applies to its use in southern Arizona.

Construction: The branches were used for walls of temporary shelters.

Household: The branches were fashioned into brooms.

Medicine: For dogs or cats with fleas, ticks, or mange, the green sprigs were cooked in about a gallon of water and the animal washed with the liquid, although by the 1980s it was no longer used for this purpose. The same treatment was used for horses and cattle.

Arizona: [Tucson], Cortaro and Silverbell roads, 15 Jan 1953, *Caldwell 158*.

Sonora: Miramar, *Felger 85-514*.

Bebbia juncea var. **aspera**

Common names: MASO KUTA (deer stick); *hierba ceniza*; sweetbush

Arizona and Sonora. Rounded, bushy perennials, the stems and leaves slender and rather rough to the touch. Leaves sparse and quickly drought-deciduous. The flower heads are fragrant, yellow, and lack rays. The pappus bristles are feathery. Flowering at almost any time of the year.

Common on arid rocky slopes and the desert floor.

Medicine: Pieces of the plant are boiled with cinnamon (*Cinnamomum verum*, Lauraceae), and the tea consumed to counteract fever. It is also used as a "female medicine."

Arizona: [Tucson], Tumamoc Hill, *Turner 68-148*.

Sonora: Southeast of Pótam, *Felger 88-624*.

***Carthamus tinctorius**

Common names: KARTAMO; *cártamo*; safflower

Sonora. Sturdy annuals, usually with a single thick, whitish main stem. Leaves firm, the edges with spine-tipped teeth or spineless. Flower heads thistle-like, 4 to 5 cm wide, with large, bright orange-yellow flowers. Fruits ("seeds") white and plump, and without a pappus.

Safflower, of Old World origin, is a common oil-seed crop in Sonora and has been grown in the Yoem Bwiara. Occasional plants are seen along roadsides resulting from seeds falling from passing trucks.

Sonora: 4 km south of road to Microondas Las Avispas on México 15, Sierra Libre, *Van Devender 95-323*.

Chloracantha spinosa [*Aster spinosus*]

Common names: NAOWO (a core word); spiny aster

Arizona and Sonora. Shrubby perennials with nearly leafless, green, spinescent stems, and flower heads with white rays and a yellow center.

Sometimes forming thickets in seasonally wet places with clayish soils in roadside ditches, agricultural waste places, and along riverbanks, especially in southwestern Sonora.

Household: Bundles of the stems have been fashioned into brooms.

Arizona: Tucson, Santa Cruz Valley, 24 Sep 1903, *Thornber 34*.

Sonora: Road to Las Guásimas, *Felger 85-1154*. Pótam Viejo, *Felger 85-1438*.

Encelia farinosa

Common names: HACHI'IHTIA (sneeze), KOPALKIN, KOPAL OUWO (copal plant), TOOKO HUYA, TOROKO HUYA (gray or light blue plant); *incienso, rama blanca*; brittlebush

Arizona and Sonora. Rounded, brittle-stemmed bushes. The leaves are often whitish and clustered at the ends of thick twigs. The flower heads are raised on slender, branched stalks; the daisy-like flower heads have bright yellow rays and yellow or brownish-purple centers.

Hachi'ihtia refers to hay fever caused by this plant, and this name was also given for *Verbesina encelioides*. *Toroko* refers to the usual, grayish color of the plant. The descriptive name *toroko huya* is also used for several other plants including *Abutilon palmeri* (Malvaceae), *Krameria* (Krameriaceae), and *Solanum erianthum* (Solanaceae).

Asteraceae, *Encelia farinosa*, Pima County, Arizona. Near Sierra Del Aguila, 19 Mar 2017 (SR).

Widespread and often abundant, especially in arid, lowland habitats from the coast to mountain foothills in the Yoem Bwiara, and in upland areas in southern Arizona.

The stems exude a clear yellow to amber-colored, sticky resin called *chu'ukam*. The name *kopalkin* relates to the copal-like resin. When heated it becomes plastic and was used by Sonoran Desert people as a glue or sealant, for medicinal purposes, and as incense (Uphof 1968; Felger and Moser 1985; Felger 2007).

Ceremony and incense: The resin is used as incense in ceremonies and devotions both at homes and in churches.

Medicine: As a remedy for flu, the flowers are dried, placed in the nostrils and the aroma inhaled. The bark (or outer portion of the stem) is boiled and the tea consumed to counteract a stomachache.

Arizona: Tucson, Tumamoc Hill, 11 Apr 1917, *Shreve 5160*.

Sonora: San Carlos, *Felger 85-502*. 3 mi east of Mex Hwy 15 (Mapoli on map), east of Las Guásimas, *Felger 88-617*.

Encelia halimifolia

Common names: KOPAL OUWO (copal plant)

Sonora. Perennial bushes with grayish-green foliage. Flower heads solitary or several in an inflorescence; the daisy-like flower heads have yellow rays surrounding a brownish-purple disk.

Dunes and silty-sandy soils near the coast, such as the delta region of the Río Yaqui.

Ceremony **and medicine**: The plants produce a yellowish resin, or *copal*, used medicinally and as incense. The uses are presumably the same as for *Encelia farinosa*.

Sonora: 2.4 mi inland from Chiinim, *Felger 88-567*. Kuubwa'e Kawi, 1 km west of Kompuertam, *Felger 89-80 & Molina*.

Ericameria, see **Xylothamia**

Haplopappus, see **Isocoma**, **Xylothamia**

*Helianthus annuus

Common names: TAA'ATA VITCHU (looking at the sun, sun gazer); *mirasol*; sunflower

Arizona and Sonora. This is the common sunflower, a warm-season annual, often 1 to 2 m tall, with few to many large flower heads bearing bright yellow rays. The flower heads turn through the day, tracking the sun, as indicated by the Yoeme and Spanish-language names.

Common agricultural weed and seen along roadsides.

Agriculture and gardens: Sometimes cultivated as a commercial crop in the region and as an ornamental. Commercially grown sunflowers have a single, very large flower head.

Arizona: Tucson, Saint Mary's Road, 16 Sep 1935, *Shreve 7467*.

Sonora: San Carlos, roadside, 13 Aug 1985, *Felger*, observation. 2 mi west of Cd. Obregón on Mex Hwy 15, *Van Devender 93-22*.

Asteraceae, *Helianthus annuus*, Santa Cruz County, Arizona. 23 Aug 2014 (SC).

Isocoma tenuisecta

Common names: YERBA PAHMO; *hierba de pasmo*; burroweed, golden burroweed

Arizona and northern Sonora. Small shrubs or subshrubs. Flowers yellow.

Common in southern Arizona.

Medicinal: Teresa Amarillas was visiting in Marana in 1992 when Richard and Felipe brought her some plants from the Santa Cruz river near Yoem Pueblo. She identified *Isocoma tenuisecta* as *yerba pahmo* and told us it is an important medicinal plant.

Bernaldo Valencia was traveling with Felipe in 1977, and at a rest stop about 10 miles east of Globe, Arizona, he saw plants he called *yerba pahmo*. He was excited to see the plants and he gathered bundles of *yerba pahmo* to take home for medicinal purposes in Tucson.

Arizona: 5 mi north of Globe on AZ Hwy 60, *Darrow 6 Jun 1943*. Tucson, floodplain of Santa Cruz River by Silverbell Road, *Molina 17 Jan 2007*.

***Lactuca sativa**

Common names: LECHUUWA; *lechuga*; lettuce

One of the store-bought vegetables in Arizona and Sonora. Sometimes grown in Yoem Bwiara home gardens as a cool-season vegetable.

***Matricaria chamomilla** [*Chamomilla recutita; Matricaria recutita, M. suaveolens*]

Common names: MANSANIA, SI'IYA; *manzanilla*; chamomile

Arizona and Sonora. Aromatic annuals with small, rounded, yellow flower heads.

Si'iya is often grown in home gardens in the Yoem Bwiara.

Beverage and medicine: This well-known herbal tea is purchased in *boticas* (herbal shops) and supermarkets on both sides of the border. Tea made from the flower heads is taken to alleviate fever. *Si'iya* is regarded as a "cool" plant. Mostly, however, the tea is just an enjoyable drink. In Mexico it is widely known as *té de manzanilla* and in the United States as chamomile tea.

Arizona: 2–3 km southeast of Pan Quemado, Avra Valley, 21 Mar 2003, *Wiens 2003-9*.

Sonora: Guaymas, purchased in the marketplace as *té de manzanilla*, used for "*colicos, dolor de estomigo*," the plants said to come from "Cócorit" near Cd. Obregón, 20 Mar 1975, *Felger 75-110*.

Palafoxia linearis

Common name: VAI MANSANIA (sea chamomile [*manzanilla*])

Sonora. Annuals or short-lived perennials. The leaves and stems are dull grayish with coarse hairs, and the leaves elongated and sometimes semi-succulent. The flowers are whitsh to pink.

Sandy soils near the shore. *Palafoxia arida* in Arizona is a closely related and similar-appearing plant.

Sonora: Playa del Sol, *Felger 85-1128B*. Chiinim, *Felger 88-561*.

Pectis

Common names: WO'I SI'IYA (coyote chamomile); *manzanilla de coyote*; desert chinchweed, cinchweed

The three *Pectis* species listed here are small annuals growing and flowering with summer rains, except *P. coulteri*, which can flower almost any time of the year. These plants are pungently aromatic and dotted with conspicuous oil glands on the leaves and flowering bracts (phyllaries). The leaves are opposite, narrow, and with one to several pairs of bristles near the leaf base. The flowers are bright yellow.

Medicine: *Wo'i si'iya* is said to be "good for headaches and fever." It is prepared by boiling all parts of the plant, and the person drinks the liquid.

Pectis coulteri

Sonora. The ray corollas are broader than those of *Pectis rusbyi*. Common across most of the Yoem Bwiara.

Sonora: Las Guásimas, *Felger 85-274*. 1 mi northeast of Mex Hwy 15 at Pitahaya junction, *Felger 88-123*.

Pectis papposa var. **papposa**

Arizona and Sonora. Flower stalks usually shorter than the adjacent leaves. Seasonally abundant across much of the Sonoran Desert, although generally not common in the Yoem Bwiara.

Arizona: [Tucson], Tumamoc Hill, *Wilder 06-258*.

Sonora: Corral, Yaqui River side, 1939, *Gentry 4745*. Northwest side of Río Yaqui near Esperanza, *Van Devender 94-466*.

Pectis rusbyi [*Pectis palmeri*]

Common name: *manzanilla del campo*

Arizona and Sonora. The flower heads are on slender stalks longer than the leaves.

Common in the Yoem Bwiara and one of the most abundant hot weather wildflowers in the southern part of the Sonoran Desert. It often replaces *Pectis papposa* in less arid habitats and regions. *Pectis rusbyi* is generally larger, more upright, and has longer flowering stalks, larger flower heads, and different pappus bristles than *P. papposa*.

Arizona: Near Sells, 16 Aug 1931, *Kearney 8039*.

Sonora: Guaymas, 23 Oct 1939, *Gentry 4682* (DES, NY).

Perityle californica

Common names: MANSANIATA SAILA (chamomile's younger brother); California rock-daisy

Sonora. The name indicates that the plants look like *mansania* (chamomile, *Matricaria chamomilla*). Cool-season annuals, the plants variable size, often with delicate stems. Ray and disk flowers yellow.

This species is seasonally common in the Yoem Bwiara, often carpeting the landscape in yellow. Sandy plains and rocky slopes.

Medicine: The plants are boiled and the tea consumed to alleviate fever.

Sonora: Las Guásimas, *Felger 85-282*. Kuubwa'e Kawi, 1 km west of Kompuertam, *Felger 89-82 & Molina*.

Perityle microglossa var. microglossa

Common name: MANSANIATA SAILA (chamomile's younger brother)

Sonora. The same name is given to *Perityle californica* and they are used in the same manner. Cool-season annuals. Flower heads small; rays white, the disk yellow.

Common across the Yoem Bwiara, including shaded arroyo banks, irrigation ditches, and the riverbed and banks of the Río Yaqui.

Medicine: The plants are boiled and the tea consumed to alleviate fever.

Sonora: Las Guásimas, *Felger 85-294*. Cuesta Alta, *Felger 85-1368*. Kompuertam, *Felger 88-594*.

Pluchea sericea [*Tessaria sericea*]

Common names: *cachanilla*, *chamiso*; arrowweed

Arizona. Tall, slender shrubs with silvery foliage and pink flowers. The foliage is fragrant, and this kind of fragrance is called *winhuva*, a term for perfume, a sweet smell, or fresh fragrance.

Several people in Waalupe (Guadalupe) said this plant is known as *chamiso*, which is the Hispanic common name for a number of different shrubs.

This species is characteristic of bottomlands, often with a shallow water table and alkaline or saline soils, and often forms dense thickets along river floodplains.

Construction: At Waalupe, Pedro Alvarez showed Felipe a *pahko heka* (ramada) roofed with the freshly cut, leafy *chamiso* brush gathered nearby.

Religion: In about 1976, José Matus from Waalupe told Felipe that on Chandler Road, south of Guadalupe, there was a big area of this *chamiso* (also see Rea 1997). There was a shrine there in remembrance of a man who lost his life in that place.

Arizona: Santa Cruz River bank south of Valencia Road, 15 Jan 2004, *Mauz 24-132*. Tempe, 3 May 1935, *McLellan 404* (ASU). Mesa City, 16 Jun 1892, *Toumey 609*.

Porophyllum gracile

Common names: MASO KUTA (deer stick); *hierba del venado*; odora

Arizona and Sonora. Short-lived perennials, glabrous, bluish-green, and dotted with dark-colored oil glands. When the plants are bruised or damaged, they produce a very pungent odor. The foliage is sparse and the leaves slender and thread-like. Flowers dull white.

Common in Arizona and thinly distributed in southern Sonora.

Medicine: This species is used in the same manner as *Porophyllum pausodynum*. *Porophyllum gracile* is an important medicinal plant in traditional Seri culture (Felger and Moser 1985).

Arizona: Tumamoc Hill, 18 Apr 1917, *Shreve 5170*.

Sonora: Rancho San Antonio (north side of Sierra El Aguaje), 23 Jan 2001, *Sánchez-Escalante 01-8* (USON).

Porophyllum pausodynum

Common name: MASO KUTA (deer stick)

Sonora. Semi-shrubby perennials, usually sparsely branched, often 1 to 1.5 m tall. This species can be distinguished from *Porophyllum gracile* by its larger size, larger stems, broader and greener leaves, and larger flower heads in dense clusters. If one brushes against the plant it produces a sweet-smelling aroma, known as *winhuva* (sweet-smelling).

Common, mostly on rocky slopes. This species is known only from west-central Sonora including the Sierra El Aguaje, Sierra Libre, and Sierra Bacatete.

Medicine: To alleviate a fever, the stems and leaves are boiled, the tea consumed, and one then covers up with a blanket to sweat out the fever. It is considered a "warm" medicine. Small pieces of the larger stems are used to make a tea taken by women for menstrual difficulties. Edward Palmer reported that that the plant was used as a remedy for headache in the Guaymas region in 1887 (Watson 1889).

Painter (1986:55) reported: "*Maso kuta* is boiled in water and used for bathing, to alleviate deer sickness caused by the witchcraft of a deer, skin disease, or menstrual pain." It is not certain if Painter was referring to *maso kuta* in Arizona, in which case it would be *Porophyllum gracile*, but most likely the medicinal uses of both species would be interchangeable.

Sonora: 1 km west of Kompuertam, *Felger 89-84 & Molina*. Guaymas, *Palmer 279 in 1887* (NY).

Asteraceae, *Porophyllum pausodynum*, San Carlos. 21 Dec 2010 (SC).

*Sonchus oleraceus

Common names: CHINIITA, KORAI; *chinita*; common sow-thistle

Arizona and Sonora. Annual herbs, the larger leaves are in a basal rosette. Pale yellow, dandelion-like flower heads are borne on hollow flower stalks.

This is a common garden and agricultural weed, and sometimes grows in natural habitats. It is a worldwide weed of Mediterranean origin.

Korai grew as a weed in gardens and yards in Marana. It is described as being about two feet tall and having yellow flowers. Some people would leave it in their gardens and others would remove it.

Food: The young, tender leaves are gathered and cooked as a wild green. It is sometimes encouraged or protected in home gardens for such use. In January 1994 at Marana, Meregilda Ochoa showed us *chiniita* in her garden and said it is used for greens.

Arizona: Tucson, Santa Cruz [River] bottoms, 30 May 1903, *Thornber 466*.

Sonora: Cañón Nacapule, *Felger 85-253*.

Thymophylla concinna [*Dyssodia concinna*]

Common names: WO'I SI'IYA (*wo'i*, coyote + *si'iya*, the word for *Matricaria chamomilla*; coyote chamomile); *manzanilla de coyote*; dogweed

Arizona and Sonora. Annual spring wildflowers. The plants are aromatic and the flower heads have white rays and yellow centers.

Mostly on sandy or gravelly soils, and also on rocky bajadas and hills. Seasonally common from southern Arizona to the northern part of the Yoem Bwiara including the Sierra Bacatete.

Arizona: Tucson, foot of Cat Mountains, 1 Apr 1913, *Thornber 916*.

Sonora: Ruin of cuartel, Sierra Bacatete, *Felger 89-141 & Molina*. 6 mi northwest of Guaymas, 28 Feb 1932, *Shreve 6133*.

Verbesina encelioides

Common names: HACHI'IHTIA (sneeze), TAA'ATA VITCHU (looking at the sun, sun gazer); *mirasol*; golden crownbeard

Arizona and Sonora. These coarse annuals are mostly seen during warmer months. The plants are

A. Vicente Molina collecting *yerba pahmo* (*Xylothamia diffusa*, Asteraceae) at Chiinim. 13 Dec 1988 (WS). B. *Yerba pahmo* from Chiinim, on house wall in Pótam. 15 Dec 1988 (WS).

stinky, and the foliage is gray-green. The daisy-like flower heads are bright yellow and 3.5 to 5 cm wide.

Hachi'ihtia refers to hayfever caused by this plant, and the same was also given for *Encelia farinosa*. *Taa'ata vitchu* refers to the flower heads turning through the day, tracking the sun, as indicated by the Yoeme and Spanish-language names; the same name is also given to the common sunflower, *Helianthus annuus*.

Verbesina encelioides is a widespread garden and agricultural weed, and also occurs in other disturbed areas

Medicine: Johnson (1962:261) reported that "*ha'chihita*" is ground and put in the nostrils to cure a head cold (catarrh); although this information might also refer to *Encelia farinosa*.

Arizona: Tucson, 18 Oct 1941, *Benson 11068* (DES).

Sonora: La Huerta (near Bahía Algodones), *Felger 85-348*.

*Xanthium strumarium

Common names: KAME'EROI; *huicholi*, *mata de cadillo*; cocklebur

Arizona and Sonora. Large annuals with broad leaves. The female flowers develop into large burs with hooked spines.

This common weed occurs in disturbed places and sometimes in natural areas along washes and canyons. The pesky burs tangle in clothing and the hair of livestock, especially horses' tails. It is a worldwide weed in warm-temperate regions.

Medicine: To relieve any kind of cramps, but especially menstrual cramps, dried burs are parched or toasted and ground into a powder and applied externally with olive oil. The powder can also be taken internally for the same malady, but should be taken in small quantities. The powder is mixed with boiling water, allowed to simmer and the liquid is imbibed when it is clear. One person said it is "Good for cramps: parch the seed burs and grind; drink the tea or apply on afflicted area." Some people keep a big jar of the burs for medicine.

Arizona: 5 mi south of Tucson, 11 Oct 1939, *Benson 9820*.

Sonora: Guaymas, *Felger 85-429*. Rancho Bacatetito, 13 Mar 1989, *Felger & Molina*, observation.

Xylothamia diffusa [*Ericameria diffusa; Haplopappus sonoriensis*]

Common names: YERBA PAHMO; *hierba de pasmo*

Sonora. Slender-stemmed shrubs, gland-dotted and resinous. Leaves slender. Flower heads small and crowded at stem tips; flowers bright yellow, visited by large numbers of flies.

Alkaline/saline soils of the coastal plain, coastal dunes and strands, and rocky slopes near the shore.

Medicine: In December 13, 1988, we were at Chiinim on the Sonora coast with several men from Pótam. They all collected bundles of *yerba pahmo* to take home. Bunches of leafy stems were tied with twine and hung on house walls for later use. Another time, about 6 km east of Isla Lobos on the road from Pótam, Teresa Amarillas asked Felipe to stop the car so that she could gather *yerba pahmo*. There are extensive stands of the plant in that region. As a remedy for infection, the leafy stems are boiled, and the inflicted area or entire body is washed with the resulting liquid. The plant also can be boiled and one bathes in the water to treat body aches and pains.

Sonora: Chiinim, *Felger 88-560, Molina, & Steen* (ARIZ, USON). Las Guásimas, *Felger 88-607*. District of the Yaqui River, *Palmer 11 in 1869* (GH).

AVICENNIACEAE, see ACANTHACEAE

BATACEAE – SALTWORT FAMILY

Batis maritima

Common names: VAI MUUNI (water bean); saltwort

Sonora. Succulent perennials; leaves glassy, yellowish-green and rounded in cross section. Male and female flowers are on different plants, and the minute flowers are borne in succulent, cone-like structures.

Coastal, forming mats at tidally-inundated mangrove margins and on muddy soils of inlets, bays, and tidal flats. It often grows with *Sesuvium portulacastrum* (Aizoaceae), which is given the same Yoeme name.

Medicine: The plant is boiled and the tea consumed as a remedy for stingray wounds.

Sonora: Guaymas, Isla Pajaros, 7 Nov 2006, *Búrquez-Montijo 2006-38-ABM*. Bahía San Carlos, 2 Apr 1982, *Sanders 2516* (ASU, UCR).

BIGNONIACEAE – BIGNONIA FAMILY

Plants in this family often have large, showy flowers, and several species are grown ornamentally. The African tulip tree (*Spathodea campanulata*) is a large tropical tree grown in southern Sonora, with large, brilliant red-orange flowers. Jacaranda (*Jacaranda mimosifolia*), cultivated for its blue-violet flowers, grows easily in lowland Sonora. Catclaw vine (*Macfadyena unguis-cati*) is often seen covering walls and buildings in Arizona and Sonora; it has large, bright yellow flowers in spring.

Chilopsis linearis

Common names: *mimbre*; desert willow

Arizona and northern Sonora. Winter-deciduous trees with hard wood. Leaves long and slender, flowers pink or whitish, fruits narrow and elongate. Mostly along riverbanks and larger washes. A horticultural variety with purplish flowers is grown in Arizona.

Masks: Yoeme artists in Tucson, including José Guadalupe ("Lupe") Flores, occasionally used the wood for pahko'ola masks (Kolaz 1985, 2007).

Arizona: St. Mary's Road near Anklam Road, 29 Jun 1990, *Rondeau 90-108*. Tucson, *Toumey 3 Aug 1891*.

***Crescentia alata**

Common names: AYAL; *jícaro, tecomate*; gourd tree

Sonora. This small tree has unique, shiny green leaves, most of which are in the shape of a cross. The flowers and fruits are borne directly on the trunks and larger limbs. The large bat-pollinated flowers emit a musty odor at night. The fruits develop into hard-shelled, rounded gourds.

Ayal is occasionally cultivated in the Yoem Bwiara. Edward Palmer noted in 1887: "It is cultivated at Guaymas, under the name of 'ayal,' for shade and for medicinal properties of the fruit, which is filled with water and the liquid afterwards taken as a remedy for contusions and 'internal bruises'" (Watson 1889: 66).

Crescentia alata is found from southernmost Sonora through the American tropics.

Household: The gourds are fashioned into utensils and gourd rattles in many tropical regions.

Sonora: Guaymas, *Palmer 85 in 1887* (US).

Tabebuia impetiginosa [*Handroanthus impetiginosus; Tabebuia palmeri*]

Common names: SEMALULUKUT KUTA (hummingbird stick), TOOWO; *amapa, amapa rosa*

Sonora. Trees with hard wood. The large leaves are palmately compound with five stalked leaflets, appearing with summer rains and shed soon after the rains cease. These trees produce masses of bright pink flowers on the dry mountain slopes during winter and early spring.

Basilio (1890 [1634]:227) recorded *toohuo* as the name for "*amapa*, árbol," which in Yoem noki would be *toowo*.

Amapa rosa is a common tree of the tropical deciduous forest in the mountains east of the Yoem Bwiara.

Cultivation: *Amapa* is occasionally cultivated in southern Sonora. We saw several in home gardens in Pótam.

Medicine: Wagner (1936:83) reported, "Smallpox is treated with a tea made from *amapa* (dye wood). The wood is scraped in water causing it to turn to a light orange color. The tea is drunk half a cupful at a time. It has a delightful cooling taste."

Sonora: 35 mi northeast of Cajeme on road to Tesopaco, 3 Mar 1933, *Shreve 6149*.

Tecoma stans var. **angustata**

Common name: Arizona yellow bells

Arizona and Sonora. Low shrubs in its northern range; the plants, leaves, and flowers are smaller than those of variety *stans*.

Known from a few localities in the Guaymas region and common in hills and mountains near the eastern margin of the Yoem Bwiara, likely also in the Sierra Bacatete. It occurs along the desert border region of the United States, including uplands of southern Arizona, and south into the Sierra Madre Occidental.

Sonora: San Carlos, north side of Takalaim, rare, *Felger 85-371*.

Bignoniaceae, *Tecoma stans* var. *angustata*, San Carlos. 26 Aug 2017 (SC).

***Tecoma stans** var. **stans**

Common names: LORIO; *flor de fortuna, lluvia de oro, palo de arco*; yellow bells

Sonora. Large shrubs and small trees. Leaves to about 25 cm long, odd-pinnate with large leaflets. Flowering prodigiously much of the year, the flowers bright yellow, and in large terminal panicles.

This is one of the most commonly cultivated flowering trees in southern Sonora. Variety *stans* is native from eastern Mexico to South America.

Gardens: This beautiful, yellow-flowered tree is often grown in gardens in the Yoem Bwiara.

Medicine: Teresa Amarillas, from Pótam, told us that a powder can be made from the dried flowers and applied to sores. One may also stir a flower in a glass of water, remove the flower, and drink the liquid to treat diarrhea.

Sonora: San Carlos and Miramar, cultivated, 29 Oct 2015, *Felger & Carnahan*, observation.

BOMBACACEAE, see MALVACEAE

BORAGINACEAE – BORAGE FAMILY

Phylogenetic studies of the formerly broadly interpreted Boraginaceae have resulted in certain genera being recognized in different families: *Cordia* (see Cordiaceae), *Heliotropium* (see Heliotropiaceae).

Cryptantha and **Johnstonella**
Common name: SEVOA'A

Arizona and Sonora. Small, cool-season annuals. The plants have irritating, glassy hairs called *sevoam*, which is also the term for glochids of chollas (*Cylindropuntia*) and prickly-pears (*Opuntia*, Cactaceae). The flowers are small and white. Three species occur in the Yoem Bwiara and a greater diversity occurs in southern Arizona.

Cryptantha maritima

Arizona and Sonora. One of the most common cryptanthas in western Sonora.

Sonora: San Carlos, *Felger 85-229A*.

Johnstonella angustifolia [*Cryptantha angustifolia*]
Common name: narrow-leaf cryptantha

Arizona and Sonora. This is the most common and widespread cryptantha in the Sonoran Desert and reaches its southernmost limits in the Yoem Bwiara.

Arizona: Tucson, *Toumey 20 Apr 1894*.

Sonora: Kompuertam, *Felger 89-79 & Molina*. Guaymas, *Palmer 169 in 1890* (UC). 4 mi east of Empalme, 8 Mar 1933, *Shreve 6191*.

Johnstonella grayi var. **cryptochaeta** [*Cryptantha grayi*]

Sonora. The plants are much smaller and more delicate than *Johnstonella angustifolia*. Widespread in the Yoem Bwiara.

Sonora: Las Guásimas, *Felger 85-281*. Rancho Bacatetito, *Felger 89-152 & Molina*.

BRASSICACEAE (CRUCIFERAE) – MUSTARD FAMILY

Nearly all of the mustards in western Sonora and across most of the Sonoran Desert are cool-season annuals. Most mustards in the region begin life with the leaves in a basal rosette.

***Brassica oleracea**
Common names: REPOOYO; *berza*; cabbage

One of the store-bought vegetables in Arizona and Sonora. In the Yoem Bwiara often purchased at village stores and sometimes grown in home gardens as a cool-season vegetable.

***Brassica rapa** var. **oleifera**
Common names: canola, field mustard, rape mustard

Robust winter–spring annuals with bright yellow flowers. This mustard is sometimes grown in Sonora as a seed crop for canola oil and occasionally is seen in weedy places.

Sonora: Campo El Baguo, 30 km northwest of Pótam on Mex Hwy 15, cultivated in irrigated field, seeds harvested for canola oil, 4 Mar 2006, *Van Devender 2006-246*.

***Brassica tournefortii**
Common names: KAUPO HI'U (*kaupo*, in-the-mountains + *hi'u*, wild-greens; mountain wild greens), MOHTAASA; *mostasa cimarona*, *quelite cimarona*; Sahara mustard

Arizona and Sonora. Weedy, winter–spring annuals; highly variable in size. The leaves and stems have coarse hairs, and the leaves are rough to the touch. The flowers are small and pale yellow.

Native to the Old World, now widespread in arid regions worldwide. A serious invasive species across much of the Sonoran Desert.

Food: At Marana, some said it is considered to be good to eat as greens ("*buena para comer como quelite*"). However, most people said it is *chivu* (bitter; see *Sinapis arvensis*).

Arizona: Saguaro National Park West, *Bertelsen 90-2918*.

Sonora: 1 km north of Estero Soldado, roadside, 13 Mar 1985, *Felger 85-645*.

Descurainia pinnata
Common names: AASAM, HUYA AASAM, PAMIITAM (from the common name in Mexico); *pamita*; tansy mustard

Arizona and Sonora. Plants erect and often slender, with soft white hairs and much-divided, lacy leaves. The flowers are whitish to pale yellow and 1.5 mm wide; the fruits are narrowly club-shaped and 4 to 7 mm long. The numerous, tiny seeds are light brown and become mucilaginous when wet.

Widespread and seasonally common in natural and disturbed areas across the Sonoran Desert and in southern Sonora.

Beverage and medicine: The seeds are used to make a refreshing, cooling drink. The seeds are harvested when the fruit is "hard" (dry and firm). The seeds are soaked in water and the resulting liquid is consumed as a stomachic. This drink is "good for the stomach, it is not used so much now; it is used if your stomach was hot, just soak the seeds in water and drink that liquid. You can sift the seeds out or can drink it with the seeds, or you can chew the seeds if you want. Older men, *yoem yoyo'owe*, used to gather this plant by the handful" (Teresa Amarillas, 1987 in Marana).

Pamita tea is given to infants because it "cleanses the stomach, sort of like a mucilage or high-fiber cleansing." Another common use is to put a seed into the eye to absorb a foreign object; a use similar to that of chia (*Salvia hispanica*, Lamiaceae). The seed becomes mucilaginous and coats the foreign object, and tears can wash it out.

Arizona: Tucson Mountains, 16 Mar 1941, *Shreve 10028*

Sonora: Kompuertam, *Felger 89-73 & Molina*. Rancho Bacatetito, Sierra Bacatete, *Felger 89-165 & Molina*.

***Eruca vesicaria** subsp. **sativa** [*Eruca sativa*]

Common names: KONI WOKI (crow foot); *rúcula*; arugula, garden rocket

Arizona and Sonora. Flowering stems arising from a basal rosette of dark green leaves. Petals pale yellow or white, with purple veins. Seeds mucilaginous when wet.

This mustard was documented in the Yoem Bwiara region in 1912. It is native to the Mediterranean region and has become an invasive weed in parts of southwestern Arizona. Arugula has become popular in United States markets as a spicy salad green.

Arizona: Tucson Farms, 22 May 1913, *Thornber 7379*.

Sonora: Ontogata, Yaqui Valley, introduced weed, *Mackie in 1912* (UC).

***Raphanus sativus**

Common names: RAVANO; *rabano*; radish

One of the store-bought vegetables in Arizona and Sonora. Sometimes grown in home gardens as a cool-season vegetable.

***Sinapis arvensis**

Common names: MOHTAASA; *mostasa*; charlock

Arizona and Sonora. Robust winter–spring annuals, often with relatively large leaves and bright yellow flowers; mature flowering plants can reach about 1 m in height.

In Arizona and Sonora it is sometimes common in agricultural areas and occasionally seen at roadsides where the bright yellow flowers stand out prominently. Native to the Mediterranean region.

Food: In Arizona, the leaves of tender young plants are gathered in winter after good rains. The leaves are taken home, rinsed, briefly boiled, and then taken out and fried with onions, tomatoes, and garlic, and seasoned with black pepper and salt. In Sonora, *mohtaasa* and other greens are often eaten with beans and corn tortillas.

Arizona: Tucson, *Toumey 1 May 1892*.

Sonora: Guaymas airport, roadside, *Felger 85-423*.

***Sisymbrium irio**

Common names: KONI WOKI (crow foot), WIKIT WOKI (bird foot); *pamita*; London rocket

Arizona and Sonora. Plants mostly erect. Leaves smooth and lobed, the early leaves in a rosette, the stem leaves well-developed but reduced upwards. Flowers small and yellow, the fruits slender and elongated.

Widespread in the lowland desert and thornscrub, especially in disturbed sites, often along washes and as an urban and agricultural weed. Cattle avoid it and it is often common in grazed areas. Native to the Old World.

Food: In Arizona and Sonora it is picked while tender and young, and cooked as greens in the same manner as *mohtaasa* (*Sinapis arvensis*). The tops (the flowering portion) and older leaves are bitter, therefore the lower parts of younger plants are used.

Arizona: Tucson, freight depot, 9 Feb 1915, *Thornber 7030* (ARIZ, UNM).

Sonora: Las Guásimas, *Felger 85-280*. Near Kompuertam, 11 Mar 1989, *Felger & Molina*, observation. Rancho Bacatetito, 13 Mar 1989, *Felger & Molina*, observation.

BROMELIACEAE – BROMELIAD FAMILY

***Ananas comosus**

Common names: NOAI, PINYA; *piña*; pineapple

Basilio (1890 [1634]:183), cited *noai* for *"piña, fruta conocida,"* indicating it was cultivated in the Yoeme or Yoreme region. Pineapple is native to the Caribbean region. *Pinya* is the name used nowadays,

Hechtia montana

Common names: EKONIA; *aguamita, mescalito*

Sonora. Perennial rosette plants forming dense and sometimes extensive colonies. Leaves tough, semi-succulent, silvery green, and bearing sharp, recurved marginal spines. Male and female flowers occur on different plants; flowering stalks reach about 1 m tall; flowers small and rather inconspicuous. Flowering in summer.

Once abundant on the rock ledges around Bahía San Carlos, and common on many open or partially barren rocky hills and mountains in the Sierra Bacatete and Sierra El Aguaje.

José María Jaimez said this plant is like an agave (*kuu'u*) but smaller, and grows at the ridge crest of a mountain "east of Vícam," known as *Chinipove*, at the south end of Sierra Bacatete. Chinipove is said to be impassible because these spiny plants grow so close together. When the Yoemem were in the mountains during the wars, the federal soldiers sometimes could not get to them because of these plants.

Bromeliaceae, *Hechtia montana*, San Carlos. 12 Dec 2013 (SC).

Food: Guarijío people in southeastern Sonora make use of *Hechtia montana* as a food plant. The leaves are trimmed and the center (*cabeza*) of the plant is roasted like a miniature agave (Felger and Yetman 2000).

Sonora: Sierra Libre, *Felger 85-814*. Bahía San Carlos, 3 Sep 1989, *Sanders 9174* (UCR).

Tillandsia

These plants are epiphytes (growing on another plant, usually a tree, but not parasitic). The silvery-gray color of many of the species is due to the stalked (peltate) scales. The scales readily trap minute amounts of water, and the scales then become transparent so that the leaves appear green.

The two species in western Sonora occur along the coast southeast of Guaymas, where maritime dew drenches the vegetation during many winter–spring nights. Additional species occur in the mountains of southeastern Sonora.

Tillandsia exserta

Common names: WIVIS KUU'U (*wivis* agave); *quiqui*

Sonora. These bromeliads have silvery-gray rosettes of leaves and are found on many kinds of trees. The larger leaves reach 50+ cm long and often hang down in curls. The flowering stalks are 30 to 140 cm tall and bear many tubular, deep violet flowers visited by hummingbirds. Flowering may occur from October through December.

Common along the coast from the vicinity of Las Guásimas southward. A small, northernmost outlier population occurs near the shore at San Carlos near Estero Soldado.

Gardens: People in the Yoem Bwiara sometimes place this bromeliad in trees around their homes as an ornamental plant.

Medicine: This plant is said to be "good for earaches." At Las Guásimas we learned that plants of either *Tillandsia exserta* or *T. recurvata* are parched, the ashes mixed with olive oil, and the paste applied in the ear with fingers as a remedy for earache. Mateo González said it is also used to treat measles: "Boil the plant and drink the tea until cured."

Sonora: 15 mi southeast of Guaymas, 18 Feb 1953, *Blakley B1645*. West of Estero Soldado, abundant in a large *Forchhammeria* tree, also seen in *Bonellia*, *Felger 84-533*. Las Guásimas, *Felger 88-612*.

Tillandsia recurvata

Common names: WIVIS KUU'U (*wivis* agave); *quiqui*; ballmoss

Sonora and limited areas in southern Arizona. The plants form rounded clusters of many small shoots with slivery-gray, 2-ranked leaves less than 10 cm long. Flower stalks slender, usually less than twice as long as the leaves, with one to several small blue flowers, flowering during summer rains.

Coastal plain from the vicinity of Las Guásimas southward, often growing with *Tillandsia exserta*. Near Las Guásimas it is found on shrubs and trees such as *Fouquieria* and *Parkinsonia praecox*.

Medicine: Used medicinally in the same manner as for *Tillandsia exserta*.

Sonora: East of Las Guásimas, *Felger 85-392*. South of Peón, *Turner 61-58*.

BURSERACEAE – FRANKINCENSE FAMILY

Trees and shrubs with resin ducts containing aromatic triterpenes and ethereal oils. Old World members of the family are the source of frankincense, made from the gum of *Boswellia carteri* and related species, and myrrh, made from *Commiphora myrrha* and related species; both genera, especially *Commiphora*, are related to *Bursera*.

Bursera

Trees and shrubs. The flowers are white or yellowish, small, and solitary or in small inflorescences. The fruits are small, 1-seeded, and have leathery valves (exocarp or husk) that fall away to reveal a brightly colored, aril-like pulp partially to fully covering the seed.

The aromatic foliage and resin, or *copal*, of various species of *Bursera* has a long history of medicinal and religious use in Mexico. The wood is soft and lightweight, easy to carve, and usually does not split during drying. An attractive, red-flowered mistletoe, *Psittacanthus sonorae* (Loranthaceae), is parasitic on burseras in the Yoem Bwiara.

Burseras in Sonora are generally called *torote*, a name applied to various semi-succulent (sarcocaulescent) shrubs and trees including certain species of *Jatropha* (Euphorbiaceae). Basilio (1890 [1634]:227) listed the word *toro* for "*torote, árbol*," indicating a species of *Bursera*. The Spanish term *torote* might be derived from the Piman or Uto-Aztecan name, which is similar to the Yoeme name. The thick-trunked desert burseras are sometimes called elephant-trees, a name also applied to *Pachycormus discolor* (Anacardiaceae) of the Baja California Peninsula.

Medicine: Wagner (1936:84) reported, "for coughs, a tea is made from the bark of the *torote* tree. The tea is boiled, and a half cupful taken three times a day for six or eight days." Although he did not identify the species, it is likely to be *Bursera fagaroides*.

Bursera fagaroides var. **elongata** [*Bursera confusa*, in part]

Common names: SAA TOORO (*saa*, from *sawai*, yellow; yellow *Bursera*); *torote, torote amarillo, torote blanco, torote de vaca, torote de venado*; fragrant elephant-tree

Sonora. This small tree resembles *Bursera microphylla* but usually becomes larger and has a larger and generally thicker trunk, and the leaves are also larger and the leaflets broader. The outer bark peels away in papery sheets during dry seasons. The leaves, present only during the summer rainy season, are not as aromatic as those of *Bursera microphylla*.

Common throughout the Yoem Bwiara, from the coastal plain to the higher mountains. When leafless, this tree and *Jatropha cordata* (Euphorbiaceae) may be difficult to distinguish. In general, this *Bursera* forms a spreading tree about as wide as tall and the *Jatropha* trees tend to be taller than wide.

Ceremony and incense: The resin is aromatic and used at home and at churches as *kopal* (*copal*; incense) for blessing and ritual food for the saints and departed ancestors. The incense is also used to cleanse the air in a home (see *Bursera microphylla*).

Harvesting: The bark from large *saa tooro* trees has been used as a honey container. The bark is peeled off in large sheets and tied together into a kind of bag. The honey does not spill or leak from the bag.

Masks: The wood has often been used for making a pahko'ola mask.

Medicine: Tea prepared from this plant is said to be "good for a cough." It is prepared by boiling a piece of the stem or bark in about a liter of water with cinnamon (*canela*, *Cinnamomum verum*, Lauraceae).

Paper: The papery bark was used like paper for writing letters, for instance to the governor and to officials in Mexico (probably early twentieth century and earlier).

Smoking: The peeling, papery bark was used as cigarette paper, especially in times of shortage during the Mexican-Yaqui wars.

Sonora: 2.5 km north-northwest of Torocobampo, Sierra Bacatete, 14 Dec 1988, *Felger & Molina*, observation. Guaymas, 23 Oct 1939, *Gentry 4718* (NY). Southeast of Peón, *Turner 61-47*.

Burseraceae, *Bursera fagaroides*, Los Anegados, Sierra El Aguaje. 9 Oct 2016 (SC).

Bursera laxiflora

Common names: CHUKUI TOORO (black tooro, black torote); *copalquín*, *torote prieto*; red-bark elephant-tree

Sonora. Large shrubs or small trees. The bark is dark red-brown to gray and does not peel away. The leaves are produced after rains and are smooth, once- or twice-divided into small segments, and moderately aromatic when crushed.

Common and widespread at all elevations in the Yoem Bwiara.

Masks: The wood is sometimes used to make a pahko'ola mask.

Medicine: This plant is said to provide a good remedy for cough, including whooping cough. Any part of the plant, but especially the bark or leaves, is boiled and the tea consumed. It is not available in Arizona and some of the elders say they think about the plant when they have a cough. Sometimes they obtain it in the mountains around Hermosillo, but there is more of it in the Yoem Bwiara, especially in the vicinity of Vícam Pueblo. The tea is bitter.

Chukui tooro is highly regarded for treating a body rash. One consultant related that, "this tree is bitter, therefore good for a rash." The bark is mashed on a metate, the mashed bark put in water and the afflicted area bathed in the liquid.

Torote prieto (probably *Bursera laxiflora*) is sold in herbal markets (*mercados*) in Sonora. Tea made from the bark is said to be "good for diabetes."

Sonora: Cañón Bacatete, 17 Dec 1988, *Felger & Molina*, observation. Guaymas, 22 Oct 1939, *Gentry 4680*. In ravines near Guaymas, *Palmer 280 in 1887* (GH).

Burseraceae, *Bursera laxiflora*, San Carlos. 11 Nov 2014 (SC).

Bursera microphylla

Common names: SAA TOORO (yellow *Bursera*), TOORO; *torote*; elephant-tree

Arizona and Sonora. The term *saa* comes from *sawai*, the word for yellow. Small trees or large shrubs. The trunk and limbs are thick and semi-succulent, and the bark peels away in thin, papery sheets during dry seasons. The wood is soft and the branches and twigs flexible. The leaves, appearing after rains, are smooth and divided into slender leaflets arranged like a feather (pinnately compound). The sap and crushed leaves are highly aromatic.

In Arizona it is found in scattered, mostly remote desert localities such as Organ Pipe Cactus National Monument and desert mountains south of Phoenix. It is common and widespread in the desert in western Sonora, where the northern part of the Yoem Bwiara marks its southern limit. *Bursera fagaroides* and *B. microphylla* are closely related, and occasional plants in the Guaymas region appear to be hybrids. Cultural uses of the two species seem to be interchangeable.

Incense: The plant produces *kopal* (*copal*) that is burned at night as incense to produce an aromatic smoke. *Hecha una goma que se utiliza para hacer humo en la noche; mucho de la gente creen que limpia la casa* (Cruz Matus, 1985 in Guaymas). ("the plants produce a gum that is used to make smoke at night; many people believe it cleanses the house").

Medicine: This plant is used for medicines in a manner similar to that of *Bursera fagaroides*.

Arizona: Maricopa County, south side of South Mountain, *Fertig 29385* (ASU).

Sonora: Road to Las Guásimas, *Felger 90-604*. Guaymas, 24 Oct 1939, *Gentry 4709* (NY).

Burseraceae, *Bursera microphylla*, Bahía San Pedro. 13 Dec 2014 (SC).

CACTACEAE – CACTUS FAMILY

Cacti, ranging from small "pincushions" to large columnar species, are prominent features of the landscape in the Yoem Bwiara and southern Arizona deserts. About 100 species of cactus occur in the state of Sonora (Paredes et al. 2000) and 28 native species are documented for the Yoem Bwiara.

Cactus spines are arranged in unique clusters called areoles. The fruits of most of the species in the region are edible, although some are hardly worth the bother. The large, delicious fruits of the columnar cacti have been important food resources and were used in wine making.

Columnar cacti are significant elements in the landscape, economy, and conservation of natural resources in southwestern Sonora (Semotiuk et al. 2017).

Carnegiea gigantea [*Cereus giganteus*]
Common names: SAUWO; *saguaro*; saguaro/sahuaro

Arizona and Sonora. These giant columnar cacti tend to be larger and more common in southern Arizona than in the Yoem Bwiara, and therefore were more widely used in Arizona. The large white flowers appear in May and June in Arizona. The large, essentially spineless fruits ripen during the very hot, dry weather of early summer before the monsoon rains. Flowering and fruiting occur about one month earlier in Sonora than in Arizona. The fruits have a thick, inedible rind and small, edible black seeds embedded in sweet, juicy, red pulp. The ripe fruits split open to reveal the red pulp and inner fruit wall, and from a distance may look like red flowers instead of fruits.

The dry, woody ribs of the saguaro skeleton are called *akwo*. Fruits on the plant are called *sauwo taakam* (*taaka* is the name for any fruit). Naturally fallen fruits on the ground are *sopi'ichim*, a term that is specific for the saguaro.

Saguaros in the Yoem Bwiara generally occur on south-facing rocky slopes, the open desert, and the coastal plain southeast of Empalme.

Saguaro and *cardón* (*Pachycereus pringlei*) are the largest cacti in the Sonoran Desert and often occur together in western Sonora. Saguaros tend to initiate branches higher off the ground than *cardones* and the mature saguaro stems are yellow-green rather than bluish-green (glaucous). Young plants may be difficult to distinguish.

Construction: Saguaro ribs were used for house walls and roofing, mostly in Arizona, and also for fences. In the early 1900s, people at Marana gathered saguaro ribs in the foothills of the nearby Tortolita Mountains for house building. Around 1914, a man gathering saguaro ribs in that area for his house was killed by a *rinchi* (ranger). After that incident the people no longer gathered in the Tortolitas and shifted their gathering to the Tucson Mountains.

In Marana, *akwom* (the woody ribs), were sometimes fashioned into trellises for roses (see *Rosa*, Rosaceae).

Woody ribs of *Carnegiea gigantea*, Pima County, Arizona. 4 Feb 2005 (SR).

Food and harvesting: From at least the early 1920s until the mid-1940s, entire families would venture into desert foothills and mountains to gather the fruits. It was common for families to set out in horse-drawn wagons and camp among the saguaros to harvest the fruit, sometimes camping for a week. People from Marana camped at the north end of the Tucson Mountains because even though the Tortolita Mountains were closer, they feared the "rangers" would shoot at "trespassers."

The saguaro fruit-gathering pole is called *hiabwai*, and the process of picking or pulling down the fruit is called *hiabwa*. Similar gathering poles are used for harvesting fruit from *aki* (*Stenocereus thurberi*), *huvahe* (*Vitex mollis*, Lamiaceae), and *mako'ochiini* (*Pithecellobium dulce*, Fabaceae). The *hiabwai* is made from two long saguaro ribs or two long pieces of *vaaka* (cane or *carrizo*, *Arundo donax*, Poaceae), tied together with wire, plus a short cross-piece tied near the top. The hiabwai is similar to saguaro fruit-gathering poles made by other Sonoran Desert people including the Tohono O'odham and Comcaac (Seris). The fruits are gathered during the cooler parts of the day, mostly in the early morning, and sometimes again in the late afternoon. Men, women, and children participated in the saguaro fruit harvests in Arizona.

The fruit is eaten fresh, either picked from the plant or from freshly fallen fruits. Arizona Yoemem prepared the fruit in ways similar to Tohono 'O'dham methods. The fresh fruit is a special treat. Cooked fruit is made into syrup for later use. Felipe's uncle, Juan Luis Garcia, called the saguaro fruit jelly *sito'im*, which is also the term for honey.

The people talk to the saguaro, orally or mentally, before picking the fruit. They ask permission and thank the saguaros for a good year of fruit.

Utilitarian: In Arizona the main arm of the *taravia*, the traditional rope twister, was made from a piece of saguaro rib, and the smaller pieces were made from a different, harder wood such as tamarisk (*Tamarix aphylla*). Taravia is also a Yoem noki word for spindle.

Arizona: Tucson, 4 May 1940, *Benson 10011.*

Sonora: Rancho Bacatetito, 13 Mar 1989, *Felger & Molina*, observation. Vicinity of Guaymas, 10 Mar 1910, *Rose et al. 12615* (US).

Cylindropuntia

Common names: CHOA; *cholla, choya*; cholla

There are two readily recognized major groups of *Opuntia* relatives in the Yoeme regions of Arizona and Sonora: Prickly-pears (*Opuntia*) have stems made up of flattened stem segments, or "pads." Chollas (*Cylindropuntia*) have stems made up of more-or-less cylindrical segments, or "joints." Most chollas are classified as *Cylindropuntia*, formerly a subgenus of the genus *Opuntia*. Hybrids between cholla species are sometimes encountered.

Choa is the general term for cholla. People and

animals growing up in the desert learn to be careful of chollas. The spines have tiny, backward-facing barbs that tear flesh when you pull them out. The larger spines on chollas and prickly-pears are called *wicham*. The glochids—the tiny, highly irritating and readily detaching spines—are called *sevoam* (*alguátes* in Spanish).

Medicine: Rosalio Moisés described his grandmother's use of cholla in Hermosillo, most likely *Cylindropuntia fulgida* and/or *C. thurberi*: "She often had to cure Abelardo, who was subject to the throat ailment that has killed so many Yaquis. When he lost his voice, María would gather chollas, pound them until she had a soapy substance that she put in water, and give the mixture to Abelardo for three days, and his voice would be all right again" (Moisés et al. 1971:22).

Cholla seeds are soaked in water (the liquid becomes gelatinous). This liquid is said to be good for stomach troubles. For a remedy for dysentery, ripe cholla fruit is cleansed of its spines and boiled in water. After boiling, it is stirred to break up the clumps, and the liquid drained off. This liquid is consumed and in a short time the dysentery will be cured.

Cholla root is a popular remedy among the Yoemem and others in Sonora. At markets in Mexico, such as the Mercado Municipal in Guaymas, small packages of the roots are sold as a remedy for pain, inflammation, and kidney problems including kidney stones, to be prepared in the same manner as for *chicura* (*Ambrosia ambrosioides*, Asteraceae): "*Dolor, inflamación, infermedades de los riñones y cálculos; preparadora igual que la chicura, y se toma igual que la chicura.*"

Cylindropuntia arbuscula [*Opuntia arbuscula*]
Common names: SEVII; *tasajo*; pencil cholla

Arizona and Sonora. Shrubby chollas, often 1 to 1.5 m tall and with a short but well-developed woody trunk. Stems moderately slender, green, with widely spaced spine clusters (areoles). Larger spines one per areole, variable in size, the stems often with at least some long spines with golden-yellow papery sheaths. The flowers are yellow-green. The fruits are fleshy and remain green even when ripe, and often persist for about one year.

Often on fine-textured soils in swales and plains, and sometimes on bajadas or lower slopes. Especially common on the coastal plain of the Yoem Bwiara and extending into the Sierra Bacatete.

Food: The fruits are used in the same manner as for *Cylindropuntia thurberi*.

Arizona: Tucson, 6 mi west of Silverbell Road on Ina (Picture Rocks) Road, 3 May 1990, *Baker 7820* (ASU).

Sonora: Cañón Bacatete, 17 Dec 1988, *Felger & Molina*, observation.

Cylindropuntia fulgida [*Opuntia fulgida*]
Common names: SEVE'E CHOA; *choya, velas de coyote*; chain-fruit cholla, jumping cholla

Arizona and Sonora. This tree-like cholla is often 0.5 to 3 m tall. The trunks are seldom straight, and often have several major branches from about mid-height. The flowers are about 4.5 cm wide, and pinkish-purple. The fruits are green and fleshy, and form perennial, hanging chains. The fruits (seeds) are nearly always sterile. The plants propagate by the readily detaching joints (stem segments) as well as the fruits, which readily form roots and develop into new plants.

Cactaceae, *Cylindropuntia fulgida*, Santa Cruz County, Arizona. 24 Jul 2017 (SC).

There are two varieties. Variety *fulgida* is abundant in southern Arizona and in Sonora southward to the Guaymas region. This variety is characterized by stems densely covered by relatively large spines with inflated, papery sheaths. Variety *mamillata* has fewer and shorter spines that do not obscure the stem surface and have tight-fitting spine sheaths. Variety *fulgida* is not known south of the Guaymas region, whereas var. *mamillata* ranges southward to Sinaloa. Plants of var. *mamillata* in the Yoem Bwiara are mostly about 1.5 m or less in height, although photos from the mid-twentieth century show plants that appear to be much larger.

Beverage and food: Like most fleshy-fruited chollas, the fruit is edible. It is roasted in hot ashes to burn off the spines and then eaten.

Gum exuding from the stems dries into a hard, black mass called *choa vooram* (*vooram* is an untranslatable core word). This gum apparently forms at the site of a wound, such as from a hole made by certain beetles. The gum is roasted and ground to a coarse flour, which is mixed with water to make a thick, atole-like drink called *vaahipo'im* (to drink as waters; *vaa* = *vaa'am*, waters + *hipo* = *hi'ipo*, to drink + *im* = plural indicator). This cool, refreshing beverage is made in an olla or bucket. It is popular and can be served iced. Children playing the role of the Apostles on Holy Thursday in the Easter ceremonies are given this vaahipo'im.

Don Ignacio Amarillas in Pótam showed us how the cholla gum is cooked. The spines are cleaned away, then fist-sized wads of gum are crushed into small chunks. On the earthen stove in the kitchen, the gum is roasted in a frying pan over a small fire while being constantly stirred. After roasting, it is ground into a kind of flour. Don Nacho and other people in Pótam are always looking for choa vooram to sell in Las Guásimas.

Medicine: Since the late 1980s, vaahipo'im has become a popular beverage to control diabetes.

Variety *fulgida*:

Arizona: Tortolita Mountains, 14 Sep 2013, *Lindley 283*.

Sonora: 32 mi west of Sonoita, 22 Mar 1969, *Pinkava 1550* (ASU).

Variety *mamillata*:

Arizona: 21 mi SE of Florence, 1 Jun 1981, *Baker 3770* (ASU).

Sonora: Guaymas, *Brandegee 12 May 1892* (UC).

1 km west of Estero Soldado, sandy flat, *Felger 85-891*. Kopas, 15 Mar 1989, *Felger & Molina*, observation.

Cylindropuntia leptocaulis [*Opuntia leptocaulis*]
Common names: *siviri*; desert Christmas cactus

Arizona and Sonora. Slender-stemmed chollas to 1.5 m tall. Flowers pale yellowish to cream, 1.5 to 2 cm wide, opening late in the afternoon and remaining open in the early evening. Fruits bright red.

Widespread and common in southern Arizona and in Sonora.

Arizona: Tucson Mountains, 15 Apr 1991, *Baker 8236* (ASU).

Sonora: 11.5 mi east-southeast of Empalme, 8 Mar 1993, *Baker 10368* (ASU). Guaymas, *Brandegee 12 May 1892* (UC).

Cylindropuntia thurberi subsp. **versicolor** [*Cylindropuntia versicolor*; *Opuntia thurberi* subsp. *versicolor*, *O. versicolor*]
Common names: SEVII; *siviri*; staghorn cholla

Arizona and Sonora. Shrubby chollas, often reaching 1.5+ m tall, the stems (joints) turning purple-brown during cooler or drier months. Flowers greenish-yellow in Sonora and variously colored (yellow, orange, red, greenish, or bronze) in the Tucson region. These variably-colored flowers

Cactaceae, *Cylindropuntia thurberi*, San Carlos, 9 Apr 2019 (SC).

probably result from extensive, former hybridization. Flowering March and early April. Fruits fleshy, yellow-green even when ripe, and usually persistent until the following year.

Basilio (1890 [1634]:224) recorded "*siviri, cierta planta cactiforme,*" which might also be *Cylindropuntia arbuscula*, *C. leptocaulis*, or *C. thurberi*.

Widespread in the Yoem Bwiara, on the coastal plain, lowland bajadas and hills, and mountains. Also common in the Tucson region.

Food: In Sonora the fruits are eaten fresh or prepared as *vannaim* (*atole* in Spanish).

Medicine: Pieces of the stem are roasted or charred on a grill or on hot coals, cooled, and applied directly to sores. It is also used with *Forchhammeria* and *Vallesia* for sores that are slow to heal (see *Forchhammeria*, Resedaceae).

Arizona: Tucson Mountains, 15 May 1991, *Baker 8250*.

Sonora: Guaymas, road to Cerro El Vigía, *Baker 4043* (ASU). San Carlos, *Sanders 2500*. 14 mi northwest of Guaymas, *Shreve 7306*.

Echinocereus leucanthus [*Wilcoxia albiflora*]
Common name: NOONO

Sonora. This delicate cactus has slender stems densely covered with short spines. The root system has a cluster of succulent, tuberous roots, blackish on the outside and fleshy and whitish inside. The flowers open during the daytime and are 3.5 to 4.5 cm wide and white, with a green stigma. Flowering occurs in April. The fruits are slender, spiny, and not eaten by people.

The same name is used for *Peniocereus*.

Locally common in the vicinity of Las Guásimas. The plants often grow through the small shrubs of *Lycium californicum*; the cactus stems seem to mimic those of the *Lycium*. It is otherwise known from similar habitats in southwestern Sonora (Martin et al. 1998) and northwestern Sinaloa. Much of the potential habitat has been converted into agriculture or altered by woodcutting and clearing.

Medicine: The tuberous roots are prepared as a remedy for headache in a manner similar to that for *Peniocereus striatus*.

Sonora: Las Guásimas, *Felger 85-380*; *Voss 29 Dec 1968*.

Ferocactus
Common names: ONO'E; *biznaga*; barrel cactus

These three names are applied to each of the species listed below. Basilio (1890 [1634]:221) recorded *onnore*, a Yoreme term, as the word for *biznaga*.

Barrel cacti in the Yoeme lands in Sonora and in southern Arizona generally have only a single and very thick stem. These large cacti have stout spines and fleshy yellow fruits with a tart, lemon-like flavor.

Food: The fruits were eaten fresh as a minor food resource or snack.

Ferocactus emoryi [*Ferocactus covillei*]
Common names: *biznaga*; Emory barrel cactus

Arizona and Sonora. The spine clusters (areoles) are made up only of stout spines (there are no bristle-like spines); the spines are much larger on juvenile plants than on mature ones, and the central spines are straight or only slightly curved near the tip. The plants in southern Arizona and northern Sonora have red or red-orange flowers while those in Sonora south of Guaymas have yellow flowers; flowering is in summer.

In the Yoem Bwiara this species occurs on plains, bajadas, and rocky slopes.

Sonora: Sierra Libre, road to Microondas Avispas, flowers yellow to red, *Felger 85-811*. Rincón del León, flowers yellow, *Felger 85-910*. Guaymas, *Jones 26 Jan 1927* (RSA).

Ferocactus herrerae [*Ferocactus wislizeni* var. *herrerae*]
Common names: *biznaga*; spiral barrel cactus

Sonora. Stems to 1 m or more in height, often leaning southward, and with conspicuously spiraled, deep, and narrow ribs. Spine clusters with stout, flattened, hooked central spines, and also bristly radial spines. Flowers orange and reddish; flowering late summer or early fall, after *Ferocactus emoryi* has finished flowering.

This species is common along the coastal plain southeast of Guaymas, most often on fine-textured alluvial soils. Much of its habitat has been converted to agriculture. In cultivation it is fast growing as compared to the other barrel cacti in the region if provided with ample water during the hot Sonoran Desert summer.

Sonora: Las Guásimas, *Felger 85-271*. Cochorit, *Felger 85-397*.

Cactaceae, *Ferocactus herrerae*, Tórim. 1 Jun 2019 (PB).

Ferocactus wislizeni
Arizona barrel cactus

Arizona and Sonora, north of Hermosillo. The thick stem of this large barrel cactus generally leans southward as it grows. The spine clusters (areoles) have both stout (central) and bristly (radial) spines; the central spine is flattened in cross section and hooked at the tip. The flowers are orange to orange-red. This barrel cactus is common in southern Arizona.

Food: In Arizona, some people made barrel-cactus candy. Marcelina T. Flores related that some of her best memories are from daylong trips with her parents and grandparents in the hills west of Old Pascua. They went to gather a certain clay to make pottery and took "along a big pot, which was used to boil water and sugar in the making of candy from barrel cactuses." Barrel-cactus candy apparently was not a traditional Yoeme food.

Emergency water: The top of the plant was cut off, and the highly succulent central part of the stem cut up. One could then suck on pieces of the stem to obtain emergency liquid.

Narrative: Anselma Tonopuame'a told Felipe that when you are young and in love, and leave your girlfriend or boyfriend, and have not seen her or him for a long time, you start to get like lonely. You should not overdo it, because if you are in the desert a barrel cactus will turn into the image of

Cactaceae, *Lophocereus schottii* var. *australis*, Sierra Bacatete. 11 Sep 2008 (JS).

your friend and you will get happy and run to the cactus, thinking the image is your friend. Then you will hug your friend and will realize it is a barrel cactus. This is to help you learn not to think about far-away love.

Arizona: Sabino Canyon, *Benson 1 Mar 1940*. Fresnal Canyon, Baboquivari Mountains, 8 Apr 1926, *Peebles SF-111*.

Lophocereus schottii [*Pachycereus schottii*]
Common names: MUSEO; *músaro*, *sina barbona*, *sinita*; senita

Sonora. The distinctive columnar cactus is often about 2 to 2.5 m tall in the Yoem Bwiara. The dull green stems have 6 to 10 stem ribs. The lower and upper parts of the stems are very different from each other (dimorphic). The sterile (lower) part of the stem is relatively thick, with fewer ribs, widely spaced areoles, short and stout spines, and does not produce flowers. These thicker stems are also more succulent, with more mucilage, than fertile, upper stem portions. The fertile (upper) portions of the stems are relatively more slender, less succulent, with a higher number of stem ribs,

Cactaceae, *Lophocereus schottii* var. *schottii*, El Sahuaral. 30 May 2005 (JS).

and have a shaggy mane of long, slender and twisted spines, and the areoles are close together (or confluent), and produce flowers. The flowers are 4 to 4.5 cm wide, whitish to pale pink, opening at night and closing early the following morning. Flowering and fruiting occur during hot weather, especially when humid. The fruits are 2.5 to 4 cm wide, red, with a sweet, juicy pulp, and have small black seeds. It is often difficult to find the fruits before the birds and ants have eaten or carried away the pulp and seeds.

Basilio (1890 [1634]:219) recorded *múseo* as the term for *pitahaya barbona*, meaning "bearded" because of the densely spiny stem tips.

Common in open desert and plains, usually on flat terrain on fine-textured soils. Occasional plants or small colonies seen on rocky hills and mountains may result from bird-disseminated seeds. Museo is especially common near Loma Vahkom, a village north of the Río Yaqui near Vahkom (Bácum).

The senita cactus occurs in a few places in Arizona adjacent to the Mexican border, such as in Organ Pipe Cactus National Monument. This species ranges southward in western Sonora to northwestern Sinaloa, and extends through most of the Baja California Peninsula. The northern populations of senita, in Sonora and Baja California, have thicker stems and fewer stem ribs than populations farther south.

Food: The fruits are a minor food resource or snack: "The *museom* cactus fruit is also eaten fresh" (Moisés et al. 1971:158).

Medicine: This is a significant medicinal plant. *Sinita* is highly esteemed in Mexico, by the general population as well as the Yoemem. The stems are widely used for medicinal purposes, especially ones with five ribs. Five-ribbed stems are the sterile (lower) growth, and are thicker, "softer" and more flaccid and much juicier than the fertile (upper) stems that have a higher number of ribs. The five-ribbed *músaro* stems are esteemed in treating diabetes, stomach ulcers, and cancer, and also as a remedy for falling hair (loss of hair, "*evita la caída del pelo*") and blood circulation ("*circulación de la sangre*"). For example, a cross section of the stem, about 2 cm thick, is cooked in about one liter of water until it boils. This liquid is then consumed through the day as one would drink water ("*se pone en un litro de agua hasta hervir, y se toma como agua de diario*"). The fresh as well as dried, star-shaped cross-sections of *músaro* stems are sold in herbal shops and *mercados* in Mexico, including in Mexico City and Tijuana.

Song: A song concerning an old, dying senita (said to be drying up), called *museo waake* (museo drying-up), is instrumental music played on the flute and drum, but no words are sung. Bernaldo Valencia played this song for us at his home in Tucson.

Sonora: Vicinity of Rancho Bacatetito, 13 Mar 1989, *Felger & Molina*, observation. 44 mi by road (Mex Hwy 15) southeast of Guaymas, 27 Dec 1961, *Lindsay 3213* (SD). Cañón Nacapule, 15 May 1982, *Parfitt 3037* (ASU).

*Lophophora williamsii

Common name: peyote

This famous little cactus is native to the Chihuahuan Desert Region and does not occur on the western, Pacific watershed of Mexico. We found no evidence of traditional use of peyote among the Yoemem.

Beals (1945:195, 196) reported, "Peyote was unknown to my informants in any form...Hrdlička [1908:244] states peyote was well known to the Opata and Yaqui, but this seems doubtful. The use of charred cactus on wounds may have misled him, although peyote is early mentioned for the Opata for the same purpose."

Mammillaria bocensis
and
Mammillaria johnstonii

Common names: YOTUI KOVA (old-person's head); *biznaguita*

Sonora. Stems globose, solitary or clustering, with multiple stems and milky sap. The center of each stem or "head" has woolly hairs from which sharp spines emerge. One consultant said, "You feel like touching it but it has sharp spines." The spines are stout, short, straight (not hooked), and spreading. The flowers are about 2 cm wide, and the perianth segments ("petals") pale straw-yellow to red-brown with a pinkish midstripe. The fruits are red.

Mammillaria bocensis is common on the coastal plains southward from the vicinity of Las Guásimas to northwestern Sinaloa. This species resembles the closely related *M. johnstonii*, which occurs from the vicinity of Las Guásimas to San Carlos and the Sierra El Aguaje.

Mammillaria bocensis: Sonora: Las Guásimas, *Felger 85-272*. 25 mi southeast of Guaymas on dirt road (west of Mex Hwy 15), *Voss 594* (SD).

Mammillaria johnstonii: Sonora: Bahía San Carlos, tuff ledges about bay, 8 Jul 1921, *Johnston 4373* (CAS).

Mammillaria grahamii

[*Mammillaria grahamii* subsp. *sheldonii*, *M. inaiae*, *M. microcarpa*, *M. oliviae*, *M. sheldonii*, *M. swinglei*]

Common names: CHIKUL AAKI (mouse organpipe), CHIKUL HU'I (mouse penis); *cabeza de viejo*; fishhook or pincushion cactus

Arizona and Sonora. There are several species of similar-appearing pincushion mammillarias in the Yoem Bwiara and in southern Arizona, although *Mammillaria grahamii* is the common one. These are small, globose or short-stemmed cacti, with hooked and/or straight spines in spine clusters densely covering the stems. The small but attractive whitish or pink flowers are mostly borne in a crown (a ring of flowers) near the tip of the stem. The fruits are fleshy, without spines, red to orange, and have small black seeds.

Food: The ripe fruits are eaten fresh as snacks.

Medicine: The plant with the spines removed is placed in hot coals. The roasted stem is spread on sandals, which are tied onto the feet with a cloth as a remedy for *wok vetiam* (burning feet; perhaps athlete's foot fungus).

Arizona: Pinal County, Sacaton Mountains, *Peebles 23 Nov 1940*.

Sonora: Nacapule Spring, 4 Jan 1995, *Felger & Schneider 95-62*. San Carlos Bay, 13 Jan 1969, *Kimnach & Lyons 629* (HNT). Sierra Bacatete, Rancho El Álamo, *Sánchez-Escalante 2007-07-03* (USON).

Mammillaria yaquensis [*Mammillaria thornberi* subsp. *yaquensis*]

Common names: CHIKUL AAKI (mouse organpipe); Yaqui fishhook cactus

Sonora. Stems many, flaccid, with additional branches arising almost anywhere on the parent stems. Central spines unusually stout and strongly hooked, readily catching on fingers and clothing, which pulls off the whole stem. Flowers are white and rose pink. Fruits red.

Common on alluvial soils of the coastal plains from near Empalme and Las Guásimas southward to northwestern Sinaloa. Often growing beneath spiny shrubs or small trees and chollas.

Sonora: Junction road to Guásimas and Mex Hwy 15, *Foster 19 Sep 1965*. 41 mi [on Mex Hwy 15] southeast of Guaymas, 27 Dec 1961, *Lindsay 3215* (SD).

Opuntia

Common names: NAAVO; *nopal* (the plant), *tuna* (the fruit); prickly-pear

Prickly-pears have flattened (laterally compressed) stem segments, or "pads" (cladodes). The areoles bear glochids and may or may not have actual spines. Areoles of young, developing pads, flower buds, and flowers have small, green, succulent, and conical leaves, which fall away with age (although small and deciduous, these are true leaves, and are also seen on chollas, *Cylindropuntia*). The flowers are relatively large, diurnal, and have sensitive stamens (the stamens close inward when touched). The fruits are mostly fleshy. The seeds are whitish, hard, and relatively large.

Naavo is the general term for prickly-pear; the fruit is called *taaka*, which is the general term for fruit. *Navo Ho'ara* (home of the prickly-pears) is the town of Navojoa in Sonora.

Beverage and food: Pérez de Ribas (1645; Reff et al. 1999:90) wrote, "Their wine was made from various native plants and fruits such as pitahaya or prickly pear." The fleshy fruits are edible, but the relatively large, bony seeds are not eaten. *Navo sitoi'm* is prickly-pear jelly made from the fruit, and probably most often from *Opuntia ficus-indica* (see *Stenocereus thurberi* for preparation; also Moisés et al. 1971:158). Prickly-pear fruits are considered a "hot" food (see *Opuntia engelmannii*).

Opuntia bravoana

Common names: NAAVO; NAKKAIM (ears), TAHIWECHIA NAAVO, TAIWECHIA NAAVO (fever prickly-pear)

Sonora. Shrub-sized prickly-pear. Pads bright green, sometimes purplish at the areoles and on the margins (edges) of the pads. Flowers yellow and relatively large, appearing in spring and early summer. Fruits red-purple, fleshy, and edible.

This prickly-pear occurs at San Carlos and Guaymas, especially on small islands and island-like sites, and is common from the vicinity of Las Guásimas southward to Sinaloa.

Sonora: 2.5 mi east of Vikam Suichi, 12 Mar 1989, *Felger & Molina*, observation. 2.4 mi east from shore at Chiinim, 13 Dec 13, 1988, *Felger & Molina*, observation. 7 km northwest of turnoff to Pótam, on Mex Hwy 15, 2 m tall, 4 m diameter, *Reina-G. 2004-519* (ASU).

Opuntia engelmannii var. engelmannii

Common names: NAAVO, TAHIWECHIA NAAVO, TAIWECHIA NAAVO (fever prickly-pear); *nopal*; desert prickly-pear

Arizona and Sonora north of Hermosillo. This prickly-pear has pads to about 30 cm long and light-colored spines. The flowers are yellow but become pale orange on their second day. The fruits are juicy and red-purple when ripe.

This prickly-pear is widespread in southern Arizona and northern Sonora.

Food: The fruits are edible. However, prickly-pears fruit are considered a "hot" food and therefore eaten only occasionally and then only in small quantity, because it is said to cause a fever if one eats the fruit regularly.

Arizona: Tucson Mountains, 1 Dec 1939, *Benson 9872*.

*Opuntia ficus-indica

Common names: TOSAI NAAVO (white prickly-pear); Indian-fig prickly-pear, tree prickly-pear

Arizona and Sonora. This is the large, common, cultivated prickly-pear, reaching about 4 m tall with a thick, woody trunk. It has large, green pads that have glochids but often few or no spines. The flowers are 6 to 10 cm wide and yellow to orange. The fruits can be 5 to 10 cm long. It was domesticated in southern Mexico where it is grown on a large scale.

Cultivation: This prickly-pear is often grown in gardens in Arizona and Sonora, and in the Yoem Bwiara it is sometimes cultivated in small fields. It is extensively farmed farther south in Mexico.

Food: The young, tender pads are eaten as a vegetable called *nopales* in Mexico and southwestern United States. Nopales are prepared in various ways, often with tomatoes, onions, garlic, and red or green chilies. The fruits are delicious and are eaten fresh. The pads and fruits are sold in supermarkets, especially in Sonora, and also harvested from home gardens.

Arizona: Sabino Canyon, several individuals (possibly clones) locally, 3 May 2003, *Baker 15330* (ASU).

Sonora: Tórim, cultivated, 1 Jun 2019, *Blystone*, photo.

Opuntia gosseliniana

Common names: NAAVO, NAKKAIM; *duraznillo*; Sonoran purple prickly-pear

Sonora. Plants to about 1 m high. The pads become purplish during winter and spring; the purple color fades with humid summer weather or spring rains. The larger spines are 5 to 12 cm long, and young plants have bristly spines. The flowers are bright yellow, appearing mostly in March and April, and fruits are pinkish or purplish and moderately fleshy.

This prickly-pear is common in the Yoem Bwiara; on exposed rocky slopes, and lowland desert flats and plains. It ranges across most of Sonora.

Food: The fruit is eaten when ripe. To remove spines (glochids) from the fruit, a leafy branch (such as *sita'avao*, *Vallesia glabra*, Apocynaceae) can be used to roll the fruit around on the ground. The fruit is washed in water and may be eaten fresh or boiled (the seeds are not eaten and are removed before cooking).

Sonora: Vicinity of Las Guásimas, 16 Mar 1989, *Felger & Molina*, observation. Cochorit, along coast several miles south of Empalme, *Heed 26 Dec 1984* (ASU). Ensenada de San Francisco [San Carlos], 30 Mar 1937, *Remple 313*.

Opuntia santa-rita

Common names: NAAVO, NAKKAIM; *duraznillo*; Santa Rita purple prickly-pear

Arizona. The pads take on a purplish color in cool, dry seasons (similar to *Opuntia gosseliniana*). Plants of this species are distinguished mostly by having fewer and much shorter spines or with glochids only, as compared with *O. gosseliniana*. Use of the fruits is the same for both species.

This cactus is found in the desert foothills of mountain ranges in southernmost Arizona and northeastern Sonora.

Arizona: [Tucson], west of Silverbell Road on Ina Road, 2 May 1990, *Rondeau 90-10*.

Pachycereus pecten-aboriginum

Common names: ECHO; *etcho*

Sonora. *Echo* is a tree-like columnar cactus 4 to 6 (8) m tall and has a well-formed trunk and large branches in the form of a candelabrum. Taller plants occur in Yoreme lands and elsewhere in the tropical deciduous forest beyond the Yoem Bwiara. Immature (sterile) or lower portions of the stems have rigid, very sharp, large, and stout spines. Mature (fertile) or upper portions of stems have dense tufts of short, slender, bristly spines. The flowers, borne on the upper part of the stems in spring, are white, 8 to 10 cm wide, opening in the evening and closing about midday. The fruits ripen in late May and June, before the summer rains, and are densely covered with long, golden-yellow spines; the fruits split apart at maturity, revealing scant, bright red-purple pulp, and shiny blackish seeds. The seeds are 4 to 5 mm in diameter and are apparently the largest of all the columnar cacti, at least in northern Mexico.

Basilio (1890 [1634]:183, 211) presented "*echo, pitahaya cardón*" as well as "*etzo, cardón.*" Many

Cactaceae, *Pachycereus pecten-aboriginum*, San Carlos. 10 Mar 2019 (SC).

authors, including early Jesuits, used the name *cardón* for *Pachycereus pecten-aboriginum*. We use the name *cardón* for *Pachycereus pringlei*, although farther south in Mexico the term *cardón* is widely used for *P. pecten-aboriginum* as well as many other large columnar cacti. The Yoreme name is *etcho*.

This large cactus is abundant in the thornscrub and tropical deciduous forest in Sonora and across much of the Yoem Bwiara. The plants tower above most of the thornscrub. Echo is relatively fast growing. The fruits have less pulp than those of the *cardón* and *aaki* (organpipe cactus, *Stenocereus thurberi*).

Construction and fencing: The dry, woody ribs have been used in house walls. Cut sections of the stems readily produce roots and are made into very effective living fences: "Natural fences in the form of cactus growth are pruned from the ever present and abundant thorny vegetation" (McMillan 1936:76). This note probably refers to stems of both *echo* and *aaki*.

Food: The fruit husk is taken off and the red fruit pulp is eaten fresh, or cooked in water and as a beverage. The fruit pulp is also made into jam. The seeds are said to be "somewhat like squash seeds" (meaning they can be prepared like squash seeds) or ground and made into a buttery oil or paste. The ground seeds are also made into a beverage (*atole*). Although undoubtedly a major resource in the past, by the late twentieth century *echo* fruits were seldom harvested.

Household: The fruit is made into a hairbrush. The stiff, bristly spines are burned away from about two-thirds of the fruit, the portion of the fruit where it will be held. The remaining spines are then clipped short, and the fruit serves as a brush (Holden 1936c:71). This is a widespread use in northwestern Mexico and is reflected in the name *pecten-aboriginum*, meaning "aboriginal comb." It is said that you should not comb your hair at night because you will die young.

Song: In a song called *echo moela* (old echo), the cactus laments its imminent death: "The old *echo* is crying because it won't be green anymore. It is dying, dying. The song says that the *echo moele* talks like this" (Bernaldo Valencia, 1994). This instrumental tune, played on the flute and drum, has an associated story but no words are sung.

Sonora: Sierra Bacatete, Rancho El Álamo, 8.3 km (recta) al NE de Vícam, 12 Sep 2008, *Sánchez-Escalante*, observation. Southwest of Empalme, *Soule & Krizman 24 Aug 1964.*

Pachycereus pringlei

Common names: HAWESOWI; *cardón, sahueso*

Sonora. This is one of the most massive cacti anywhere and the largest cactus in the Sonoran Desert, often 8 to 10 m tall in the Guaymas region and reaching greater sizes farther north such as in the vicinity of Bahía Kino. The trunk is short and often enormous, with many massive branches arising from below 2 to 3 m above the ground. The stem surfaces are glaucous (pale bluish) or grayish-green. The flowers are white, 7 to 10 cm long and about as wide, and appear from March to May. As with saguaro, the flowers open in the evening and remain open part of the following day depending on temperature. The fruits, which ripen in early summer, are about the size of a peach. The fruits are covered with dense, felt-like hairs, and vary among individual plants from spineless to spiny. At maturity the fruits split open to reveal a vividly colored crimson-purple to pinkish-white pulp filled with small black seeds. The pulp is sweet and delicious. The shiny black seeds are 3 mm in diameter.

Basilio (1890 [1634]:183) reported *hahueso*, a term for *pitahaya cardón*, which seems similar *hawesowi*.

This cactus is relatively scarce in the Yoem Bwiara except in the northernmost portion. *Cardón* occurs on arid sites such as rocky south-facing slopes at low elevations and partially saline soils of the coast plain. A notable grove occurred just northwest of Empalme at a place aptly called *cardonal* and extremely dense stands cover several small rocky islets in Guaymas harbor. *Cardón* ranges from the coastal plain near Las Guásimas northward along the Sonora coast to Puerto Lobos and inland to Pitiquito. It occurs on most of the islands in the Gulf of California and through much of the Baja California peninsula.

Food: Rosalio Moisés wrote, "Some cactus fruits are eaten fresh. The *ahuesim* does not grow in the Yaqui Valley, but there are lots [of] plants scattered between the low hills near the sea at

Guaymas. The red *ahuesim* fruit is thought to be very good when ripe" (Moisés et al. 1971:158). We can presume that people with access to *hawesowi* harvested the fruit in substantial quantity.

Medicine: In Sonora and elsewhere in Mexico the stems are used medicinally in the same manner and for the same purposes as *músaro* (*Lophocereus schottii*). In November 1984, Richard saw pieces of stems for sale in the *Mercado Municipal* in Guaymas. He was told it is used for cancer and the same purposes as for *músaro* and prepared in the same manner: "*Para cáncer y mismas enfermedades que el músaro, se prepara igual que el músaro, también hasta hervir.*"

Sonora: Guaymas, 11 Feb 1903, *Coville 1666* (US); *Jones 26 Jan 1927* (RSA). 1 mi east of San Carlos Bay, on hill, 12 Jan 1966, *Kimnach & Lyons 604B* (HNT).

Peniocereus

Common names: NOONO; *reina de la noche*; desert night-blooming cereus

The populations of these unusual, slender-stemmed cacti are rather extensive, although plants may occur widely dispersed, patchy, or scattered.

The plants are cryptic, but with careful searching in the right habitats you can locate them. They are not rare in the Yoem Bwiara but are much less common in Arizona. They occur in a wide variety of habitats and soils.

Peniocereus greggii var. transmontanus

Common names: NOONO; *reina de la noche, sarramatraca*; desert night-blooming cereus

Arizona and northern Sonora. This well-known but seldom-seen cactus has a single, large, tuberous root and slender, grayish-purple stems. The large, white, nocturnal flowers are powerfully fragrant—each flower opening only one night, followed weeks later by a red fruit. Because it is such a privilege, there is great excitement when one sees the flowers open.

Plants in central and southwestern Arizona and adjacent Sonora are mostly seen on valley floors and bajadas.

Medicine: In Marana the tuber was collected and used for various medicinal purposes such as treating a headache. For example, Juan Luis Garcia collected the tuberous root, washed it, sliced it like a potato, and then strung it up to dry for later use. The root, fresh or dried, is diced, rubbed or ground into flour, and boiled; the flour or powder can also be dried and stored for future use. The moistened flour is massaged or otherwise applied to the skin for various ailments. The cool root slices reduce fever and bring down feverish body temperatures. Some say the root is toxic if taken internally, causing the body to swell or puff up.

Arizona: Tucson, mesas, 10 Jun 1905, *Thornber 5363*.

Peniocereus marianus

Common names: NOONO; *sacamatraca, sarramatraca*; Sonoran night-blooming cereus

Sonora. The root system has several large, tuberous roots clustered closely together. The stems can be 1.5 to 2 m long (or even to 4 m in canyons), often growing through shrubs and small trees, and are grayish-maroon, with 4 or 5 ribs or angles. The flowers are nocturnal, white, fragrant, and about 7 cm wide. Flowering occurs in late February and March and the fruits can be found from early summer to at least late July. The fruit pulp and exterior are bright red.

This unusual cactus is widespread and sometimes fairly common across the Yoem Bwiara, from the coastal plain to mountains including canyons and rocky slopes.

Peniocereus marianus can be distinguished from *P. greggii* by having larger and often longer stems, several rather than one tuberous root, and somewhat smaller flowers. They do not occur together.

Food: The fruit pulp is sweet and edible, but the fruits are seldom numerous and thus are

Cactaceae, *Peniocereus marianus*, Cerro El Vigía. 26 Feb 2019 (SC).

mostly a snack food. José María Jaimez said the tuberous root is edible and that he has eaten it.

Sonora: Las Guásimas, *Felger 85-285*. 2.5 mi by road east of Vikam Suichi, 12 Mar 1989, *Felger & Molina*, observation.

Peniocereus striatus [*Neoevansia striata; Peniocereus diguetii; Wilcoxia striata*]
Common names: NOONO; *cardoncillo, sacamatraca, sarramatraca*

Sonora. The gray-brown stems, about as slender as a pencil (4.5 to 8 mm in diameter), often scramble through spiny shrubs such as *Lycium* (Solanaceae) and *Citharexylum* (Verbenaceae); the stems seem to mimic those of their nurse plants. The root system has numerous potato-like tubers strung on slender connecting roots. The spines are small, bristle-like, and fall away from older stems. The flowers open during the night and are about 8 cm long and wide, which seems remarkably large for such slender stems. The fruits, about 5 cm long, are rounded, red, and juicy when ripe. Flowering is in summer and the fruits ripen in late summer and early fall.

Although rather cryptic, it is a fairly common across a broad span of habitats through the Yoem Bwiara, including coastal plain, inland bajadas, and rocky slopes.

Food: The potato-like roots are said to be tasty and are eaten fresh, and can be salted to add flavor. The fruit is eaten fresh, mostly as a snack.

Medicine: Every part of the plant is used as medicine. The stems or roots are mashed and then boiled. The resulting liquid is consumed to alleviate body aches. The tuberous root can be sliced like a potato, salted, strung and hung up to dry, and stored for future use. A sliced root may be ground, mixed with other herbs, sometimes mixed with olive oil, and applied to the body, somewhat like a salve, to alleviate "body-swelling." To further help with the healing one should wear clean, white cotton clothing such as a tee-shirt.

To treat headaches, the *paapas* (a term for potatoes and also the potato-like roots of this cactus) are sliced into thin pieces, placed on the temples and forehead, and tied on with a white cloth. If one cannot find *noono*, then an ordinary potato can be used for the same purpose (see *Solanum tuberosum*, Solanaceae).

Sonora: Playa del Sol, *Felger 85-733*. Northwest of Empalme, *Krizman & Soule 24 Aug 1964*. 41 mi southeast of Guaymas (on Mex Hwy 15), 27 Dec 1961, *Lindsay 3214* (SD).

Cactaceae, *Peniocereus striatus*, Las Cocinas, Sierra El Aguaje. 2 Oct 2018 (SC).

Stenocereus alamosensis [*Rathbunia alamosensis*]
Common names: SINA; *sina*; snake cactus

Sonora. This cactus forms dense tangles of slender, snaking, and arching stems armed with sharp, stout spines. The flowers are red and tubular, 7 to 10 cm long, and are open during the daytime. The fruits are red, juicy, and relatively small—about the size of *sinita* (*Lophocereus schottii*) fruit. The young or juvenile plants have very slender stems creeping across the ground, usually beneath a spiny shrub or tree. These worm-like plants have small, bristly spines, and do not look like the adult plants. These juvenile plants are called *sevii nawia* (cholla weak).

Basilio (1890 [1634]:224) cited *sina* as "*pitahaya que llaman 'nacido'*."

Sina is widespread across the Yoem Bwiara, from the coastal plain to lower rocky slopes.

Food: The fruit is eaten fresh from the plant, mostly in June and July. Although some say the fruit is very sweet, others say it is bitter.

Song: There is a song for the flute called *sina wakia* "dry sina." This song is not vocalized but there is an associated story: *muki'iseka weye* "ready to die/dry up (on its way to death)." The plant is saying that it doesn't catch any more rainwater. "The *sina wakia* talks like that" (Bernaldo Valencia, 1994).

Sonora: Rancho Bacatetito, 13 Mar 1989, *Felger & Molina*, observation. Bahía San Carlos, 14 Jan 1966, *Kimnach 619* (HNT). 27 mi southeast Guaymas, Mex Hwy 15, 26 May 1975, *Lehto 18576* (ASU). Sierra Bacatete, Rancho El Álamo, 8.3 km (recta) al NE de Vícam, 130 m, 12 Sep 2008, *Sánchez-Escalante*, observation.

Cactaceae, *Stenocereus alamosensis*, Rancho Aguajito, San Carlos. 10 Mar 2015 (SC).

Stenocereus thurberi [*Lemaireocereus thurberi*]
Common names: AAKI; *pitaya/pitahaya*, *pitaya dulce* organpipe cactus

Sonora and parts of southwestern Arizona. This large columnar cactus is one of the most conspicuous plants throughout the Yoem Bwiara. It generally has several short, stout trunks, and attains 3 to 8+ m in height with many arms. The stems have 12 to 19 shallow ribs. The white to pinkish flowers open at night and close around dawn. Peak flowering occurs in early summer and the fruits mostly ripen in July and early August. The fruits are about 5 to 7 cm wide (not including spines) and covered with sharp spines. Spines on unripe, green fruits are tenacious and difficult to remove. When the fruits are fully ripe the spines tend to fall away or are easily scraped away. The fruit husk ("skin") is red, relatively thin, and often does not split open when ripe. The pulp of mature fruits is sweet, juicy, bright red, and delicious. The seeds are numerous, black, and small (2 to 2.5 mm long).

The organpipe cactus ranges northward to parts of southwestern Arizona but does not occur in the Tucson region. Plants in the desert of northern Sonora and southern Arizona are trunkless or nearly so. As you go southward in Sonora the organpipes become taller and more massive, especially south and east of the desert, such as in the Yoem Bwiara. In the thornscrub and tropical deciduous forest of southern Sonora the trunks are often substantial, and branching occurs well above the ground.

Pérez de Ribas (1634; Reff et al. 1999:88) wrote:

> The *pitahaya*...is an unusual plant in European terms...Its branches are like green grooved candles growing straight up from the trunk, which grows short and robust, making a very striking crown. It has not a single leaf...Its flesh [the fruit] is similar to that of the fig, but softer, the flesh is very white in some plants, red or yellow in others. It is very delicious, particularly when grown in the dry soil along the seashore...where it rains very little.

The description of white or yellow fruits might refer to another species of columnar cacti, *Stenocereus montanus*, *saguira* or *pitaya colorado*, in what is now southeastern Sonora and northern Sinaloa (Felger et al. 2001). "Dry soil along the seashore," however, might refer to fruits of *Pachycereus pringlei* that can vary in color (Felger and Moser 1985).

Basketry: Beals (1945:11) illustrated and described a "carrying basket used in gathering pitahaya fruit" made with split canes (*Arundo donax*, Poaceae) and "mesquite bark" strips (see *Prosopis glandulosa*, Fabaceae).

Construction, fencing, and household: The woody "skeletons" (dry, woody vascular cylinders) are split into slats and woven into walls for houses, which are plastered with mud. These slats are also used for fencing. In March 1989, we saw two young

Cactaceae, *Stenocereus thurberi*, Pima County, Arizona. 2 Aug 2013 (SR).

Cactaceae, *Stenocereus thurberi*, and Devlin Houser. Rancho Aguajito, Sierra Santa Ursula. 28 Dec 2019 (MB).

Yoeme cowboys at a ranch north of Vícam weaving split organpipe slats to form a house wall. We also saw slats of split organpipe ribs and sections of whole green stems used in irrigation-ditch weirs (see *Lysiloma divaricatum*, Fabaceae).

The woody slats have also been used to make beds: "We slept on beds made of pitahaya wood covered with *petates*" (Moisés et al. 1971:14).

Deer songs: A number of songs feature the *aaki*. Here is one of our favorites:

Akita vampo nachine
husama yoliyoliti awa hiluke
Akita vampo nachine
husama yoliyoliti awa hiluke
(repeated two more times)

Near the place of the Organpipe Waters,
I am rubbing antlers in an enchanted brown way.
Near the place of the Organpipe Waters,
I am rubbing antlers in an enchanted brown way.

Ayamansu seyewailo huyata naisukuni
yo vaa bwibwikola weyeka
Senu yo vai vakuliau su yepsaka
Husama yoliyoliti awa hiluke
Akita vampo nachine
Husama yoliyoliti awa hiluke

Over there in the middle
Of the flower-covered wilderness,
Walking alongside the enchanted water,
Arriving to one enchanted fresh branch,
I am rubbing antlers in an enchanted brown way.
Near the place of the Organpipe Waters,
I am rubbing antlers in an enchanted brown way.

Food and beverage: The fruit continues to be harvested as a favorite food. As with the saguaro and other plants, the people speak to the *aaki* at harvest time. The fruit is picked with *hiabwia* (poles) made from a large stalk of *vaaka* (cane or *carrizo*, *Arundo donax*, Poaceae) or two pieces of cane tied together to make a taller pole, or a saguaro (*Carnegiea gigantea*) rib. A stick is lashed crosswise near the end of the pole. This cross-piece is made from several shrubs that have hard wood that does not impart a bitter, or unpleasant taste to the fruit (e.g., *Cordia parvifolia*, Cordiaceae, or *Lycium* including *L. californicum* and *L. brevipes*, Solanaceae). The fruit-gathering pole is called *hiabwai*.

For home use, the fruit has often been gathered by women and children, but men may also participate in the harvest. The women take buckets into the desert to carry the fruits home. The fruit is eaten fresh together with the small seeds. The fruit husk is edible (Felger and Moser 1985), but apparently most people other than the Seris did not bother with the husk. The husks might be fed to animals such as pigs or chickens. The fruits are often cooked, in which case they are boiled into syrup and the seeds are strained out. When the syrup is served it is called *sito'im* (also the term for honey).

The fruit was the most significant sources for wine. Pérez de Ribas (1645; Reff et al. 1999:90) wrote, "Their wine was made from various native plants and fruits such as pitahaya or prickly pear."

Writing about the Yoem Bwiara in the 1930s and 1940s, Rosalio Moisés reported:

> During June, July, and August everyone goes out to gather pitahaya. With a long, sharpened carrizo cane you spear the fruit, which may grow fifteen or twenty feet above the ground. At home the fruit is peeled, put in an *olla* without water, and cooked. The women decide when it has reached the jelly stage by dropping a little in a cup of water. The sweet pitahaya jelly, which we call *sitoim*, can be kept for a long time. This sort of jelly can be made from other cacti, but that made from pitahaya is the best. (Moisés et al. 1971:158)

Pitaya fruit is also harvested for sale, and sometimes sold in Sonoran markets or directly along Mexico Highway 15. Such fruits are harvested mostly by men and youths. The fruits are harvested around dawn and taken to markets and sold the same day, since they quickly spoil in the hot and humid summertime weather.

Yetman (2006:46) provided an account of the commercial harvesting of *pitaya* fruit in Sonora: "Most years, natives of the region and throughout Sonora gather the fruits and sell them in local markets. Yaquis regularly congregate near a checkpoint along the International Highway, where vehicles must stop for inspection. There they offer freshly gathered and despined fruits... every July and August."

Sonora: Bacatete, Cuartel, 13 Mar 1989, *Felger & Molina*, observation. 27 mi southeast of Guaymas, Mex Hwy 15, 26 May 1975, *Lehto 18572* (ASU). Cañón Nacapule, 15 May 1982, *Parfitt 3036* (ASU).

CANNABACEAE – HEMP FAMILY

***Cannabis sativa**

Common names: SIARI VIIVA (green cigarette, green tobacco); marijuana

Arizona and Sonora. This warm-season annual has been grown illicitly as an economically significant crop. Occasional feral plants are seen along such places as riverbanks, irrigation canals, and even roadsides.

Smoking: A marijuana cigarette is called *siari viiva* (green cigarette). Writing about the 1930s and 1940s, Rosalio Moisés reported:

> Many Yaquis like marijuana better than mescal or tequila. Nearly all of the Yaquis who returned from Central Mexico [after the deportations] used marijuana. In Torim twelve boys between the ages of ten and eighteen died of marijuana, and eight others went crazy. [Rosalio's claim of boys dying from marijuana use seems exaggerated, or perhaps *Datura* was involved.] Many times the Mexican government has sent orders to the general in Vicam Switch to destroy all the marijuana. Every time such orders came, Mexican soldiers had to search all over the Yaqui territory for marijuana; they never

> found any because the Yaquis plant it out in the heavy brush or among carrizo. The Mexican soldiers did not really want to find marijuana plants, because a lot of them like to use it too. Mexicans often came to Vicam Switch with marijuana cigarettes to sell for one and a half pesos each. A lot of marijuana is grown around Hermosillo and at San Ignacio, north of Magdalena. (Moisés 1971:203)

Weaving: Pérez de Ribas (1645; Reff et al. 1999:92) reported, "The women also practiced the art of spinning and weaving cotton or other fibers such as Castilian hemp or pita." (Reff et al. 1999:92). Castilian hemp is a form of *Cannabis* selected for fiber production; pita is agave fiber.

Celtis pallida subsp. **pallida** [*Celtis tala* var. *pallida*]
Common names: KUNWO; *cúmero*, *garambullo*; desert hackberry

Arizona and Sonora. Large, rambling, briar-like thorny shrubs to 4+ m tall with hard, flexible wood. The leaves are small and rough-surfaced, and the margins usually toothed. The small, orange fruits have a thin, fleshy, moderately sweet, edible pericarp. The fruits are often abundant in early fall and variously available at other seasons. Desert hackberry is common in Arizona, especially along washes. In the Yoem Bwiara it is widespread in washes, arroyos and canyons, bajadas, and desert plains, and less common on rocky slopes.

Some of the non-food uses and names might apply to either or both of the two *Celtis* species.

Food: The fruits were made into *vannaim* (a kind of *atole*, gruel or pudding). The fruits were one of the food resources used by people in the mountains during the Mexican-Yaqui wars. In 1989, José María Jaimez, of Kompuertam, recalled that, "Eating all these fruits of the desert, our elders fought with the Yoim [Mexicans]." Teresa Amarillas, visiting in Yoeme Pueblo in October 1987, said the fruits are sweet and the birds also like them.

Hunting and weapons: Bows for hunting and warfare were made from *kunwo* wood. A member of the *Wiko'i Yau'ura* (Bow Leaders' Society), carries a bow made from kunwo (Spicer 1980:182).

Utilitarian: The carrying yoke, or shoulder yoke, called a *palanca* in Spanish, was often made from *kunwo* wood. It was used by men to transport water containers and other heavy items. The wood was also used for making axe handles.

Arizona: 2.4 mi by road south of Sasco Road on east side Cerro Prieto, Samaniego Hills, *Felger 88-502*.

Sonora: Cuesta Alta, Río Yaqui, *Felger 85-1411*. Cañón Nacapule, *Felger 92-1016*.

Celtis reticulata
Common names: KUMARO; *cúmero*; canyon hackberry

Arizona and Sonora. Trees often 5 to 10 m tall with a well-developed trunk. The leaves are scabrous (with short, rough hairs). In Arizona, this tree has unique, irregular, warty bark on the trunk and major limbs, and the leaves are winter-deciduous. In southern Sonora, the bark is gray and smooth, and the leaves are essentially evergreen. Flowers inconspicuous; the fruits orange to red-brown, rather hard and less than 1 cm wide.

Basilio (1890 [1634]:208) indicated *cumro* as the word for "*vainoro, árbol*."

This tree grows along canyons and larger washes in southern Arizona and Sonora. It occurs in Sierra El Aguaje and Sierra Libre, and presumably in the Sierra Bacatete.

Arizona: Santa Cruz Valley, Tucson, 15 Nov 1902, *Thornber 2375*.

Cannabaceae, *Celtis pallida*, Pima County, Arizona. 15 Sep 2013 (SR).

Sonora: Cañón Nacapule, *Felger 96-82*. Ejido Punta de Agua, arroyo, northeast of Sierra Bacatete, 5 Oct 2011, Van Devender, observation (MABA/SEINet 2020).

CANNACEAE – CANNA FAMILY

***Canna indica** [*Canna coccinea, C. discolor, C. edulis*]
Common names: COLA DE PERICO (parrot tail); *caña de las Indias, lengua de dragon*; canna lily

Sonora. Herbaceous perennials with red flowers, grown for the edible, starchy root. José María Valenzuela and María Valenzuela grew rows of canna in semi-shade in their home garden in Kompuertam.

Horticultural forms with large, variously colored flowers, called *cola de perico*, are grown as ornamental plants such as in Las Guásimas and Pótam.

Sonora: Kompuertam, 14 Mar 1989, *Steen, Felger, & Molina*, photos. Las Guásimas, cultivated in home garden, called *cola de perico*, 16 Dec 1988, *Felger, Molina, & Steen*, observation. Pótam, cultivated in garden, 16 Dec 1988, *Felger, Molina, & Steen*, observation

CAPPARACEAE – CAPER FAMILY

Atamisquea emarginata [*Capparis atamisquea*]
Common names: HUVAK VENA (*huuva*, stink + *k*, past tense + *vena*, like; like a skunk); *palo hediondo, palo zorillo*

Sonora. Woody shrubs with leathery leaves, dark green above and silvery to brownish below with a dense covering of scales. The flowers are cream white. The fruits, about 1 cm long, are bright red inside and have a strong, spicy odor.

Common in western Sonora, including the coastal plain in the Yoem Bwiara.

Medicine: The leaves or leafy twigs are mashed or ground and applied to aching or sore legs or sores that are slow to heal ("when sores don't heal"). These may include infected areas that result from lacerations, cuts, scrapes, or other wounds. *Huvak vena* is also used to relieve a toothache. Bernaldo Valencia said the leaves are boiled and applied directly to the afflicted tooth or placed on the face next to the tooth.

Sonora: East of Chiinim, 13 Dec1988, *Felger*, observation. East of Las Guásimas, *Felger 89-196 & Molina*.

Capparidaceae, *Atamisquea emarginata*, Pima County, Arizona. 27 Jun 2006 (SR).

Forchhammeria watsonii, see RESEDACEAE

CAPRIFOLIACEAE (*Sambucus cerulea*), see VIBURNACEAE

CELASTRACEAE – BITTERSWEET FAMILY

Tricerma phyllanthoides [*Maytenus phyllanthoides*]
Common names: PASEO VEAK VENA (*veak*, a while ago + *vena*, like; it seems like a while ago) *mangle dulce*

Sonora. Medium-sized to large shrubs with evergreen, succulent leaves. The fruits are small, fleshy, and red.

Paseo is a general term for mangroves. Basilio (1890 [1634]:138) cited *beracbena* as an "*árbol de marisma*." (marsh tree).

These shrubs are abundant along the inland margins of mangroves and near the shore in coastal thornscrub. The Seris ate the fleshy, red fruits (Felger and Moser 1985).

Sonora: Estero Soldado, *Felger 85-541*. East of Chiinim, 13 Dec1988, *Felger & Molina*, observation.

Celastraceae, *Tricerma phyllanthoides*, Bahía San Pedro. 5 Feb 2015 (SC).

CHENOPODIACEAE, see AMARANTHACEAE

COCHLOSPERMACEAE – COCHLOSPERMUM FAMILY

Amoreuxia palmatifida

Common names: SAAWA; *saiya*

Arizona and Sonora. Perennial herbs with thick, tuberous, and succulent roots often 20 to 30 cm long. The stems are often 15 to 30 cm tall and bear palmately lobed leaves. The flowers, 5 cm across, have bright orange petals with large maroon blotches. The fruits are 3 to 4 cm long, egg-shaped and green until drying as thin-walled capsules. The seeds are 5 mm long, kidney-shaped, and blackish. The new shoots usually appear with the first summer rains, and flowers and fruits are found during hot, humid weather. The flowers open in the early morning, dotting the landscape bright orange, and wither with daytime heat. The plants are dormant from fall through winter and spring.

Saawa is widespread in the Yoem Bwiara, on plains, hills and mountains. It is especially common in the Sierra Bacatete and some areas of the coastal plain. It is widespread in Sonora and extends into southern Arizona near the Mexican border.

Food: Every part of the plant is edible—the roots, young leaves, buds, fruits, and seeds—and has a good flavor. The thick, tuberous roots can be eaten fresh or variously cooked, such as boiled with meat or roasted on coals. The green, unripe fruits are reported to be sweet. The succulent roots have been a significant Yoeme food resource and are sometimes seasonally sold in local Sonoran markets. Some people still harvest the plants in Sonora. Saawa was a significant food for people hiding out in the mountains during the Mexican-Yaqui conflicts. Men relied on saawa when they went into the mountains, sometime for a week or so at a time.

Beals (1945) called it, "sawa, a root resembling a sweet-potato" and "a mainstay in war times." He thought that *sayámme* might be the same plant and reported that the root was called *salas*. *Samawaaka* is a mountain in the Bacatete range (*Vakatetteve Kawim*) southwest of the Hill of the Rooster (*Totoitakuse'epo*). Tomas Martinez thought that *Samawaaka* came from word *saawam*. Families made trips to this mountain to collect *saawam*.

Arizona: East edge of Nogales, Sierrita Mountains, 6 Sep 1940, *Benson 10404*.

Sonora: Miramar, *Felger 85-894*. 0.5 mi W of Hwy 15, 12 mi SE of Guaymas and 1 mi south of Cruz de Piedra, 26 Aug 1980, *Sanders 1800* (UCR).

Amoreuxia gonzalezii is similar to *A. palmatifida*, differing in part by larger fruits and round rather than kidney-shaped seeds. It is edible in the same manner as *A. palmatifida*. It ranges from south-

Cochlospermaceae, *Amoreuxia palmatifida*, Santa Cruz County, Arizona. 10 Aug 2017 (SC).

central Arizona to Sinaloa, and is likely to occur in mountains along the northeastern margin of the Yoem Bwiara.

COMBRETACEAE – COMBRETUM FAMILY

Laguncularia racemosa

Common names: SIARI PASEO (green mangrove); *mangle blanco*; white mangrove

Sonora. Large evergreen shrubs or small trees. The leaves are opposite, smooth, semi-succulent, and green on both surfaces. The flowers are small and white.

Siari paseo grows in mangrove esteros and inlets. Seedling development begins while the fruit is still on the tree, and the seedling emerges rapidly after the fruit falls and becomes stranded in shallow tidal water. This is one of the three common mangrove species along the coast of Sonora.

Sonora: Estero Soldado, *Felger 85-874A*. Bahía de San Carlos, mangrove swamp, 19 Jul 1970, *Lehto 16987* (ASU).

Combretaceae, *Laguncularia racemosa*, Estero Soldado. 30 Oct 2015 (SC).

COMPOSITAE, see ASTERACEAE

CONVOLVULACEAE – MORNING-GLORY FAMILY

***Convolvulus arvensis**

Common names: *correhuela*; bindweed

Arizona and Sonora. Perennial vines from rhizomes; flowers pink or white. Troublesome weeds in agricultural fields, disturbed habitats, and home gardens. Native to Europe.

Arizona: Tucson, Santa Cruz Valley, 12 Apr 1916, *Thornber 7522*.

Sonora: Campo El Baguo, 30 km northwest of Pótam, irrigated field, *correhuela*, *Van Devender 2006-250* (USON).

Cuscuta

Common names: *fideo*; dodder

Several species of these small parasitic vines occur in Arizona and Sonora. The stringy stems are orange or yellow. The leaves are reduced to minute scales and the flowers small and white.

Ipomoea arborescens

Common names: HESE'I; *palo santo*; tree morning-glory

Sonora. Distinctive trees with thick, smooth, whitish trunks and very soft, spongy wood. The leaves appear with the summer rains and are quickly shed after the rains cease. The attractive white flowers, shaped like typical morning-glory blossoms, appear in winter and spring.

These unusual trees grow in and around the larger mountains of the Yoem Bwiara. It is common southward and eastward from the Yoem Bwiara and as well as north of Hermosillo.

Sonora: Microondas mountain overlooking Guaymas, 11 Jan 1982, *Daniel 1960* (ASU). Vicinity of

Convolvulaceae, *Ipomoea arborescens*, north of Hermosillo. 11 Dec 2013 (SC).

Cañón Bacatete, 17 Dec 1988, *Felger & Molina*, observation.

***Ipomoea batatas**
Common names: KAMOOTE; *camote*; sweet potato

Arizona and Sonora. Perennial vines, with tuberous roots. Corollas 4 to 7 cm long, funnelform, lavender or sometimes white, with darker color inside.

Sweet potatoes are purchased from stores. Also cultivated in gardens and occasionally found outside of cultivation in southern Sonora, especially in densely vegetated disturbed areas.

Beverage: *Teswiin*, an alcoholic beverage, was fermented from sweet potatoes. It was just for personal use at home, not for ceremonies; a similar beverage was made using barley, wheat, or corn (see *Hordeum vulgare*, *Triticum aestivum*, and *Zea mays*, Poaceae).

Sonora: Guaymas, *Chan & Folkner 25 Apr 1960*. Road to Las Guásimas, in shrubs and trees along drainageway, petals pink with magenta center, rare, 9 Oct 1985, *Felger 85-1160*.

***Ipomoea carnea** subsp. **fistulosa** [*Ipomoea fistulosa*]
Common names: MANTO; *gloria de la mañana, palo santo de Castilla*; bush morning-glory

Sonora. Multiple-stem shrubs. The leaves have prominent leaf stalks and moderately large, ovate leaf blades with pointed tips. The flowers are rather large and pink.

It is cultivated as an ornamental plant in the Yoem Bwiara.

Medicine: Manto is used to relieve headache and fever. Mentholatum, lard, Vaseline, or olive oil is spread on a leaf and the oily side of the leaf is applied to the forehead and tied on with a white cloth, and replaced as needed. It is classified as a cool medicine because it relieves a fever.

Sonora: Pótam, 13 Dec 1988, *Felger & Molina*, observation.

Ipomoea seaania
Common names: Yaqui morning-glory

Sonora. Shrubs 1 to 4 m tall. Flowers showy, corollas 7 to 8 cm wide, white with a maroon band at the base. Flowering January to March.

This unique morning-glory, known only from hills and mountains in the vicinity of Bahía San Carlos, should be sought in mountains in the Yoem Bwiara. The species name derives from *Sea Ania*, the Flower World. This species and *Ipomoea arborescens* belong to a group of about 11 species of arborescent morning-glories in Mexico and Central America (Felger and Austin 2005).

Sonora: 1 km north of Bahía San Carlos on old road to Bahía Algodones, broad canyon, *Felger 85-301*.

Convolvulaceae, *Ipomoea seaania*, San Carlos. 26 Mar 2013 (SC).

***Ipomoea tricolor**
Common names: TRONPIO; *trompillo*; morning-glory

Ornamental vines with tricolored blue-purple-white flowers grown as ornamentals in Arizona and Sonora. The name *tronpio* also includes other morning-glory vines.

CORDIACEAE – CORDIA FAMILY

The plants in this family were formerly included in Boraginaceae.

Cordia parvifolia
Common names: WOTOVO; *vara prieta*; desert cordia, littleleaf cordia

Sonora. Shrubs to about 2 m tall with slender, hardwood branches and dark-colored bark. The relatively small leaves have grayish-white hairs and are rough to the touch. The shrubs flower at various seasons following rains except in mid-winter, producing masses of pure white flowers. The flowers (corollas), each 2.5 to 3 cm wide, open an hour or so after dawn and fall off with midday or afternoon heat.

Abundant in the open desert, plains and rocky slopes of hills and mountains; characteristic of arid habitats. On hot summer mornings, ground squirrels (Sciuridae) often sit in the branches gobbling the flowers. This drought-resistant shrub is cultivated in southern Arizona.

Deer song: *Wotovoli Seewa* (cordia is blooming) is a song of the *wotovo*. In song language the name becomes *wotovoli*, an affectionate term and the white flowers represent the tail of the white-tailed deer. This song is from Guadalupe Molina of Vikam Suichi, recorded by Felipe at his grandfather's house in Marana in 1979:

wotovoli seewa	wotovoli is blooming
wotovoli seewa	wotovoli is blooming
wotovoli sewatakai	wotovoli really is blooming
wotovoli seewa	wotovoli is blooming
wotovoli seewa	wotovoli is blooming
wotovoli sewatakai	wotovoli really is blooming

(repeated three more times)

Ayamansu seywailo huyatanaisukuni	
	Over there in the middle of the flower-covered wilderness
uhyolisi weyekasu seewa	[wotovoli] is standing beautifully blooming
wotovoli seewa	wotovoli is blooming
wotovoli seewa	wotovoli is blooming
wotovoli sewatakai	wotovoli really is blooming

Food harvesting: The wood of this shrub (or that of *Lycium*, Solanaceae) is used for the cross section of the organpipe cactus fruit-picking stick (see *Stenocereus thurberi*, Cactaceae). These woods are used because they are hard and do not impart a bitter taste to the fruit.

Ritual: In the *Wiko'i Yau'ura* (Coyote Warrior Society, also called the Bow Leaders' Society), the corporal (soldier) uses *wotovo* wood as a staff. Similarly, *wotovo* is used for the staff of office for the leadership of the Yoeme community—for the *ya'ura*, or leader (official).

Sonora: 10–15 mi east of Empalme along Hwy to Ciudad Obregón, coastal plain, 5 Sep 1974, *Gentry 23441* (DES). Guaymas, *Palmer 174 in 1890* (NY).

Cordia sonorae

Common names: POMAHE; *palo de asta*

Sonora. Small, hardwood trees. The leaves are dark green, tough, and usually smooth. A profusion of sweet-scented, white flowers can be seen January to April except during years of severe drought. The flowers, 3 to 3.5 cm wide, become satiny brown and keep their shape after drying and falling.

Basilio (1890 [1634]:222) cited *pómahau* as the word for "*palo de hasta*." *Asta* is the Spanish term for a staff.

Arroyos, canyons, foothills, and mountains in the Yoem Bwiara, but not in the drier parts of the region.

Tools: The wood was made into axe and shovel handles.

Sonora: San Carlos, 2 Feb 2014, *Carnahan SC 920*. 8 mi by road northeast of Vikam Suichi, 12 Mar 1989, *Felger & Molina*, observation. Sierra Bacatete, Rancho El Álamo, 8.3 km (línea recta) al NE de Vícam, 12 Sep 2008, *Sánchez-Escalante*, observation.

Cordiaceae, *Cordia parvifolia*, south of Tastiota. 6 Nov 2018 (SH).

CRUCIFERAE, see BRASSICACEAE

CUCURBITACEAE – SQUASH FAMILY

This family includes squashes, gourds, and melons. Most are vines and the flowers are unisexual (male and female flowers may occur on the same plant or on separate plants). *Kama* is the general term for any kind of squash or pumpkin, both the plant and the fruit, and also the name for crocodile.

Certain kinds of squash (*Cucurbita*) have been grown in the Yoem Bwiara since pre-contact times, and melons (*Cucumis*) since Spanish colonial times. Pérez de Ribas (1645; Reff et al. 1999:87–88) reported, "Plants brought from Castilla grow well in these regions...Melons do so well that you rarely find one that is not pleasing." He also noted, "Among the maize they also sow various kinds of squash, which are sweet and tasty. Some are cut into slices and dried in the sun; these last for most of the year."

Members of the gourd family are frost-sensitive. Squashes and pumpkins are planted in the Yoem Bwiara in February after the danger of frost has passed.

Apodanthera palmeri

Sonora. Perennials with trailing stems. Leaf blades palmately lobed, the lobes mostly toothed. Male and female flowers on the same plant; flowers bright yellow. Fruits 3–4 cm long, striped and mottled dark and light green, thin-skinned, and many seeded.

Sonora from the vicinity of Carbó southward to the Sinaloa.

Food: In 1877, Palmer reported "Fruit edible, with the taste nearly of muskmelon, ripening in September (282)" (Watson 1889:51).

Sonora: Plains about Guaymas, *Palmer 282 in 1887* (UC, US, NY).

***Citrullus lanatus** subsp. **lanatus**

Common names: SAKOVAI, SAKVAI; *sandia*; watermelon

Arizona and Sonora. Watermelon is native to Africa and Asia. It was brought to the New World in early Spanish colonial times, and presumably was cultivated in southern Sonora even before 1634, when Basilio (1890 [1634]:223) reported "*Sacobari.*—Badea, sandía." *Sacobari* is a Yoreme term, although linguistically close to the Yoeme name. *Badea* comes from Classical Arabic (Real Academia Española 2019).

Agriculture: Watermelon cultivation was described for the Yoem Bwiara by Studhalter 1936:122–123).

> The Yaquis are apparently just as fond of watermelons as are other Indians in the Southwest. This plant is not extensively grown in the fall of the year, but constitutes one of the major crops noted in the spring. The seeds are planted almost exclusively in pits (at least such was the case during the very dry season of 1934), these being ten to fifteen feet apart...Corn and watermelons are often planted together in the same pit. Planted in January and maturing in May, the vines are said to grow to a length of 12 or 15 feet. Five or six seeds are planted in each pit, which is at first vertically walled and with a flat bottom, like a broad U; as the vines grow in length, the pit is scraped down to the form of a shallow wide-open V. We were told of at least two varieties, a spherical and an oblong type. Seeds for the next crop are stored in tequila bottles.

Watermelons were often planted from the end of November to mid-January and would be harvested from May to August. Rosalio Moisés planted watermelons on November 27, 1946. On December 7, he and his wife made paper covers to protect the seedlings. In February they "uncovered all the watermelon plants because frost danger was past" (Moisés et al. 1971:203). In late April they moved from their house to a ramada they built at the garden so they could protect their crop from animals and thieves. "By the end of May the watermelons were ripening. We stayed in the ramada until August 14. Our garden was in such a sandy spot that big wagons could not get in to pick up a load. As a result, I did not sell many watermelons" (Moisés et al. 1971:205).

In March 1989, we saw Yoeme farmers growing fields of watermelons east of the highway and northeast of Pitahaya.

Medicine: Mateo González said, "dry the rind, burn it, grind and apply to a sore."

***Cucumis anguria**

Common names: HI'IKRE (from Johnson 1962); *melón de coyote*; bur gherkin, West Indian gherkin

Sonora. Annual vines with deeply lobed leaves.

Flowers yellow, the fruit usually with soft prickles, and edible especially when young, although somewhat bitter.

There are no records for this species in the Yoem Bwiara, although it is naturalized nearby in the Yoreme region. Johnson identified *hi'ikre* as *Cucumis anguria*, although it might instead be *Cucumis melo* var. *dudaim*.

Laundry and shampoo: Johnson (1962:263) listed *hi'ikre* as a vine that is used to wash clothes or the head (shampoo); presumably the fruit was used.

***Cucumis melo** var. **dudaim**

Common names: VAA MINAI (watermelon), WO'I MINAI (coyote melon); *melón de coyote*, *meloncillo de coyote*; coyote melon, dudaim melon, stink melon

Sonora. Annual vines. The stems and petioles have rough, prickly hairs; the leaf blades are broadly ovate and sometimes shallowly lobed. The flowers are yellow and the fruits fleshy, yellow and smooth when ripe and highly fragrant. The plant is most vigorous during the summer and fall, but is also seen at other seasons.

Widespread Yoem Bwiara where it thrives in roadside ditches, the margins of savanna swamps, in hedgerows and irrigation ditches, and in remnant gallery forest along the Río Yaqui. Native to southern Asia.

Food: The fleshy fruits are sometimes edible. Consultants in Sonora said some *wo'i minai* are sweet while most are bitter.

Medicine: Yetman and Van Devender (2002:180) noted that two Yoreme men "report that the gourds are boiled, then crushed. The wash is rubbed in the hair to kill lice. Vicente says that he learned this practice from Yaquis."

Sonora: Cuesta Alta, Río Yaqui, *Felger 85-1365*. Southeast of Pótam, hedgerows and canal bank, *Felger 88-591*.

***Cucumis melo** var. **reticulatus**

Common names: MINAI; *melon*; cantaloupe

Sonora. Cantaloupes have been planted in the Yoem Bwiara after the danger of frost had passed, often in February (Moisés et al. 1971).

Pérez de Ribas (1645; Reff et al. 1999:88) reported that melons were popular among the Yaquis. These probably were varieties of *Cucumis melo*, likely grown from seeds introduced by the Jesuits.

Cucurbita argyrosperma var. **callicarpa**

Common names: AYA'AWI, KAMA, KUTA KAMA (wood squash); *calabaza*; cushaw squash

Arizona and Sonora. *Kama* is also the term for crocodile (*Crocodylus acutus*, Crocodilia), perhaps due to comparison of the crocodile hide and rind and stalk of this squash.

This large, robust-growing annual forms long-running stems with large leaves. The large, rather irregularly shaped squash is unmistakable; it is pale green with whitish to yellowish stripes and speckles and can have a tough rind, and the very thick, corky stalk (peduncle) is unique. It has been cultivated in the Sonoran Desert region since ancient times. It differs from *sewalka* (an unidentified native squash, *Cucurbita*) by the corky stem (peduncle) and in being hard and elongated at the stem end, whereas *sewalka* is rounded and has a softer rind.

Cucurbitaceae, *Cucurbita argyrosperma* var. *callicarpa*, Marana Yoeme Pueblo. Ca. 2019 (L. Maahs).

Basilio (1890 [1634]:203) recorded *aiahui* as the word used for "*calabaza llamada arota*."

The *Cucurbita argyrosperma* complex includes four varieties, native and wild or cultivated, from the southwestern United States to Central America. Variety *callicarpa* is the large, cultivated cushaw squash of the southwestern United States and Mexico.

Although there are corn songs, no songs are known for squash and beans.

Agriculture, food, and gardens: This squash was a mainstay in the traditional diet. In Pótam, Antonia Amarillas described it as being "*mahauta vena*," and that it "looks like a turtle, the head and neck long and slender, and the rest enlarged." She knew it from the 1930s. By the 1980s it was apparently no longer grown in the Yoem Bwiara, but was still commonly grown in the Yoreme region.

Various reports of pumpkins grown for food in the Yoem Bwiara probably refer to this indigenous squash (e.g., Moisés et al. 1971:199).

Some seeds would be saved for planting in the next season. It was planted in early March with corn and beans, after cold weather had passed. It could be planted again around the time of summer rains. It could be planted alone, but was traditionally grown in cornfields. The plant eventually overtakes the whole cornfield—it takes over the whole ground but does not affect the corn. Both share the water. The squash foliage covers the ground and keeps it from drying too fast. Usually there is a fence around the garden to keep out animals such as goats and dogs.

In Marana in the mid-1970s, Juan Garcia would cut up this squash and string the pieces to dry, for later use. The strings of squash were hung up in a ramada or inside the house. Some squashes would be cooked fresh. It was eaten every day; the squash would be the main meal, eaten with tortillas and coffee, and sometimes also with beans. The squash was also made into soup or fried. "You can eat the rind and all of the younger, smaller and tender squashes. But the rind of the larger, older squashes, is not eaten." The whole squash could be stored for an entire season.

Beliefs: People should not point at squash plants, especially when the plants are flowering. It is said, "If we point at the plant it will not produce." This holds for all plants, but especially squash since it is the most shy of all plants—other kinds of plants can tolerate being pointing at but not the squash.

Cucurbita digitata

Common names: TETA'AHAO; *calabacilla*; coyote gourd

Arizona and northern Sonora (not in the Yoem Bwiara). A common and often conspicuous trailing vine from a tuberous root with 5-parted leaves. The flowers are yellow, and the female flowers develop into hard-shelled, round gourds about 9 cm in diameter.

Medicine and shampoo: The inside of the gourd, while still fleshy, was used as a hair shampoo and conditioner to maintain a healthy scalp and to keep the hair black and shiny. It was also used to get rid of dandruff and lice. The gourd was cut open and boiled to get a gel-like shampoo.

Washing: The gourd was used as clothing bleach and to further clean the laundry. The still-green gourd was chopped in half and put in boiling wash water. It was being used in this manner in the 1950s when people in Arizona washed clothes in metal washtubs over a fire. The clothes were dunked up and down in the boiling water with a *carrizo* (*Arundo donax*, Poaceae) stirring stick. There was a lot of laughter among some elders in Marana when they called it "Indian soap."

Arizona: Tucson: West Branch of the Santa Cruz River, 19 Jul 2001, *Mauz 21-43*; Santa Cruz River Valley, 22 Sep 1903, *Thornber 24*.

Unidentified Cucurbita

Common name: SEWALKA

Sonora. A pumpkin or squash cultivated in Sonora, but more commonly in the Yoreme region; it was apparently not grown by Yoemem in Arizona. Antonia Flores said it is like a round squash or pumpkin.

Ibervillea sonorae [*Maximowiczia sonorae*]

Common names: KAU CHAANI (mountain chaani); *guarequi*, *wereke*; cow-pie plant

Sonora. The large swollen stem-base or caudex resembles not much more than a large, smooth cow-pie. The vines of this plant, which emerge from the top of the caudex just before or with the first summer rains, rapidly grow up through shrubs and trees, and wither soon after the rains cease. The leaves are often bluish-green (glaucous) and deeply lobed. Male and female flowers occur on the same plant. The fruits are globose to ovoid and fleshy, often 5 to 8 cm long, green and white and become yellow or red when fully ripe in late summer and fall. The fruit is not palatable to humans (Felger and Moser 1985).

Chaani is an untranslatable core word.

This plant is common in the Yoem Bwiara and does not occur in Arizona.

Medicine: Preparations made from the caudex are used to heal sores (wounds, insect bites, carbuncles). The caudex is sliced into thin sections, which are dried and ground to a fine powder. Applied to sores, this material is said to "dry up infection."

Cucurbitaceae, *Ibervillea sonorae*, Rancho San Antonio, Sierra El Aguaje. José Jesús Sánchez-Escalante (right) and Guadalupe Aguilar-Villavicencio. 31 Jan 2001 (RF).

Guarequi is harvested from the wild, such as in the vicinity of Pótam, and sold in regional markets as living but dormant plants ("roots," actually the caudex). It is popular in Arizona and Sonora. Most medicinal herb vendors, such as in Cd. Obregón, proport it as a remedy for diabetes and advertise it as *wereke*. Although used for various maladies, it is has been popular as a medicine for sores and stomach ulcers. A small piece of the caudex is cut off and heated in water until it boils and the liquid saved to be consumed through the day in place of drinking water ("*Un trozo chico en un litro de agua hasta hervir, tomarla como agua de diario*"). The caudex or pieces of it are also pounded like *carne machaca* (dried meat) and then cooked in water. It is also used to treat wounds: "It is a remedy for wounds, it is very bitter; pound the root and apply it to the leg [or other afflicted area] (*Es un remedio para una herrida, muy amargo; machaca, se pone en la pierna*)."

Beliefs: Rosalio Moisés provided names of some plants used in witchcraft (Moisés et al. 1971:20; see *Pleradenophora bilocularis*, Euphorbiaceae). He mentioned "some ground-up roots of the plant called *ochani*," which probably is *kau chaani*.

Sonora: Playa del Sol, *Felger 85-908*. Guaymas, *Palmer 283 in 1887* (US).

***Lagenaria siceraria**

Common names: VISA'E; *bule*, *guaje*; bottle gourd, calabash gourd

Arizona and Sonora. These robust annual vines produce large gourds that are popular as utensils and especially as musical instruments. *Visa'em* are grown in gardens in Arizona and Sonora. *A'ookos* is the entire gourd (Johnson 1962:247). Felipe tells us that the term *a'ookos* is seldom heard nowadays except in Vícam. Basilio (1890 [1634]:204) used the term *arcosi* for *bule* (arcosi is likely a Yoreme rather than Yoeme term).

The bottle gourd was domesticated in the Old Word, perhaps more than 14,000 years ago. Subspecies *asiatica* was domesticated in Asia and brought to North America by Paleoindians and was present in southern Mexico as early as 10,000 years ago. Europeans brought cultivars of subspecies *siceraria* to the Americas; this subspecies was domesticated in Africa. Multiple cultivars of the Old World subspecies have largely replaced New World cultivars (Erikson et al. 2005).

Beliefs: *Bwehe'im* (half gourds) used for deer singing are very powerful and not just anyone should grow them for this purpose. For this reason, such visa'em would be grown away from a home garden, in the wild, such as along a river or wherever water collects. Gourds grown by a non-Yoeme neighbor, however, would not be subject to the power within the visa'e as it grew, and these neighbors might supply the gourds for Yoemem. The practice of growing ceremonial gourds away from the home was generally observed in Sonora, but by the latter part of the twentieth century visa'em were grown in home gardens.

Gardens: Visa'e has long been grown in home gardens. In Arizona it is planted in spring, after frost, and the gourds are harvested in fall before frost. In Sonora, it can be planted in spring and again with summer rains, and the gourds ripen in late summer or early fall. Visa'e is often grown next to small mesquite trees or fences that serve as support for the vines.

In the 1970s in Marana, several Yoemem were lucky that their gourd plants produced copious gourds that other Yoemem came to buy for ceremonial use. By the early 1980s people at Marana were no longer growing visa'e. Since then people have been looking elsewhere, for example, obtaining gourds from Native Seeds/SEARCH in Tucson, or at swap meets. Yoemem in Sonora were still growing visa'em in 2015.

Household and utilitarian: A half gourd was used as a dipper for drinking water (e.g., Holden 1936c:70). Basilio (1890 [1634]:147) cited the words *bueha* for a "*calabaza que sirve de vaso*" (a gourd cup), and *arocosi* for a "*calabaza que sirve de cántaro*" (a gourd pitcher). Gourds were fashioned into water bottles and especially canteens, *vaa nuu'um* "water containers," essential items for venturing into the *monte*. A wooden or corncob stopper was fitted into a hole cut around the stem end of the gourd. The canteen was supported by strips of rawhide or agave twine.

Music: A half gourd, bweha'i, is used as a resonator for the musical rasper and water drum of the deer singers. The water drum is made from a gourd larger than one used as a resonator. For use as a water drum, a half gourd is hardened and waterproofed by boiling it in lime-water. When a pahko is about to start, animal fat is often rubbed on the inside of the gourd to further ensure that it will be water-resistant. It floats inverted in a water-filled large clay bowl, which more recently has been replaced by a metal bowl.

The visa'e gourd is also fashioned into *ayam*, the hand-held gourd rattles used by the deer dancer and by the *matachinim*. Pieces of discarded, broken glass are put into a visa'e gourd and shaken to clean and scrape the inside. Various small objects are put inside the cleaned gourd to produce the sounds, these generally are different kinds of seeds, but some people may prefer small pebbles or beads. The pebbles are usually from an ant hill. Painter (1986) reported that small, smooth, store-bought beads are put in the ayam in Arizona. In Sonora, seeds from the balloon vine (*Cardiospermum corindum*, Sapindaceae) or precatory bean (*Rhynchosia precatoria*, Fabaceae) are often used, and in both Arizona and Sonora the seeds of fan palms (*Washington*, Arecaceae) might be used. A wooden handle is fitted into the *ayam*.

Arizona: Desert Botanical Garden, Plants and People of the Sonoran Desert Trail, *Davis 13 Sep 2002* (DES).

Sonora: Río Mayo, Arroyo Aduana between La Aduana and Mexico Highway 15, east of Minas Nuevas, tropical deciduous forest, single very large vine to 3 m in trees, flowers white, open at dusk, young fruit and elongated gourd, not near houses, 11 Oct 1992, *Van Devender 92-1204*.

*Luffa aegyptiaca

Common names: WAHAROM; *estropajo*; loofah, sponge gourd

Arizona and Sonora. Warm-season annual vines, sometimes 6 to 8 m long. Cultivated in gardens and sometimes feral (growing wild, without being planted) in the Yoreme regions, such as near Navojoa. This species is native to Asia.

Food and household: The young, fleshy fruits are edible, and the dry interior of mature fruits serve as bath sponges and to scrub pots and pans.

Medicine: The fruits are used as a remedy for sores and earache. To treat sores, the whole fruit is parched, then ground and applied to an afflicted area. To relieve earache, the fruit is boiled, and the resulting liquid used as eardrops.

Sonora: Near Río Mayo, Pueblo Viejo in Navojoa, *Van Devender 95-1130* (USON).

Luffa quinquefida

Common names: WAHAROM; *estrapajo de coyote*; coyote loofah

Sonora. Robust, summer-growing, annual vines. The flowers are yellow. The fruits are 5 to 7

Found in low wet places such as roadside depressions and the savanna-like swamps on the coastal plain of the Yoem Bwiara.

Sonora: Las Guásimas, *Felger 85-298*. Road to Playa del Sol, *Felger 85-1112*. East of Chiinim, small dune in saline mud flats, *Felger 88-577*. Near Mori (Yaqui country), heavily brushed coastal plain, 24 Oct 1961, *Gentry 4742*.

Tumamoca macdougalii [*Ibervillea macdougalii*]

Common name: Tumamoc globe-berry

Arizona and Sonora. Stems vining from tuberous roots. New growth appears with summer rains, or just prior to onset of the monsoon; the stems and leaves quickly perish when the rains cease. Male and female flowers variously on the same or different plants. Fruits globose, 1 cm wide, glabrous and fleshy, dark green when immature and bright red when ripe.

Common on fine-textured silty-sandy soils in coastal plains southeast of Guaymas to northwestern Sinaloa, and locally less well-known in southern Arizona.

Tumamoca is usually visible only when in leaf or bearing ripe fruits. On the coastal plain southeast of Guaymas, the delicate stems are strung like spider webs on briar patches of *Lycium californicum* (Solanaceae). The vegetative stems might be overlooked because of their resemblance to *Ibervillea sonorae*. *Tumamoca* is most readily distinguished from *Ibervillea* by its below-ground tuberous roots, smaller leaves generally with narrower segments, and smaller, edible fruits. *Tumamoca* often occurs together with *Ibervillea* and both flower at the same time. The fresh fruits served the Seris as a minor food resource (Felger and Moser 1985:388).

Cucurbitaceae, *Tumamoca macdougalii*, Pima County, Arizona. 23 Aug 2015 (SC).

Sonora: Playa del Sol, *Felger 85-904*; road to Playa del Sol, 0.2 mi N of beach, *Felger 85-1126*. Road to Las Guásimas, *Felger 85-1154B*.

CYPERACEAE – SEDGE FAMILY

Most sedges in Arizona and Sonora are wetland plants or grow on seasonally wet ground.

Cyperus esculentus var. **leptostachyus**

Common names: KOONI SAAKI (crow roasted); *cebollín*, *coquillo amarillo*; yellow nutgrass

Arizona and Sonora. This relatively large sedge forms a small nut-like tuber at the ends of rhizomatous roots.

Common in wetlands, often in mud at edges of ponds. The Yoeme names and uses for the two sedges known as nutgrasses are probably interchangeable (see *Cyperus rotundus*).

Food: The small root tubers (*bolitas en la raíz*) are eaten fresh. The fresh root tubers are also toasted or parched (as one would do for coffee beans) and then ground into flour and prepared as *saktusi* (*pinole*).

Medicine: Flour from the ground root tubers is applied as a remedy for sores.

Arizona: Tucson, west branch of Santa Cruz, *Mauz 22-68*.

Sonora: Guaymas, *Palmer 203 in 1887*.

***Cyperus rotundus**

Common names: VAA VA'AKO; *cebollín*, *coquillo morado*; purple nutgrass, purple nutsedge

Arizona and Sonora. This sedge resembles *Cyperus esculentus* but differs in having rhizomes that produce chains of several small tubers (smaller than those of *C. esculentus*). *Cyperus rotundus* is a common agricultural weed.

Medicine: The small root tubers are used in the same manner as for *Cyperus esculentus*. Painter (1986:55) reported that vaa va'ako (nut grass) "is ground and put on sores. It also imparts energy to living things, such as a pahko'ola dancer or a racehorse."

Arizona: Tucson, University of Arizona campus, 23 Oct 1952, *Parker 8194*.

Sonora: Miramar, Hotel Playa de Cortes, *Felger 20151128-1*. Pótam, Colonias Yaquis, *López-E. 23 May 1985* (USON).

Eleocharis macrostachya

Common names: VAA VASO (water grass); *zacate del agua*; mountain spikerush

Arizona and Sonora. Grass-like perennials with tough rootstocks and shiny green stems. Emergent from shallow water or at margins of waterholes or ponds, and along irrigation ditches.

Animal food: It is eaten by horses and cattle.

Arizona: Santa Catalina Mountains, Sabino Canyon, 24 Aug 1914, *Thornber 7445*.

Sonora: Kompuertam, *Felger 89-71 & Molina*.

EBENACEAE – PERSIMMON FAMILY

*Diospyros sonorae

Common names: HUPSI; *guayparín*; Sonoran persimmon

Sonora. This potentially large tree forms a dense crown. The fruit has fleshy, soft, blackish-brown pulp that is slightly sweet and astringent.

Guayparín is found in southern Sonora and northwestern Sinaloa. Part of the distribution pattern suggests naturalization from purposeful introduction (Felger et al. 2001). We have not seen it in the Yoem Bwiara.

Basilio (1890 [1634]:207) listed "caurara, *guaiparime*." Johnson (1962:267) identified *hupsi* as "*guayparín* (*árbol*)." *Hupsi* is a name widely used in the Yoem Bwiara for *papache borracho*, *Randia thurberi* (Rubiaceae). Although the *papache* and *guayparín* are very different in appearance, both produce fruits with soft, blackish, edible pulp that has a similar taste and texture.

Food: The fruit is used by the Yoremen and Yoemem undoubtedly knew of the fruit.

EUPHORBIACEAE – SPURGE FAMILY

Many members of this family, such as *Euphorbia* and *Jatropha*, can be highly toxic. Some of the medicinal uses reported here might be dangerous if one is not experienced with quantities and other practices.

Cnidoscolus palmeri

Common names: *ortiguilla*, *quemadora*, *zumaque venenoso*

Sonora. Small shrubs with stout, stinging hairs. These hairs can inflict a painful rash, although it tends to only last about half an hour. The plants form potato-like tuberous roots. The leaves are 2 to 5 cm long and about as wide, relatively thick, crisped ("crinkled" on the edges), with prominent milky-white veins. The flowers are small and white.

This unique shrub occurs locally in mountains and hills in the vicinity of Guaymas, often on steep slopes and cliffs. The peeled, tuberous roots, fresh or cooked, were an important staple of the Seris (Felger and Moser 1985). Yoeme consultants in the Guaymas region were unaware that the roots are edible.

Medicine: The Yoemem drank tea made from the leafy stems of *quemadora* as a remedy for fevers ("*para calenturas*"; Cruz Matus, Guaymas, 1985).

Sonora: Guaymas, *Felger 84-175*; *Palmer 302 in 1887* (GH).

Croton texensis

Common names: TAI PUSI (with-sun eyes); *tortolita*; Texas croton

Arizona and Sonora. Annuals to about 1+ m tall, openly-branched, with slender stems and rather sparse foliage. Occasional weeds in the Río Yaqui valley.

Beliefs: It is said that if you touch this smelly plant it will dry up.

Arizona: About 10 mi northwest of Tucson, banks of Santa Cruz River, 3 Nov 1944, *Gould 2902-a*.

Sonora: Northeast of Tórim, *Felger 88-596*.

Euphorbia

Common names: EE OONA (ant salt), KOAPA'IM, MAHKOAPA'I (deer koapa'i; referring to the white-tailed deer, *Odocoileus virginianus*); *golondrina* (swift or swallow, bird); desert spurges

Arizona and Sonora. There are 17 species of spurges in the Yoem Bwiara, three of which are shrubs (*Euphorbia cymosa*, *E. lomelii*, and *E. tomentulosa*), while the remainder are low-growing, annual or perennial herbs. All have milky latex ("sap"). The unique flower-bearing structure, the cyathium, includes small, unisexual flowers without sepals or petals; the cyathium often has petal-like appendages attached to round or oval glands. The cyathium superficially resembles a single "normal" flower.

The concept of *koapa'im* and *ee oona* refers to more than a dozen species of small spurges. Included here are *Euphorbia abramsiana*, *E. albomarginata*, *E. arizonica*, *E. capitellata*, *E. florida*, *E. hirta*, *E. hyssopifolia*, *E. incerta*, *E. pediculifera*, *E. petrina*, *E. polycarpa*, *E. prostrata*, *E. setiloba*, and *E. trachysperma*.

Medicine: Although euphorbias have been widely used medicinally, we caution that these plants are generally regarded as toxic, especially if ingested, and the milky juice can cause severe eye pain and injury. There are older reports of euphorbias used as remedies for poisonous snake bites and other toxic bites. We also caution that these traditional herbal remedies are not a substitute for life-threatening causes and emergency medical care.

These plants have been used to heal sores and bruises. The green plant is boiled, flattened, and then bound to the sore, or it can be fried in lard with a little salt and used in the same manner. The plant may also be washed with water, dried, and ground, or chopped up green if there is urgency for use if the dried, ground herb is not available. These euphorbias have been used similarly for a spider bite or carpenter bee sting.

Cruz Matus (Guaymas, 10 March 1985) said Yoemem use *golondrina* (we were looking at *Euphorbia polycarpa*) medicinally, and that it is an excellent remedy for wounds ("*es muy bueno para un golpe*"): "wash the leg or other afflicted part, wash it well, put it [the plant] in water, boil it, cool it, and use a cloth (*trapo*) soaked in the liquid and soap to wash the afflicted part (*es muy buena*)." One can also boil the plant and use the liquid to wash a sore (infection) on the feet; also "boil and wash feet for sore or tired feet" (Alfonso Flores, 17 Dec 1988; see *Euphorbia polycarpa*).

Euphorbia abramsiana [*Chamaesyce abramsiana*]

Arizona: Marana, *Streets 25 Oct 1939*.

Sonora: 8 mi by road northeast of Vikam Suichi, 12 Mar 1989, *Felger 89-100 & Molina*.

Euphorbia albomarginata [*Chamaesyce albomarginata*]

Arizona: Santa Cruz Valley near Tucson, *Pringle 2 May 1881*.

Sonora: Mex Hwy 15, 0.2 mi NW of Pótam junction, *Felger 85-1263* (ARIZ, RSA, USON).

Euphorbia arizonica [*Chamaesyce arizonica*]

Arizona: [Tucson], Saguaro National Park West, *Bertelsen 89-505*.

Sonora: Guaymas, island in harbor, *Palmer 321 in 1887* (GH).

Euphorbia capitellata [*Chamaesyce capitellata*]

Arizona: Tucson, Tumamoc Hill, *Carter 16 Mar 1932*.

Sonora: Cañón Bacatete (Bacateve), Sierra Bacatete, *Felger 89-140 & Molina*.

Euphorbia cymosa [*Euphorbia colletioides*, *E. plicata*]

Common names: HUYA TONO'OARA (plant with many joints/bending parts); *candelilla*, *jumete*

Sonora. Shrubs, often 1.5 to 3 m tall, with copious milky sap, flexible green stems, and sparse foliage that is drought-deciduous. The flowering structures are small and white.

Mountain canyons and higher elevations in the Sierra Libre and expected in the Sierra Bacatete.

Medicine: In 1994, Bernaldo Valencia recognized the plant at the Arizona-Sonora Desert Museum. He said the plant or stems are boiled and "one bathes with it...just for the body and also for blood circulation."

Sonora: Sierra Libre, *Felger 85-828*; Cañón La Pintada, *Van Devender 3 Jan 1984*.

Euphorbia florida [*Chamaesyce florida*]

Arizona: Tucson, Tumamoc Hill, *Wilder 06-261*.

Sonora: Strand at Estero Soldado, *Felger 84-405*.

***Euphorbia hirta**

Arizona: Fresnal Canyon, Baboquivari Mountains, 1 Sep 1945, *Gould 3251*.

Sonora: Río Yaqui at Mex Hwy 15, *Van Devender 94-461*.

Euphorbia hyssopifolia [*Chamaesyce hyssopifolia*]

Arizona: Tucson Mountain Park, *Bertelsen 89-516*.

Sonora: Guaymas, *Palmer 82 in 1887*.

Euphorbia incerta [*Chamaesyce incerta*]

Sonora: Strand at Estero Soldado, *Felger 85-873A*.

Euphorbia lomelii [*Pedilanthus macrocarpus*]
Common names: KANDELIA, KANTEELA; *candelilla*; slipper plant

Sonora. This unusual plant has few-branched, greenish-gray, thick succulent stems that produce copious milky latex when cut, and few, small, quickly deciduous leaves. The strange flowering structures are bright red and about 2.5 cm long; the shape is reminiscent of a miniature slipper.

Fairly common on the coastal plains of the Yoem Bwiara.

Toxicity: The plant is considered poisonous, causing diarrhea if eaten or even put in the mouth.

Sonora: South of Peón, *Turner 61-31*.

Euphorbia pediculifera [*Chamaesyce pediculifera*]
Common name: Louse spurge

Arizona: Anklam Road, Tucson Mountains, *Shreve 8132-A*.

Sonora: Strand at Estero Soldado, *Felger 84-404*.

Euphorbia petrina [*Chamaesyce petrina*]

Sonora: Between Chiinim and Pótam, coastal dune, *Felger 88-562*.

Euphorbia polycarpa [*Chamaesyce polycarpa*]
Common name: *golondrina*

Arizona and Sonora. Annuals to small perennials. Flowering structures (cyathia) small, with white, petal-like appendages or sometimes without appendages.

Euphorbiaceae, *Euphorbia polycarpa*, San Carlos. 26 Jun 2013 (SC).

One of the most widespread and common euphorbias in the region, found in many habitats.

Medicine: Boil the plant, and then use it to wash feet when sore or tired (Alfonso Flores, Pótam, 17 Dec 1988).

Arizona: [Tucson], Saguaro National Park West, *Bertelsen 17 Apr 1989*.

Sonora: 3 mi east of Mex Hwy 15 (Mapoli on map), east of Las Guásimas, 17 Dec 1988, *Felger 88-618 & Molina*. Guaymas, *Palmer 96 in 1887*.

Euphorbia setiloba [*Chamaesyce setiloba*]

Arizona: Tucson, Tumamoc Hill, *Wilder 06-247*.

Sonora: 8 mi by road northeast of Vikam Suichi, *Felger 89-101 & Molina*.

Euphorbia tomentulosa [*Chamaesyce tomentulosa*]

Sonora. Shrubs often 1+ m tall. Leaves about as wide as long with serrated margins. Floral structures white, in dense clusters at ends of branchlets.

Abundant and widespread in the Yoem Bwiara, mostly in arid, rocky sites, also bajadas, and along washes and canyons; from near sea level to mountaintops. This is one of the few shrubby species of subgenus *Chamaesyce* section *Anisophyllum* in Sonora.

Sonora: Guaymas, 21 May 1939, *Gentry 4716*.

Euphorbia trachysperma [*Chamaesyce trachysperma*]

Arizona: Pinal County, Sawtooth Mountains, *Mauz 25 Aug 1998*.

Sonora: Mex Hwy 15 south of road to Pitahaya, *Felger 85-204*.

Jatropha

Sapo is the generic Yoeme term for *Jatropha* shrubs (not to be confused with the Spanish word for toad). Basilio (1890 [1634]:223) recorded *sapo* as the term for "*sangre de drago*." Latex ("sap") from deep cuts in the lower stems of some jatrophas is blood-red, hence the common name *sangrengado* in Mexico, a corruption of *sangre de drago* (dragon's blood). Damaged stems, especially those of *J.*

cuneata, produce copious fluid that dries clear on clothing, but when oxidized after laundering becomes a permanent, blood-like, red-brown stain. Mateo González pointed out that *sapo* juice produces a red stain.

Jatropha cinerea, *J. cardiophylla*, and *J. cordata* are host plants for caterpillars of the large saturniid moth *Rothschildia cincta* (Lepidoptera: Saturniidae). The cocoons of this moth are used to make the *tenevoim* leg-rattles; they are most often collected on *J. cinerea*. Cocoons found on *Jatropha* are called *sapo tenevoim*.

Caution: Although we supply accounts of how the sap or other parts of the plants were used in traditional medicine, we advise that various *Jatropha* species are known to be highly toxic. Elders cautioned that jatropha is toxic if ingested.

Jatropha cardiophylla

Common names: KAU SAPO (mountain sapo); *palo sangrón*, *sangrengado*; limberbush

Arizona and Sonora. Small to medium-sized shrubs with sparingly branched, red-brown, flexible stems, producing glossy-green, heart-shaped leaves. The plants leaf out and bloom just before or during the summer rainy season, and become leafless soon after the end of the summer rains. The flowers are relatively small and inconspicuous, white and bell-shaped. Sap from deep cuts in the lower stems is blood-red.

This limberbush is common in the hills and lower mountains near Tucson and is uncommon in the Yoem Bwiara.

Dye: We were told that the roots are used for a reddish dye (in Arizona), apparently augmented with roots of *Calliandra eriophylla* (Fabaceae) and *Krameria bicolor* (Krameriaceae).

Medicine: The viscid, clear juice (sap) from cut stems of this species and *Jatropha cinerea* was used as an eye drop to "cleanse" the eyes. We do not know more details of how the sap was used, and caution that various jatrophas are known to be highly toxic.

Arizona: Waterman Mountains, *Van Devender 88-444*.

Sonora: 8 mi by road northeast of Vikam Suichi, 12 Mar 1989, *Felger & Molina*, observation. Sierra Bacatete, Arroyo El Álamo, 6.8 km (línea recta) al NE de Vícam, 11 Sep 2008, *Sánchez-Escalante*, observation

Jatropha cinerea

Common names: SAPO; *palo sangrón*, *sangrengado*; ashy limberbush

Sonora. A desert shrub with smooth, light-colored bark and broad leaves. The whitish or grayish, dull and non-peeling bark and relatively dull-colored leaves distinguish it from the other *sapo* species in the region. Cut branches produce copious clear to pale yellowish latex that is intensely staining. The plants may produce leaves at various seasons, especially after the summer rains. This is the most common and widespread limberbush species in western Sonora.

Medicine: To treat sores, the bark is stripped from the plant, fermented with alcohol, and the liquid used to wash the sores. Mateo González said it is a remedy for *makoam* (vitiligo). As with *Jatropha cardiophylla*, the sap is used as eyedrops.

Caution: We do not know more details of how this plant was used, and various jatrophas are highly toxic.

Sonora: Rincón del León, *Felger 85-912*. 5 mi northeast of Pitahaya junction, 13 Mar 1989, *Felger & Molina*, observation. Guaymas, 22 Oct 1939, *Gentry 4672* (UC).

Euphorbiaceae, *Jatropha cinerea*, San Carlos. 24 Aug 2017 (SC).

Kau sapo seewa — Mountain limberbush is flowering/blooming

Kau sapo taaka — Mountain limberbush is giving fruit
Kau sapo seewa — Mountain limberbush is flowering/blooming
Kau sapo taaka — Mountain limberbush is giving fruit

Haisane mamachika wana — How do I look/appear
weeka seewa — as I stand there flowering/blooming

Kau sapo seewa — Mountain limberbush is flowering/blooming
Kau sapo taaka — Mountain limberbush is giving fruit
Kau sapo seewa — Mountain limberbush is flowering/blooming
Kau sapo taaka — Mountain limberbush is giving fruit

Jatropha cordata

Common names: KAU SAPO (mountain sapo); *papelio*

Sonora. Small trees, often with a well-formed trunk and a high, thin crown bearing leaves only during the summer rainy season. The leaves are smooth and shiny green. The bark becomes papery and peels away in thin sheets during dry seasons.

Euphorbiaceae, *Jatropha cordata*, Espinazo Prieto, north of Hermosillo. 15 Oct 2008 (JS).

Widespread in southern Sonora, especially in the foothills and mountains along the northeast side of the Yoem Bwiara.

In their leafless, dry season aspect, these trees are strikingly similar to *Bursera fagaroides* (Burseraceae), generally differing by their narrower stature, whereas *Bursera* generally has a larger, spreading crown.

Deer song (below): The flowers and fruits of this plant feature in a deer song. Felipe heard Miguel "Miki" Maaso sing this song at a ceremony in the 1980s and wrote the words on a piece of paper. The first stanza is repeated 4 to 6 times.

Masks: One Tucson maskmaker said that *sapo* is used in the Yoem Bwiara to carve pahko'ola masks. This is the only *Jatropha* in the region that has trunks large enough to provide wood for carving a mask.

Sonora: Cañón Nacapule, *Felger 85-858*. Near Chiinim, 13 Dec 1988, *Felger & Molina*, observation. 3 km north of Totoitakuse'epo (Hill of the Rooster), 13 Mar 1989, *Felger & Molina*, observation. Sierra Bacatete, Rancho El Álamo, 7.9 km (línea recta) al NE de Vícam, 12 Sep 2008, *Sánchez-Escalante*, observation.

Jatropha cuneata

Common names: TOROTE; *matacora, sangrengado*; limberbush

Dry regions of western Sonora and southwestern Arizona. Desert shrubs with flexible stems and small leaves that are longer than wide and mostly clustered on stout, short branchlets. The sap is blood-like and intensely staining. The leaves may be produced at almost any season following rains and the small flowers appear with the summer rains.

Generally found on hot, dry slopes in the Yoem Bwiara.

Leather tanning: This jatropha has been used to tan hides: "Para curtir cueros. La raiz, da una tinta colorada, machaca y ponen en agua con los cueros" (Cruz Matus, Guaymas, 10 Mar 1985; To cure leather. The root gives a red dye, crush and put it in water with the leather).

Sonora: Cañón Bacatete near the cuartel, 17 Dec 1988, *Felger & Molina*, observation. Kuubwa'e Kawi, 1 km west of Kompuertam, 12 Mar 1989, *Felger & Molina*, observation. Guaymas, 5–11 Jun 1897, *Rose 3749* (US).

Jatropha macrorhiza

Common names: KOOWI SAAWA (*koowi*, pig + *saawa*, the word for *Amoreuxia palmatifida*; pig saiya); *jicamílla* (little *jícama*); ragged jatropha

Arizona. *Jatropha macrorhiza* is an herbaceous perennial, dying back to the ground in winter and drought. The relatively large, tuberous root is poisonous. The shoots with their large leaves appear in late spring or summer; the leaves are deeply and irregularly toothed. Because its palmately-divided leaves resemble *Amoreuxia palmatifida* (Cochlospermaceae) in their general aspect, *Jatropha macrorhiza* seems a probable identification for *koowi saawa*.

Usually found above the desert in Arizona. In Sonora it does not extend into the Yoem Bwiara.

Toxicity: Tomas Martinez told Felipe that *koowi saawa* grows in Arizona and is like *saawa* (*Amoreuxia*, Cochlospermaceae) but has bitter fruit, and if you eat the fruit diarrhea will develop. Rosario Vakame'eri-Castillo told Felipe that eating it makes you hallucinate, saying "A tiny stick will look like a big moving snake." While Rosario's information seems characteristic of *Datura* poisoning (see *D. discolor*, Solanaceae), Tomas' description points to *Jatropha macrorhiza*, a highly poisonous plant.

Arizona: Santa Rita Mountains, Box Canyon, 17 Sep 1965, *Barr 65-396*.

Pedilanthus macrocarpus, see **Euphorbia lomelii**

Pleradenophora bilocularis [*Sapium biloculare*; *Sebastiania bilocularis*]

Common names: HOYO KUTA (poison stick), KAU HOWO (mountain poison); *hierba de la fleche*; Arizona jumping-bean, arrow-poison plant

Arizona and Sonora. These large, multiple-stemmed shrubs produce copious milky latex ("sap"). The leaves are glossy-green to reddish, longer than wide, and have serrated edges. The inconspicuous female flowers produce 2-lobed seed capsules.

It is common in western Sonora and southwestern Arizona, in desert and thornscrub.

Beliefs: Rosalio Moisés provided a lively discussion of *hoyo kuta* (he called it the *howocuta* bush) as an ingredient in witchcraft. The time frame was about 1907, when he was a boy in Sonora:

> Pedro Charavan and I used to go dove hunting in the brush. Once we came upon an *olla* sitting in the fork of a mesquite tree. We knew just what it was, although we had never seen one before. Older Yaqui women sometimes talked about these witching *ollas*. The witch who placed the *olla* there was Luisa Sosa, the fat wife of Pedro Sosa. She often offered cups of pinole to boys, but we never accepted anything from her because we knew she was a witch.
>
> The way the witching *olla* was prepared by Luisa, or by any witch wanting to use one, was to take a little *olla* about eight or nine inches in diameter and put in all sorts of thorns, needles, thread, poisonous herbs, *toloache* (jimsonweed), the milk of the *howocuta* bush, some leaves from the Don Juan plant, and maybe some ground-up roots of the plant called *ochani*. The *olla* is then put out in the brush, up in the fork of a tree. In two or three days it starts to boil, and on the sixth night itexplodes with colored lights like firecrackers. A wind blows the odors and whoever smells them gets sick.
>
> When Pedro Charavan and I found the *olla*, it was boiling strongly, giving off a terrible stench. I am sure it was almost ready to explode and that Luisa Sosa was planning to use it that night. We threw it in a nearby creek. The next day Luisa lay face down on the floor of her house all day long with her face covered. She cried a lot and was sick for three days. She never did anything to us. (Moisés et al. 1971:20)

In Rosalio's narrative, *toloache* is *Datura discolor* and the Don Juan plant is *Nicotiana glauca* (both in the Solanaceae), and *ochani* is probably *kau chaani*, *Ibervillea sonorae* (Cucurbitaceae).

Toxicity: This shrub is widely regarded as being highly poisonous and is carefully avoided. One man said it is "very bad, you will get diarrhea if you sit under the shrub, especially if you fall asleep under one." The latex (sap) is toxic and is well-known as an arrow poison. The sap was smeared on arrow points to be used for combat and hunting. Some men said this method was not used in hunting, such as for deer, because the meat would be poisoned; there is, however, contrary information.

This arrow poison was greatly feared. In the early seventeenth century, Pérez de Ribas (1645; Reff et al. 1999:90, 93) wrote:

> Most of these arrows are smeared with an herb so venomous that, as long as it is somewhat fresh, there is no antidote or remedy that can save the life of anyone who has been wounded, no matter how little venom has entered any part of the body.
>
> Because they normally use poisoned arrows (even for hunting), it takes no longer than twenty-four hours for an animal to die...They find the place where the animal has dropped by looking up to the sky, where the *zopilotes* [vultures]...are flying in circles.

This arrow poison was made from the sap of *Pleradenophora bilocularis*. While the sap is certainly toxic, the strength of the venom might be exaggerated. The Seris also used a much-feared arrow poison made from this shrub, and they also feared the plant (Felger and Moser 1985).

Arizona: Pima County, Ajo, *Vorhies 18 Apr 1924*.

Sonora: Wapari, southeast of Pótam, *Felger 88-627*. Guaymas, *Jones 25 Jan 1927* (RSA).

***Ricinus communis**

Common names: KEVENIA; *higuerilla*; castor bean

Arizona and Sonora. Sparingly branched perennials that can grow very quickly to almost tree-like stature. The large, dark glossy-green leaves are deeply lobed and can be 50 cm or more across. Some forms have reddish-bronze foliage.

In southern Sonora mostly in weedy, disturbed places, especially along irrigation canals and roadsides. Native to the Old World but long naturalized in North America.

Gardens: The red leaved form is occasionally grown in home gardens as an ornamental plant.

Medicine: Kevenia is sometimes used to alleviate a headache, fever, or sore throat. For this purpose, a fresh leaf is tied on the head with a white cloth or scarf, or sometimes lard is applied to the leaf before it is tied on the head. The same procedure is employed to relieve overall body fever, but the leaf is applied to the chest, stomach, and areas of the back.

Euphorbiaceae, *Ricinus communis*, San Carlos. 18 Dec 2012 (SC).

Toxicity: All parts of the plant are extremely poisonous if ingested.

Arizona: Tucson, *Yatskievych 22 May 1981*.

Sonora: Bahía San Carlos, *Felger 85-549*. Las Guásimas, 16 Dec 1988, *Felger & Molina*, observation. Rancho Bacatetito, 13 Mar 1989, *Felger & Molina*, observation.

Sapium biloculare, *Sebastiania bilocularis*, see **Pleradenophora bilocularis**

Tragia jonesii [*Tragia amblyodonta* (misapplied), *T. scandens* M.E. Jones (not *T. scandens* Linnaeus)]
Common names: NATA'E; *ortiga*, *quemador*, *rama quemadora*; noseburn

Sonora. This small vine is widespread in the Yoem Bwiara, including canyons, bajadas, rocky slopes, and plains, and especially riparian and semi-riparian sites. It has stinging hairs like those of a nettle. A similar species, *Tragia nepetifolia*, occurs in southern Arizona.

Basilio (1890 [1634]:178, 205) gave *natare* and *batare* as words for *ortiga* (literally "nettle," but a word used for plants with stinging hairs), possibly indicating this *Tragia* species.

Sonora: Palo Parado, Río Yaqui, *Felger 85-1400*. Guaymas, *Jones 26 Jan 1927* (RSA).

FABACEAE (LEGUMINOSAE) – LEGUME or BEAN FAMILY

Acacia sensu lato

The classic genus *Acacia* has been segregated in a number of smaller genera, three of which are included in the book. *Acacia* sensu stricto has been limited to Australia.

Acacia cochliacantha,
see **Vachellia campechiana**
Acacia constricta,
see **Vachellia constricta**
Acacia farnesiana,
see **Vachellia farnesiana**
Acacia greggii,
see **Senegalia greggii**
Acacia occidentalis,
see **Senegalia occidentalis**
Acacia willardiana,
see **Mariosousa heterophylla**

Albizia sinaloensis [*Hesperalbizia occidentalis*]
Common names: HOOSO; *joso*, *palo joso*

Sonora. Large, handsome, trees often 15+ m tall, with a tall trunk, smooth yellowish-white bark, and feathery leaves.

Hooso grows along the banks of the Hiak Vatwe and in river towns in the Yoem Bwiara including Pótam.

Boats: Dugouts were made from large river trees; sometimes from *hooso* but mostly from *avaso* (*Populus mexicana*, Salicaceae).

Ceremony and masks: Pahko'ola masks are sometimes carved from *hooso* wood. The wood is often carved into the foundation of the *maso kovo*, the deer-head headdress of the deer dancer.

Household: The wood is used to make the *vatea*, a heavy dough-kneading bowl, and the *alteesa*, a canoe-shaped kneading trough with legs. In Kompuertam on March 11, 1989, José Maria Jaimez's mother showed us a *vatea*, about 60 cm wide, made there by her brother before the Mexican–Yaqui wars. She was using it to knead dough. (She did not want to sell it; although we did not ask, someone earlier had asked to buy it.) The wood also is used to make large spoons for stirring and cooking, and wooden dishes for All Souls' Day.

Fabaceae, *Albizia sinaloensis*, Suaqui Grande, southeast of Hermosillo. 30 Apr 2015 (JS).

Weapons: The wood has been used to make rifle stocks.

Sonora: Cañón Bacatete (Bacateve), Sierra Bacatete, *Felger 88-632, 89-131 & Molina*. Kompuertam, *Felger 89-76 & Molina*.

***Arachis hypogaea**
Common names: KAKAWAATE; *cacaguate*; peanut

One of the store-bought food items in Arizona and Sonora.

Caesalpinia cacalaco, see **Tara cacalaco**
Caesalpinia gilliesii, see **Erythrostemon gilliesii**
Caesalpinia palmeri, see **Erythrostemon palmeri**

Caesalpinia pulcherrima

Common names: TABWIKOSEEWA (around-the-sun flower, flowering around the sun); *tavachín*; red bird-of-paradise

Arizona and Sonora. A slender-stemmed, medium-sized shrub beset with prickles. The leaves are large and divided into numerous small leaflets. The flowers are large and brilliant red-orange with long stamens.

Cultivated and naturalized or native in the northern part of Sierra El Aguaje and in eastern Sonora.

Gardens: It is commonly grown in gardens in Arizona and Sonora.

Arizona: Tucson, 30 May 1903, *Thornber 303*.

Sonora: 8 mi by road northeast of Vikam Suichi, 12 Mar 1989, *Felger & Molina*, observation.

Calliandra californica

Common name: TAVACHIN, *tabuchín*

Sonora. Small shrubs. Leaves tardily drought-deciduous, with many small leaflets. Flowers bright red and showy. Flowering March and probably at other seasons.

Hills and mountains in the San Carlos region and in canyons in the Sierra El Aguaje. Also grown in southern Arizona as a garden and landscape plant.

Sonora: Takalaim, *Felger 85-362*.

Calliandra eriophylla

Common names: *huajillo*; fairy duster

Arizona and Sonora. Dwarf woody shrubs with firm but flexible stems; often propagating by root sprouts. Leaves drought-deciduous, with many small leaflets. Flowers pink to nearly white, sweet-scented in the morning.

Dye: Yoemem in Arizona use the roots to augment red dye of *Krameria bicolor* (Krameriaceae) and/or *Jatropha cardiophylla* (Euphorbiaceae).

Arizona: Tucson Mountains, 30 Apr 1905, *Thornber 5203* (ARIZ, NY).

Sonora: Rincón del León, *Felger 85-918*. Guaymas, *Palmer 293 in 1887* (NY).

*Cicer arietinum

Common names: KAAVANSA, KAAVAPSA; *garbanzo*; chickpea, garbanzo bean

Arizona and Sonora. Introduced from the Old World. *Kaavansam* is the modern term for garbanzo beans, and *kaavapsam* is an older name used by elders in Marana in the 1960s.

Agriculture and food: In the early 1930s, the garbanzo bean was "rather an important crop at Vicam village (Studhalter 1936:123).

Its importance was no doubt enhanced by the active interest of ex-president Obregón, who is said to have made himself the *garbanzo* king of Mexico, growing a large part of the entire Mexican crop on his estate just south of the Yaqui River and controlling the distribution of the remainder.

The Yaquis call the plant 'Spanish bean.' It is planted in November and ripens in May. Grown in the same soil as wheat, the plants not only furnish in their seeds some of the protein eaten by the Yaquis, but in addition their tender tips are broken off, boiled, and eaten as greens. They are grown in plowed fields of fair extent (one-half to three acres).

Their harvesting, as described to us, is of particular interest, reminding one of the post-cultural care of wheat. The entire plants are pulled up and permitted to dry from one-half to one day on a hard smooth dirt floor, such as is used for the threshing of wheat. The dried plants are now raked away and the pods winnowed from the seeds during a wind by a very large wooden shovel. Seeds are stored in sugar or flour sacks to be eaten at any time during the year.

Coursetia glandulosa

Common names: SAAMO, SAMO; *sámota*; samota

Arizona and Sonora. Unarmed shrubs, the stems and branches slender and flexible. Leaves once-pinnate, the leaflets with a short projection from the tip. Flowers pale yellow and white with faint red tinges. Pods densely glandular-pubescent and constricted (narrowed) between the seeds.

Widespread in arroyos and canyons and on rocky slopes. The stems are sometimes encrusted with ant-tended, orange-colored lac produced by the scale insect *Tachardiella fulgens*. This lac was used as an all-purpose adhesive in many regions of Sonora (Felger and Moser 1985).

Arizona: Saguaro National Park, Tucson Mountain District, King Canyon Trail, 18 Nov 1988, *Rondeau 88-83*.

Sonora: Cañón Nacapule, *Felger 85-256*. Cañón Bacatete, 17 Dec 1988, *Felger & Molinia*, observation.

Fabaceae, *Coursetia glandulosa* with lac insects (*Tachardiella fulgens*) tended by Pseudomyrmex ants, Cañón del Nacapule. 26 Jan 2016 (SC).

***Delonix regia**

Common names: TAVACHIN; *flamboyán*; flame tree, royal poinciana

Sonora. Ornamental trees with spectacular red-orange flowers and enormous bean-like woody pods. Widely cultivated in the essentially frost-free lowlands of Sonora including the Yoem Bwiara.

Sonora: Koasepe, southeast of Kompuertam, 14 Mar 1989, *Felger*, observation.

Desmanthus bicornutus

Common names: HÍCHUIQUIA; *dais*; bundle-flower

Sonora. Small unarmed shrubs with slender stems, and drought-deciduous leaves. Flowers white. Pods slender, in digitate clusters. Common roadside weed or understory plant.

The Yoeme names for the two *Desmanthus* species are probably interchangeable.

Sonora: Guaymas, *Palmer 642 in 1887* (NY). Bacum Station near Río Yaqui, *Pennell 20211* (US).

Desmanthus covillei

Common names: HIITEPOA, SITE'EPOA; *dais*; Coville's bundle-flower

Sonora. Small shrubs similar to *D. bicornutus* but generally with smaller leaves and fewer pods per cluster. Widespread in Sonora in washes, arroyos, and rocky slopes.

Animal food: This plant is eaten by the *maaso* (white-tailed deer).

Sonora: Guaymas, *Gentry 4675*. 10–15 mi E of Empalme along Hwy to Cd. Obregón, *Gentry 23438*. Sierra Bacatete, Rancho El Álamo, 8.3 km (línea recta) al NE de Vícam, 12 Sep 2008, *Sánchez-Escalante*, photos.

Erythrina flabelliformis

Common names: CHIRIKOOTE; *chilicote*; coral bean

Arizona and Sonora. In the foothills and mountains of southeastern Sonora, *chirikoote* develops into a tree reaching 8 to 10 m tall with a very thick trunk, often more than 60 cm wide. In mountains closer to the desert it is a much smaller tree. In southern Arizona coral bean is found in the mountains at elevations above the desert, and ranges into the lower oak zone. Coral bean in Arizona is merely a shrub, growing among rocks; it is frost-sensitive, and the stems are repeatedly frozen back.

The bright red, showy flowers appear on leafless stems in early summer, and the pods ripen later in the summer. The seeds are bright red, about 12 to 18 mm long, and are reported to be poisonous. The leaves quickly fall after the end of the summer rains and the plants remain leafless until the summer rains of the next year.

Chilikoote Kawim (Coral-bean Mountain) is the traditional Yoeme name for Cerro El Vigía at Las Guásimas.

Adornment: The seeds are sometimes strung as necklaces.

Masks: Some men say that the lightweight wood is excellent for mask making, but that it occurs too far away and for that reason is seldom used. The wood is very soft and does not split upon

Fabaceae, *Erythrina flabelliformis*. Upper left: Santa Cruz County, Arizona. 29 May 2013 (SC). Right: Cerro El Vigía. 25 Oct 2019 (SC). Lower left (seeds): Pima County, Arizona. 4 May 2021 (SR).

drying. Some older pahko'ola masks were made from this wood, and occasional contemporary masks have been carved from the wood.

Arizona: Sabino Canyon, Oct 1916, *Thornber 9207*.

Sonora: [Las Guásimas], Cerro El Vigía, *Martin 7 Jan 1979*.

***Erythrostemon gilliesii** [*Caesalpinia gilliesii*]
Common names: KAU TAVACHIN (mountain *tavachín*), TABWIKOSEEWA (around-the-sun flower, flowering around the sun); *tabachín amarillo*; yellow bird-of-paradise

Arizona and Sonora. Slender, spineless shrubs. The flowers are large and showy with yellow petals and long, red stamens.

Often cultivated and sometimes naturalized in Arizona and Sonora. Native to South America.

Gardens: This shrub is grown in home gardens in Arizona and Sonora.

Arizona: [Tucson], Tumamoc Hill, 15 Dec 1969, *Burgess 7611*.

Sonora: Sierra Bojihuacame southeast of Cd. Obregón, *Gentry 14517*.

Erythrostemon palmeri [*Caesalpinia palmeri*]
Common names: KUME'A OUWO (grinding tree); *palo piojo*

Sonora. Unarmed, hardwood shrubs with bright yellow flowers and pods that burst open with a loud noise. *Palo piojo* ("louse stick") refers to the conspicuous whitish lenticels on the stems.

This is one of the more common and widespread shrubs of the region, often found on open plains and rocky slopes, and in arroyos and canyons.

Sonora: 8 mi by road northeast of Vikam Suichi, 12 Mar 1989, *Felger & Molina*, observation. Guaymas, 1 Apr 1897, *Maltby 184* (NMC); *Palmer 70 in 1887* (NY, US).

***Glycine max**
Common names: SOOYA; *soya*; soybean

Sonora. Soybeans, native to the Old World, have been grown as an agricultural crop in the Yoem Bwiara. Soybeans were often planted after wheat was harvested in late spring or early summer, often in a system of rotation involving maize, soybean, and wheat.

Haematoxylum brasiletto
Common names: HUCHAHKO; *brasil*; brazilwood

Sonora. Large, spiny shrubs or small trees with

very hard wood. The stems are irregularly fluted and the heartwood reddish. The stems begin to develop their distinctive fluted ridges at about 2.5 cm in diameter, and develop additional ridges as they enlarge, forming living sculptures. The twigs often zigzag and thorn-tipped. The leaves mostly have 2 to 4 pairs of leaflets. The leaflets are mostly broadest above the middle, may be notched at the tip, and have well marked parallel veins. The flowers are bright yellow and the pods small and flattened.

Brasil is widespread across the Yoem Bwiara, on the coastal plain as well as in the foothills and mountains. Pérez de Ribas (1645; Reff et al. 1999:84) wrote, "Most of the province is flat but full of dense thickets and native trees, including some red brazilwood [*palo colorado del Brasil*]." Basilio (1890 [1634]:218) cited *mápau* as the term for *palo-colorado*.

Throughout much of Latin America, this and another closely related species (also called brazilwood) have served as major sources of dye and medicine.

Agriculture: The planting stick, *wi'ika*, was carved from a strong huchahko branch.

Dye: *Haematoxylum* wood has been used as a source of reddish dye.

Hunting and weapons: Arrow points, made from the heartwood, were used for hunting large animals and for warfare. Pérez de Ribas (1645; Reff et al. 1999:90–91) wrote, "Some also use a kind of pike or pointed pole...they made the point or sometimes the whole weapon from brasilwood. These pikes are used by Indians who have a role similar to that of our captains."

Medicine: Huchahko has been an important Yoeme medicinal plant. A common use is an infusion prepared by chipping a piece of wood off a small stem and placing this chip in water where it is left for a time. The infusion is consumed to "strengthen and purify the blood," treat jaundice, and alleviate sourness or bitterness of the mouth. Infusions of the wood can be used as a refreshing bath. Painter (1986:55) reported, "Huchahko wood, made into a drink, stops pain and eases tired muscles." A small branch is boiled and one drinks the liquid to improve circulation of the blood ("*circulación de sangre*").

Fabaceae, *Haematoxylum brasiletto.* Left: Cañón del Nacapule, 30 Dec 2011 (SC). Right: Cañón del Nacapule, 16 Dec 2012 (SC).

In 1887, Edward Palmer recorded that *Haematoxylum* was used as a dye, and the young wood as a remedy for jaundice (Watson 1889). Also see *Arundo donax* (Poaceae) for use of huchahko wood to treat gunshot wounds. A combination of huchahko heartwood, mesquite leaves, and *vavis* (*Anemopsis californica*, Saururaceae) is soaked in water and the infusion consumed to relieve fatigue from sun and overexertion ("burning body," a fever, when too hot and tired from working in the hot sun). Wagner (1936:84) reported, "fainting spells are treated with medicine made of Brazil wood and mesquite [*Prosopis glandulosa*] leaves."

Rosalio Moisés described his grandmother's curing practice in Sonora: "One thing that María did for just about every sick person who came to her was to shave Brazil wood in water. The sick person drank part and left part in the glass. If the water in the glass remained red, the person could be cured, but if the water turned gray, nothing could be done and she told them she could not cure them" (Moisés et al. 1971:18).

Music: This is the preferred wood for the notched musical rasper and rasping stick. Especially since the mid-twentieth century, people from Sonora have supplied the Arizona communities with huchahko raspers.

Religion and ritual: *Huchahko* is a spiritually significant plant. A cross about 4 cm wide made of huchahko wood is placed on the top of the harp to ward off evil. Small crosses (or less often small arrows) made from huchahko wood serve as amulets. Such crosses are sewn into small satin or cloth bags and kept on one's person to ensure good luck, provide positive energy, and ward off evil (Painter 1986:39, 42). Roman Borbón gave Richard a gift of one these small satin bags with huchahko.

The maso yi'ireo (white-tailed deer dancer) uses a stick about two or so inches long in the back of his headdress (see *Olneya tesota*), the preferred wood being huchahko, but if one forgets or does not have brazilwood, then any other stick will suffice. For example, Felipe says, "they sometimes break a pencil and use that, or a popsicle stick...innovation if you don't have what you should."

In describing village governors, Spicer (1954:65) wrote, "each carries a cane, an extremely important symbol of office. This is a brazil wood cane stick about two feet long, capped with silver at one end."

Sonora: Miramar, *Felger 85-899*. Cañón Bacatete, 17 Dec 1988, *Felger & Molina*, observation.

Havardia sonorae [*Pithecellobium sonorae*]

Common names: WOKOHNA; *jócono*; Sonoran ebony

Sonora. Multiple-trunk, thicket-forming small trees. The wood seems to be hard but apparently deteriorates rapidly. The branches have painfully sharp, curved, paired spines that persist to form horizontal ridges or bands on the gray bark. The gray-green leaves are twice-divided with small leaflets. The flowers are mimosa-like, white, and with strong, sweet fragrance.

Basilio (1890 [1634]:213) recorded "hocona, *gato, árbol*." The word *gato* (cat) is often used in Mexico for legume trees with curved spines like a cat's claw.

Common on desert and thornscrub flats across the Yoem Bwiara. At first it seemed strange that there was not wider use of this woody legume. Later, however, we came to realize that its lack of popularity is due the poor quality of the wood and to the abundance of mesquite (*Prosopis glandulosa*) and ironwood (*Olneya*).

Utilitarian: The wood was used to make carrying yokes.

Sonora: Mapoli (name on map), 3 mi northeast of Mex Hwy 15 on road to Cañón Bacatete (Bacateve), *Felger 88-619*. Guaymas, 21 Oct 1939, *Gentry 4670I* (NY); *Palmer 58 in 1887* (NY).

Indigofera suffruticosa

Common names: CHIIHU; *añil*; indigo plant

Sonora. Unarmed shrubs reaching 2 to 3 m tall. The leaves are soft and once-pinnate with 9 to 14 leaflets. The flowers are minute and salmon-pink and orange. The pods are about 2 cm long, relatively thick and curved, and in dense, persistent clusters.

Widely separated small populations of these shrubs are sometimes encountered near old ranches in the mountains from the Guaymas region southward. Chiuhu was apparently once common along the Río Yaqui. The foliage and upper stems are occasionally damaged by cold weather. It is more common southeast of the Yoem

Bwiara, where it seems to be native (Yetman and Van Devender 2002). This shrub occurs from lowland Mexico to South America and the West Indies, and is naturalized in many parts of the Old World and Pacific Islands.

A site on the Río Yaqui was called El Añil. In the 1880s, the Yaqui leader Cajemé "was able to persuade his men to build a fort at a place called El Añil between Potam and Vicam on the south side of the Yaqui River" (Spicer 1980:147).

Dye: The indigo plant is the most widespread indigenous source of blue (indigo) dye in western Mexico and many other regions of Latin America. It was widely grown and harvested across much of Sonora as a source of indigo dye (*añil*). There are early reports of blue cotton fabrics, the blue color undoubtedly resulting from indigo dye. For example, Pérez de Ribas (1645; Reff et al. 1999:91, 329) mentioned "a blue cotton cape" and "blue cotton thread" and reported Yoeme officials wearing blue capes adorned with mother-of-pearl shells during early encounters with Spaniards. In 1887, Edward Palmer recorded that *añil* was "used by the Indians for dyeing palm leaves. Guaymas" (Watson 1889:46). In 1902, Hrdlička (1904:65) reported, "The only clothing of native weave now to be seen among the Yaquis is the faja and the white serape, the latter ornamented with one or two broad stripes in pale blue...but even these garments are scarce." He also stated, "they gather the indigo which is produced in abundance on both rivers (Yaqui and Mayo), and prepare the color" (Hrdlička 1904:64).

Both the Yoemem and Yoremem call the plant *chiju* (Spanish orthography) and the Yoreme continue to use the foliage to produce blue dye (Yetman and Van Devender 2002). Felipe heard about chiihu as a dye plant in the Yoreme lands in the 1990s from Doña Rosa Matus, a blanket weaver at Saniel, in the Río Mayo Region. She called the plant *chiju*, and said it grew along the small wash that ran to the east of the village. She also remarked that the goats eat the plant.

Chiihu was wild-harvested in the Yoem Bwiara; it was also cultivated but the practice was lost and not taken up again after the Mexican–Yaqui wars.

Yoemem in Arizona described chiihu as a plant used for indigo (blue) dye in Sonora. Elders in Marana said that many people used indigo to dye cloth. For example, in the 1950s the people in Marana preferred this indigo color, called *teweli* (dark blue). They used commercial dye instead of the indigo plant, which does not occur in Arizona. During that time they dyed their clothes in big metal wash tubs over an open fire.

Sonora: Rancho La Huerta, 2 km inland from Ensenada La Manga, [NW of San Carlos], garden irrigated with water from canyon-spring, *Felger 85-347*. San José de Guaymas, *Felger 85-474*.

Fabaceae, *Indigofera suffruticosa*, San Carlos. 17 Dec 2016 (SC).

***Leucaena leucocephala** subsp. **glabrata**

Common names: TAVACHIN; *guaje*; white popinac

Arizona and Sonora. Unarmed shrubs or small trees, essentially glabrous. Leaves twice-pinnate, with nectaries on the leafstalks. Flowers in small, dense, globose heads, white to yellowish. Pods flattened, dehiscent, with a thickened rim. Flowering at any season, especially April through June.

Grown in gardens in Sonora and often weedy in empty lots and along roadsides. It is also grown in southwestern Arizona and sometimes becomes weedy. Native to southern Mexico and Central America.

Food: The tender green seeds and young pods are important in traditional southern and central

Mexican cuisine but are little used in Sonora. It is grown in the tropics worldwide for lumber and animal feed.

Arizona: Apache Junction on AZ 88, *Reina-G. 2004-995.*

Sonora: Koasepe, 3 km southeast of Kompuertam, 14 Mar 1989, *Felger & Molina*, observation. Empalme, 28 Jan 1992, *Hughes 1568* (NY). Guaymas, Cerro El Vigía, 28 Apr 1975, *Turner 75-27.*

Lysiloma divaricatum [*Lysiloma microphyllum*]

Common names: VAMYO; *mauto*

Sonora. Trees with hard wood and without spines or thorns; the limbs and branches are strong and flexible, and usually rather slender. The bark is thick and woody, and often peels away in strips. The foliage is feathery green during the summer rainy season and then becomes gradually deciduous. Leaves twice-divided (bipinnate); stipules leafy, larger than the leaflets, and soon deciduous. Flowers small and white, in globose heads. Pods flattened and dry, the rim separating from the rest of the pod.

Basilio (1890 [1634]:218) identified *mahu* as "*mauto* (*árbol*)."

Common along larger arroyos, canyons, and mountains, especially at higher elevations.

Fabaceae, *Lysiloma divaricatum*, Cañón del Nacapule. 13 Dec 2013 (SC).

Construction: Vamyo poles have been incorporated in walls and roofing of traditional homes in the Yoeme Bwiara. Branches and limbs are used for the *chapa kari*, a pitched-roof temporary shelter or hut, and as roofing for ramadas. In March 1989, in watermelon fields north of Vícam, we saw irrigation-ditch weirs made of black plastic sheets between *vamyo* poles, split organpipe-rib slats, and sections of whole green organpipe cactus stems.

Medicine: The bark and leaves are mashed in water and the liquid is applied to burns, as often as necessary, to relieve discomfort ("keep pouring onto the burned area").

Sonora: Kuubwa'e Kawi, 1 km west of Kompuertam, 12 Mar 1989, *Felger & Molina*, observation. Guaymas, *Palmer 640 in 1887* (NY).

Lysiloma watsonii

Common names: MACHAO; *tepeguaje*; feather tree

Arizona and Sonora. These large shrubs or small trees have dense, evergreen crowns of feathery leaves. Flowers small and white, in globose heads. It is apparently not common in the Yoem Bwiara and mostly grows in mountain canyons. It is widely cultivated in southern Arizona, where it is native in a few canyons in the Rincon Mountains.

Basilio (1890 [1634]:218) as well as Johnson (1962:274), identified *máchao* as "*tepehuaje, árbol.*" *Macham* is the term for thigh. The Guarijío name is *machaquí* (Yetman and Felger 2002) and the Mayo (Yoreme) name is *macha'aguo*, "man's thigh" (Yetman and Van Devender 2002).

Arizona: Tucson, University of Arizona, Campus Agricultural Center, *Johnson 15-014.* Rincon Mts, *Thornber 7138.*

Sonora: Guaymas, 5 mi southeast of Mex Hwy 15 at Puente El Tigre, *Felger 85-1531.*

Mariosousa heterophylla [*Acacia willardiana; Mariosousa willardiana*]

Common names: NAWI'O; *palo blanco*; Willard acacia

Sonora. This thin, wispy tree is unarmed and has paper-like, peeling, white bark, especially in dry seasons. The trunk and limbs are mostly slender and flexible, and the wood is hard. The smaller branches are notably slender and flexible. The leaves are long, stringy, and drooping.

It is common in Yoem Bwiara regions, often growing on hot, exposed rock slopes, ridges, and dry arroyos.

Mateo González and others pointed out the usefulness of the stems because they are strong and flexible.

Basketry: Bernaldo Valencia told us that nawi'o was used to make baskets.

Hunting: A pole made from the stem or trunk is used to make a snare for small game, such as cottontails.

Music: The wood from stems about 4 or 5 cm in diameter is considered excellent for making the *hiponia* (drumstick) for the *tampaleo* drum.

Paper: Don Alfonso Leyva-Flores said that the bark of the nawi'o was used as paper for writing, which was done with mesquite charcoal: "*Wame itom yoyo'owam waka nawi'o veata hiosiapo sauwan. Matue vea hiohten*" (Our ancestors used the bark of the nawi'o as paper. They wrote with charcoal). He was talking about his grandparents, so that would have been in the nineteenth century.

Sonora: 2 mi north of Guaymas, 29 Dec 1951, *Blakley B-820* (DES). Guaymas, *Palmer 628 in 1887* (US).

Fabaceae, *Mariosousa heterophylla*, San Carlos. 11 Nov 2017 (SC).

*Melilotus albus

Common names: TEEVO; *trébol agrio*; white-flowered sweet clover

Arizona (and Sonora perhaps in the Yoem Bwiara). The name derives from the Spanish *trébol.* Spring annuals with white flowers. Often weedy in gardens and fields. It is native to Eurasia.

Food: The plants are cooked as greens. In 1994 in Marana, Meregilda Ochoa showed us the plant in her yard and said it is "edible and really good."

Arizona: Tucson, *Thornber 2 Aug 1905.*

Sonora: Hermosillo, *Reina-G. 2005-854* (USON).

*Melilotus indicus

Common names: TEEVO; *trébol agrio*; yellow sour clover

Arizona and Sonora. The name derives from the Spanish *trébol.* Spring annuals with yellow flowers. Often weedy in gardens and fields, and sometimes in wetland places. It is native to the Mediterranean region.

Arizona: Tucson, Church Wash Diversion, 21 Feb 2003, *Mauz 23-3.*

Sonora: San Carlos, *Felger 85-403.* Near Kompuertam, 11 Mar 1989, *Felger & Molina*, observation.

*Mucuna pruriens

Common names: MASO PUUSIM (deer eyes); *ojo de venado*; sea bean

Large, vining shrubs in the tropics worldwide. The large, rounded seeds occasionally wash ashore along the Sonora coast. The smooth, brown seeds have a black band around the middle, making it look like a miniature hamburger. The seeds are sold commercially for jewelry.

The seeds are bought from Mexican herbal stores in Arizona and Sonora. Vendors offer them on Saint Francis feast days in Magdalena, Sonora.

Beliefs: The large seed is pierced and strung as a necklace to ward off evil thoughts and evil energy, or *puhtuana*, meaning someone would give an evil eye. Painter (1986:42) mistakenly called "*maso pusim*...dried eyeball of a deer." In Arizona, before the 1970s these necklaces were only used for babies and young children. Since then these necklaces are strung with beads on either side of the seed and are also sometimes worn by adults.

Olneya tesota

Common names: EHEA; *palo fierro*; ironwood

Arizona and Sonora. Ironwood is one of the

most conspicuous landscape features of the Sonoran Desert and arid coastal thornscrub in the Yoem Bwiara. It usually develops into a small tree, often with several massive trunks. The trees are long-lived. Masses of pale, violet-pink flowers are usually produced in May and develop into short, thick pods in June and July. Each pod bears 1 or 2 (occasionally 3 or 4) seeds. The wood is unusually hard and heavy and takes a fine polish.

Beliefs: The maaso or deer dancer uses a stick in the back of his headdress (see *Haematoxylum brasiletto*). For example, Glafiro Perez reported, "I place a large white piece of cloth that is folded like a triangle on my head. It crosses in the back, then ties in the front. Under the cloth in back, I place a small ironwood stick about two inches wide with a cross carved on it. This symbolizes the strength of the plant world, and it helps to keep the deer head on top of my head while I dance" (Perez and Vance 1993: 49).

Beverage: Yoemem living in Hermosillo used the seeds to make to make a coffee substitute.

Music: In Arizona, the wood occasionally has been used to make the *sena'asom* (disk rattle) frame because *sita'avao* (*Vallesia*, Apocynaceae), the preferred wood, does not occur in Arizona.

Arizona: Tucson Mountains, 2 Jun 1958, *Caldwell 164*.

Sonora: Rancho Bacatetito, 13 Mar 1989, *Felger & Molina*, observation. Cañón Nacapule, *Felger 92-1026*.

Parkinsonia

Common names: SIARI KUTA (green stick; the general term for palo verde); palo verde

Arizona and Sonora. Trees with green or greenish bark.

Mask: A pahko'ola mask from the early twentieth century, formerly at San Xavier, and now at the Arizona State Museum, is reportedly made from palo verde wood.

*Parkinsonia aculeata

Common names: VAKA'APO, VAKA'APOA, VAKA'APORO; *bagota, retama;* Mexican palo verde

Arizona and Sonora. This large, green-barked tree is fast growing and has sharp spines at the leaf bases. The leaves are long and stringy with tiny, drought-deciduous leaflets. Masses of bright yellow flowers are produced in early summer.

This is one of the most commonly cultivated trees in Arizona and Sonora. It is often weedy and sometimes naturalized through much of Sonora where it seems to be spreading as natural vegetation gives way to disturbed habitats. It is especially common along irrigation canals, arroyos, riverbanks, and roadsides in southern Sonora. It is probably native to Central America and southern Mexico.

Medicine: The leaves are used in remedies to relieve coughs and rid oneself of phlegm or mucus. For coughs, the leaves are twisted into a circular form and sucked much like one would a cough drop.

Arizona: Tucson, Grant Road just west of Silverbell Road, *Johnson 89-56*.

Sonora: Potam, 13 Dec 1988, *Felger & Molina*, observation. Las Guásimas, home garden, 16 Dec 1988, *Felger & Molina*, observation.

Fabaceae, *Parkinsonia aculeata*, San Carlos. 28 Mar 2013 (SC).

Parkinsonia florida [*Cercidium floridum*]

Common names: TE'OWE; *palo verde azul*; blue palo verde

Arizona and Sonora. Trees generally with a well-formed trunk and a distinctive bluish-green color to the branches and leaves. The leaves are small and there are spines at the twig nodes. Masses of yellow flowers are produced in spring.

Mostly along washes and arroyos and on the coastal thornscrub plains of the Yoem Bwiara. In Arizona it is sometimes called *cho'i*, the name for *Parkinsonia praecox*.

Medicine: It was used medicinally, but we do not have specific information.

Music: The gum or resin, called *chu'ukam*, is used as rosin for the violin bow.

Washing: The bark is stripped away and the wood cut up, mashed, and used as laundry soap.

Arizona: Tucson, West Branch channel and Enchanted Hills Wash, 8 Apr 2004, *Mauz 24-55*.

Sonora: Guaymas, *Carter 5594a*. Kompuertam, 11 Mar 1989, Felger, observation. Rancho Bacatetito, 13 Mar 1989, *Felger*, observation. San Carlos Bay, *Shreve 6554*.

Parkinsonia microphylla [*Cercidium microphyllum*]
Common names: WO'I VA'AM, WO'I VOA'AM (*wo'i*, coyote + *voa'am*, hairs/furs); *palo verde*; foothill palo verde

Arizona and Sonora. Widespread and common in southern Arizona deserts and ranging southward to the northern part of the Yoem Bwiara.

Medicine: Mateo González said, "boil it and let it sit, then drink the tea, warm or cool, to treat diabetes, cancer, body infections, kidney, heart, and *imparto*."

Arizona: Tucson Mountains, *Thornber 706*.

Sonora: Cañón Nacapule, *Felger 92-1054*. Guaymas, Isla Almagre Chico, 27 Sep 2006, *Suarez-Gracida 2006-120-CGSG*.

Parkinsonia praecox [*Cercidium praecox*]
Common names: CHOI, CHO'I; *brea*, *palo brea*; brea palo verde

Sonora. Small trees with green bark, straight spines at the twig nodes, and a horizontal, spreading canopy. The leaves are rather small and the flowers golden yellow. Mass flowering is seen from February to April. The pea-green bark, open, spreading growth habit, and pale green foliage are distinctive.

Palo brea is abundant in the desert and thornscrub across the Yoem Bwiara, and thrives on overgrazed lands and disturbed sites. It is also commonly cultivated in southern Arizona.

Adhesive: The "gum" is made into a paste and used like ordinary glue for repairing broken items, or to glue paper together, etc. The name *palo brea* "is derived from the fact that the waxy substance coating the bark, after being scraped from the branches and melted by heat, is used as a 'gum' for gluing together leather objects and furniture; thus it is used just like *la verdadera brea*" (Carter 1974: 30).

Medicine: The *ota kuta* (stem wood) is said to be "good for diabetes."

Wagner (1936:85) reported, "sometimes abortions are produced by mixing the resin of the *brea* tree with tallow and making a paste. This is rubbed on the abdomen. When this remedy is effective it is probably due to the severe rubbing instead of the paste."

Music: The gum or resin, called *chu'ukam*, is used as rosin for the violin bow.

Utilitarian: The "gum"(resin) is made into a paste and used like ordinary glue for repairing broken items, or to glue paper together, etc. The name *brea* "is derived from the fact that the waxy substance coating the bark, after being scraped from the branches and melted by heat, is used as a 'gum' for gluing together leather objects and furniture; thus it is used just like *la verdadera brea*" (Carter 1974:30).

Sonora: Cuartel at Cañón Bacatete, 13 Mar 1989, *Felger & Molina*, observation. South of Peón, *Turner 61-35*.

Fabaceae, *Parkinsonia praecox*, Sierra Mazatán. 17 Apr 2003 (JS).

Phaseolus acutifolius
Common names: *tépari*; tepary bean

Arizona and Sonora. Annual vines. Wild and cultivated teparies are distinctive and indigenous to northwestern Mexico and southwestern United States (Nabhan and Felger 1978). Basilio (1890 [1634]:161–162) recorded the names of seven different kinds of bean (*muni*), possibly both wild and cultivated, including: *iori munim* for "*frijoles pequeños de la tierra*"; *iol couuni* for "*frijoles pinta de la tierra*"; *selaim* for "*frijoles pequeño*"; plus *subae munim*, *tosa selaim*, and *tosali munim* (white beans).

Fabaceae, *Phaseolus acutifolius*, Pima County, Arizona. 8 Sep 2014 (SR).

Wild Teparies

Common names: HUA MUUNI (wilderness bean), HUA SE'ELAIM (wilderness teparies), SUVA'U MUUNIM (quail beans)

Arizona and Sonora. Native teparies are common in the mountains in the Yoem Bwiara in Sonora and in Arizona in areas above the desert. They grow with warm-weather rains. The flowers are small and pink. The pods and seeds are smaller than those of cultivated teparies. The seeds (the edible beans) are usually gray-speckled and resemble gravel, although apparently sometimes also other colors. Unlike cultivated teparies, the pods are readily dehiscent—opening explosively when dry and ripe, flinging out the seeds (Nabhan and Felger 1978).

José María Jaimez said he likes wild teparies and there are four names for them. He recalled three of the names: *hua muuni* (wilderness bean), *hua se'elaim* (wilderness teparies), and *suva'u muunim* (quail beans).

Food: Wild teparies were eaten in former times, especially by people in the mountains. José María Jaimez told us, "The older people in the mountains ate these beans and lived a healthy life." Wild teparies are gathered when the pods are ripe and mature, usually in August. (The pods would be gathered in the early morning, while still cool and damp, since dehiscence generally occurs with drier and warmer air later in the day.) The seeds are separated from the small pods by beating them with a *karoote* (wooden club). Wild teparies are prepared as food in the same manner as common cultivated beans.

Arizona: [Tucson], Saguaro National Park West, Sweetwater trail, *Bertelsen 89-563*.

Sonora: Near San Carlos Bay, flood-bottom with shrubs and ipomoeas, flowers bright pink, 24 Oct 1939, *Gentry 4728*.

Domesticated Teparies

Common names: HESEIM (brown tepary; cultivated), SE'ELAI (white tepary; cultivated)

The domesticated teparies resemble the common bean (*Phaseolus vulgaris*) but have somewhat smaller leaves and smaller seeds (the edible beans). There are teparies with different-colored seeds.

The tepary was domesticated and cultivated in the Sonoran Desert region. It has long been a major crop in the Yoem Bwiara. Pérez de Ribas (1645; Reff et al. 1999:87) reported, "All these people use the *frijol*." References to "beans" in pre-contact times in northwestern Mexico and southwestern United States would be to teparies rather than to the common bean (*Phaseolus vulgaris*).

Food: Teparies are esteemed by Yoemem and Yorim alike and considered nutritionally better than the common pinto bean.

Cultivated teparies in the Yoem Bwiara are planted from January to March, "at the same times that you plant regular beans." We were told that Yorim like to buy teparies from the Yoemem. The white teparies are usually preferred, although there are also black and brown teparies "but nobody likes them." It is said that, "*se'elaim* will make you fart all night."

*Phaseolus vulgaris

Common names: MUUNI; *frijol*; common bean, pinto bean

The common bean was domesticated in Mesoamerica and has been grown in northern Mexico at least since early Spanish colonial times. Basilio (1890 [1634]:224) recorded *siqui muni* as the word for *frijol colorado* (red bean). (Studhalter 1936:122):

> Several varieties of beans are grown by the Yaquis including *frijoles*, string beans, and lima. The first type is the most common. All the varieties are grown either in plowed fields or in pits; the former may range up to two acres in extent. There is both a spring and a fall crop. Sometimes beans are threshed and winnowed in the same manner as...described for *garbanzos* [see *Cicer arientum*]. The seeds are stored in sacks or in mat bins, and occasionally in underground pits.

Seeds of the common bean are substantially larger than tepary seeds. A number of different kinds of *muunim* have been grown. *Mun sikiri* and *sikili muunim* (red bean), the traditional red bean (actually pink) is delicious. It was also called *mun roosam* (pink beans). This variety apparently has not been grown in the Yoem Bwiara since the mid-1980s and then only in home gardens. This bean, however, is available from Mexican vendors in Ciudad Obregón. *Mayokovam* (*asufrado* in Spanish) is light brown and tastes like the regular pinto beans; it is not speckled like the pinto bean. Mayokovam are more common from Navojoa and southward. *Ehootam* are green beans (*ejote* is Nahuatl term)

Juanita Paula Valle said that in the Marana–Tucson area, *sikiri/sikili muunim*, also called *mun roosam*, were common in the 1930s and 40s, and that pinto beans came in during the 1950s.

Although there are corn songs, no songs are known for beans and squash.

Food: Beans have long been a traditional Yoeme staple. Beans are sorted to remove any rocks or small dirt clods, rinsed, and sometimes soaked for hours. A big pot is filled with water, a handful of salt added in the form of a cross, simultaneous with a prayer. The beans are put into the pot, again added in the form of a cross and with a prayer. If one bean should fall out during the sorting or while being added to the water, it is picked up immediately, rinsed and thrown into the pot, because they say it cries, it does not want to be left behind. It is considered an orphan. *Mun posoim* "bean posole" has long been a traditional food in Arizona and Sonora. Beans are cooked (boiled), meat is added, plus hominy and sometimes other ingredients such as spinach, and salted. *Mun nohim* are bean tamales.

Medicine: Juanita Paula Valle said that the beans called *sikiri muunim* (*mun roosam*) were crushed and used for relief of headaches. The crushed beans are applied to forehead and temples by being tied on with a white cloth.

Wagner reported several treatments for rabies, the accuracy of which seems questionable. He wrote:

> Another remedy when *fresno* [*Fraxinus*, Oleaceae] is not available is made of beans. About a pint of beans is roasted and then ground very fine. This bean meal is mixed with cold water. The patient is made to drink the mixture until he vomits. The next day the treatment is repeated. If the patient is not better it is repeated again on the third day." (Wagner 1936:85)

Piscidia mollis

Common names: HO'OPO, HOPO, MOORA; *palo blanco*; fish-poison tree

Sonora. Distinctive legume trees of moderate size, with white bark. The leaves are made up of several large, oval leaflets that are tough and velvety haired, often imparting a silvery color. The pea-like flowers are cream-white, and the seed pods are woody and prominently winged.

In the Yoem Bwiara east of Mexico Highway 15; arroyos and plains, and in the Sierra Bacatete.

The leaves contain toxic compounds, which have been used as fish poison in the Río Mayo region (Gentry 1942). However, we do not know such use among the Yoemem.

Adornment: The seeds are used for rosary necklaces.

Deer song: A song in the *Maso Me'ewa* (Killing of the Deer) ceremony compares the leaves (actually the large leaflets) of *ho'opo* to mountain lion ears (Evers and Molina 1987:145; see *Puma concolor*, Felidae).

Medicine: Preparations made from this tree are used in the treatment of measles and "blood infection." The leafy herbage is ground, toasted in ashes, and the still-warm ashes are rubbed on the body.

Tools: The wood is excellent for making axe handles.

Sonora: Road to Torocobampo in Sierra Bacatete, *Felger 88-585*. 3 mi east of Mex Hwy 15 (Mapoli on map), east of Las Guásimas, *Felger 88-620*.

*Pisum sativum

Common names: PEONASIM; *chicharos*; peas

Peonasim are cultivated peas as well as commercial, store-bought peas. Peas are of Mediterranean origin.

Agriculture: Studhalter (1936:123) reported that in Sonora, "English peas are grown in very much the same manner as are the *garbanzos*."

*Pithecellobium dulce

Common names: MAKO'OCHIINI; *guamúchil*; Manila tamarind, monkey pod

Sonora. This is a large tree with light gray bark. Paired spines at the base of each leaf persist and become broad and enlarged on the major limbs and trunk as horizontal bands. The leaves have several large, paired leaflets and the flowers are white and mimosa-like. The pods, which ripen in May and June, are curved or coiled, and partially split open to reveal black seeds embedded in either pinkish or white, soft, spongy pulp. The scientific name translates as "sweet monkey-ear," referring to the fruits.

Basilio (1890 [1634]:218, 205) recorded *macochin* as *guamuchil*, and *bacochini* for "*guamuchil, árbol*."

The trees are seen at nearly every settlement or town in southern Sonora and seemingly naturalized trees are sometimes seen along irrigation ditches and the Río Yaqui. This species is native to the hot lowlands of Mexico to northern South America, and is probably native only as far north as central Sinaloa. The pods are sold in quantity in Sonoran markets.

Cultivation: This is one of the most commonly cultivated trees in the Yoem Bwiara.

Beverage, food, and harvesting: The sweet, fleshy pulp is eaten fresh or made into a beverage. In the lower Río Yaqui and Río Mayo valleys, a long bamboo or carrizo fruit-gathering pole is often seen leaning against a guamúchil. This pole is used to harvest the pods. A transverse stick is tied near the top of the pole, much like the saguaro-fruit gathering pole (see *Carnegiea gigantea*, Cactaceae).

Sonora: Cuesta Alta, Río Yaqui, *Felger 85-1389*. Las Guásimas, 16 Dec 1988, *Felger & Molina*, observation.

Prosopis

Common names: HUU'UPA; *mesquite*; mesquite

Huu'upa is the noun and *hu'upa* is the adjective/modifier. The noun drops one "*u*" when it serves as an adjective. *Hu'upa sawa* (mesquite leaf) is one example. However *hu'upa* as a noun is a variant in common use.

Three species of native mesquite occur in the Yoem Bwiara and southern Arizona: *Prosopis articulata*, *P. glandulosa* var. *torreyana*, and *P. velutina*. These are large shrubs to large trees, and often spinescent.

Mesquites are widespread in southern Arizona and through much of Sonora on desert and thornscrub plains as well as in valley bottoms and drainageways, and they thrive in disturbed habitats. Mesquites can be quite fast growing, especially in southern Sonora where there is deep, rich soil and sufficient soil moisture. They are readily spread by cattle and thrive in grazed areas—the indehiscent pods (they do not split apart to release the seeds) with the seeds embedded in a leathery pit are adapted for dispersal by large mammals; in this regard, cattle have replaced the long-extinct Ice Age native mesquite eaters (Felger 1977, 2007; Felger et al. 2001; Hodgson 2001).

The trunks of even well-grown trees are often crooked or irregular. The roots are pliable and of rather soft wood, while the wood of the trunks and branches is hard, heavy, and strong although somewhat brittle, and very resistant to decay. The bark is rough and usually dark brown to blackish.

Chu'ukam refers to the sticky or gooey gum ("sap") from a mesquite tree. White gum (from the phloem) often oozes from wounds and black pitch (from the xylem) drips onto the bark or the ground, likewise from injuries.

Mesquites are winter deciduous, although often gradually so. The brilliant green new foliage appears with warm weather in March and April—typically after all danger of frost has passed. The new growth can be exceedingly rapid and there is great variation in leaf size.

Chunahko is a term specific for mesquite flowers. The flowers are small and yellowish, and crowded into cylindrical, spike-like inflorescences. Mass flowering occurs soon after leaf out in spring and the pods ripen in early summer, typically before the first summer monsoon. Additional flowering and pod development may occur later in the summer and early fall.

Hu'upa taakam (mesquite fruit) is the term for the pods; the Mexican term is *péchita*. Mesquite pods are relatively large, slender and elongated, tan when ripe and often suffused with red, and indehiscent (they do not split apart to release the seeds). The pods tend to hang on the tree until they ripen, at which time they fall to the ground. The pod typically consists of four parts: (1) the exocarp, the tough outer husk or "skin"; (2) the mesocarp, the sweet, carbohydrate-rich pulp (lacking or reduced in *Prosopis articulata*); (3) the endocarp, the leathery-bony "pit" that encloses each seed; (4) the seeds, tightly held inside the endocarp and released after the endocarp is broken or abraded, such as by being tumbled
down a gravelly wash, pounded in a mortar or by the feet of animals, or passed through the gut of an animal. Each pod has multiple seeds, but the number can be highly variable.

Very large mesquite trees (*bwe'u hu'upa*, large mesquite), once plentiful along riverine floodplains and larger arroyos and washes, have become scarce in Arizona and Sonora. *Wohnaiki hu'upam* (eight mesquites) is the name of a place where eight large mesquites stood near Tórim. In the late 1980s we were told that the Yoim (Mexicans) cut them down.

Hu'upa sania or *sania hu'upa* (small mesquite grove) refers to an area of small mesquites—lush, green mesquites that may become tree-sized. In contrast, *saniloa* or *sanila* is an area of mesquites that remain as shrubs, as if the mesquites are stunted (see *Prosopis articulata*). In saniloa the world is said to be different in some way, perhaps associated with, but different from Yo Ania. It would not be an urban area like the towns or cities. It would be a place one must not go far by oneself, especially at night.

One day in December 1988, we were returning from the Sierra Bacatete. The road was muddy due to recent rains and our van got stuck in the mud, and it took a while to get started again, and then it got dark. It was an area with stunted, shrubby mesquites. We could not find the road out to the highway until late at night after a number of wrong turns. Richard suggested sleeping in the van until morning light, but Ignacio Amarillas and others from Las Guásimas would not do so. Don Nacho said, "Remember what I told you this morning, compadre, the world [this wilderness world, the Huya Ania] is playing with us."

Cultural knowledge and use of mesquite: Huu'upa is the most common and useful tree in the Yoem Bwiara and likewise was important for Yoemem in southern Arizona. Velvet mesquite (*Prosopis velutina*) and western honey mesquite (*P. glandulosa* var. *torreyana*) both have pods with edible mesocarps, and the cultural uses are essentially indistinguishable. The following discussion may apply to both species. Non-food uses in Sonora might also include the bitter mesquite (*P. articulata*).

Fabaceae, *Prosopis velutina* pods, Ajo area, Pima County, Arizona. 4 Aug 2006 (SR).

Adornment: Mesquite leaf stalks have been used after ear piercing to keep the hole from closing. The person must rotate the leaf stalk daily. If mesquite is not available then red thread should be used instead.

Basketry: Beals (1945:11) illustrated "A carrying basket or wakal, used in gathering pitahaya fruit. Made of canes twined with mesquite-bark strips. The bottom is made of mesquite-bark strips interwoven; the carrying strap is of canvas. Diameter, 10 inches; height 10½ inches." The mesquite strips probably were made from the inner bark or roots.

Beliefs: Shutler (1977:189) reported:

> Two plants, in particular, have supernatural powers above the natural powers of plants to nourish and cure. These are *hu'upa* (mesquite), and *hiyakvivam* (tobacco). *Hu'upa* grows both in Sonora, in the Yaqui homeland, and it Arizona, and although all *hu'upa* has natural curing powers, only the *hu'upa* which grows in the sacred Yaqui territory has mysterious power to detect and vanquish witchcraft. A piece of *hu'upa* will protect its owner against witches, especially if it is cut in the shape of a cross. The *hu'upa* wood, moreover, will glow at night to warn of an approaching witch.

Man using a long pestle to pound mesquite pods. 1923 (unknown photographer).

There are many sayings or beliefs involving mesquite, some of which are listed here:

> Don't put the tip (small end) of a mesquite log in a fire because your wife will have a hard time giving birth. (Instead, put the large end into the fire).
>
> When there is a long dry spell, a *mata* (grinding stone) is leaned against the *tevat kuus* (house or patio cross made of mesquite wood) to help bring rain.
>
> If you pee in a body of water you will not be capable of building a mesquite fire.
>
> Never touch or gather firewood that has been hit by lightning, because it has power that can make you sick (Estefana Garcia told this to Felipe in Marana when he was 11 years old).

Beverage and food: In earlier times the sweet pulp from the fruit was a major food resource.

Pérez de Ribas (1645; Reff et al. 1999: 87–88) wrote: "They also use a type of *algarrobilla*, or carob bean, which comes from a native tree called *mesquite*. These beans are ground and made into a drink with water, and because they are somewhat sweet, they are for these Indians what chocolate is for Spaniards." He also noted, "Their wine was made from various native plants...Sometimes the wine was made from the carob bean of the mesquite tree" (Pérez de Ribas 1645; Reff et al. 1999:90).

Beals (1943:20) listed mesquite as one of the most important wild foods gathered in former times by the Yaquis. The pods were gathered by the entire family. Ripe pods were picked from the trees and carried home in buckets. The pods are rinsed and dried on a rooftop. When dry, the pods are put in a large cottonwood or mesquite mortar (*huchu'ukia*) and pounded with a large pestle (*hiponia*). The huchu'ukia is made from a mesquite stump or cottonwood stem, and the pestle is made from the core of the mesquite (José María "Chema" Jaimez, 12 March 1989, Kompuertam). The resulting fine flour was made into *hu'upa vannaim* (mesquite *atole*; vannim may be from Spanish for bean pods, *vainas*).

Another way to make mesquite *atole* is by boiling the ground pods in water. When the pods are soft, the fiber and "seeds" are removed, and the remaining floury liquid is mixed with white or brown sugar and white flour to make a pudding. Any part of the mesquite left over from pounding or cooking, or in fact any food or medicine left over, is taken to the east and offered to plants by throwing in the form of a cross. Nowadays some people throw coffee or medicine, etc., at the foot of the house yard cross.

People who came to Arizona from Sonora in the early twentieth century harvested mesquite pods much as they had in Sonora. For example, in the early twentieth century, Geraldo ("Lalo") Alvarez and his "family lived in the desert at Kingsley Ranch near Arivaca" and among the wild harvests they made "*atole* from mesquite taakam pechitam [pechitas or pods]" (Molina et al. 2003:4).

The red-orange or clear brownish-white gum (resin), called *chu'ukam*, served as chewing gum, used mostly by children. José María Jaimez called it *huu bwania* (mesquite's wax).

Charcoal, fuel, lumber, and woodcutting: *Maatu* is the term for charcoal, or *karvoonim* from the Spanish. (Matus is a common Yoeme surname.) In the nineteenth century, mesquite charcoal was used for writing on bark paper, such as from *palo blanco* (*Mariosousa heterophylla*).

In describing the Yoem Bwiara shortly after the mid-twentieth century, Jane Holden Kelley (Moisés et al. 1971:xii) wrote, "The omnipresent mesquite has gained a new importance in the modern cash economy because of the great commercial demand for fence posts, charcoal, and firewood."

Charcoal making has long been a means for Yoeme men in the Yoem Bwiara to earn cash and subsistence. From the 1930s to the mid-twentieth century, small-scale charcoal production supplemented mesquite wood-cutting (e.g., Moisés et al. 1971).

In the later decades of the twentieth century, charcoal making, *carbonim hoowa*, became an important economic activity. Charcoal work brings quick income between the time of farm loans and crop harvests and sales, when there is no other significant income. Many of the men work at making charcoal. They help each other in the making of it. In Arizona and Sonora people were well aware that charcoal-making was destructive and not sustainable. In Marana in 1987, Juan Luis Garcia told Felipe that they were destroying all the big mesquite in Sonora (in the Yoem Bwiara) for the purpose of charcoal-making and he didn't like it.

In 1989 we saw charcoal production near Kompuertam: There is a large oven-like, cone-shaped, earthen mound and smoke is seeping out of it. The mesquites in the vicinity, about 5 m tall, have just been cut. It looks like the mesquites here certainly grow fast, but there is a general lack of large trees away from the houses. Cutting of mesquite for charcoal accelerated in subsequent decades.

In an interview in late May 2019 in Pótam, Juana Lugo-Osuna told Felipe that the men could not find enough mesquite ramada (*heka*) posts for the 2019 Holy Trinity Pahko (celebration). The men traveled far and wide looking for the posts not only for the pahko ramadas, but also for the kitchen and dining areas. She said the people who make charcoal to sell on the market are cutting down the mesquites to earn a living. She said this has been going on for many years. Juana said the people are poor and this is the only way they can make a living. She was also concerned about the *vaaka* (*Arundo donax*, Poaceae).

Mesquite has been the most important and preferred wood for cooking and heating in Sonora. In Arizona it was the most extensively used cooking and heating fuel until about the 1970s. Only solid dead wood was burned (Holden 1936c:71).

Eulalia (Alvarez) Suarez, of Old Pascua, said that she "preferred mesquite wood for cooking the family meals. The family would go to gather mesquite wood weekly as far as Ruthrauf Road, north of the city" (Molina et al. 2003:67). Obtaining firewood in the Tucson region became increasingly difficult as the city grew. Rosario Vakame'eri-Castillo and his family were in a pickup truck in the 1930s and had gathered firewood for the house when a rancher in the Tortolita foothills accosted them at gunpoint and made them unload all of the wood, and told them never to come back again.

Holden (1936a:51) reported that during the Gloria in Tórim, "Piles of dead, solid mesquite wood had been stacked near the new arbor earlier in the day to furnish fires for the night. Shortly after sundown a number of fires were lighted, the military society having then by far the largest one."

Procuring firewood in the Yoem Bwiara sometimes was extremely time-consuming. Firewood was usually gathered by women and sometimes by boys. Holden (1936b:71) wrote that as the supply of suitable wood near settlements was used up, the women had "to go farther and farther from their houses for it. It is not unusual for a woman to walk a mile or more, break up a bundle of wood, tie it with a leather thong, balance the load on her head, and carry it home."

Men in Arizona and Sonora often cut mesquite wood as a means for income. Rosalio Moisés tells of cutting mesquite in 1933: "In the United States, wood was sold by the cord. In the Río Yaqui, it was sold by the *carga*, or load. Mexican truckers who came into the woods to pick up the *cargas* paid one peso a *carga*" (Moisés et al. 1971:142). During these times men in the Yoem Bwiara often worked at cutting mesquite wood for firewood, fence posts, and railroad ties. It was hard work and the wages or money earned was meager but essential during those hard times.

Rosalio Moisés' experiences in 1944 further illustrates the nature of the mesquite cutting in the Yoem Bwiara and how the worker would be poorly paid (Moisés et al. 1971:195).

> After Easter I went back to the wood forest, cutting seven-foot mesquite posts. I was paid fifty centavos a post, of which Miguel Toledo took thirty centavos because he had the contract for the posts. The axe was in pretty bad shape, so I could only make twenty-five posts a day. We had to walk seven miles out to the place where we cut the posts and seven miles back every night, even though there was good mesquite much closer. The governor of Torim and Miguel Toledo told us, 'If you folks want to cut some posts, you better go over past *Paros Hackia* (Rabbit [Jackrabbit] Creek).'
>
> I was making about five pesos a day and walking fourteen miles through terrible heat. Miguel Toledo was very happy because he did not have to work at all. He just sat around the *guardia* all day and got rich off of our hard work. Such easy money for him. We cut enough posts (seven thousand) to fulfill Miguel's contract in two months, finishing on July 18, 1945. On the nineteenth, I started making [railroad] ties, and four days later I went back to the wood forest.

Construction: Large, forked trunks and major branches are made into posts for ramadas at homes and the *pahko heka* (ceremonial ramada), and especially for the traditional homes (mostly in Sonora). The more slender, straight poles are used as roofing. In Sonora, one often sees mesquite poles, much taller than the roofs, leaning against houses. The poles are stacked against the roofs to preserve them, instead of lying on the ground where termites, other insects, and rot can damage them.

Mesquite harvesting should be done during the time of the full moon. It is said that if one should get the wood other than during the full moon, termites will damage the posts (Vicente Baltazar, 1980).

Household: The three-branched mesquite post or tripod, *hu'upa vahim taka'ariakame* (mesquite with a 3-forked branch), is seen at every household holding a large, unglazed water vessel or olla. This olla is called *va'achia* when holding drinking water on the tripod mesquite-stand. Slow evaporation through the porous wall of the olla cools the water.

Medicine: Anselma Castillo Tonopuame'a said that bark from the east side of a young mesquite is used for headache: "Cut a strip the width of your forehead and tie it onto forehead with a white cloth." People usually use the fresh green bark. Similar information, from Tórim, was given by Wagner (1936). (See *Anemopsis californica*, Saururaceae, and *Haematoxylum brasiletto* and *Vachellia farnesiana*).

Tea made from mesquite gum and elderberry (*Sambucus cerulea*, Viburnaceae) is taken to alleviate a cold.

Mesquite flowers, *chunahkom*, have been used as a remedy for pink eye. Painter (1986:55–56) reported that, "mesquite leaves mashed in water are effective as eye drops" and "ashes of the preferred pine, mesquite, and cholla may be used in massage for broken bones, sprains, or cramps." To alleviate irritated, pink, or sore eyes, mesquite flowers are soaked overnight in water and eyes rinsed with the water.

Wagner (1936:82–85) cited several medicinal uses for mesquite:

> Gas relief is obtained from a brew made from mesquite bark. The bark must come from the tree on the side of the rising sun. A strip of bark is beaten into a pulp and then steeped in water. The liquid is drunk as needed.
>
> Cathartics [laxatives] are made from various herbs. One very commonly used is a tea made from the macerated bark of mesquite twigs.
>
> Fainting spells are treated with medicine made of Brazil wood [*Haematoxylum brasiletto*] and mesquite leaves.
>
> For ant bites the Yaquis use the wax from the mesquite tree. In the last few years this practice has been modified by the use of commercial glue, which is smeared over the bite.
>
> A remedy for headache, is made from mesquite leaves. A quantity of these is mashed into a pulp. This is mixed with water and urine, made into a poultice and applied to the forehead.

Music: Wilder (1963) reported that in Arizona in the 1940s the deer singers' rasping sticks were made from mesquite wood, although catclaw acacia (*Senegalia greggii*) was more often used.

Religion and ritual: The many kinds of wooden crosses, especially in the Yoem Bwiara, were almost always made of mesquite wood. Spicer (1980:65) reported, "In front of each [house] was a wooden cross, made of mesquite limbs peeled and at right angles to each other." This household cross is usually about 1 m tall. Boundaries between the Wohnaiki Pweplom, the eight traditional towns, were marked with three mesquite crosses, the largest of the three often about 2.5 meters tall (Spicer 1980:121, 215; Evers and Molins 1992). When the crosses rot away the villagers make new ones.

Shampoo: Mesquite leaves are pounded into a green mass and used as a shampoo, giving the scalp a fresh scent.

Toys: Rosalio Moisés told of children using mesquite thorns as darts: "We would take a branch of the Don Juan plant, slash it, put it on the ground, and throw mesquite thorns at it, like darts" (Moisés et al. 1971:22; see *Nicotiana glauca*, Solanaceae).

Fabaceae, *Prosopis glandulosa*, Tórim. 12 Apr 2018 (SC).

Prosopis articulata

Common names: HUU'UPA, SANEAL; *mesquite amargo*; bitter mesquite

Sonora. The leaves may have one or two pairs of pinnae, each bearing many leaflets. The flowers tend to be yellowish orange. The major distinguishing feature is the pod, which has little or no pulp (mesocarp) and therefore not used as a food source.

This mesquite is found near the coast from

Bahía Kino to southwestern Sonora and perhaps northwestern Sinaloa, as well as through much of the Baja California peninsula, especially the southern half of the peninsula.

The concept of *saniloa/sanila*, an area of stunted mesquites might be related to this mesquite. *Sanéa* is the Yoreme (Mayo) name for this species (Yetman and Van Devender 2002:215). The Yoreme village of Saneal (Spanish) would be *Saniel* in the Yoreme language.

Sonora: Guaymas, 24 Oct 1939, *Gentry 4689; Palmer 197 in 1887* (GH, NY, US); *Rose 1225* (NY). Bacum Station, near Río Yaqui, 7 Sep 1935, *Pennell 20204* (NY).

Prosopis glandulosa var. **torreyana** [*Prosopis juliflora* var. *torreyana*]
Common names: HUU'UPA; *mesquite*; western honey mesquite

Arizona and Sonora. Western honey mesquite is distinguished by its leaves with one pair of pinnae, and often relatively large and widely spaced leaflets. The leaves are generally glabrous (smooth, without hairs) or nearly so, and the leaflets usually more than twice as long as wide.

Variety *torreyana* is the common, widespread mesquite of the Yoem Bwiara. It is most numerous and largest along drainageways, riverine floodplains, and low-lying, coastal plains with fine-textured soils. It grows in near the shore, along washes, in the plains, at ranches and towns, and at roadsides.

Variety *torreyana* ranges northward along the Sonora coast to southwestern Arizona. It is also widespread elsewhere in arid regions of southwestern United States and northwestern Mexico. Variety *torreyana* is replaced by var. *glandulosa* east of the Continental Divide.

Basilio (1890 [1634]:215) cited *huupa* as the word for "*mezquite, árbol.*" Vast *bosques* of *mezquital* (mesquite forests) once covered much the coastal plains of southwestern Sonora. We were told that mesquites and willows along the Río Yaqui were larger and more beautiful before the river dams were built. The trunks were big and there were lots of very large mesquites. One man said, "these trees were more beautiful before the dams because there was more water, more water bringing nutrients."

El árbol de la horca, the hanging tree, is a large mesquite tree in Vícam that was used for executions. It was still standing in the late 1990s.

In 1772, Miguel del Barco reported: "In Yaqui and other parts of that coast these seeds [of mesquite] have a good flavor, and the people eat them. But in [Lower] California they are all bitter, and that is why only beasts eat them. In Loreto and in one or another mission some sweet mesquites can be found, but these came originally from the area of the Yaquis, or from that coast. They were taken from there to [Lower] California either as seed or when they were very small and grew there fairly well. This transplant was made after the conquest had begun, because before there was not a single one which was not bitter." (Barco 1980:122; also see Felger 1977 and Hodgson 2001). Miguel del Barco and the translator obviously confused "seeds" for pods.

Arizona: Phoenix, New River wash, *Landrum 11393*.

Sonora: Cañón Bacatete, 17 Dec 1988, *Felger & Molina*, observation. Guaymas, 24 Oct 1939, *Gentry 4690*. Cerro Cruz de Piedra, east of Empalme, *Martin 28 Dec 1988*.

Prosopis velutina [*Prosopis juliflora* var. *velutina*]
Common names: HUU'UPA; *mesquite*; velvet mesquite

Arizona and Sonora. Velvet mesquite is distinguished by leaves with two pairs of pinnae, and the leaves and young stems are pubescent with small hairs. The leaflets are generally shorter than those of *Prosopis glandulosa*.

Velvet mesquite replaces honey mesquite across much southern Arizona, including the Tucson region, and in Sonora northward and inland from Guaymas.

Bwe'u Hu'upa (large mesquite) was a Yoeme settlement in Tucson on the west bank of the Santa Cruz River south of Irvington Road. The families moved to Pascua, Barrio Libre, and Marana.

Arizona: Silverbell Road, northwest of Tucson, 30 May 1945, *Gould 3186*. Tucson, *Toumey 20 Jun 1891*.

Sonora: Vicinity of Hermosillo, 4 Mar 1910, *Rose 12362* (NY).

Fabaceae, *Prosopis velutina*, Santa Cruz County, Arizona. 24 Jul 2020 (SC).

Rhynchosia precatoria

Common names: CHAMPUUSI (*cham*, a core word + *pusi*, eye), SANTA PUUSIM (holy eyes); *ojo de chanate*; precatory bean, rosary bean

Arizona (scarce in Santa Cruz County) and mostly found in Sonora. A robust vine with small, shiny, red-and-black seeds. It is common along canal hedgerows and the Río Yaqui. The name *santa puusim* refers to Santa Lucía, the patron saint of the eyes.

Ojo de chanate might be translated as grackle eyes; *chanate* is the Great-tailed Grackle or another icterid bird (Icteridae).

Medicine: As a remedy for headache (*punsadam vetchi'ivo* "for headaches," from Spanish "*punzada*," for sharp piercing pain), the ground seeds are placed on the forehead and temples over the aching area as a poultice. It is held in place in the usual manner employing a white cloth. (A white cloth, or white color signifies purity and healing.)

Music: The seeds are put in the *ayam* gourd rattles used by the deer dancer and the matachinis (see *Lagenaria*, Cucurbitaceae), as well as into *tenevoim* cocoon rattles (see Saturniidae, Lepidoptera).

Arizona: Tumacacori Mountains, north of Rock Corral Canyon, 17 Mar 2014, *Carnahan SC 406*.

Sonora: Cuesta Alta, *Felger 85-1413*. Potam, canal bank, *Felger 88-555*. Kompuertam, *Felger 89-77 & Molina*.

Fabaceae, *Rhynchosia precatoria*, Alamos. 15 Dec 2017 (SC).

Senegalia greggii [*Acacia greggii*]

Common names: HU'UPA KEKA'ALA (mesquite with mange, dog mange); *uña de gato*; catclaw acacia

Arizona and northern Sonora. Shrubs or small trees with an irregular trunk, and twigs with small sharp, recurved spines. Flowers cream-colored in spike-like racemes (inflorescences longer than wide). The wood is strong and hard, and the heartwood reddish.

The same name was given for *Condalia globosa* (Rhamnaceae).

Common in southern Arizona, especially along desert washes, and in Sonora mostly north of Hermosillo.

Music: If one cannot obtain *huchahko* (*Haematoxylum brasiletto*), the deer-dance rasper stick is made from straight branches of this acacia. In Arizona, during the first half of the twentieth century most musical raspers were made from this acacia.

Hunting and Weapons: The wood was used for arrow points.

Arizona: Silverbell Road, 6 mi northwest of Tucson, 30 May 1945, *Gould 3183.*

Senegalia occidentalis [*Acacia occidentalis*]
Common names: TESO; Mexican catclaw acacia

Sonora. Trees resembling *Senegalia greggii* but characteristically forming larger trees with a well-formed trunk, the leaves often somewhat larger, and the flowers are in rounded clusters (heads) rather than elongated spike-like racemes. The two species can difficult to distinguish.

This tree is widespread in southern and eastern Sonora, often along washes.

Masks: Yoeme and Yoreme men living in Hermosillo sometimes used the wood to carve pahko'ola masks. A small mask made for Roman Borbón as young boy in the 1940s was carved from wood obtained at a *teso* grove along the Río Sonora near Hermosillo.

Hunting and Weapons: Presumably the wood was used for arrow points, the same as for as with *Senegalia greggii*.

Sonora: San José de Guaymas, *Felger 85-491.* 3 km north of Totoitakuse'epo (Hill of the Rooster), 13 Mar 1989, *Felger*, observation. San Carlos, 20 Mar 1934, *Shreve 6550.*

Senna covesii [*Cassia covesii*]
Common names: KAU OHASEN (mountain *hojasen*); *daisillo*, *hojasen*; desert senna

Arizona and Sonora. Common herbaceous perennials with pale yellow flowers. Leaves with 2 to 4 pairs of leaflets.

Mostly on open, arid sites; scattered on desert plains, bajadas, and lower slopes.

Medicine: A tea made from the plant is consumed to alleviate problems associated with blood in the urine. This tea is also esteemed for controlling fever.

Arizona: Tucson, mesas, *28 Aug 1903, Thornber 190.*

Sonora: Ensenada San Francisco [San Carlos], *Dawson 6335.* San Carlos, *Felger 85-670.*

Sesbania herbacea
Common names: VAEKIO; *bequilla*; river-hemp

Sonora and also in Arizona, such as at Eloy and Marana in wet soil around irrigation. These giant annual herbs often reach 4 m tall in the Yoem Bwiara. Abundant in depressions between irrigated fields.

Construction: The plant was used as roofing for traditional houses, to cover the earthen layer on the roof.

Food: The seeds are ground and used like butter or lard: "*Se usa la semilla para manteca.*"

Household and utiliarian: The whole plant is ground to make an oily liquid that is washed on the hard-packed dirt floors of houses in Sonora to hold down the dust.

Arizona: Experimental Farm, Sacaton, 22 Aug 1928, *Kearney 5653.*

Sonora: Palo Parado, Río Yaqui, *Felger 85-1355.* Guaymas, *Gentry 4678.*

***Tamarindus indica**
Common names: TAMARINDO; *tamarindo*; tamarind

Sonora. Large trees native to the Old World, cultivated in tropics worldwide for shade, lumber, and especially the sweet pulp of the pods, which is used as a flavoring and food.

The trees are sometimes grown in southern Sonora.

Sonora: Las Guásimas, cultivated, 16 Dec 1988, *Felger & Molina*, observation.

Fabaceae, *Tamarindus indica*, El Tular, Miramar. 17 Feb 2018 (SC).

***Tara cacalaco** [*Caesalpinia cacalaco*]
Common name: *cascalote*

Large shrubs or small trees. Trunk and major branches with large, woody spines (unique among woody legumes in Sonora, and resembling those of *Ceiba aesculifolia*), the smaller branches with rose-like spines. Flowers bright yellow. Pods constricted between seeds. Native to Sinaloa and occasionally cultivated, such as at Pótam, and rarely escaped or scarcely naturalized in nearby roadside ditches.

Sonora: Pótam, cultivated as ornamental tree, several seen in town, 5–6 m tall, *Felger 85-1492*.

Vachellia campechiana [*Acacia cochliacantha*]
Common names: KOOWI TAMI (pig teeth); *güinolo*; boat-spine acacia

Sonora. Large shrubs or small trees with hard wood. The paired spines at the base of the leaves are mostly broad and boat-shaped. Flowers yellow-orange in globose heads. Pods thick and moderately flattened, and not constricted between seeds.

Common in the coastal thornscrub, also in foothills, and especially in disturbed habitats.

Kowireepa (pig earring) was listed by Johnson (1962:271) as "a very hard wood like mesquite, with what they make arrow points." This may be *Vachellia campechiana*.

Fabaceae, *Vachellia campechiana*, San Carlos. 21 Dec2 012 (SC).

Sonora: Northeast of Cuartel at Bacatete, 13 Mar 1989, *Felger & Molina*, observation. 7 mi northeast of Cajeme, 2 Mar 1934, *Shreve 6141*. 2.3 mi northwest of Guaymas, 13 Aug 1958, *Turner 155*.

Vachellia constricta [*Acacia constricta*]
Common names: CHUKUI KUTA (black wood); *mezquitillo*; white-thorn acacia

Arizona and Sonora, north of the Yoem Bwiara. Woody shrubs with straight, white spines, yellow flowers in globose heads, and slender pods (*taakam*) that are narrowed between the seeds. The twigs or branches are dark, and said to be even darker "inside."

Widespread in southern Arizona and in Sonora generally north of the Guaymas region. It was well known to Yoemem in Arizona before the 1970s from places such as the Avra Valley south of Marana. Felipe learned of this plant from his uncle, Juan Luis Garcia, who pointed one out in the Tucson Mountains. It is described as having taakam pechitam [*pechitas* or pods], "ousi huuva [*muy apestoso* or very stinky]" bark, and dark twigs or branches said to be even darker inside. It was also pointed out that *kuka* (*Vachellia farnesiana*) does not have dark stems.

Music: The wood is said to be good for making the scraper (*hirukia aso'ola*) for the musical rasper.

Arizona: Silverbell Road, northwest of Tucson, 30 May 1945, *Gould 3184*.

Sonora: 15 mi south of La Palma, between Guaymas and Hermosillo, *Wiggins & Rollins 235*.

Vachellia farnesiana [*Acacia farnesiana*]
Common names: KUKA; *huisache*, *vinorama*; sweet acacia

Arizona and Sonora. Shrubs or trees rarely more than 5 m tall. The flowers are bright yellow-orange, in globose heads, and very fragrant. It has huuva (*apestoso* or stinky) bark called *chukui kuta* (black wood) and small, thick pechitam (pods or *pechitas*).

Basilio (1890 [1634]:208) reported *cuca* as the word for *vinorama*.

Mostly in disturbed habitats; roadsides and along arroyos, washes, and canyons, also in hedgerows in agricultural areas. In 1990, we saw a sweet acacia in Pótam about 13 m tall with a well-developed, thick trunk. It was planted next to a house by a Mexican woman who once lived there.

Sweet acacia, widespread in the New World, has been introduced to many parts of the world and the flowers have been a staple in the perfume industry in France and elsewhere.

Medicine: Wagner (1936:82) reported a remedies for headache "is made from the bark of *cuhuca (huisache)*."

While visiting in Marana in 1987, Teresa Amarillas saw *kuka* and told us that as a remedy for hiccups, "you boil twelve flowers [meaning twelve of the flower heads] and drink the tea (*Dose seewam va'ampo vakne heoktiata vetchi'ivo*)."

Perfume: Mateo González said his father made perfume from *kuka* flowers. The flowers were put in alcohol in little bottles.

Arizona: Baboquivari Canyon, Baboquivari Mountains, 3 Nov 1931, *Peebles 8373.*

Sonora: Potam, tree 15 m tall, planted in yard, *Felger 88-553*. Rancho Bacatetito, 13 Mar 1989, *Felger & Molina*, observation. Guaymas, *Palmer 305 in 1887*.

***Vigna unguiculata** subsp. **unguiculata**

Common names: YOI MUUNIM, YORI MUUNIM (Mexican beans); *guisante negro de ojos*; black-eyed pea

Native to the Old World, grown in Arizona and Sonora.

Basilio (1890 [1634]:217) recorded *iorimuni* as the term for "*frijol buche negro*" (black crop bean), possibly meaning this or another cultivated bean.

Food: *Yoi muunim* (*yori muunim*) are cooked on New Year's Day to ensure a prosperous and healthy new year.

FAGACEAE - OAK FAMILY

Quercus emoryi

Common names: KOOWI VEYOOTAM (pig acorns), KUSIM (the acorns or *bellotas*); *bellota*; Emory oak

This tree is common in the mountains at elevations above the desert in southern Arizona and northern Sonora. *Kusi* is the term used in east-central and southeastern Sonora for another black oak, *Quercus albocincta*. The name *koowi veyootam* indicates the acorns were fed to pigs.

Food: The acorns are often sold on the streets in northern Sonora, and locally harvested in southern Arizona. Presumably Yoemem in parts of Sonora and Arizona made use of the tasty acorns.

Arizona: Santa Catalina Mountains, Sabino Canyon, 4 Jun 1916, *Thornber 8199.*

Sonora: 7 mi south of Nogales, *Parker 1 Jan 1953.*

FOUQUIERIACEAE - OCOTILLO FAMILY

The three ocotillo species in Sonora and Arizona are called *mureo* and used in similar manners.

Fouquieria diguetii

Common names: MUREO; *palo adán*; Adam's tree

Sonora. Large shrubs or small trees with red-orange flowers. Inflorescences relatively compact, usually longer than wide; the flower stalks

Fouquieriaceae. A. *Fouquieria macdougalii*, Cañón del Nacapule. 16 Dec 2012 (SC). B. *Fouquieria diguetii*, San Carlos. 30 Nov 2015 (SC). C. *Fouquieria splendens*, west of Sierra del Águila. 19 Mar 2017 (SR).

(pedicels) 2 to 6 (13) mm long.

Widespread throughout the lowlands of the Yoem Bwiara, along the coast, on rocky slopes, desert flats, and alluvial plains.

This ocotillo resembles *Fouquieria macdougalii*, but it is not as tree-like, generally has smaller leaves, smaller flower clusters (inflorescences), and shorter flowers and flower stalks. It is used in the same way as *F. macdougalii*. *Fouquieria diguetii* somewhat resembles *F. splendens* but is more shrub- or tree-like and the flowers are darker red.

Sonora: North-northeast of Las Guásimas, 16 Dec 1988, *Felger & Molina*, observation. Guaymas, Isla Almagre Chico, 27 Sep 2006, *Suarez-Gracida 2006-113-CGSG*. Guaymas, Île d'Ardilla, Jun 1866, *Thiébaut 1105* (P). S of Peón, *Turner 61-34*.

Fouquieria macdougalii

Common names: MUREO; *ocotillo macho*; tree ocotillo

Sonora. Large shrubs or small trees with a thickened, twisted trunk and lower limbs. The bark is often waxy yellow-brown and peeling in dry seasons. The spiny branches bear small leaves and the flowers are bright red. Inflorescences relatively loose and open, usually as wide as or wider than long, the flower stalks (pedicels) are (3) 5 to 30 mm long; the flowers are bright red.

Thornscrub and desert; plains, hills and mountains throughout the Yoem Bwiara. Usually a shrub in the desert and larger and more tree-like in the southern thornscrub and tropical deciduous forest.

Construction: The branches are used for fencing and walls for the traditional wattle and daub house.

Medicine: Fluid or nectar (*kauwo*, milk) from the flowers is squeezed into the eyes as eyedrops for allergies, infections, and pink-eye. As a remedy to heal sores the bark is left to soak in water and the sore then bathed with the resulting water. Liquid from bark soaked in water is also used as a remedy for dandruff.

Washing: Bark mashed in water produces foam and has been used as laundry soap.

Sonora: Road to top (Cerro El Vigía), N of Guaymas, 11 Jan 1982, *Daniel 1954* (ASU). Cañón Bacatete, 17 Dec 1988, *Felger*, observation. South of Peón, *Turner 61-33*.

Fouquieria splendens

Common names: MUREO; *ocotillo*; ocotillo

Arizona. Shrubs with wand-like, mostly long, straight, spiny branches arising from an essentially trunkless base. The small leaves are produced after rains. Red-orange flowers appear at the stem tips in spring.

Widespread in the Sonoran Desert. The southern limits in Sonora are on the exposed slopes and ridges in the Sierra El Aguaje north of San Carlos. It is abundant in southern Arizona, and the cultural uses are for southern Arizona.

Construction: The straight branches were used for walls in the traditional wattle and daub house. The branches are also used in roofing for ramadas and as fencing. When ocotillo canes are planted in the ground, the stems form roots and grow into a new plant, forming a living fence.

Medicine: This ocotillo is used in the treatment of dandruff, similar to that described for *Fouquieria macdougalii*.

Arizona: Tucson Mountains, 18 Apr 1947, King Canyon, *Kurtz 5*.

Sonora: Cañón Nacapule, *Felger 85-542*.

GERANIACEAE – GERANIUM FAMILY

*Erodium cicutarium

Common names: ALFIDILLO; *alfilerillo*; filaree, stork's bill

Arizona. Common, cool-season annuals with rosettes of leaves that are hairy and somewhat fern-like. The flowers are small and lavender-pink. The plants are described as *bwiapo naikim sisimeme* (on-the-ground spreads one-that-goes, or one that spreads on the ground). Filaree is common in southern Arizona. It is native to Europe and has been in North America for centuries. It is valued as forage for goats.

Arizona: Silverbell road at north end of Tucson Mountains, 25 Mar 1980, *McLaughlin 2441*. Tucson, 27 Mar 1917, *Shreve 133*.

GRAMINEAE, see POACEAE

HELIOTROPIACEAE – HELIOTROPE FAMILY

The plants in this family were formerly included in Boraginaceae.

Heliotropium angiospermum
Common name: VAA KO'OKO'I (water chile)

Sonora. Herbaceous to semi-woody or shrubby perennials to about 1.5 m tall. The leaves are broad, thin, and hairy; the flowers are white and 1 to 1.5 mm wide.

Mostly along streams and riverbanks including the lower *Hiak Vatwe* (Río Yaqui).

Sonora: Palo Parado, Río Yaqui, *Felger 85-1401*. 2 km southeast of Torim, *Felger 88-601*. Stream at Rancho Bacatetito, *Felger 89-157 & Molina*.

Heliotropium curassivicum [*H. curassavicum* var. *oculatum*]
Common name: Alkali heliotrope

Arizona and Sonora. Perennial herbs and annuals; Plants glabrous, semi-succulent, bluish glaucous. Flowers white with a yellow center becoming purple with age. Alkaline soils, brackish water at estero margins, and an urban and agricultural weed.

Sonora: Cuesta Alta, Río Yaqui, *Felger 85-1375B*. Guaymas, 1 Apr 1897, *Maltby 191*. Bahía San Carlos, *Urry 1349*.

KRAMERIACEAE – RATANY FAMILY

Krameria
Common names: KOSAWE (Sonora), KOSAWI (Arizona); *cósahui*; ratany

Three species occur in the Yoeme regions of Arizona and Sonora. The fruits are small, spiny burs. They are all used in the same manner medicinally. *Krameria erecta*, because of its smaller size and smaller stems and roots, was probably seldom, or less often, utilized.

Krameria bicolor [*Krameria grayi*]
Common names: desert ratany, white ratany

Arizona and Sonora, north of the Yoem Bwiara. Sprawling, often leafless or nearly leafless shrubs with many straight, slender, spinescent twigs, and magenta-purple flowers. The wood is red-brown and rather hard, and the roots rather thick and also red-brown. This species is common in the desert in southern Arizona and northern Sonora.

Dye: In Arizona the roots are used for a reddish-brown dye, augmented with roots of *Calliandra eriophylla* (Fabaceae), and used with roots of *Jatropha cardiophylla* (Euphorbiaceae).

Arizona: Tucson Mountains, 23 Apr 1903, *Thornber 927*.

Krameria erecta [*Krameria parvifolia*]
Common name: range ratany

Arizona and Sonora. Dwarf shrubs with tough, knotty stems, small leaves, and magenta-purple flowers.

Rocky hills and plains with open, sparse desertscrub.

Arizona: Saguaro National Park, Tucson Mountain District, King Canyon Trail, 22 Nov 1988, *Rondeau 88-104*.

Sonora: 2.5 km north-northwest of Torocobampo, Sierra Bacatete, 14 Dec 1988, *Felger & Molina*, observation. E of Empalme, 5 Sep 1974, *Gentry 23439*.

Krameria sonorae
Common names: KOSAWE; Sonoran ratany

Sonora. Openly-branched shrubs usually 1 to 2 m tall, with slender, mostly smooth and nearly leafless stems, and pale pink flowers.

Beverage: A pleasant-tasting tea is made from the larger stems as well as, or especially, from the roots. It is often mixed with *canela* (cinnamon). Armando Felix of Arizona said it is better than coffee, and that he grew up drinking this tea and did not drink coffee.

Krameriaceae, *Krameria sonorae*, San Carlos. 30 Nov 2014 (SC).

Medicine: *Kosawe* is an important medicinal plant. The roots or stems (or any part of the plant) are made into tea as a remedy for iron-poor blood (*mal de sangre*). One may take it for one or two months, or as long as desired. It is consumed as one does tea or coffee. Kosawe tea, best made from the green (fresh) plant, is considered a good remedy for coughs. Teresa Amarillas told us that the stem only is used as a remedy for cough, and that kosawe also is used to alleviate fevers. Carmen Liowe Garcia said that kosawe and cinnamon together help prevent diabetes.

Sonora: 3.5 mi by road northeast of Mex Hwy 15 road junction to Torim, *Felger 88-582*. East of Las Guásimas, *Felger 88-608*.

LAMIACEAE (LABIATAE) – MINT FAMILY

Condea albida [*Hyptis albida, H. emoryi*]

Common names: VIVINO; *salvia*; desert lavender

Arizona and Sonora. Abundant and widespread desert shrubs often 1.5 to 3 m tall. The foliage is grayish-white, and the leaves and flowers are fragrant. The flowers are small, lavender-blue, and can occur any time of the year. It often grows along larger washes as well as rocky slopes.

Basilio (1890 [1634]:205) recorded *bibino* as the word for *salvia*, probably referring to this species.

Household and medicine: Bundles of this plant are used to freshen the air in the house because of its *winhuva*, sweet fragrance. When in the mountains or out in the country, people often collect bundles of this plant to take back to their homes to be used for the fragrance and for medicinal purposes.

Lamiaceae, *Condea albida*, San Carlos. 21 Dec 2012 (SC).

The plant is used to make a tea to alleviate fever and for *heka'uk* ("*viento que pega*"; see *Nicotiana glauca*, Solanaceae). As a remedy for insomnia, a handful of leaves are cooked in water and the liquid or tea consumed. It is said to make one very calm ("*hace muy mancito*"). Edward Palmer reported in 1887 that it was used as a remedy for rheumatism (Watson 1889).

Arizona: Silverbell Mountains, 23 Feb 2001, *McLaughlin 8900*.

Sonora: Cañón Bacatete (Vacateve), Sierra Bacatete, *Felger 88-623*. 8 mi by road northeast of Vikam Suichi, *Felger 89-91 & Molina*.

*Mentha

Common names: YERVA BWENA, YERVA WENA; *hierba buena* (good herb); common mint

Arizona and Sonora. Several kinds of mint are cultivated in home gardens, especially peppermint (*Mentha piperita*) and spearmint (*M. spicata*). These well-known herbs are esteemed for their highly aromatic, pleasant-smelling leaves. The leaves are chewed to freshen the breath after eating.

Medicine: Tea made from mint is said to be "good for everything (*sirve para todo*)," including headache, fever, and stomach cleansing. The tea is consumed after it colors the water.

Yerva bwena, taken as a tea, has been used in treatment of infants and children, and is said to be good for stomach cleansing of infants. This tea was given as a preventative to children as a cool beverage during summer. It is boiled and cooled, then taken as a tea to alleviate croup, to cleanse the stomach of acid reflux, for heartburn, or to alleviate gases from eating or drinking the wrong foods.

Sonora: Kompuertam, home garden, 13 Mar 1989, *Felger & Molina*, observation.

*Ocimum basilicum

Common names: ALVAAKA; *albahaca*; sweet basil

Arizona and Sonora. This Old World native is often cultivated in gardens for the aromatic leaves, which are used as seasoning and for medicinal purposes.

At Marana in January 1994, Meregilda Ochoa and Luis Ochoa described this plant and called it *alkanflor*.

Beliefs: This plant is used to ward off negative spirits. Just growing it in the yard, near the house; it provides protection from negative power because it is a powerful plant.

Medicine: A small ball or wad of the leaves is placed in the ear together with *ruura* (*Ruta graveolens*, Rutaceae) to alleviate an earache. One person said, "it is a garden plant, 2 to 3 feet tall, smells like Vicks Vapor or mint, and is used for earaches and other pains."

Sonora: Las Guásimas, cultivated in garden, 16 Dec 1988, *Felger & Molina*, observation. Potam, Oct 1995, *David Shorter*, photo.

Salvia columbariae

Common names: CHIIYA; *chia*; chia

Arizona. A winter–spring annual with basal leaves and small blue flowers in dense clusters. It has been used and named in the same manner as S. *hispanica*.

Arizona: Saguaro National Park, Tucson Mountain District, Picture Rocks Pass, 6 Apr 1982, *Starr 54*.

*Salvia hispanica

Common names: CHIIYA; *chia*; chia

This medicinal herb is purchased at Mexican herbal shops and health food stores.

Medicine: The seeds, placed in water, soon become gelatinous. These soaked seeds are used for stomach problems and also as an energy drink.

Vitex mollis

Common names: HUVAHE (spider), KAU HUVAHE (mountain spider); *uvalama*

Sonora. Large shrubs or small trees, rarely large trees in riparian habitats. Leaves divided into 3 to 5 large leaflets. The flowers are violet and white and have a far-reaching, pungent fragrance. The fruits are 1.5 to 2 cm wide, rounded, with a single seed surrounded by soft, fleshy pulp that turns blackish when mature. The fruits ripen from June through November.

Basilio (1890 [1634]:215) listed "*húbari*" as "*obalama, árbol.*" *Húbari* is probably a Yoreme term comparable to *huvahe* and *obalama* seems comparable to *uvalama*.

Found in canyons and arroyos, mostly in the mountains and foothills in the Yoem Bwiara, especially in the Sierra Bacatete, and is sometimes planted in home gardens.

Beverage: The fruit is made into a beverage and drunk as a coffee substitute.

Food: The fruits are harvested in the wild as well as from cultivated trees. The fruits, which are sometimes sold in local markets in Sonora, are eaten fresh and also cooked with *panela* (brown sugar). Some fruits are sweet and some are spicy, even on the same plant. The seeds are not eaten. The fleshy, soft fruit pulp (mesocarp) has the color, consistency and taste of cooked prunes. Bernaldo Valencia said the male trees have spicy fruits and the female trees have sweet fruits. (The flowers have both male and female parts and, as far as we know, are not unisexual.)

Gardens: Huvahe is sometimes cultivated, such as in Pótam, as an ornamental shade tree and for the fruit.

Harvesting: When discussing the saguaro we were told that the same kind of fruit-picking pole was used for the *huvahe* tree (see *Carnegiea gigantea*, Cactaceae).

Medicine: The fruit was boiled and the tea used as a remedy for fever. A decoction of the fruit and leaves has been used as a remedy for diarrhea.

Sonora: Kompuertam, *Felger 89-75 & Molina*. San Carlos, 9 Jan 2015, *Carnahan SC 921*. 3.4 mi (by road) northeast of Ortíz, *Turner 69-49*.

Lamiaceae, *Vitex mollis*, San Carlos. 10 Mar 2017 (SC).

LAURACEAE – BAY FAMILY

***Cinnamomum verum**

Common names: *canela*, cinnamon

Food: Cinnamon may be added to wheat *atole* at a *pahko* or baptism. It is also put in chocolate drinks at weddings.

Medicine: This spice is added to a number of herbal remedies, for example see *Bebbia juncea* (Asteraceae), *Bursera fagaroides* (Burseraceae), *Krameria sonorae* (Krameriaceae), and *Larrea tridentata* (Zygophyllaceae).

LEGUMINOSAE, see FABACEAE

LINACEAE – FLAX FAMILY

***Linum usitatissimum**

Common names: *lino*; common flax, flax, linseed

Annuals with blue flowers, the stem fibers are used to make linen. A single population was found on the coastal plain in the Yoem Bwiara. Native to the Old World.

Weaving: Cultivation of flax was established by Jesuits in early colonial times for weaving on newly introduced Spanish looms for cotton and wool. Hu-DeHart (2018:100, 102) reported that around 1784, the Franciscan Padre Valdéz, "to increase both the quantity and quality of raw materials for the new weaving industry, taught Yaquis how to cultivate the *lino*, or flax seed." The early flax cultivation apparently did not persist.

Sonora: 26 mi S of Empalme along Hwy 15, just north of Pitahaya, low, hot, coastal flats, dried, cracked clay of a seasonal pond near the road, surrounded by thorn forest with *Prosopis*, *Parkinsonia*, *Opuntia*, *Lemaireocereus*, etc., bed of pond with *Sida hederacea*, *Cressa truxillensis*, etc., *Sanders 2 Apr 1982* (UCR).

LORANTHACEAE – SHOWY MISTLETOE FAMILY

Another mistletoe, *Phoradendron californicum*, belongs to the family Santalaceae (including Viscaceae).

Psittacanthus calyculatus

Common names: CHICHIHAM; *toji*

Sonora. This large, tropical mistletoe has broad, dark green, and leathery leaves, bright orange flowers, and the ripe fruits are bright red. It grows on cottonwood (*Populus mexicana*) and willow (*Salix gooddingii*, Salicaceae) trees along the Río Yaqui.

Sonora: Cuesta Alta, *Felger 85-1358*. Yaqui River crossing near Corral, 2 Mar 1933, *Shreve 6136*.

Struthanthus palmeri

Common names: CHICHIHAM; *toji*

Sonora. This mistletoe grows on many different host trees, especially mesquite (*Prosopis*, Fabaceae). The stems hang down in loose, leafy spirals. The flowers small and white, and the fruit fleshy and bright red. It is common throughout the Yoem Bwiara.

Loranthaceae, *Struthanthus palmeri*, San Carlos. 10 Mar 2015 (SC).

Edward Palmer reported these plants "growing on fig, lime, mesquite, and other trees" near Guaymas in 1887 (Watson 1889:73). The Sonoran plants, previously known as *Struthanthus haenkeanus*, are now known as *S. palmeri*.

Sonora: Cuesta Alta, *Felger 85-1412*. Near the Cuartel, Sierra Bacatete, *Felger 88-628*. Guaymas, Jun 1887, *Palmer 199 in 1887* (NY). South of Peón, *Turner 61-29*.

LYTHRACEAE – LOOSESTRIFE FAMILY (includes Punicaceae)

***Punica granatum**

Common names: KANNAO; *granada*; pomegranate

Arizona and Sonora. Small trees cultivated in home gardens for the edible fruit. Pomegranates, native to the Old World, were brought to Mexico in early Spanish colonial times.

Sonora: Las Guásimas, 16 Dec 1988, *Felger & Molina*, observation. Kompuertam, 14 Mar 1989, *Felger & Molina*, observation.

Lythraceae, *Punica granatum*, Maricopa County, Arizona. 2 Jun 2015 (SR).

MALPIGHIACEAE – ACEROLA FAMILY

Malpighia emarginata [*Malpighia umbellata*]

Common names: KAU SIKRO'OPO (*kau*, mountain + *sikro'opo*, the word for *Lycium*, Solanaceae); *granadilla*

Sonora. Shrubs 1 to 2 m tall, with rigid, woody branches, gray bark, and dark green leaves. Flowers pink. The ripe fruits are red and fleshy. *Malpighia* fruits are rich in vitamin C.

There is one record for it in the Guaymas region and we found it on volcanic hills near the Río Yaqui in the Yoem Bwiara. This species is more common eastward and farther south in Sonora and ranges southward in western Mexico.

Food: The fruits, eaten fresh, are juicy and sweet with a plum-like flavor.

Sonora: Kuubwa'e Kawi, a lava rock hill, 1 km west of Kompuertam, *Felger 89-81 & Molina*.

MALVACEAE – MALLOW FAMILY

(includes Bombacaceae and Sterculiaceae)

Current botanical information places the Kapok Family (Bombacaceae, including *Ceiba*) and the Cacao Family (Sterculiaceae, including *Guazuma*) in the Mallow Family.

Abutilon incanum

Common name: TOOKO HUYA, TOROKO HUYA (gray or light blue plant)

Arizona and Sonora. Shrubs with slender branches and velvety hairs. Leaves usually gray-green. Flowers yellow-orange with large maroon spots at the center.

Widespread, especially along drainageways and on brushy hillsides.

Household: The stems are tied together to make brooms.

Arizona: Tucson, West Branch of the Santa Cruz River, 18 Aug 2001, *Mauz 21-111*.

Sonora: 8 mi by road northeast of Vikam Suichi, *Felger 89-85 & Molina*. Guaymas, *Jones 26 Jan 1927* (RSA).

Abutilon palmeri

Common names: RIPTIA, TOOKO HUYA, TOROKO HUYA (gray or light blue plant)

Riptia is from *ripte*, meaning blind, due to dust or sickness, such as "pink eye." Likewise, the usual Spanish language name is "*mal de ojo*" for various mallows (see *Sphaeralcea*).

Southwestern Arizona and Sonora. Sparsely-branched shrubs or subshrubs. Densely hairy, the leaves velvety. Flowers pale orange, on branches raised well above the foliage.

It is common in the Yoem Bwiara, especially in low-lying places, roadside depressions, drainageways in the coastal plain, and upper bajadas.

Arizona: Pinal County, Picacho Mountains, 14 Mar 1992, *Wiens 92-114*.

Sonora: Potam Viejo, *Felger 85-1434*. East of Chiinim, *Felger 88-571*.

Ceiba aesculifolia subsp. **aesculifolia** [*Ceiba acuminata*]

Common names: POCHOOTE, VAUWO; *pochote*; kapok tree, silk-cotton tree

Sonora. Distinctive trees of subtropical areas. The trunks are studded with large, conical, woody thorns. The leaves, usually present only during the summer rainy season, have 5 to 8 elongated leaflets arranged like fingers on a hand. The large, whitish flowers are primarily bat-pollinated and appear on the leafless twigs in May and June. The fruits are large, green, and slightly fleshy before ripening, and open explosively at maturity to release seeds embedded in fluffy kapok fiber. These trees develop many, very large, tuberous roots, which are tough, fibrous, and very succulent.

Vauwo is a core word, although the first part of the word seems to be derived from *vaa'am* (water) and *vau* can also mean "to the water."

Growing in mountains along the northeastern side of the Yoem Bwiara and in tropical deciduous forest in eastern and southern Sonora.

Malvaceae, *Ceiba aesculifolia*, Álamos. 1 May 2014 (SC).

Food and emergency water: The young and tender green fruit is said to be edible fresh. The young tuberous roots were used for emergency food and are described by elders as being rather like potatoes: "Eat the roots when there is nothing else to eat—when you are hungry you eat anything. [The roots are] peeled and sliced like a cucumber. Eat the young root, not when it is full size" (Bernaldo Valencia, Tucson, 1996). The oily seeds are also edible.

After digging out the root, one would just suck on the raw root to obtain the watery juice. During the Mexican–Yaqui wars, Bernaldo Valencia lived with his elders in the Sierra Bacatete. In 1996, in Tucson, he told us that he grew up drinking liquid from the roots of this tree when water was otherwise unavailable:

> It has roots like a potato. Use roots about this big [he indicates about 25 cm long and about 6 cm wide]. The little plants, small trees, do not have tuberous roots, only the bigger trees. Use a knife or machete to dig them, about 60 cm deep, the ground is not too hard when the tree is growing [summer rainy season]. It is moving the earth so it is not too hard to dig. Scrape the root, and mash with a rock but not too hard, and suck it like a watermelon. It was used during the wars. You can get enough water for all day—there is a lot [of it] over there [in Sonora]. The new trees have a lot of water, not the old ones, the young tender tree, but not real small ones. The old trees don't have much water. You can get this water any time of year, whenever you are there.

Sonora: 2.5 km north-northwest of Torocobampo, Sierra Bacatete, *Felger 88-586 & Molina*.

***Ceiba pentandra**

Common names: *ceiba*; kapok tree

Sonora. *Ceiba pentandra* is an enormous tree cultivated in tropical climates worldwide. It is grown in the Yoem Bwiara and Guaymas, and elsewhere in southern Sonora as far north as Hermosillo.

Sonora: Tórim, 1 Jun 2019, *Blytsone*, photo.

Malvaceae, *Ceiba pentandra*, Hermosillo. 2 Jun 2019 (PB).

***Gossypium hirsutum**

Common names: CHIINI; *algodón*; cotton

Arizona and Sonora. Shrubby plants with yellow flowers, growing as annuals in cultivation. A seasonal fishing camp at Estero Algodones, on the coast west of Ráhum is called *Chiinim*. *Chin sookam* is a cotton plant from a previous year or years.

Basilio (1890 [1634]:134) indicated the plural *chinim* as the typical word in use, noting, "*Comunmente se expresan en esta lengua por el plural los nombres indeterminados.*"

Malvaceae, *Gossypium hirsutum*, north of Guaymas. 18 Nov 2018 (SC).

Cultivation, clothing, and weaving

Cotton was grown in the Yoem Bwiara in pre-contact times. Pérez de Ribas (1645; Reff et al. 1999:92) documented the native use cotton:

> The women also practiced the art of spinning and weaving cotton...They made mantas from it, but they did not use looms...Instead, these mantas were woven by means of very laborious handiwork, placing stakes in the ground for use in producing cloth...Women covered their bodies from the waist down with a cotton manta.

He also reported, "Now...they weave large mantas, for the Yaqui women are great weavers" (Reff et al. 1999:329).

The Yoemem used horizontal looms instead of vertical looms like the Navajo use, said Felipe Molina. The Yoreme still use the horizontal looms.

In 1902, following and amidst decades of upheaval in Yoeme culture, Hrdlička (1904:65) observed, "The manufacture of cotton and woolen fabrics has greatly declined. The only clothing of native weave now to be seen among the Yaquis is the faja and the white serape, the latter ornamented with one or two broad stripes in pale blue and natural brown or black; but even these garments are scarce."

In the 1930s, Studhalter (1936:120) was surprised to find "an absence of cotton," due to the recent warfare. Later in the twentieth century, cotton became an important commercial crop in the Yoem Bwiara and across much of the rest of Sonora and Arizona. In the late twentieth century, however, cotton production became less important in the Yoem Bwiara. In Pótam in December 1988, Guillermo Amarillas said, "*algodón anda aqui pero no mucho, el banco no nos da buen ayudo, falta ayudo* (there is cotton here but not much, the bank does not give us much help, they fail to help)." As usual, the bank dictated agricultural policy.

During the twentieth century, until around the 1970s, many Yoemem in Arizona worked in the cotton fields and industry. *Chin pua* (picking cotton) and *chin saako* (a bag cotton is put in) are among the terms developed in the cotton fields. In the early twentieth century, entire families including children worked in the fields. Although it was hard work, it was a far better life than in the war-torn homeland in Sonora or slavery in the Yucatán henequen fields before 1910.

Guazuma ulmifolia

Common names: AIYA; *guásima*; picklenut

Sonora. A medium-sized tree with a stout, well-developed trunk, and soft wood. The fruits are 2.5 to 4.5 cm long, oval, dark brown, cone-like, hard and woody, and covered with short, blunt prickles.

Often common along the larger washes and the banks of the Río Yaqui, but much more common to the south and east of the Yoeme lands. The northernmost Yoeme town in Sonora is called

Waasimam or Las Guásimas, where these trees grow in a large wash near the town. Basilio (1890 [1634]:203) reported *ahia* as the word for *guásima*.

Food: The fresh, tender green fruits become mucilaginous when chewed and can relieve thirst. The flavor is bland and not sweet. The mature fruits and seeds were ground and used to make tortillas and *atole*, or sometimes roasted and ground as a coffee substitute.

Malvaceae, *Guazuma ulmifolia*, Álamos. 17 Dec 2017 (SC).

Masks: Some artists carved pahko'ola masks from *aiya* wood.

Medicine: The bark is used as a remedy for kidney ailments.

Music: The wood is used for the frame or rim of the *tampaleo* drum. Aiya is the preferred wood for drum making, but it is becoming increasingly difficult to obtain. Tampaleo drums were made in Arizona, but by the end of the twentieth century the young people no longer knew how to make them, so they were purchased from Sonoran Yoemem.

Sonora: Las Guásimas, *Felger 85-1160*. Palo Parado, *Felger 85-1407*. Cañón Bacatete (Vacateve), Sierra Bacatete, *Felger 88-634 & Molina*.

Hibiscus tiliaceus, *see* **Talipariti tiliaceum**

*Hibiscus rosa-sinensis

Common names: Chinese hibiscus, hibiscus

Arizona and Sonora. Non-native shrubs grown in gardens in the Yoem Bwiara for their large, showy flowers; also sometimes in southern Arizona in areas with little danger of frost.

Arizona: Tucson, University of Arizona campus, 21 Jul 1982, *Starr C44*.

Sonora: Near Kompuertam, 11 Mar 1989, *Felger*, observation.

*Malva parviflora

Common names: MALVA; *malva*; cheeseweed

Arizona and Sonora. Winter–spring annuals with nearly circular leaf blades and small, white flowers.

In Arizona and the Yoem Bwiara it is a common garden and agricultural weed, and also occurs along arroyos and canyons and in low-lying places such as cattle ponds and floodplains; generally not common among natural vegetation. Native to Eurasia and naturalized around the world.

Medicine: The plants are boiled, and the liquid used as an enema as a remedy for stomachache and constipation. The liquid is also used as a refreshing bath.

Arizona: Tucson, *Toumey 1 May 1892*.

Sonora: San Carlos, *Felger 85-407*. Near Kompuertam, 11 Mar 1989, *Felger*, observation.

Sphaeralcea coulteri

Common names: HEOKO KUTA (*heoko*, slimy or mushy + *kuta*, wood), SEVOA'ARA; *mal de ojo*; annual globe-mallow

Arizona and Sonora. Widespread spring wildflower, sometimes carpeting the desert in fields of orange.

Abundant on sandy or fine-textured and poorly-drained soils, arroyo margins, plains and slopes, and often in disturbed, weedy habitats.

Arizona: Saguaro National Park, Tucson Mountains, west of Sandario Road, 1 Apr 1990, *Bertelsen 90-74*.

Sonora: San Carlos, *Felger 85-325*. Kompuertam, *Felger 89-74 & Molina*. 8 mi by road northeast of Vikam Suichi, *Felger 89-89 & Molina*.

Sphaeralcea emoryi

Common names: HEOKO KUTA (slimy wood), OCHOKO KUTA (greasy stick/wood), TOOKO HUYA, TOROKO HUYA (gray or light blue plant), VUU NAKA (*vuuru*, burro/donkey + *naka*, ear); *mal de ojo*; globe mallow

Arizona and Sonora. Bushy, short-lived perennials or annuals with reddish-orange flowers.

Common in southern Arizona and northern Sonora, and apparently not common in the Yoem Bwiara.

Medicine: The root is mashed and put in boiling water, and then allowed to simmer. After it has cooled, the liquid is consumed to alleviate colic, and also to treat diarrhea.

Arizona: Tucson, Tumamoc Hill, 4 Mar 1983, *Bowers 2586*.

Sonora: San Carlos, *Felger 85-689*. Southeast of Potam, *Felger 88-590*. 12 mi south of Ortíz, *Shreve 27 Nov 1933*.

Sphaeralcea

A medicinal herb called *negrita* that is sold in Mexican herbal shops has been identified by Moore (1990) and others as one or more species of globe mallow (see *Sambucus cerulea*, Viburnaceae).

Sphaeralcea

Common names: RIPTIA (see *Abutilon palmeri*); *mal de ojo*; globe mallow

*Talipariti tiliaceum var. tiliaceum [*Hibiscus tiliaceus*]

Common names: CHIINI (cotton); *majagua del mar*; sea hibiscus, wild cotton tree

Sonora. A small tree with thick, woody trunks and limbs. Foliage dark green, forming a rounded crown of rather large, shiny green leaves. Flowers large and bright yellow with a red center, fading orange and sometimes red.

As the Yoeme name indicates, it somewhat resembles a very large cotton plant. It is cultivated in tropical, mostly coastal, regions around the world, and is native to the Old World. It is grown in yards for shade in southern Sonora.

Gardens: A cultivated tree in the Amarillas' yard in Pótam is said to have been brought from "the south" in Mexico. Don Nacho Amarillas called it *chiini*. We also saw it planted at a few other places in the Yoem Bwiara.

Sonora: Bahía San Carlos, sprawling tree 4.5 m tall × 6 m wide, 14 Feb 2017, *Carnahan 2258*. Potam, Oct 1995, *David Shorter*, photo.

Malvaceae, *Talipariti tiliaceum*, San Carlos. 14 Feb 2017 (SC).

MARTYNIACEAE – DEVIL'S CLAW FAMILY

Three species in two genera feature in Yoeme culture. These plants grow during warm or hot weather, especially summer rains. The plants vary considerably in size and some may become quite large. The leaves and stems are somewhat fleshy, and the surfaces are viscid-sticky due to gooey, glandular hairs. The fruits develop into unusual woody capsules.

*Martynia annua

Common names: TAMKO'OKOCHI (*tam*, teeth + *ko'oko*, pain or hurt + *chi*, on; painful teeth); *uña de gato*, *aguaro*; cat's claw.

This large annual plant produces a strange woody fruit 2 to 3.5 cm long, with a pair of grasping (recurved) hooks 3 to 5 mm long at the distal end.

Martynia annua occurs south of the Yoem Bwiara and is often common in the Yoreme regions of southern Sonora. It is occasionally cultivated in Arizona and probably in Sonora.

The fruits are generally purchased from herbal stores in Arizona and Sonora, and in Sonora also at local *mercados* (markets). Felipe tells us, "Native Seeds/SEARCH sold *tamko'okochi* when I worked there in the late 1990s."

Beliefs: The fruit is used as a necklace pendant to ward off negative energy. For example, in the 1990s, a Tucson artist made glass bead necklaces that had one of the capsules (he called it *tamko'okochi*) attached to the center as a pendant. He filed off the small claws, because they would catch on clothing. He said these necklaces were worn for protection from witches and curses.

Medicine: The fruits have been used in a remedy to control diabetes. Wagner (1936:83) reported, "For hiccough a tea made from the seed of the *torito* plant is used. The seeds from one pod are ground and steeped in water. The patient sips the liquid until he gets relief. The torito plant grows south of Sonora in Sinaloa." Torito in this case is probably *Martynia*.

Sonora: San Bernardino, Río Mayo, *Arguelles 18 Sep 1958* (DES).

Martyniaceae, *Martynia annua*, east of Álamos. 15 Dec 2017 (SC).

Proboscidea altheifolia [*Proboscidea arenaria*]
Common names: AWA'ARO (has horns); *cuernitos, gato, torito, uña de gato*; desert unicorn-plant, golden devil's claw

Arizona and Sonora. Herbaceous perennials from a thick, tuberous root. The plants often begin growing and flowering after the weather turns warm in mid-spring. The flowers are yellow and the distinctive devil's claw capsules split apart upon drying to form two long, curved hooks. The devil's claw capsules are often found long after the plants are dead and dry. Usage is similar to that of *Proboscidea parviflora*.

Common in dry plains and washes in southern Arizona and also in the Yoem Bwiara.

Food: The seeds are edible in the same way as those of *Proboscidea parviflora*. Felipe tells us, "we used to eat the seeds of the *awa'aro* in Marana in the 1960s."

Arizona: Avra Valley, 30 mi west of Tucson, 23 Jul 1953, *Parker 8293*.

Sonora: Estero Soldado, sand soil of strand, *Felger 85-879A*. Cañón Bacatete, 17 Dec 1988, *Felger & Molina*, observation.

Proboscidea parviflora subsp. **sinaloensis**
Common names: AU'ORI; *aguaro, cuernitos, torito, uña de gato*; devil's claw, unicorn-plant

Arizona and Sonora. Annuals, the plants sometimes becoming large. The flowers are lavender with purple, white, and yellow markings. The fruits are similar to those of *Proboscidea altheifolia* although often larger and generally with longer claws.

This devil's claw grows with summer rains and is often found in brushy areas and sandy-gravelly or clay soils along larger arroyos and washes. These are wild and often weedy plants.

Martyniaceae, *Proboscidea parviflora*, San Carlos. 10 Oct 2016 (SC).

Food: The fresh seeds are peeled and the white inside portion is eaten, mostly as a snack. The seeds are said to be oily or like lard. The young, tender fruits are sometimes cooked as a vegetable.

Arizona: West Branch of the Santa Cruz River, floodplain, 7 Aug 2001, *Mauz 21-106*.

Sonora: San Carlos, 10 Oct 2016, *Carnahan 2059*. Canyon 2 mi northwest of San Carlos Bay, 3 Sep 1989, *Sanders 9185* (UCR).

MELIACEAE – CHINABERRY FAMILY

*Melia azedarach

Common names: PIOCHA; *piocha*; chinaberry tree

Arizona and Sonora. Common cultivated trees native to Asia with glossy, deep-green leaves divided into many leaflets with toothed margins. After the leaves fall in the winter, the round, hard, yellow-brown fruits persist. The leaves and fruits are poisonous if ingested in large quantities.

Masks: Rosario Vakame'eri-Castillo once experimented with the wood for mask making, but it split and was not suitable.

Toxicity: It is said that the fruits cause retardation if eaten by a child.

Arizona: Tucson, Tumamoc Hill, *Turner 83-4*.

Sonora: Las Guásimas, cultivated, 16 Dec 1988, *Felger & Molina*, observation.

MORACEAE – MULBERRY FAMILY

Dorstenia drakena

Common names: NAOTO'ORIA (together laid out); *baiburilla*

Sonora. Perennial herbs with a short, thick stem. One to three large leaves are produced during the summer rainy season and quickly fall away after the rains cease.

Common in the mountains of southeastern Sonora.

Moraceae, *Dorstenia drakena*, Río Aros. 17 Jul 2007 (SJ).

Medicine: The dried stems are used as a remedy for dizziness. They are obtained from herbal shops.

Wagner (1936:82) reported "three remedies for headache. One is from *bailburilla* root, dried, beaten to a powder, and mixed with tallow forming a salve."

Sonora: Onavas, 14 Oct 1986, Rea 1282 (UCR). Álamos, Río Cuchujaqui, *Wilder 06-290*.

*Ficus benjamina [*Ficus nitida*]

Common names: YUKATEEKO; *yucateco*; Benjamin fig

Sonora. Large trees commonly grown for their dense shade in gardens, along streets, and near buildings. The Spanish and Yoeme common names indicate it is from Yucatán, although this species is native to the Orient.

*Ficus carica

Common names: CHUUNA; *Higuera*; fig

Arizona and Sonora. The common, cultivated fig was brought to Mexico from Spain and introduced into Sinaloa and Sonora by Jesuit missionaries. Pérez de Ribas (1645; Reff et al. 1999:88) reported that, "Plants brought from Castilla grow well in these regions, especially orange and fig trees."

Basilio (1890 [1634]:210) listed *chuna* as the word for *higuera*.

Food: Figs are grown in home gardens in Arizona and Sonora.

Sonora: Kompuertam, cultivated, 14 Mar 1989, *Felger & Molina*, observation.

Ficus insipida [*Ficus radulina*]

Common names: KAU CHUUNA (mountain fig); *chalate*

Sonora. These trees grow 10 to 15+ m tall, forming a massive trunk and large limbs, smooth bark, gnarled and buttressed trunks (roots), and a dense, spreading crown. The leaves are 12 to 30 cm long and the lower surfaces are rough like sandpaper (scabrous). The figs are 2.5 to 3 cm in diameter, mottled green and white, and borne singly in the leaf axil. This species is readily distinguished by its large and often rough-surfaced leaves, exceptionally long stipules (4.5 to 8 cm long, seen as the sheath covering a new shoot or

Moraceae. **A.** *Ficus insipida* (left) and *F. pertusa* (right), Cañon del Nacapule. 24 Jan 2018 (SC). **B.** Moraceae, *Ficus insipida*, Rancho Bacatete. 1 Apr 2009 (MB).

growth bud), and solitary figs.

Basilio (1890 [1634]:165) cited *chunan* as the term for "*higo de la tierra*" (wild fig).

A spectacular grove of large *kau chuuna* trees grows in the Cañón Bacatete in the Sierra Bacatete. These trees form a closed canopy over a stream trickling even during dry seasons. The ground is covered with the dry, crackling, fallen leaves. We were told that *kau chuuna* grows everywhere in these mountains where water runs. There are other groves in scattered riparian canyons in the Sierra El Aguaje north of San Carlos and east of the Yoem Bwiara. This tropical species ranges southward to South America.

Food: "In the mountains the people ate one of the mountain figs, *kau chuuna*; it is sweet and ripens almost the same as the ordinary fig" (José María Jaimez, 13 Mar 1989).

Narrative: Mateo González said that it flowers at midnight and if you see the flower you will have good luck. (Like all figs, however, the individual flowers are minute and inconspicuous, lining the inside of the fig and would not be recognized as flowers by non-biologists.)

In the story of *Malinero'okai*, "As she traveled she contemplated...occasional fronds of *kauchunam*. These enchanted trees reminded her that she and her husband had passed by here two weeks before" (Giddings 1959:41–42).

Sonora: Cañón Nacapule, *Felger 84-17*. Cañón Bacatete (Bacateve), riparian arroyo below the Cuartel, *Felger 89-139 & Molina*.

Ficus palmeri and Ficus petiolaris

Common names: *tescalama*; cliff fig

Sonora. Mostly shrubs and small trees, sometimes becoming large trees. Root and stem bark whitish. Leaves broadly ovate to heart-shaped, the stipules often 15 to 25 mm long. Figs about 1.5 cm in diameter, paired but one may fail to develop or fall off.

These strange trees grow on cliffs, sheer canyon walls, and mountain rock. The roots grasp the rock and cascade down over the surface, and if they reach moist soil the plant develops into a tree, otherwise it remains dwarfed as a shrub.

In the more arid the Yoem Bwiara, such as near Guaymas and San Carlos, the plants have the aspect of *Ficus palmeri*, while in more favorable, inland mountains they resemble *F. petiolaris. Ficus palmeri* is widespread in the Gulf of California desert region, and *F. petiolaris* is characteristic of thornscrub and tropical deciduous forest regions beyond the desert. *Ficus petiolaris* generally has heart-shaped leaves, often with conspicuous pink veins, and is glabrous or essentially so. In contrast *F. palmeri* tends to have ovate leaves with yellowish white veins and pubescent twigs and peduncles.

Food: The figs are eaten, usually cooked, by the Seris (Felger and Moser 1985:348). Presumably some Yoemem made use of the small figs.

Ficus palmeri: Sonora: San Carlos, small riparian canyon SW of Cañón El Alacrán (La Navaja), 15 Dec 2015, *Carnahan 1628*. Cañón Nacapule, *Felger 85-1326* (ARIZ, USON).

Ficus petiolaris: Sierra Bacatete, Rancho El Álamo, Arroyo El Álamo, a 7.9 km (línea recta) al NE de Vícam, 12 Sep 2008, *Sánchez-Escalante*, observation.

Moraceae, *Ficus palmeri*, near Cañón La Navaja, Sierra El Aguaje. 15 Dec 2015 (SC).

Ficus pertusa

Common names: NAKA'APULI (ear lobe), NAKAPURI; *nacapule*, *nacapuli*; Central American banyan

Sonora. Large shrubs to large trees. The leaves are shiny green and smaller (5 to 10 cm long) than those of the other native figs, and the stipules are mostly 1 to 2 cm long. The figs are about 1 cm in diameter and usually paired (one may fail to develop or fall off).

Pérez de Ribas (1645; Reff et al. 1999:86) described a large, banyan-like fig tree:

> ...another natural wonder...a tree often found in the valleys of Sinaloa and in other warm places. This tree, which has a very large crown, is called *tucuchi* in the language of the region. Its fruit is a small, sweet fig, some of its branches are very long and large, extending so far that they could no longer sustain themselves without forked branches to hold them up...some trunks that are separate from the main trunk of the tree grown straight up out of the earth and under the tree branches.

Although the most common banyan-like fig in southern Sonora and northern Sinaloa is *Ficus cotinifolia*, this description might be that species and/or *F. pertusa*.

This tropical–subtropical species reaches its northern limits in riparian canyons in the Sierra El Aguaje and is more common in mountains southeast of the Yoem Bwiara.

Food: The small figs are edible.

Sonora: Tórim, tree 30 m crown, planted in the plaza, 1 Jun 2019, *Blystone*, photo. Cañón Nacapule, *Felger 84-601*. Guaymas, *Palmer 92 & 646 in 1887* (CAS).

Moraceae, *Ficus pertusa*, Sierra El Aguaje. 23 Feb 2019 (SC).

*Morus alba

Common names: MOORA; *mora*; white mulberry

Arizona and Sonora. Cultivated trees native to China with rather large, often lobed leaves. Male and female flowers are on separate trees. Commonly grown in Tucson and sometimes in the Yoem Bwiara, occasionally becoming feral.

Food: The fruits are good tasting.

Arizona: Tucson, West Branch of the Santa Cruz River, 29 Jul 2001, *Mauz 21-72*.

MORINGACEAE – MORINGA FAMILY

***Moringa oleifera**

Common names: *moringa*; horseradish tree, moringa

Small, fast-growing, slender trees to 5 meters. Leaves more than 30 cm long with numerous leaflets, smelling like horseradish. Flowers fragrant, cream-colored, 2 to 3 cm long. Fruits are long, slender, hanging capsules ("beans").

Small moringa groves are grown in home gardens in the Yoem Bwiara. Also cultivated elsewhere in southern Sonora and in the tropics worldwide for food and nutritional supplements. Native probably to India.

Food: Nearly every part of the plant is edible, especially the leaves.

Sonora: El Tular near Estero de Miramar, feral shrubs on refuse piles, 14 Feb 2018, *Carnahan 2868* (ARIZ, USON). Las Guásimas, cultivated small trees, 1 Jun 2019, *Blystone*, photos.

MUSACEAE – BANANA FAMILY

***Musa ×paradisiaca** [*Musa sapientum*]

Common names: PLATANO; *platano*; banana

Sonora. Bananas are sometimes grown in home gardens in the Yoem Bwiara.

Sonora: Kompuertam, *14 Mar 1989*, *Felger & Molina*, observation. Potam, Oct 1995, *David Shorter*, photo.

MYRISTICACEAE – NUTMEG FAMILY

***Myristica fragrans**

Common name: nutmeg

Medicine: Wagner (1936:85) reported, "hydrocephalus (water on the brain) is treated with nutmeg. The nutmeg is ground into a powder and rubbed on the temples." This and some other medicinal uses reported by Wagner seem questionable.

MYRTACEAE – MYRTLE FAMILY

***Psidium guajava**

Common names: WAEVAS, WAIVAS; *guayava*; guava

Sonora. Cultivated shrubs and small trees with simple, smooth leaves and bark that peels in jigsaw puzzle-like pieces. The egg-sized fruits are yellow. Native to Central America.

Food: The highly valued, aromatic fruit is eaten fresh.

Sonora: Kompuertam, 11 Mar 1989, *Felger & Molina*, observation. Potam, Oct 1995, *David Shorter*, photo.

NYCTAGINACEAE –
FOUR-O'CLOCK FAMILY

Boerhavia

Common names: MOCHI, MOOCHI; spiderling

Arizona and Sonora. Annuals or short-lived perennials. Moochi may refer to any of several common *Boerhavia* species.

Medicine: Wagner (1936:83) reported, "For scarlet fever the yellow root of the *mochi* plant is used. It is beaten to a pulp and steeped in cold water. The liquid is placed where the dew will form on it overnight. Beginning next day it is given to the patient three times a day for three days. During the time the patient eats no food."

Boerhavia coccinea [*Boerhavia caribaea*]

Common name: scarlet spiderling

Arizona and Sonora. Plants perennial but commonly flowering in the first season; roots thickened, and becoming moderately tuberous. Often forming sprawling herbs 1 to 2 m across. Flowers bright reddish-purple in sub-umbellate clusters. Fruits sticky from glandular-hairs.

Common weed in disturbed habitats including roadsides, and spreading into natural areas.

Food: José María Jaimez told us that people used to eat the "seeds" (fruits, or technically, the anthocarps). The seeds were boiled and mashed on a grinding stone (metate).

Arizona: Tucson, West Branch of the Santa Cruz River, 8 Aug 2001, *Mauz 21-84*.

Sonora: Road to Playa del Sol, *Felger 85-1119*. Southeast of Rancho El Tigre, *Felger 85-1465*. Kompuertam, 14 Mar 1989, *Felger & Molina*, observation.

***Bougainvillea spectabilis** and hybrids

Common names: MASA'ASAI, BUGAMBILIA; *bugambilia*; bougainvillea

These large, spectacular flowering vines are widely cultivated in Sonora and lower elevations in Arizona for their spectacular masses of purple, orange, pink or even white inflorescences. The colorful floral bracts appear like petals, surrounding actual flowers.

Sonora: Kompuertam, cultivated, 11 Mar 1989, *Felger & Molina*, observation.

Pisonia capitata

Common names: SUTU'URA (one with fingernails), VAHEWO; *garabata*, *vainoro*

Sonora. Large, briar-like shrubs with rigid branches armed with recurved thorns, the short twigs also sometimes thorn-tipped. These thorns are especially sharp and rigid. Male and female flowers are on separate plants. The flowers are dark red-purple, fragrant, and in small clusters. The fruits are club-shaped, about 1 cm long, and sticky with thick-stalked glandular hairs that adhere to feathers, fur, and clothing.

Nyctaginaceae, *Pisonia capitata*, Álamos. 1 May 2014 (SC).

Riparian canyons and larger, densely vegetated arroyos and washes; common in the Sierra El Aguaje and Sierra Bacatete and coastal thornscrub from the vicinity of Vícam southward.

Rosario Vakame'eri-Castillo told Felipe that *sutu'ura* was a "shrub and has many thorns, the plants are fairly big, and the thorns point inward, making it easy to go in, but impossible to get out." He said it grows in the Guaymas region.

Sonora: 1 mi northeast of Mex Hwy 15 at Pitahaya junction, *Felger 88-124*. Guaymas, *Palmer 175 in 1890*.

Salpianthus macrodontia

Common names: TAAHUMAI; *guayavilla*

Sonora. Shrubs 1 to 2 m tall, with long tuberous roots. Flowers small, purplish and white, drying brown and persistent in clusters.

Sierra Bacatete, near the Río Yaqui, and occasional along roadsides.

Sonora: 4 km N-NE of Las Guásimas, 16 Dec 1988, *Felger 88-604, Molina, Steen, & Flores* (ARIZ, USON).

NYMPHAEACEAE – WATER LILY FAMILY

Nymphaea elegans

Common names: KAPO, KAPO SEEWA; *capomo*; water lily, tropical royal-blue water lily

Sonora. Aquatic perennials with thick rootstocks deeply buried in mud, and floating leaves with red-brown mottling. Flowers large, blue, and showy.

Growing and flowering during hot weather in flooded roadside ditches and pools in swampy areas on the coastal plain.

Song: This plant is the subject of many Yoeme love songs, such as *Kapo Seewa* (seewa is flower). In 2006 and 2007, the Tucson radio station *Caliente* (FM 102.1) often played the popular *Flor de Capomo* in Spanish or *Kapo Seewa* in Yoem noki.

Sonora: 1.7 mi on Mex Hwy 15 southeast of Pitahaya (Belém, Río Yaqui), shaded pond, *Felger 85-1268*. 7 km north of road to Potam on Mex Hwy 15, *Meyer 26 Oct 1990*.

Nymphaeaceae, *Nymphaea elegans*, Rancho los Charcos, southeast of Moctezuma. 8 Sep 2018 (GY).

OLEACEAE – OLIVE FAMILY

Fraxinus velutina

Common names: *fresno*; velvet ash

Arizona and Sonora east and north of the Yoem Bwiara. These large trees are found in riparian canyons in southern Arizona and northern and eastern Sonora, and are sometimes grown as shade trees.

Medicine: Wagner (1936:85) reported several Yoeme remedies for rabies, obtained in an interview at Tórim: "The one by which most store is placed is a tea made from the bark and leaves of the *fresno* tree. The tree is said to grow in the mountains. Only *atole* is given the patient to eat. If he becomes irrational he is tied up until the medicine has its effect."

PALMAE, see ARECACEAE

PAPAVERACEAE – POPPY FAMILY

Argemone gracilenta

Common names: TATCHI'INA; *cardó*; prickly poppy

Arizona and Sonora. Annuals or perennials, often to 1 m or more tall. Plants spiny nearly throughout, and with yellowish sap. The flowers are large with white petals and numerous yellow stamens.

This prickly poppy is common along dry washes and in disturbed desert places.

Dye: Johnson (1962:286) reported that *táči'ina* was used to dye textiles yellow ("*cardo, planta muy espinosa que da tinta amarilla para teñir textiles*"). The reference might be for *Argemone gracilenta* and/or *A. ochroleuca*. Both occur in Arizona and Sonora.

Arizona: Pima County, near Coyote Mountains, 13–14 Apr 1963, *Gentry 19983*.

Sonora: 5 mi northeast of Pitahaya, *Felger 89-125 & Molina*.

PASSIFLORACEAE – PASSION-VINE FAMILY

Passiflora arida [*Passiflora foetida* var. *arida*]

Common names: MASTAOKA; desert passion-flower

Sonora. Perennial bushes, or vines often growing over shrubs, the herbage densely white-woolly (not sticky glandular), sometimes partially drying brownish. Flowers white, about 5 cm wide. Fruits to 3 cm long, ovoid, soft, and green to yellow-green. Flowering and fruiting almost any season.

Often common along the banks of the Río Yaqui as well as the open desert and thornscrub.

Food: The fruits are eaten fresh.

Sonora: Palo Parado, Río Yaqui, *Felger 85-1408*. Miramar, *Felger 92-1015*. Guaymas, *Hastings & Turner 64-19*.

Passifloraceae, *Passiflora arida*, La Balandrona, Sierra El Aguaje. 10 Oct 2000 (JS).

PEDALIACEAE – SESAME FAMILY

***Sesamum indicum** [*Sesamum orientale*]

Common names: ORHOLIINIM; *ajonjolí*; sesame

Sonora. Warm-season annuals native to tropical Asia. The flowers are white or lavender-tinged and about 2.5 cm long. The fruit capsules produce numerous small seeds.

A crop plant and occasionally seen along roadsides growing from seeds falling from trucks.

Agriculture: Sesame is an important cultivated crop in Sonora and has been a major crop in the Yoem Bwiara since the 1990s.

PETIVERIACEAE

Rivina humilis

Common names: WO'I KO'OKO'I (coyote chile); *chile de coyote*; bloodberry, rouge plant

Arizona and Sonora. Bushy perennials, often more than 1.5 m tall and growing through other shrubs. Stems and leaves often reddish-green, the leaves ovate and thin. Flowers white to pink, about 5 mm wide. Fruits 3 to 4 mm wide, fleshy, red, and yielding a red dye.

Riparian and semi-riparian habitats in Sonora, mostly brushy arroyos, canyons, and banks of the lower Río Yaqui.

HAICOCOA

Wagner (1936:84) reported, "pinkeye is treated with the juice of *haicocoa* berries. The berries are about the size of buckshot and of a deep red color. The juice is mashed out and smeared on the eyes, temples, forehead and cheeks." *Haicocoa* seems to be *Rivina humilis.*

Sonora: Cañón Nacapule, *Felger 85-1168*. Palo Parado, Río Yaqui, *Felger 85-1402.*

Petiveriaceae, *Rivina humilis*, La Pintada, Sierra Libre. 6 Oct 2007 (JS).

PHYTOLACCACEAE (*Rivina*), see PETIVERIACEAE

PHYTOLACCACEAE (*Stegnosperma*), see STEGNOSPERMACEAE

PLANTAGINACEAE – PLANTAIN FAMILY

Plantago ovata var. **fastigiata** [*Plantago insularis, P. insularis* var. *fastigiata*]

Common names: MUMSA; *pastora*; Indian-wheat, woolly plantain

Arizona and Sonora. Winter–spring annuals, highly variable in size depending on soil moisture. The plants in the Guaymas region are generally small in comparison with their usual size farther north. The leaves are simple, elongated, and silky-silvery pubescent. Each plant may have a half-dozen or more slender flowering stalks, 10 to 15 cm long, bearing numerous, small yellow- to red-brown seeds.

Mostly in open desert habitats, washes, and roadsides. This is one of the more common winter–spring annuals of the Sonoran Desert and reaches its southern limits in the Guaymas region.

The New World populations, long known as *Plantago insularis*, are closely related to the Old World *P. ovata* and they are regarded as a single species. The seeds are edible and cultivated forms yield commercial psyllium seed (fiber supplement).

Arizona: [Tucson], Silverbell Road west of Santa Cruz Wash, 25 Mar 1980, *McLaughlin 2442.*

Sonora: Guaymas, road to Cerro El Vigía, *Felger 85-435*. San Carlos, *Felger 85-641.*

POACEAE (GRAMINEAE) – GRASS FAMILY

Vaso is a general term for grass and the name for many of the common native grasses. Basilio (1890[1634]:230) gave *vasoparia* for *zacatales*, a field of grass. *Vahu* is a term that describes the act of gathering and cutting of grasses (Basilio 1890 [1634]:230) and can also be used with other kinds of plants.

Grasses were used for roofing for the traditional dome-shaped house and for temporary houses or shelters in the mountains or country. Various kinds of large grasses were tied in bundles and then tied onto the traditional arched house frame, or on the frame of a *chapa kari*, "*casa de campo*," a kind of hut shaped like a tent with a pitched roof. These huts are put up in the field or mountains such as when one is herding goats. These grass roofs readily shed rainwater.

Profound cultural changes followed Jesuit introductions of cereal crops, notably wheat (*Triticum aestivum*), barley (*Hordeum vulgare*), and oats (*Avena sativa*).

***Arundo donax**

Common names: VAAKA; *carrizo*; cane, giant reed

Arizona and Sonora. This giant, bamboo-like grass has leafy stems reaching 3 to 5 m tall. New shoots emerge from thick, perennial rhizome-rootstocks throughout the warmer months, but especially in spring, and rapidly grow to full height. The "roots" (rhizomes and roots) are called *vakaroroa*. The mature stalks are overtopped with a large, feathery inflorescence or panicle, called *vaka moa*. *Vaaka* is one of the most important plants in traditional Yoeme culture (e.g., Felger and Molina 2019).

Vaaka was harvested along the Santa Cruz River in southern Arizona. In Arizona access to vaaka in the latter part of the twentieth century became more limited than in the past, and it became necessary to ask permission to harvest it from the owner of the property.

Vaaka occurs along the lower Río Yaqui and is often common along irrigation ditches. The people say that in earlier times vaaka grew thickly along the banks of the Hiak Vatwe (Río Yaqui), that it grew big and tall, arching over the river. Felipe tells us, "In the 1970s, my cousin and his wife Francisca, from Potam would visit my grandmother and talk about the vaaka. Francisca said when she was young in the 1930s, the vaaka along the river banks arched over the river. That must have been a beautiful sight."

Spicer (1954:11) described "cane-breaks which make up the dense, often hard-to-penetrate bush characteristic of the banks of the Yaqui River along the lower sixty miles of its course." He also showed a 1942 photo of people in a path through tall carrizo, with the following captioned:

> Thick, high stands of carrizo once lined the banks of the Yaqui River. The cane was used for mats, house walls, bird cages, knives, spoons, and many other articles. Its exploitation by outsiders for commercial furniture was a political issue among Yaquis during the 1940s and 1950s. After the Obregón Dam was built to develop irrigated agriculture, the river ceased to flow in its lower course near the towns, and this important resource was much diminished. (Spicer 1980:9)

Vaaka has become increasingly scarce along the river as the river flows have diminished. In an interview on 31 May 2019 in Pótam, Juana Lugo-Osuna told Felipe that her son was expecting visitors for the Holy Trinity Feast Days but had a hard time looking for vaaka to cover his ramada roof. Felipe tells us, "He did bring in a truck load as we listened to Juana's report. On our way out, Juana's son came out to bid farewell and mentioned the hard times people are having looking for resources to build houses and ramadas. He said he had to go out again because he needed more [vaaka] to accommodate all the visitors from the other villages. He said the visitors will rest and sleep under the ramadas during the four days of Holy Trinity."

Botanical publications report that *Arundo* is not native in the New World, although it has long

Poaceae, *Arundo donax*, Miramar. 25 Jan 2018 (SC).

been widely cultivated and naturalized in the warmer parts of the world. The earliest known specimen of *Arundo* in Sonora was collected by Edward Palmer at the Río Yaqui in 1869. Presumably at some point during historical times *Arundo* replaced the native and similar-appearing *Phragmites* along the Río Yaqui. Their cultural uses would be interchangeable (see discussion for *Phragmites*). Basilio (1890 [1634]:148) used *baca* as the term for *caña hueca* (hollow carrizo), which could have been *Arundo* or *Phragmites*.

Arundo and *Phragmites* can be distinguished with certainty only by characters of the spikelets:

> *Arundo donax*: Lemmas with long hairs, the rachillas (spikelet stalks) glabrous.
>
> *Phragmites australis*: Lemmas glabrous, the rachillas with long hairs.

Authors such as Holden et al. (1936) and others have mistakenly used the name "bamboo" for vaaka. Actual bamboos in the region are the native *Otatea* and the cultivated *Phyllostachys*. These bamboos have rigid, woody, and rather long-lived stems.

Basketry and mats: The upper, tender and more slender part of the stem is used in making baskets (*waarim*) and mats (*hipetam*, or *petates*). For basketry, the upper portion of the stem below the panicle (inflorescence) is used. Traditional baskets were also made from willow (see *Salix gooddingii*, Salicaceae).

Nacho Amarillas-Sombra making a *waari* (basket) from *vaaka* (*Arundo donax*, Poaceae), Pótam. A 15 Dec 1988; B & C 16 Dec 1988 (WS).

Men have been traditional basket-makers, although women wove palm-fiber baskets and hats (Hrdlička 1904; Moises et al. 1971; see Arecaceae, palms) reported. Beals (1945:37) noted, "the Yaqui make almost no basketry, although the manufacture has been abandoned recently. The most common Yaqui basket around 1900 was a quadrangular shape with rounded corners woven in checker pattern...Cubical, cylindrical, and bottle shapes are mentioned." Basket making, however, thrived as a minor industry through at least most of the twentieth century.

During two days, 15 and 16 December 1988, Ignacio (Nacho) Amarillas-Sombra showed us how he makes baskets. He had an artisanal business making *waarim* (baskets). Mixing Yoem noki and Spanish, he sometimes calls them *warritos*. Only some vaaka stems serve for basket making. He selects and cuts vaaka stalks from near Pótam and brings them to the ramada at the family home. He uses selected stems tips, the upper portion of the stem below the inflorescence (panicle), about a meter or less in length, and strips away the leaves. This carrizo stem-tip stripped of leaves is called *vaka moa* (*moa*, or *flor de carrizo*, is the young growing tip or *spiga*).

After the stalks of dry vaka moa are cleaned, he lightly mashes or taps the stalks on an *avaso* block (a cross-section of a cottonwood branch). Each vaka moa is tapped in a series of light poundings all along the length of the stalk. Then holding down the pounded stalk he pulls it through, under a *mano*, splitting the stalk longitudinally. The mano is called *mata tutuha* (*mata* is a grinding stone or metate; *tutuha* is the mano rock.)

Don Nacho then starts working on the prepared vaaka splints for basket making. The work of mashing and splitting was done in the early morning. The process has split the sides of each stem. Don Nacho now halves each of the stems with a knife, and trims them longitudinally with the same knife to end up with two, flat and even-sized splints from each stalk. The material is put away for the rest of the day, as there are other things to do.

At dawn of the next day, Don Nacho has separated the vaaka stalks into four piles. Several people are helping split them into splints. Don Nacho continues working on the vaaka sticks and rattles them on the door of the room by the open kitchen to get one of the women out of bed to make coffee. It is rather cold, and the women just aren't getting out of bed. So Nacho finally is making the coffee himself and grouching about it. The blackened coffee pot hangs over the burning logs and is boiling away. After a considerable pile of vaaka splints are prepared, he starts folding over splints to make a basket start and continues weaving the basket, finishing it by late morning. A photo from the 1980s by David Burkhalter shows "Ignacio Amarillas Sombra, Yaqui basketmaker, Potam" holding a nearly finished waari (Burkhalter 1992).

Baskets of different sizes and shapes were ubiquitous in traditional Yoeme culture. For example, Holden (1936c:69) reported, "Every kitchen...had from one to ten baskets...There was always a *tortilla* basket from ten to fourteen inches in diameter and six to eight inches deep. Then there was usually a larger basket in which the eating utensils were kept....There might be other baskets, of varying sizes and shapes, for storage purposes." He also saw "a large *carrizo* basket containing about a bushel of beans" in a house in Tórim (Holden 1936c:70).

Beals (1945:11) illustrated and described a "Carrying basket, or wakal, used in gathering pitahaya fruit. Made of split canes twined with mesquite-bark strips. The bottom is made of mesquite-bark strips interwoven; the carrying strap is of canvas. Diameter, 10 inches; height, 10½ inches...Bailing wire is superseding bark." (See *Prosopis glandulosa*, Fabaceae).

Bead loom: In the 1980s, José Guadalupe ("Lupe") Flores made bead-working looms with two 4-foot lengths of carrizo, held together by two crosspieces of an unidentified wood. The carrizo grew in Flores' backyard in New Pascua. (Tom Kolaz purchased one of these looms for the Arizona State Museum in Tucson.) José Flores sold or traded his beadwork to other Yoemem.

Beliefs and medicine: Tall canes tied in a bundle (*tesia*), are often seen standing upright against a house in the Yoem Bwiara. Such bundles are considered a good omen, warding off evil thoughts called *eerim*. The bundle is there for repairs on the house or fence.

Wagner recorded some medicinal uses for *Arundo* in 1934:

> Gunshot wounds are treated in primitive but effective way. A section of bamboo [sic] having about the diameter of the wound is selected. Another piece with closed end is fitted into this, making a crude popgun. Brazil wood [*Haematoxylum brasiletto*] scrapings are placed in this 'gun,' which is then inserted and [the] wound forced full of the scrapings. These, in contact with the tissue fluids, swell, stopping [the] hemorrhage. As the wound heals this plug is extruded...Fractures are placed in as good position as possible by manipulations and then splinted with split bamboo [sic]. (Wagner 1936:86)

Painter (1986:56) also mentioned *carrizo* used as splints.

Construction, fencing, and mats: Vaaka has been extensively employed in traditional Yoeme architecture, described and illustrated in detail by McMillan (1936) and others. *Vaka chukte* (cane cutting) is the phrase used for cutting carrizo to make mats or for house construction. Vaaka is used for house walls and roofing, as well as for the pahko'ola ramada. In Sonora, especially in river bottom settlements, family compounds are often surrounded by vaaka fencing. Fencing of arrizo-walled enclosures is used to protect fruit trees such as oranges, gardens, or seedling beds from animals.

Spicer (1980:63) described Yaqui houses in early Jesuit times: "The houses which Yaquis constructed near the church and its plaza...were surrounded with cane fences." When Felipe and Richard visited the Jaimez family in Kompuertam in the 1980s, their house compound was surrounded by vaaka fencing. Near the house compound their vegetable and flower garden also was surrounded by vaaka fencing to protect it from animals. The rooms of the house compound were roofed with thatched vaaka covered with corrugated tarpaper and then covered with earth. Wildflowers, grasses, and weeds grew on the roof. Vaaka walls of the sleeping rooms were adobe plastered.

Hipetam (woven mats, or *petates*) are made from interwoven split vaaka stems. These all-purpose, strong mats form walls of traditional houses, sleeping mats and beds, and many other uses. Hipetam have been used as the foundation beneath earthen house roofs and for fencing. Rosalio Moisés wrote that in 1947, he and his wife started making petates:

> We could make three or four a day. Pancha took them to the Mexican storekeeper, who only paid forty centavos apiece. She took the 1.20 or 1.60 pesos of credit in food. Now we ate only twice a day. Making petates all the time is hard. You have to pound up the carrizo before you can weave the mats. Our hands were sore all the time. (Moisés et al. 1971:206)

made such cane containers for the family's important papers. He used cottonwood root for the caps or stoppers. Hrdlička (1904:65) reported, "On ranches each Yaqui employed keeps a personal account, which he carries in a tube made from the native bamboo [sic]. Each of these tubes is differently decorated on its surface with numerous incised figures, mostly of geometrical pattern. These figures are not strictly property-marks, yet they serve to distinguish the tubes."

Cultivation: Vaaka is grown in virtually every town and ranch in the Yoem Bwiara region, and has long been grown in household gardens in

A. Freshly cut *vaaka* (*Arundo donax*, Poaceae), Pótam. Fence and wall also of cane, and a willow house cross (*Salix gooddingii*, Salicaceae). 1 Jun 2019 (PB). B. Ramada for washing, Tórim. Cane poles and roofing (*Arundo donax*) and a corner post of Yaqui cottonwood (*Populus mexicana*, Salicaceae). 1 Jun 2019 (PB). C. Sleeping room, Tórim. Mesquite posts (*Prosopis glandulosa*, Fabaceae), wall of cane (*Arundo donax*, Poaceae), window closed with mat also of cane, and a willow house cross (*Salix gooddingii*); roofing of canes covered with plastic sheeting. 1 Jun 2019 (PB). D. Wall of cane (*Arundo donax*), Tórim. Horizontal canes chinked with mud, the vertical canes not chinked, allowing ventilation. 1 Jun 2019 (PB).

Containers: Larger cane stems were hollowed out for use as a container (like a canister or tube) to hold important papers. The open end is plugged with a clay or wooden stopper—the other end being already sealed by the natural septum in the cane stalk. Rosario Vakame'eri-Castillo in Marana Arizona. In Sonora, after cutting the vaaka in the fall, the remaining plant is often burned ("burn the *basura*") in order to improve the vigor of the new growth, "so that it comes up better."

Deer songs: Vaaka features in numerous deer songs, including the Red Quail song (Evers and

Molina 1990), which concerns the Elegant Quail (see *Callipepla douglasii*, Galliformes).

Games: Pérez de Ribas (1645; Reff et al. 1999:94) described seventeenth-century games using vaaka canes:

> It corresponds to cards or dice, but instead...they use four short reeds that have been split open, each less than a *geme* [about 15 cm] in length. Little figures and points are drawn on the reeds to indicate their value or a loss. The reeds are cast onto a small stone slab so that they bounce and fall by chance...The bet is for a string of small sea snails that they prize and wear as ornaments. They also wager bows or arrows, as well as knives and hatchets that they have obtained. These same items serve as stakes in other games, but patoli is the game they play most often.

Household: Knives and spoons were made from the vaako stems, as well as stirring sticks for use in the kitchen.

Hunting and weapons: Arrow shafts, *vaka huiwa* (cane arrow) especially the foreshaft, were made from carefully selected cane stalks. Hrdlička (1904:66–67) described bows and arrows found at the site of the Sierra Mazatán massacre of June 1902: "The arrows are stout and measure 2½ to more than 3 feet long; the shaft consists of a stout, hollow reed."

Musical instruments: The larger canes are used to make the *kusia* (flute) and water-drum stick. The kusia is a two-piece flute played by the tampaleo.

The *Wiko'i Ya'ura* (Coyote Warrior or Bow Leaders' Society) dancers use a split cane instrument, *vaka chamti* (cane split), for beating the bow in the coyote dance. The vaka chamti is made by carefully cutting slender strips into a section of cane about one foot long. The splits do not go through the septum at the stem node.

A special cane instrument, called *vaka aapa* (cane harp) is made from a meter-long length of carrizo cane. It has a long peg on one end and a single, heavy string attached. It was used in the crow dance involving the Sinaloa Crow (see *Corvus sinaloae*, Passeriformes: Corvidae). The sound of the vaka aapa is not loud and this instrument was also used by men and boys for their own enjoyment.

Religion and ritual: The *matachin koona* (matachini crown) is made from split stalks. Vaaka stalks are also made into special official canes for the matachin governors. Ceremonial arches for religious ceremonies are made from large, leafy stalks. On the outside walls of traditional houses in Sonora, such as at Pótam, one often sees small crosses of vaaka stuck into the carrizo walls.

The *pahkola moro* (pascola leader) uses a stout vaaka pole to lead the pahko'olam into the pahkola ramada and when leaving they are led out and to their quarters by the same pole. When not in use, the vaaka pole is stuck into the ramada roof. At the end of the ceremony "the *pahkolam* all danced to the *tampaleo* at the same time. When they had finished dancing to the *tampaleo* they started to bless the ground...Each *pahkola* marked a cross on the ground with the bamboo [sic] reed with which the *moro* had led him into the ramada" (Griffith and Molina 1980:41). The pahko'olam also use vaaka poles to "get power" from the harp.

Spicer (1980:171) described "*Cohetes*, the small sky rockets consisting of cane cylinders about four inches long containing gunpowder and attached to a stick or weed stem...Held in the hand by the stem, the cohete is lighted with a firebrand and flung into the air. One usually goes up about twenty-five to thirty feet, hissing as it goes, and explodes with a loud pop. Dozens of such cohetes are necessary for all ceremonies." Holden (1936b) told of many small sky rockets being set off during a funeral in Tórim.

Holden (1936b) also told of death mats made of carrizo, for babies and little children who died of whooping cough and heavy burial mats for larger adults. This use of vaaka mats was continued a number of years. Felipe saw these mats in Pótam which were burned after being used.

Shampoo: Doña Romana Sánchez said that the "roots" (*vakaroroa*), are used for shampoo.

Smoking: Pérez de Ribas (1645; Reff et al. 1999:332) wrote, "When one nation invites another to forge an alliance for war they convey the invitation by sending a number of reed canes filled with tobacco" (see *Nicotiana rustica*, Solanaceae.)

Toys: Spilt cane stalks are made into kite

frames, and also small baskets and miniature furniture for children's play. A bow made of vaaka is called *vakawiko'i* and is the name of the children's game that features this toy weapon. Children made "slingshots or blowguns...out of lengths of carrizo cane." (Moisés et al. 1971:22)

Arizona: 5 mi south of Tucson, deep channel of Santa Cruz [River], *Carter 15 Oct 1932*.

Sonora: Cuesta Alta, Río Yaqui, *Felger 85-1367*. Yaqui River, *Palmer 19 in 1869* (US).

*Avena sativa

Common names: AVEENA; *avena*; oat

Introduced by the early Jesuits. Oats are grown during cooler seasons.

*Cenchrus ciliaris [*Pennisetum ciliare*]

Common names: *zacate buffel*; buffelgrass

Arizona and Sonora. Robust perennials, inflorescences dense with purplish bristles and burs.

At Las Guásimas we were told this grass is like *teku bwasia* (squirrel tail, unidentified) or *yonso vaso* (Johnson grass, *Sorghum halepense*).

This grass, native to the warmer parts of Africa, Madagascar, and India, is the most extensively planted forage grass in Sonora. Since the mid-twentieth century hundreds of thousands of hectares of natural vegetation have been cleared and planted with buffelgrass. Since about the 1970s, it has spread into natural desert areas, especially along washes, and has become a widespread invasive weed and fire hazard. It is particularly abundant along roadsides where it is sometimes harvested for fodder.

Poaceae, *Cenchrus ciliaris*, San Carlos. 16 Dec 2013 (SC).

Arizona: Tucson, vacant lot, 17 Jan 1980, *Reeder & Reeder 7235*.

Sonora: San Carlos, *Felger 84-545*. Las Guásimas, 16 Dec 1988, *Felger & Molina*, observation.

Cenchrus palmeri

Common names: WICHA'APOI, WICHA'APOLI (*wicha*, spine + *'apoli*, not translatable); *guachapori*, *huizapori*, *toboso*; giant sandbur

Sonora. Non-seasonal annuals. The spikes bear 1 to 4 burs with sharp, stiff spines about 10 to 14 mm long. These obnoxious burs, the largest of any species of *Cenchrus*, persist half-hidden in the sand long after the plant dies.

Common, often on sandy soils, but also on rocky slopes.

Sonora: Guaymas, *Palmer 689 in 1887* (US). Estero Soldado, *Yensen 1972-11-12*.

*Cenchrus spinifex [*Cenchrus incertus*, *C. pauciflorus*]

Common names: WICHA'APOI, WICHA'APOLI; *guachapori*, *huizapori*, *toboso*; common sandbur, field sandbur

Arizona and Sonora. Annual bur grasses. The spikes bear more than 10 burs, which have sharp spines.

Generally occurring as garden and farm weeds and at scattered disturbed sites. Probably not native to the region.

Arizona: [Tucson], Fort Lowell, 7 Aug 1905, *Thornber 4278*.

Sonora: Cuesta Alta, Río Yaqui, *Felger 85-1383*. East of Chiinim, *Felger 88-573*.

*Cynodon dactylon

Common names: MO'OKO VASO (*mo'oko*, the word for *Suaeda nigra* + *vaso*, grass); *zacate bermuda*, *zacate inglés*; Bermuda grass

Arizona and Sonora. Creeping perennials with long stolons, scaly rhizomes, and short leaves.

This weedy grass is common on alkaline soils and near water in natural and disturbed habitats, including heavily grazed areas. In some places it covers the banks of the Río Yaqui. It is a common lawn grass in Arizona and many areas of Sonora and is also grown in the region for hay.

Arizona: AZ-286, 24 km south of Robles Junction, 31 Aug 1992, *Reeder & Reeder 8895*.

Sonora: Guaymas airport, roadside, *Felger 85-422*. Near Kompuertam, 11 Mar 1989, *Felger & Molina*, observation. Rancho Bacatetito, 13 Mar 1989, *Felger & Molina*, observation.

Distichlis littoralis [*Monanthochloe littoralis*]
Common names: VAA VASO (water grass); *zacate playero*; shore grass

Sonora. Wiry-stemmed, perennial saltgrass. The leaves are firm, sharp-pointed, and about 1 cm long. This is the shortest-leaved grass in the Sonoran Desert. Male and female flowers are on separate plants. The small spikelets are nearly hidden among the leaves.

Dense colonies fringe salt marshes on tidal flats and at esteros.

Sonora: Estero Soldado, 30 Oct 2015, *Carnahan 1624*. Chiinim, edge of mangroves, *Felger 89-183, Cruz, Molina, & Steen*.

***Hordeum vulgare**
Common names: SEEVA, SEEVAM; *sevada*; barley

Arizona and Sonora. Barley was introduced by the early Jesuits. It is grown in the Yoem Bwiara as a winter crop, but has seldom been planted since the late twentieth century. The grain is usually purchased in supermarkets.

Beverage: In Sonora and Arizona, *seeva* was fermented as *teswiin*, an alcoholic beverage. It was just for personal use at home, not for ceremonies. A similar beverage was made using sweet potato (*Ipomoea batatas*, Convolvulaceae), wheat (*Triticum aestivum*), or corn (*Zea mays*).

In 1980, Porfirio Yokiwa described some of the process: You have a big tin barrel on a table for fermenting the barley. A cow horn spigot at the bottom of the barrel drains into a tin bucket or pail below it, and later the teswiin is poured into available bottles. A wooden stopper was made mostly from cottonwood root. More recently the cow horn was replaced by copper tubing leading from the barrel to the bucket.

Food: *Seeva posoim* is barley soup or *posole* (from Spanish *posole*) and usually includes some meat.

***Oryza sativa**
Common names: AROSIM; *arroz*; rice

Rice has been cultivated in the Río Yaqui valley by some non-Yoeme farmers. Rosalio Moisés told of working in rice fields:

> We went to see the boss man, who said he could use us pulling weeds out of the rice fields for three pesos a day.
>
> We worked from sunrise to sunset, fourteen-hour days. All day we stood barefoot in the water, stooped over, pulling weeds. It was hot. At night we could hardly sleep because of fleas on the ground and mosquitos in the air. We stayed eight days, and it was the hardest work I ever did. Groceries were expensive. We drew our wages on February 14, 1934; each of us took ten pesos back to Torim. When we left, we said we would never again work in the rice fields of the Yaqui Valley. (Moisés et al. 1971:149)

Otatea acuminata
Common names: VAKAU, YO VAKAU (enchanted vaaka [*Arundo donax*]); *otate*; weeping bamboo

Sonora. This slender bamboo grows 4 to 5 m tall and has narrow leaves. It is native to relatively moist places in the mountains of eastern Sonora and occurs in the Sierra Bacatete. Vicente Baltazar confirmed that vakau grows up in the Vakatetteve Kawim (Tall-cane[vaaka] Mountains, the Sierra Bacatete). "*Vakatetteve* means tall bamboo [sic]....There is tall bamboo that grows by the springs up in the Vakatetteve Mountains" (Evers and Molina 1987:111). The name vakau is also used for the common cultivated bamboos (see *Phyllostachys aurea*).

Sonora: Municipio de Yécora, Arroyo Santa Ana at Son 12 (Tepoca–Cd. Obregón highway), 510 m, *Van Devender 97-1057*.

Pennisetum ciliare, see **Cenchrus ciliaris**

Phragmites australis subsp. **berlandieri** [*Phragmites communis*]
Common names: WO'I VAAKA (coyote vaaka/ cane); *carrizo*; common reed, reedgrass

Arizona and Sonora. A bamboo-like grass,

reaching 2 to 3+ m tall, or smaller when stressed. Reedgrass is the original, pre-contact cane throughout the Western Hemisphere. It still occurs in some of the more remote waterholes in the Sonoran Desert, such as on Isla Tiburón, in small, isolated patches in canyons and along the Sonora coast, and is especially abundant and vigorous along the lower Colorado River and its delta. In general, it has been replaced by the non-native *Arundo donax*.

Phragmites and *Arundo* are remarkably similar in their general appearance. *Arundo* typically produces large stems and few smaller ones. In contrast, *Phragmites* commonly produces many smaller stems as well as larger ones, and in harsh environments it may be stunted and produce only smaller stems. The two species are reliably distinguished by technical but easily-discerned differences in their spikelets (see *Arundo donax*).

Phragmites australis is native to every vegetated continent on earth. It is perhaps the world's most widespread species of flowering plant, although there are geographic genetic differences, recognized as subspecies. Presumably it was once common along the Río Yaqui but was long ago replaced by *Arundo*. *Phragmites* would have served the same purposes for which *Arundo* is now used.

Wo'i vaaka is described as being "another kind of vaaka" (*carrizo*), said to be different from the common vaaka (*Arundo*). Wo'i vaaka is said to rot quickly, and be smaller (about 1 to 1.5 m tall) than vaaka. There is a small patch of it on the lower Río Yaqui by the coast, about 15 km from Pótam at a place called *Aron Kawi* (or a close approximation of this spelling; *Aron* might be a man's name, *Kawi* is mountain). Although we have not seen specimens, it might be a small, local population of *Phragmites*. In 1988, Ignacio Amarillas (Don Nacho) and his son Guillermo told us that the only place he knew where *wo'i vaaka* occurs is on a little patch of land on an island-like place in the Río Yaqui delta near the shore southwest of Pótam. They wanted to take us there but said the road was impassable because in was in the marshy river area. This is probably the Aron Kawi site. Don Nacho described the plant as smaller than *vaaka*, with more slender stems, and that it is brittle—you cannot make mats from it like you can with vaaka.

Arizona: Tucson, Anklam Road, base of Tumamoc Hill, 27 Nov 1984, *Bowers 2980*.

Sonora: Cañón del Nacapule, 28 Nov 2015, *Carnahan 1592*. Bahía San Pedro, 27 Mar 2016, *Carnahan 1704*. Sierra Santa Ursula, Aguaje los Pelones, photos, *Bogan 4-5 Apr 2009*.

*Phyllostachys aurea

Common names: VAKALAUME (*vaaka*, cane + *laume*, not translatable); YO VAKAU (enchanted vaaka);

Poaceae, *Phragmites australis* and *Sabal uresana*, Rancho Santa Ursula, 4–5 Apr 2009 (MB).

bambú; bamboo, Chinese bamboo, golden bamboo

Arizona and Sonora. *Phyllostachys aurea* is the common, non-native bamboo grown in Arizona as well as in Sonora. *Vakau* is also the name for the native Sonoran bamboo, *Otatea acuminata*. Felipe tells us: "When people from Potam came to my house in Marana and saw the Chinese bamboo in my yard; they said it was *yo vakau* or *vakalaume*. I have seen that plant in Potam." The giant timber bamboo, *Bambusa oldhamii*, occasionally cultivated in Arizona, is also called *vakalaume*. In the 1980s, Vicente Baltazar saw Chinese bamboo at Marana and called it *vakau* and said another larger bamboo is called *vakalaume* (perhaps timber bamboo).

People say vakau or yo vakau in Pótam, depending on the person talking. Both are acceptable. *Vakau* and *vakalaume* come from the word *vaaka*, cane or reed (*Arundo donax*).

*Saccharum officinarum

Common names: SANA, YOI SANA (Mexican sugarcane); *caña de azúcar*; sugarcane

Large perennials of Old World origin. Grown in Sonora and sometimes in Arizona, mostly in home gardens, and also purchased in stores.

During the reign of the dictator Porfirio Díaz in the late nineteenth and early twentieth century, in addition to the deportations to the henequen haciendas in Yucatán (see *Agave fourcroydes*, Asparagaceae), native peoples were sent to work as slaves on sugar plantations in Oaxaca:

> There was also an increased effort to expand sugar production...in such places as the Valle Nacional in Oaxaca...Yaquis, along with Mayos, Opatas, and Pimas were regularly rounded up on the haciendas, placed temporarily in jails in Hermosillo and Guaymas, and shipped by boat to San Blas in Nayarit. From there they were herded on foot across Mexico, either to be dropped off in northern Oaxaca or pushed on to Vera Cruz for final shipment to Yucatan. (Spicer 1980:160)

Agriculture: Sugarcane was brought to Mexico in the early Spanish colonial times and grown in southern Sonora in the seventeenth century. Basilio (1890 [1634]:223) cited *sana* as the term for "*caña de comer, caña de azúcar*," and indicated *iorisana* as the word for "*caña de Castilla*" (Basilio 1890 [1634]:148).

Food: Sugarcane is used to sweeten food, especially during *Animam Mikwa* (All Souls' Day) on November 1. Pieces of sugarcane are put on the *Animam Mikwa* table as an offering for those who have a sweet tooth and favor sugarcane.

*Sorghum halepense

Common names: YONSO VASO; *zacate Johnson*; Johnson grass

Arizona and Sonora. Robust perennials, forming tough rhizomes and dense colonies. In Arizona and Sonora it is common in disturbed habitats, especially in agricultural regions, and occasionally is found in natural areas along arroyos and washes. In the Yoem Bwiara it is also found in low, seasonally wet places such as roadside ditches and coastal savannas. Johnson grass is native to the Mediterranean region.

Construction: Rosario Vakame'eri-Castillo told Felipe that when the Yoeme first came to Arizona they used Johnson grass for house roofing. They fashioned it into bundles and tied it across roof beams.

Arizona: Tucson, Anklam Road, northwest of Tumamoc Hill, 19 Aug 1983, *Bowers 2709*.

Sonora: Guaymas, *Felger 84-400*. Southeast of Potam, 15 Dec 1988, *Felger & Molina*, observation.

Sporobolus airoides

Common names: TAVA'I; *zacatón alcalino*; alkali sacaton

Arizona and Sonora. Large perennial clumping grasses commonly to 1.5+ m tall. Often in low-lying alkaline or saline soils, including on floodplains and playas. It is common on sandy soils among the coastal mudflats west of Pótam.

Tava'i refers only to this grass and derives from *vaso*, the generic term for grass.

Construction: This grass was used for thatching on house roofs. It was laid perpendicular across carrizo (*Arundo donax*) poles. The house will not leak when roofed in this manner.

Arizona: Tucson, West Branch of Santa Cruz River, 3 Sep 2001, *Mauz 21-131*.

Sonora: East of Chiinim, small stabilized dune in coastal mudflat, *Felger 88-572*.

Sporobolus cryptandrus

Common names: VAI TAVA'I (*vai*, water + *tava'i*, the word for *Sporobolus airoides*); sand dropseed

Arizona and Sonora. Tufted perennials. The base, or sometimes the entire panicle is partially enclosed by an enlarged, inflated leaf sheath; the upper part of the panicle often becomes free with spreading branchlets.

Common along the coast in the Yoem Bwiara on sandy soils of strands and beach dunes, and also widespread elsewhere.

Construction: This is one of the various perennial grasses used for roofing in Sonora.

Arizona: Tucson, West Branch of the Santa Cruz River, 17 Aug 2001, *Mauz 21-107*.

Sonora: Estero Soldado, *Felger 84-417*. Chiinim, *Felger 88-564*.

***Triticum aestivum**

Common names: TIIKO; *trigo*; wheat

Arizona and Sonora. Cool-season annuals, the spikelets are large and the varieties grown in Sonora and Arizona usually have very long, scabrous awns. Wheat plants are sometimes seen along roads in Sonora, growing from grain fallen from passing trucks.

The word *tiiko* is derived from the Spanish name.

Agriculture: Winter/spring wheat is one of the major agricultural crops in Sonora and Arizona. Wheat was an early Jesuit introduction.

The missionaries early on engaged in a policy in which, "each mission was to achieve economic self-sufficiency and stability, to eliminate totally the need to hunt and gather, in short, to reach the level of substantial surplus production" (Hu-DeHart 1981:36). Wheat eventually provided a dependable late spring-harvested crop that the Jesuits hoped would substitute for or augment wild harvests (such as mesquite), and thus encourage people to be settled at the missions. During the latter part of the seventeenth century, agriculture in the western Río Yaqui area had become highly productive. By this time the Yoemem were producing surplus livestock and wheat, which the Jesuits used to supply missions in Baja California and Pimería Alta (Spicer 1980).

Poaceae, *Triticum aestivum*, north of Guaymas. 12 Mar 2017 (SC).

Wheat planting and production in the Yoem Bwiara in the 1930 was described by Studhalter (1936:123):

> The planting of wheat is done by hand, immediately behind the plow, in November and December. The only moisture received comes from the meager winter rains. The soil at the river's edge is not suitable for wheat, the fields of which are found from one to three miles from the river, chiefly at Vicam village. Harvesting is done by hand with sickle or a small scythe. Threshing is done on a dirt floor some twenty feet in diameter, which has been wetted and made smooth and hard. Horses, which are owned by 'all the wheat farmers,' are driven over the wheat tops on the floor until the threshing is complete. If horses are not available, the tops are beaten with mesquite or other poles, after which they are removed by hand and the winnowing done during a wind with a large wooden shovel. This method is very much like those portrayed in pictures of biblical times.

In modern times, wheat continued to be a major agronomic crop in the Yoem Bwiara, thriving in irrigated fields. The "Green Revolution" had its origin from the intensive agronomic research on wheat centered in Ciudad Obregón.

Beverage: The alcoholic drink called *teswiin* was sometimes made from wheat, but might also be made from barley (*Hordeum vulgare*), corn (*Zea mays*), or sweet potatoes (*Ipomoea batatas*, Convolvulaceae). It was for home use.

Food: Dry wheat can be roasted in a large olla with hot coals, constantly turning it with a stick, such as a piece of saguaro rib. As it is turned, it pops like popcorn. It is then ground into *saktusi* (pinole). Corn and barley can be prepared in a similar manner.

Tiiko posoim (from Spanish *posole*, soup) is prepared similarly to barley posole (see *Hordeum vulgare*). In Sonora, during times of food shortage, tiiko posoim was eaten with *torim*, the indigenous tree-dwelling packrats (*Neotoma phenax*, Rodentia: Cricetidae).

Pinole is also prepared from dry flour by stirring some of the flour into a cup or dish of water or milk. Wheat flour for pinole was one of the major foods taken into the mountains when the Yoemem were fighting Mexican forces during the late 1800s and early 1900s.

The flour is often made into *atole*. It can be eaten with a spoon or by drinking. Cinnamon (*canela*, *Cinnamomum verum*, Lauraceae) is often added to atole eaten at a pahko or baptism.

Some Yoemem joined the Mexican federal army because they found out the *federales* were providing pinole and other foods to soldiers fighting the revolutionaries. One man said he was so glad because he had a lot of food. When they gave him a *mochila* (haversack, like a backpack) full of food (pinole), he was sitting down when he put on his backpack, but when he stood up he fell over backwards because it was so heavy. He laughed about it. This man lived at Old Pascua and was interviewed and recorded by Mario Flores of New Pascua in the 1960s.

Arizona: Santa Cruz wash N of Ina Road, Tucson, 9 Feb 1974, *McLaughlin 7819*. Tucson, University of Arizona, grass garden, 4 Apr 1907, *Thornber 685*.

Sonora: Mex Hwy 15, N of Guaymas, alongside new highway surface, abundant, likely seeded for erosion control, long erect awns, 11 Mar 2017, *Carnahan 2300*.

***Zea mays**

Common names: VACHI; *maíz*; corn, maize

Arizona and Sonora. Maize was the most important pre-contact crop for the Yoemem and across most of the agricultural areas of the New World. When the Spaniards encountered maize and brought it to Spain they called it *trigo de las Indias*. The English did likewise, *corn* originally being their name for wheat and other European grains. Although maize was first domesticated in southern Mexico, it has been in northern Mexico for thousands of years where countless indigenous land races were developed. In modern times, higher-yielding varieties have replaced older indigenous varieties.

Basilio (1890 [1634]:171, 205) gave *bachi* as the word for *maíz*, and *bachec* as a word meaning "*tener maíz*" (to have corn). He reported many words and phrases specific to corn and its uses.

Agriculture and gardens: Corn has been a foundation for traditional Yoeme culture. This fact was documented from the time of the earliest written records. Pérez de Ribas (1645; Reff et al. 1999: 87) wrote, "The seeds these people sow and the foods they process and eat are primarily maize."

There are Yoeme terms for maize in all its stages of development and growth. Some are listed here:

Ava choonim, corn silk.
Avai, fresh, green corn ear or *elote*.
Chiiwe, shelled corn.
Kitti, *masa de maíz* or corn *masa*.
Mako'ochia, a bundle of corn husks, such as used in the *nahi* (hoverfly, Syrphidae) and *choparao* (raccoon) dances. Johnson (1962:274) spelled it *makochiia*.
Momoi, hard, dry, mature corn.
Nao ania, corn world, or literally the corncob world, is everything having to do with corn.
Nao, *naon*, a corncob.
Sanava, cornhusk.
Sita, the young, developing ear, the emerging ear.
Siwe, the young emerging plant or seedling.
Vachia, corn kernel or seed.
Vachi momoi, ripe corn.

In the Yoem Bwiara, corn was traditionally planted around March and again in early fall. It was planted in basins irrigated with river water, or elsewhere close to a water source. Corn for family use is planted in home gardens.

In modern times most of the corn planted in the Yoem Bwiara is tied into bank loans and sold by the government. Almost all of the corn planted today is hybrid corn, but is still called *hiak vachi* (Yoeme corn) as well as *yoi vachi* (Mexican corn).

The traditional varieties, most of which have apparently been lost, produced small ears, and usually only two on a plant. The major traditional varieties from the Yoem Bwiara include:

- Chukui vachi (black corn). This is a traditional variety. Basilio (1890 [1634]:171) indicated chuculi bachi as the term for maíz negro.
- Hiak vachi. Antonia Amarillas-Flores said this corn formed small, short plants and had small ears, and produced only one or two ears. It is more frost- or cold-sensitive than modern strains. It is no longer grown.
- Kaka vachi (sweet corn). This was used mostly for making saktusi (pinole). It was grown until at least the 1930s, but is no longer grown.
- Sea vachi (flower corn). Each ear has multicolor grains such as red, purple, and white. Ornamental "Indian corn" is also called sea vachi. Like the other traditional varieties, it is no longer grown.
- Sikiri/sikili vachi (red corn). Basilio (1890 [1634]:171) cited siquili bachi as the name for maíz colorado.
- Yoi vachi (Mexican corn). This is the modern, commercial corn, which has been grown in the Yoem Bwiara since the 1940s. Each plant can produce 3 to 5 large ears. Yoi vachi is more cold-hardy than the traditional hiak vachi.

Beals (1945:9) reported, "Three kinds of maize are sown: a white maize, *báci bwéu* (maize large), *báci ilíci* (maize small), and *báci séwa* (maize flower, pinole maize). Formerly a large, yellow maize was also planted...Red ears occur occasionally and are kept for remedies."

Beverage and songs: *Vachi vino* (corn wine) was made from fermented corn kernels. Vachi vino was usually brought to the ceremonies by the people who made it as an offering for all the people present. Some said this wine is not regarded as *teswiin*. In later years, corn wine was only occasionally made and has been substituted by commercial liquor. The Corn Wine Ceremony might be part of the San Juan Pahko, held on or about June 24, or for the *Kamiino Pahko* (Our Lady of the Road, *La Señora del Camino*), between June 30 to July 4).

Porfirio Yokiwa told Felipe, "Corn is fermented in a *va'achia* [a large unglazed pottery vessel or olla]. Let it sit for a few days—someone would be guarding it so that nobody (kids) bothers it, that no one steals it. When almost ready, they sing to it, the elders, who know the songs, during the daytime." The Corn Wine Ceremony was part of the longer ceremony, performed during the day and night of the last day [of the pahko]. Men singing the corn songs, accompanied by their gourd rattles (*ayam*), would call both men and women to come and drink the *vachi vino* with them. The men would sing for the fermenting corn and men and women would dance in a circle to the music and songs. The songs may include the gourd rattle in the narratives, "the sound of the gourd rattle makes you want to move in step with the rattle." The songs usually tell about the yellow water of the corn wine, called *sawai va'am* (yellow water). "They sing about it in the songs, being there, and who is going to enjoy it. Men and women drank it, during the day and at night." The subject of one of these songs was *sisikili va'achiam* (red water-ollas). These songs are sung to bring the summer rains, calling the summer rains to the Yoem Bwiara, or similarly in Arizona.

Poaceae, *Zea mays*. Corn garden, Vicam. Jun 1987 (FM).

Deer song: Don Guadalupe Molina from Vícam sang this corn song at Yoem Pueblo in 1979, and it was recorded by Felipe:

AVAIM TE KOITANE – LET'S BREAK OFF THE CORN (EAR)

Empo yo waitavaitakai uhyoyolisi hiusime
Empo sewailo yo waitavaitakai
Sewa vam vepa uhyoyolisi hiusime

Ayamansu seyawailo huyatanaisukuni
Sewa yota vam hisam vepa uhyoyolisi hiusime
Empo sewailo yo waitavaitakai
Sewa vam vepa uhyoyolisi hiusime

You enchanted fresh corn as you are, you are beautifully sounding
You flower enchanted world enchanted fresh corn
On top of the flower water, beautifully sounding (as you are moving along)

Over there in the center of the flower covered world
Enchanted flower's water headdress, you are beautifully sounding (as you are moving along)
You flower covered world enchanted fresh corn
On top of the flower water, beautifully sounding (as you are moving along)

The "headdress of the water" is the bouncing of the water in a pottery vessel (olla, *va'achia*) as it is being carried on a woman's head. Don Lupe told how this song came about:

> A hunter woke up one morning and went to va'achia and discovered it was empty. He called out to his daughter, and she came and he asked her to get some water at the river. She agreed and went off with the *soto'i*. On the way she walked by some people who were picking corn, and she looked at them and continued to the river. When she got to the river she filled her olla and put a *moso'okia* (headring of a cloth coil) on her head and lifted up the olla and placed it on her head. She started walking back to her house and once again got to the place where the people were picking corn. She said to herself, "I think I will take a corn (ear)." She put down her olla and went into the field and broke off a fresh ear of corn. She went back to her olla and placed the fresh corn in the water. She again put the headring on and again placed the olla on her head and proceeded. As soon as she was not even ten steps away the corn in the field started to sing the song we have here.

The water vessel or olla is called *va'achia* when holding drinking water supported by the three-branched mesquite stand (see *Prosopis*, Fabaceae). At all other times the olla is called *soto'i*.

Music: The drumstick for the water drum for deer songs is "wrapped in cornhusks and tied by a spiraling cord" (Wilder 1940:45; 1963:167).

Food: Maize in one form or another featured in nearly every traditional meal, from tahkaim (tortillas) to atole, tamales, and hominy. Basilio (1890 [1634]:227, 225) listed *tusi* as a term for "*harina, masa*" (flour, dough); although "tusi" refers only to "something ground up," the word is "associated primarily if not exclusively with maize" (Merrill 2012:220). Basilio (1890 [1634]:172) also cited *bachitusi* as the specific term for "*masa de maíz*." He provided *táscari* as the name for "*tortilla, pan de maíz*" (corn tortilla; Basilio 1890 [1634]:225).

Tortillas made from corn masa would be served at almost every meal. For much of the twentieth century, many Yoeme women made and sold tortillas for their livelihood, not only in the Yoem Bwiara but in such places as Hermosillo and in Arizona, and many worked in tortilla "factories" making both corn and flour (wheat) tortillas. *Atole* made with corn flour was second in importance to tortillas, but was not consumed every day. The usual daily fare in the Yoem Bwiara would be beans and tortillas and a little bit of meat when available.

Napo vaki (ash boiled) is corn hominy. It is made in a large pot or bucket over a fire, with water, corn kernels, and white ashes of mesquite. In modern times commercial lime has been substituted for mesquite ashes. Napo vaki usually was simmered overnight, during which time the kernels become soft. The women then could make fresh corn tortillas in the morning from the napo vaki and save the rest to be used as an ingredient in *mun posoim* (bean *posole*/soup).

Rosalio Moisés wrote, "While we lived at Sasco, my grandmother sold tortillas and pinole to the Mexican workmen at the encampment. She made pinole by toasting corn in a pot; part of the corn would pop and the rest would parch. Then she ground the corn on a metate and cooked the meal with either water or milk and sugar" (Moisés et al. 1971:43).

A special food called *napa nohim* (ash tamales) was made from fresh corn; it was like a large bread—sort of like a fresh tamale but not wrapped. Felipe tells us, "It was like a round cake, baked in an oven. I ate it in Potam in the 1980s, delicious." This was a common, traditional Yoeme food, but has seldom been made since about the 1980s. Green-corn tamales, called *ava nohim*, are prepared and eaten in December.

Medicine: *Ava choonim*, corn silk, is boiled and when cooled the tea is consumed to treat *ko'okosi sise'eteko* (painful urination). Corn silk mixed with chiltepins and tobacco is used to treat earache and pain (see *Capsicum annuum* and *Nicotiana rustica*, Solanaceae).

Unidentified grass

Common name: CHO'OKO VASO (salty grass)

We were told this is a clumping or tufted grass growing in sand along the coast of Sonora. Ignacio Amarillas identified a plant of *Brachiaria fasciculata* (browntop signalgrass) as *cho'oko vaso*. Most likely other grasses are included in the concept of *cho'oko vaso*.

POLYGONACEAE – BUCKWHEAT FAMILY

Antigonon leptopus

Common names: MASA'ASAI (*masa*, wing + *sai*, older brother; older brother's wing), MASA'ASAI WIROA (*masa*, wing + *sai*, older brother + *wiroa*, vine); *san miguelito*; queen's wreath

Arizona (cultivated) and Sonora. Perennial vines from tuberous roots, often covering shrubs and trees; stems and leaves drought-deciduous. Masses of pink fllowers are produced on slender, branched inflorescences, the flowering branches terminating in tendrils. Flowering most of the year, especially during the summer rainy season. The fruit (an achene) is about 1 cm long, hard and nut-like

Common and widespread in Sonora, including arroyos and canyons, rocky slopes, riverbanks, and sometimes on bajadas.

Food: The small, tuberous roots are edible (Uphof 1968) and Yoemem in Tucson say the small tubers were cooked as a vegetable. In 1887, Edward Palmer reported "San Miguelito; its long black wiry roots develop large tubers at intervals, which have a pleasant nutty flavor and are an article of food with the Yaqui Indians" (Watson 1889:73). Guarijíos and Mayos likewise savored the tuberous roots (Yetman and Felger 2002; Yetman and Van Devender 2002).

Polygonaceae, *Antigonon leptopus*, south of Hermosillo. 26 Sep 2013 (SC).

Gardens: Queen's wreath is cultivated as an ornamental flowering vine in gardens in Arizona and Sonora.

Narrative: Giddings (1945) told of a girl named *Masa'asai* who was turned into a tobacco plant by God (see *Nicotiana rustica*, Solanaceae).

Arizona: Tucson, garden south of University of Arizona campus, *Molina 15 Nov 1987*.

Sonora: North-northeast of Las Guásimas, 16 Dec 1988, *Felger & Molina*, observation. Kompuertam, 12 Mar 1989, *Felger & Molina*, observation. Guaymas, *Palmer 59 in 1887* (Watson 1889).

Rumex hymenosepalus

Common names: KAU VATTAI (mountain *Rumex*); *cañaigre*, *hierba Colorado*; Arizona dock

Arizona. Perennial herbs with dark-colored tuberous roots to about 15 cm long, described as "beet-like." The leaves are relatively large.

The plants usually grow in seasonally moist soils and floodplains, such as along irrigation ditches near Yoem Pueblo and sandy-gravelly soils of the Santa Cruz River near Marana.

Medicine: The plant is a highly regarded medicinal plant. Preparations of the root are used to treat sore throat, loose teeth, and to clean the teeth and mouth. The roots are boiled, and the liquid is used as a gargle or rubbed on the teeth. *Kau vattai* also has been used for babies when teething. Teresa Amarillas, from Pótam, told Felipe that for "kau vattai, boil roots for sores and wash the sores with it; in Spanish it is called *yerba colorada*; also used for *vokiam* [fever blisters])".

Kau vattai is used for young children with a sickness called *sava*. This is a sickness, which is like a desire for something good to eat, but the young can't obtain it, and the tongue becomes white. They say if you eat something delicious in front of a child, it would be better that you eat it elsewhere, away from the child. If the child gets some of the food, the *sava* is less severe. The remedy is used after the tongue appears white.

People from Sonora would come to the Marana region to obtain kau vattai, and Sonoran Yoemem often asked people from Marana or Tucson to bring some of it next time they come to Sonora. In the 1980s, Teresa Amarillas would go to Marana to visit her uncle, Tomas Martinez, and ask boys to go to the river and dig up some kau vattai for her to take back to Pótam. When the boys brought the fresh, potato-like tuberous roots, she would rinse them, cut them into thin slices, and dry them in the sun. People in Pótam would go to her house to buy kau vattai.

Arizona: Tucson, *Clara Fish 30 Mar 1893*. Tortolita Mountain Park, 17 Mar 2017, *Lindley 124*. Fresnal Village, "stems roasted as wild rhubarb, *Nabhan 513*.

Polygonaceae, *Rumex hymenosepalus*, Santa Cruz County, Arizona. 25 Mar 2013 (SC).

Rumex inconspicuus

Common names: KAU VATTAI, VATTAI, VAEKA'A; *hierba colorado*

Sonora. Robust annuals, often with a large taproot; emergent from shallow muddy water or growing on wet soil. The leaf blades are 6 to 30 cm long and 2 to 10 cm wide, and more or less elliptic (oval), the leaf stalks (petioles) can be 2 to 20 cm long. Unlike *Rumex hymenosepalus*, this annual species does not produce tuberous roots.

Common in lowland areas of the Yoem Bwiara, including roadside ditches, along washes, and wetlands. This is the only species of *Rumex* known from the Yoem Bwiara.

Medicine: In Pótam, an infusion of *kau vattai* is used as a gargle as a remedy for loose teeth and also for teething babies. Cruz Matus (in Guaymas, 1985) and José María Jaimez (in Kompuertam, 13

Mar 1989) both said the vaeka'a is almost the same as *vavis* (*Anemopsis californica*, Saururaceae) but has larger leaves, and is good (as a remedy) for sunstroke or heatstroke ("*es casi igual pero tiene una hoja mas grande, sirve para las soledades lo mismo que vavis*"). The roots are said to be "good for medicinal tea" (Nacho Gonzalez-Piña and others, Las Guásimas, 2 March 1985).

Sonora: San Carlos, roadside ditch, *Felger 84-551*. Las Guásimas, *Felger 85-297*. Stream at Rancho Bacatetito, *Felger 89-164 & Molina*.

PORTULACACEAE – PURSLANE FAMILY

***Portulaca oleracea**

Common names: BWAAROM; *verdolaga*; purslane

Arizona and Sonora. Summer annuals. The plants are succulent and have flat leaves. The small, bright yellow flowers open in the morning sunlight, usually last only a few hours, then become deliquescent (dissolving or melting).

Widespread and common in weedy places around homes and on farmlands. Verdolaga is sometimes sold at swap meets in Arizona and in markets, especially in Sonora. Nearly worldwide; apparently not native in Arizona and Sonora.

Food: Much use is made of bwaarom as greens. The plants are collected, washed, and lightly boiled. Often the leaves and tender youngest stems are picked off for cooking, and the larger stems discarded. *Bwaarom* may be variously prepared, often with beans and flour to make gravy, but also with onions, garlic, tomatoes, and salt.

Arizona: Tucson, University of Arizona, *Thornber 21 Jul 1903*.

Sonora: Guaymas, *Felger 84-403*. San Carlos, *Felger 95-79*.

PRIMULACEAE – PRIMROSE FAMILY

Bonellia macrocarpa subsp. **pungens** [*Jacquinia macrocarpa* subsp. *pungens*, *J. pungens*]

Common names: TAHSI'O, TASI'O, TASSIO; *san juanico*

Sonora. Small trees with a thick trunk and a dense, leafy crown. The leaves have a sharp, terminal spine and the flowers are small and bright orange. The fruits are nut-like and hard-shelled, 2.5 to 3 cm long. There are several seeds embedded in a fleshy, gelatinous pulp. Flowering and fruiting may occur at various seasons, with peak flowering from late spring to early summer. Bernaldo Valencia (in Tucson 1994) said, "When the fruit are falling down, at that time [there are] lots of *viicham* (wasps)."

This tree is widespread through the Yoem Bwiara and is notably common near Las Guásimas.

Adornment and dye: In 1887, Edward Palmer recorded, "the flowers are used by the Indians to give a durable yellow color to the palm-leaves used in making baskets, etc., and they are also strung

Portulacaceae, *Portulaca oleracea*, San Carlos. 24 Jan 2016 (SC).

Primulaceae, *Bonellia macrocarpa*, San Carlos. 11 Mar 2015 (SC).

like beads and worn for ornament" (Watson 1889:59).

Beliefs: One or two of the fruits on a necklace can be used to ward off evil. A leafy twig or small leafy branch hung on the wall outside the door of a home protects the house from evil—a feat accomplished because of the sharp points on the leaves. Mateo González said you should not sleep under the tree because you will have nightmares, and can even go crazy.

Fishing: The mashed bark was used as a fish poison: "*Cuando los pescadores quieren matar pescados, cinco o seis de ellos se juntan y se repartan la cáscara del árbol de San Juanico. Entonces lo machacan y lo llevan al mar, y en pozo donde hay pescados, lo tiran. Se emborrachan los pescados, y ási los puedan matar*" (Johnson 1962:66).

Medicine: The fruit is mashed and rubbed on the body as a remedy for "iron-poor blood" or "poor blood," or for the same purpose one may bathe with the mashed seeds or fruits.

The fruits are regarded as a good treatment for sores and gunshot wounds. For this purpose the fruit is allowed to dry, then ground with leather scraped into the powder, and the mixture applied to the afflicted area. The fruit is also used medicinally for sinusitis, and the flowers as a remedy for earache. The flowers are put in a liter of water and the liquid placed in the ear.

Sonora: San Carlos, *Barr 62-831*. North-northeast of Las Guásimas, *Felger 88-606*.

PUNICACEAE, see LYTHRACEAE

RANUNCULACEAE – RANUNCULUS FAMILY

Clematis drummondii

Common names: CHIVA'ATO HIMSI (goat's moustache), CHIVA'ATO HIMSITA SAILA (little-brother of goat's moustache); *barba de chivata*; old man's beard, Texas virgin's bower

Arizona and Sonora. Robust perennial vines. The flowers are cream-white and the feathery fruits are in rounded clusters.

Growing in trees along river-bottom hedgerows in the Río Yaqui valley and in southern Arizona often along arroyos and canyon bottoms.

Medicine: José María Jaimez said that to treat "a blood clot in the brain" one can wash the head with an infusion of this plant.

Arizona: Tucson, Santa Cruz [River] Bottoms, *Thornber 1 May 1916*.

Sonora: 1 mi SE of Potam, near Río Yaqui, *Felger 23 Nov 1987 & Molina*, observation. Rancho Bacatetito, 13 Mar 1989, *Felger & Molina*, observation.

Ranunculaceae, *Clematis drummondii*, Sierra El Aguaje. 8 Mar 2020 (SC).

RESEDACEAE – MIGNONETTE FAMILY

Forchhammeria was formerly placed in the caper family, Capparaceae.

Forchhammeria watsonii

Common names: HI'ITO; *palo jito*; lollipop tree

Sonora. Trees of lowlands and coastal plains, 3 to 8 m tall, with a notably thick, usually solitary trunk and smooth bark. "Lollipop tree" refers to the shape of the tree and its dense, dark green crown. The leaves are nearly evergreen, narrow and thick. The flowers are small, without petals, and in dense clusters; the male (pollen-bearing) flowers are yellow and female (fruiting) flowers are purple, and are on separate trees. Flowering occurs mostly in March and April. The fruits are about 1 cm long, purplish to reddish-orange, with a sweet pulp, and ripen in May and June and sometimes again in November.

Food: The fruit is boiled and eaten. However, it is said that only certain, gifted people are able to prepare it so that it will taste good; for others it will be bitter. Mateo González said the fruits are eaten like beans, called *hi'ito vakim* (hi'ito boiled) and that the fruits were eaten by people "who used to walk in the mountains." Rosalio Moisés recalled that "The tiny red fruit the jiito tree...can be cooked like beans" (Moíses et al. 1971). Johnson (1962:263) reported that the fruit was pit-baked overnight, in the same manner as for agaves: "*la fruta se cuece en un horno en la tierra por toda una noche, (el mismo proceso se usa con el mezscal).*"

Medicine: Hi'ito fruit are mashed with stems of *sevii* (*Cylindropuntia thurberi*, Cactaceae) and fruits of *sita'avao* (*Vallesia glabra*, Apocynaceae), and the mixture applied to sores that are slow to heal.

Sonora: Cardonal near Empalme, *Felger 85-636*. Guaymas, *Palmer 167 in 1890* (ARIZ, NY). North-northwest of Torocobampo, Sierra Bacatete, 14 Dec 1988, *Felger & Molina*, observation.

Resedaceae, *Forchhammeria watsonii*, San Carlos. 11 Apr 2019 (SC).

RHAMNACEAE – BUCKTHORN FAMILY

Condalia globosa

Common names: HU'UPA KEKA'ALA (mesquite with mange, dog mange); *crucillo*; bitter condalia

Arizona and Sonora. Densely-branched shrubs often 2 to 3 m tall, with very hard wood, thorn-tipped branchlets, small leaves, and small, yellowish, fragrant flowers.

The same name is also given to *Senegalia greggii* (Fabaceae).

Widespread but seldom common in the Yoem Bwiara; lowland desert and xero-riparian habitats.

Music: The wood is sometimes used to make the musical rasper.

Weapons: The wood was used for arrow mainshafts.

Arizona: North Maricopa Mountains, Sonoran Desert National Monument, 8 Jan 2004, *Mauz 24-2*.

Sonora: Between Potam and Chiinim, small dune, *Felger 88-568*. Rancho Bacatetito, 13 Mar 1989, *Felger & Molina*, observation. Guaymas, 10 Mar 1910, *Rose et al. 12599* (US).

Sarcomphalus amole [*Ziziphus amole*, *Z. sonorensis*]

Common names: NARANHIO; *amole*, *saituna*

Sonora. Shrubs or small trees bearing stout thorns. The leaves are 4 to 10 cm long, mostly oval, and dark green. The flowers are inconspicuous and the fruits, about 1 cm wide, are red-orange when ripe.

The name, *naranhio* (from *naranja*, Spanish for orange tree) provided by Alfonso Leyva-Flores at Las Guásimas, refers to the citrus-like resemblance of the tree and its leaves.

Rocky slopes, arroyos, canyons, and fine-textured soils in low places; seldom common. Sometimes near the shore behind rocky beaches, or on slopes or alluvium near the beach.

Medicine: It is said to be a remedy "good for diarrhea, boil the bark" (Alfonso Leyva-Flores, 16 Dec 1988).

Sonora: East of Las Guásimas, *Felger 88-613 & Molina*. Cañón Bacatete (Vacateve), *Felger 88-641 & Molina*. Guaymas, *Palmer 124 (= 659) in 1887* (UC, US).

Rhamnaceae, *Sarcomphalus amole*, San Carlos. 30 Sep 2018 (SC).

Sarcomphalus obtusifolius var. **canescens**
[*Ziziphus lycioides* var. *canescens*, *Z. obtusifolia* var. *canescens*]
Common names: HUTU'UKI; *abrojo*, *marcha*; graythorn

Arizona and Sonora. Messy-looking, briar-like, grayish shrubs with thorn-tipped, zigzag branches and rather few and small leaves, or essentially leafless in dry seasons. The small flowers are followed by dark, blackish or bluish, fleshy fruits about 1 cm long.

Some people in Pótam said that *marcha* is the local Mexican name.

In southern Sonora it often found near the shore and inland, often along arroyos and low-lying, partially saline or alkaline soils, as well as desert plains.

Food: The fruits are edible either fresh or made into *atole*, although opinions differ on its palatability.

Medicine: The leaves, branches, and stems are mashed in water and the liquid poured on the body to treat a "burning (hot) body."

Arizona: Tucson, West Branch of Santa Cruz River, 19 Jul 2001, *Mauz 21-54*.

Sonora: North-northeast of Las Guásimas, 16 Dec 1988, *Felger & Molina*, observation. Near Kompuertam, 12 Mar 1989, *Felger & Molina*, observation. 6 mi northwest of Guaymas, 28 Feb 1933, *Shreve 6124*.

RIVINACEAE (*Rivina*), see PETIVERIACEAE

RHIZOPHORACEAE – RED MANGROVE FAMILY

Rhizophora mangle
Common names: NAHNAWA'ARA; *mangle colorado*; red mangrove

Sonora. Shrubs to small trees. Red mangrove is distinguished from the other Gulf of California mangroves (*Avicennia*, Acanthaceae; *Laguncularia*, Combretaceae) by prominent stilt roots, shiny green and semi-succulent leaves, and its unique "fruit." Leaves opposite and simple, leathery, and glabrous. The club-shaped, green, fruit-like seedlings, often 20 to 30 cm long, develop while attached to the tree. After dropping into the water and washing into a suitable place in warm shallow water, the roots rapidly take hold in the mud, pulling the still-leafless seedling into an upright position. The leaves and shoot then rapidly develop.

Red mangrove occurs in intertidal zones in esteros and bays; the roots are inundated daily by tidal seawater. Mangroves require alternate tidal flooding and exposure of the roots to the air and cannot tolerate stagnant water. Mangroves were abundant in all of the quiet bays and tidally-draining esteros of the Guaymas region such as Guaymas harbor, the estero at Miramar, Estero Soldado, and Bahía San Carlos. Partly because mangroves harbor the unpleasant biting flies called *jejenes* (*Culicoides* and/or *Dasyhelea*, Diptera: Ceratopogonidae), the mangroves have been largely eliminated from places where people live. Oysters (*Crassostrea*, Mollusca: Bivalvia) often form thick colonies on the stilt roots, but have been largely extirpated in the region by overharvesting.

South of Guaymas, south of the desert, the mangroves extend in a coastal band into the tropics. The demise of the Río Yaqui and Río Mayo is certainly having an adverse effect on the mangroves. Mangroves north of Guaymas occur in disjunct pockets.

Sonora: Bahía San Carlos, *Beatty Dec 1960*. Guaymas Bay, tide flats at edge of bay, common, flower & fruit, 29 Dec 1951, *Blakley B-822* (DES). Chiinim, 13 Dec 1988, *Felger & Molina*, observation. Between Empalme and Guaymas, 27 Feb 1933, *Shreve 6119*.

Rhizophoraceae, *Rhizophora mangle*, Estero Soldado. 11 Feb 2018 (SC).

ROSACEAE – ROSE FAMILY

***Cydonia oblonga**

Common names: MEMRIO; *membrillo*; quince

Quince trees are cultivated for the fruit in southern Arizona and Sonora. A popular jellied sweet, called *membrillo* in Sonora, is made from the fruit. Quince is native to the Old World.

Sonora: Kompuertam, in home garden, 14 Mar 1989, *Felger & Molina*, observation.

***Malus** hybrids

Common names: MANSAANA; *manzana*; apple

Commercial, store-bought apples. The climate is generally too hot to grow apples in lowland areas of Arizona and Sonora.

***Prunus armeniaca**

Common names: ALVERKOOKI; *albaricoque*, *chabacano*; apricot

Commercial, store-bought apricots, and sometimes cultivated in Arizona.

Food: Some apricot trees were grown at Yoem Pueblo for their edible fruit from the 1950s to 1970s, at nearby Yoeme farming communities, and also in Arizona in Eloy and elsewhere. The fruit is enjoyed by the young and old.

***Prunus domestica**

Common names: SILWEELA; *ciruela*; plum

Commercial plums are store-bought and sometimes cultivated in Arizona.

***Prunus persica**

Common names: RURAHNO (from the Spanish *durazno*); *Durazno*; peach

Arizona and Sonora. Peach trees were grown in home yards in Yoem Pueblo from the 1950s to 1970s. Sometimes the branches would break from the weight of the fruit. Peaches are sometimes also grown in the Yoem Bwiara and apparently were introduced in early Spanish colonial times. Basilio (1890 [1634]:217) listed *iurasno* as the word for *durazno*.

Beliefs: In Marana, a bone sometimes was placed on a peach tree branch as an offering to allow the tree to give more fruit.

Food: The fruit is enjoyed by the young and old.

Sonora: Las Guásimas, home garden, 16 Dec 1988, *Felger & Molina*, observation. Wiivisim, 15 Mar 1989, *Felger & Molina*, observation.

***Rosa** hybrids

Common names: ROOSAM; *rosas*; roses

Arizona and Sonora. Roses are grown in home gardens. People in Marana sometimes made rose trellises from saguaro ribs (see *Carnegiea gigantea*, Cactaceae).

Sonora: Kompuertam, home garden, 14 Mar 1989, *Felger & Molina*, observation.

RUBIACEAE – MADDER FAMILY

***Coffea arabica**

Common names: KAAPE; *café*; coffee

Arizona and Sonora. *Kafe* is a more recent term, while *kaape* is the older term.

Medicine: For sunstroke, strong, fresh, hot black coffee is given to the patient rather than an ice drink. It is said that an ice drink may make the victim go into shock and the patient is not given anything cool to drink until he or she "cools down." Use of this remedy was witnessed in Marana and Pótam.

Hintonia latiflora [*Coutarea latiflora*]

Common names: KOPALKIN; *amargo*, *copalquín*

Sonora. Slender trees to 5 m tall. The leaves are 5 to 10 cm long, follow the summer rains, and are drought-deciduous. The flowers are magnificent,

Rubiaceae, *Hintonia latiflora*, San Carlos. 24 Aug 2017 (SC).

white, very fragrant, 6 to 9 cm long, and seen during the summer rainy season. The fruits ripen in early fall.

Kopalkin grows in mountains, especially in large canyons, in the Yoem Bwiara. The geographic range extends from southern Sonora and southwestern Chihuahua to Guatemala. It should be brought into cultivation in frost-free regions, both as an ornamental landscape plant and as an agronomic crop for the bark, which contains quinine.

Medicine: Since the 1990s, many people have used it to treat high blood sugar and diabetes. A thumb-sized piece of the dry bark is broken off and boiled, and the tea consumed when it cools. The tea is very bitter (hence the name *amargo*). The bark is readily available in Mexican herbal shops but in Sonora the people often prefer to harvest it themselves, usually from *Vakatetteve Kawim* (Sierra Bacatete).

In 2000, Juana Flores-Osuna, in Pótam, had high blood sugar and her medical doctors offered her western prescription drugs, which she did not take (but did not tell her doctors). On the advice of others in Pótam she used *kopalkin*. She had sent her son to get it from the mountains. She said that it was *chiivu*, very bitter. On her next visit to her doctors, her blood sugar was conrolled and they were amazed, but she did not tell them about the kopalkin.

The bark is used as a febrifuge (fever control) and anti-malarial remedy in many parts of Mexico. The trunks, especially if near a road, are invariably scarred from bark collecting. The bark is harvested extensively in the Álamos region, made into capsules in Navojoa and sold commercially, and is likewise harvested in many other parts of Mexico (Felger et al. 2001). The tea is used as a purgative for intestinal parasites, as an energy tonic and to "restore the blood," and to reduce fevers. This tea is often used when the seasons change from hot to cool weather.

Sonora: Guaymas, *Johnston 3099* (CAS). San Carlos, *Johnston 4358* (CAS). Deep cañons near Guaymas, *Palmer 298 in 1887* (K, US).

Randia echinocarpa

Common names: HOSOINA, PIISI; *papache*, *papache borracho*

Sonora. Sprawling, irregularly shaped, spinescent shrubs with rigid, hardwood branches. The leaves are somewhat oval and hairy. The pure white flowers have gardenia-like fragrance especially strong at night (*Gardenia* is a member of the same family). Both bisexual and unisexual flowers may occur on the same plant, or male and female flowers may occur on separate plants. The distinctive, baseball-sized, hard-shelled fruits are ornamented with large, flattened, and blunt-spined projections. The fruits are almost invariably pecked or chewed open by birds or other animals.

Basilio (1890 [1634]:213) named *hosoina* as a term for "*papache, árbol*," likely indicating this species.

This *Randia* is characteristic of tropical deciduous forests in southern Sonora and likely approaches the eastern margin of the Yoem Bwiara.

Food: As with other randias, the gooey blackish pulp of the fruit is somewhat sweet and edible. However, it is said that eating too much of it causes one to become drunk, hence the name *papache borracho* also applied to several other *Randia* species in Sonora.

Harvesting: The branches of this and other *Randia* species naturally grow at right angles to the main stem. A T-shaped portion of the stem is cut out and lashed to a carrizo (*Arundo donax*, Poaceae) pole to make an organpipe cactus fruit-picker, the *hiap taka'aria* (see *Stenocereus thurberi*, Cactaceae).

Medicine: The fruits are widely sold for medicinal purposes in markets in Sonora and elsewhere in Mexico, and are used the Yoem Bwiara.

Sonora: [Herbal store], Ciudad Obregón, 19 Feb 2016, *Semotiuk 129* (USON). Huatabampo, Río Mayo Region, *Van Devender 95-1068*.

Randia obcordata

Common names: PIISI; *papache*

Sonora. Shrubs with rigid, opposite and divaricate branches (decussate, where each pair is oriented 90 degrees from the pair above and below), and often stout thorns. Leaves often broadly obovate. Flowers white, about 3 to 6 mm long, flowering in June and July with summer rains. Fruits globose, 8 to 10 mm wide, blackish, smooth and glabrous. Without the small flowers and small blackish fruits, which are ephemeral, the plants seem indistinguishable from *Randia thurberi*.

Basilio (1890 [1634]:222) indicated *pisi* as the word for "*papachito, árbol*," possibly referring to this species or to *Randia thurberi*.

Known in the flora area by only a few specimens but apparently widespread including hills, mountains, and bajadas.

Religion: The stems are used to make crosses and rosary beads in the same manner as for *Randia thurberi*.

Sonora: Northeast of Las Guásimas, *Felger 88-609*. On high gravelly mesas near Guaymas, *Palmer 648 in 1887* (US).

Randia thurberi

Common names: HUPSI, PIISI; *papache borracho*

Sonora. Spinescent shrubs with rigid, divaricate, and decussate branches, and paired, often stout spines. Leaves drought-deciduous. The flowers are white and fragrant like a gardenia, about 10 to 12 mm long. Flowering with summer rains. The fruit are globose, hard-shelled, mottled green and white, 1.5 to 2.5 cm wide in diameter, ripening at least in spring, and the mesocarp (pulp) black and gooey. The plants are similar to *Randia obcordata* except for the flowers and fruits.

Widespread in the Yoem Bwiara, especially in lowlands; bajadas, plains, and lower slopes.

The shrub is used as a substitute Christmas tree, if conifer trees are not available.

Beliefs: Cross-shaped, trimmed stems are hung anywhere inside or outside a home, for instance above a door, to ward off evil. This is one of several hardwood shrubs used to make rosary beads (see *Lycium andersonii*, Solanaceae).

Food: The sweet, gooey black pulp of the fruit is edible: "It is eaten but if one eats a lot of it, it will make you drunk (*Se come pero si se come mucho, te emboracha*)." It is said that one can safely eat three or four fruits but no more. Mateo González said eating too much of the fruit can make you vomit. (Also see *Diospyros sonorae*, Ebenaceae.)

Sonora: Cañón Nacapule, *Felger 85-871A*. Guaymas, *Palmer 299 in 1887* (US). South of Peon on Mex Hwy 15, *Turner 61-43*.

Rubiaceae, *Randia thurberi*, San Carlos. 30 Sep 2018 (SC).

RUTACEAE – CITRUS FAMILY

*Casimiroa edulis

Common names: HAPA'AWI; *chapote*, *zapote blanco*; white sapote

Sonora. This tree forms dense shade and has a thick trunk. The leaves are alternate, palmately compound, with (3) 5 large l

Arizona and Sonora. Selected, grafted varieties of sweet oranges are widely grown in Sonora and variously in Arizona. In the Marana–Tucson region, sweet oranges are often susceptible to freezing damage.

Ouwo refers to trees bearing edible fruit.

Orange trees were being grown in the western Sinaloa–Sonora region by the early seventeenth century. Pérez de Ribas (1645; Reff et al. 1999:88) reported, "plants brought from Castilla grow well in these regions, especially orange and fig trees."

Other citrus grown in Sonora include *mandarina* (tangerine), *toronha* (*toronja*, grapefruit), and *tornoha roosa* (pink or rose grapefruit).

Beliefs: It is said that a red ribbon is tied on the branches to prevent an eclipse of the sun from damaging or destroying the plant. A red ribbon also protects against *sivowame* (evil eye) or a desire for the fruit that is not fulfilled, which would cause damage because of lust. However, a red ribbon or cloth is also used to ward off birds.

Beverage: The leaves are made into a refreshing tea (from Maria Cruz Miranda, Marana, 1979).

Sonora: Koasepe, 14 Mar 1989, *Steen, Felger, & Molina*, photo (red cloth among leaves to keep birds out).

*Ruta graveolens

Common names: RUURA; *ruda*; rue

Arizona and Sonora. Aromatic perennial bushes about 1 m tall, with bluish-glaucous foliage and yellow flowers. Native to Europe, rue has long been grown in distant parts of the world and was introduced into Mexico in early colonial times.

Gardens: *Ruura* is widely cultivated in home gardens, especially in Sonora.

Medicine: Mixed with *alvaaka* (basil, *Ocimum*, Lamiaceae), rue leaves are rolled up in a small ball or the leaves are fried in olive oil until they begin to pop and crackle, then, with some oil it is wrapped in a small piece of cotton and inserted into the ear to relieve earache (Alfonso Leyva-Flores, Las Guásimas, 16 December 1988). Similarly, at Marana people would take a little bunch of leaves, roll the leaves in the fingers into a ball, apply a little bit of olive oil, and insert it into the ear. For relief of headache and fever, fresh leaves are crumbled between fingers and tied in place on the forehead with a white cloth.

Arizona: Phoenix, Desert Botanical Garden, 10 Jun 1975, *Engard 74-83*.

Sonora: Las Guásimas, *Felger 88-616*. Kompuertam, home garden, 11 Mar 1989, *Felger & Molina*, observation.

SALICACEAE – WILLOW FAMILY

Cottonwood (*Populus*) and willow (*Salix*) trees in the Yoem Bwiara and southern Arizona have soft wood and are winter deciduous. In Arizona, the leaves turn yellow in winter—the cottonwoods more brilliantly than the willows—and new leaves burst forth in late winter or early spring. In southern Sonora the leaves are winter-deciduous although only briefly so, and the leaves scarcely become yellow before falling.

Male and female flowers are on separate trees. The flowers are small, the sepals vestigial, and there are no petals. The flowers are wind-pollinated and borne on an elongated inflorescence called a catkin. The fruits are small capsules and the seeds minute, each with a tuft of long, silky hairs adapted for wind dispersal.

Great gallery forests of cottonwoods and willows lined the rivers of Arizona and Sonora prior to construction of large dams and diversions, as well as groundwater extraction, during the twentieth century. In the hot Sonoran climate, the trees are fast-growing and their biomass enormous. These trees form extensive networks of shallow roots, which are important in slowing soil erosion.

Cottonwoods and willows are often protected or planted as shade trees and for various utilitarian purposes. They readily grow from cuttings of any size. The best time to strike cuttings is in winter before the buds leaf out. The leaves and young twigs of *Populus* and *Salix* are eagerly eaten by cattle and deer, and seedlings and saplings can be eliminated by cattle. For the most part the wood is soft and not highly valued for fuel and construction, but was much used because it was so readily available.

Populus

Common names: AVASO; *álamo*; cottonwood

Cottonwoods are among the most important trees in Yoeme culture. The following cultural information applies to both Frémont cottonwood (*Populus deltoides* var. *fremontii*) and Yaqui cottonwood (*P. mexicana* subsp. *dimorpha*), except where one or the other region is specified.

On the Río Yaqui, *Populus deltoides* occupies the upper portions of the river drainage (central Sonora northward and Arizona) and, while *P. mexicana* occurs along the lower reaches (southward from central Sonora).

Avah naawa (*avas naawa*, a variant) is cottonwood root and *avas veá* is cottonwood bark; the terms *avas* or *avah*, are adjectives,

Boats: Felipe's grandfather, Rosario Vakame'eri-Castillo, and his godfather Anselmo Valencia, told about dugout canoes (*bwaktim*) in the coastal villages in Sonora, such as at Wiivisim. The dugouts were made from river trees, *avaso* or *hooso* (*Albizia sinaloensis*, Fabaceae), anything with a large trunk.

In Arizona, Rosario Vakame'eri-Castillo used cottonwood root to carve model dugout canoes, *kuta bwaktim* (wood dugout).

Construction: The wood is inferior for house construction but has been used for such purposes because it is, or was, so readily available. Roofing poles for the *heka* (*ramá* or ramada), especially in Sonora, are often made from *avaso* branches.

Fuel: The wood is of poor quality for fuel but is used for this purpose because it is (or was) readily available, especially in Sonora.

Cultivation: In the lower Río Yaqui and Río Mayo regions cottonwoods and some willows are planted around ranches, although willows seem more popular around homes and in the towns—probably because of their smaller size. Cottonwoods are readily grown from cuttings, best cut when leafless in winter.

Masks: Cottonwood root, *avah naawa*, is usually the preferred wood traditionally used by Yoeme men in Arizona and Sonora for making pahko'ola masks. The wood is lightweight, strong, and does not split upon drying. Avah naawa from *Populus mexicana* is regarded as superior to that from *P. deltoides*.

Large-diameter wood from the root became increasingly difficult to obtain in southern Arizona in the latter part of the twentieth century as the large trees were vanishing. For example, Tucson mask carvers had a hard time getting cottonwood root after about the mid-1980s. They had been accustomed to getting it in the Marana area. They often asked Tom Kolaz for cottonwood root, which he sometime brought from the Santa Cruz River near Kino Springs east of Nogales.

Wood from the limbs was sometimes resorted to for mask making in Arizona, although it is less desirable than wood from the root because it is more difficult to carve. Tom Kolaz told Richard, "You look at the inside of the mask and you can see the struggle the mask maker had with chiseling out the wood. There are minor cracks and gouges and the interior is not smooth. If he had root to carve, he would be able to make a much better mask." José Guadalupe ("Lupe") Flores, living in Tucson, carved traditional pahko'ola masks from cottonwood root. He would have relatives bring him avaso root from the Río Yaqui because he said it carved better than Arizona cottonwood root (Tom Kolaz, personal communication, 2014).

Wood from the root of Frémont cottonwood in Arizona is lighter in color than that of the Yaqui cottonwood, which is reddish-brown. When Tom Kolaz asked carvers about these differences, some of them said it is because water in the Río Yaqui is richer in minerals, etc. That coloring was one way that Tom questioned some Tucson men selling masks they said they carved in Arizona. Tom would show them the reddish-brown wood and say, "This *álamo* came from the Río Yaqui." They readily admitted the mask came from Sonora, meaning it was not made by the seller, and said, "How did you know?"

Medicine: Fresh green leaves are employed in a remedy for sores on the legs or feet. The leaves are boiled in a bucket, the water cooled, and the legs or feet are bathed with the water.

Religion: Green boughs (leafy branches) are extensively used in Yoeme and Yoreme rituals to represent the plant world during the many religious fiestas through the year. For example, during the Gloria on Holy Saturday cottonwood boughs are placed around the dance ramada. The plaza in front of the church is lined with evenly-spaced leafy cottonwood twigs to mark the edges of the sacred space for the Gloria performances. At such times spectators are not allowed to walk into the sacred space marked off by the line of cottonwood twigs. One or a few men, charged with keeping order, use a leafy cottonwood switch as a badge of authority and to gently enforce the boundaries.

Populus deltoides var. **fremontii** [*Populus fremontii*]

Common names: AVASO; *álamo*; Frémont cottonwood

Arizona. The Frémont cottonwood (*Populus deltoides* var. *fremontii*) is the largest native tree in the Sonoran Desert. It occurs along the Santa Cruz and other river systems in Arizona and northern Sonora, and is still common in certain places. However, due to destruction of the river ecosystems it has become ever scarcer. Cottonwoods are easily cultivated but are seldom planted nowadays because they require large amounts of water.

Arizona: Tucson, University of Arizona campus, in a row to the rear of South Hall, 13 Feb 1915, *Thornber 7029*

(ARIZ, ASU). Tucson, *Toumey 20 Apr 1894*.

Populus mexicana subsp. **dimorpha** [*Populus dimorpha*]

Common names: AVASO; *álamo*; Yaqui cottonwood

Sonora. This cottonwood once formed riverine forests of immense trees along the lower Río Mayo and Río Yaqui, as well as the lower Río Fuerte and other Sinaloan rivers. Pérez de Ribas (1645; Reff et al. 1999:84) wrote, "Along the banks of the rivers there are pleasant valleys full of green groves of poplars [*álamos*] that are free of thorny underbrush." Basilio (1890 [1634]:203) provided *abaso* as the word for *alamo*.

Yaqui cottonwoods once commonly reached more than 20 m in height. These trees had huge trunks, sometime to 2 m in diameter, and a broad crown with long, drooping, leafy branches. The branches and limbs have rather soft and somewhat pithy wood, while the wood of the trunks and especially the roots is considerably firmer. These trees are capable of extremely rapid growth. The trees flower in January or February, and the new leaves appear while the old ones are still being shed.

The leaves are shiny green and of two kinds (dimorphic): Leaves of mature growth are relatively large, have petioles (leaf stalks) 4 to 8.5 cm long and leaf blades 5.5 to 12 cm long, and about as broad as long (more or less triangular. Leaves of the juvenile phase (seedlings, saplings, and slender, smaller, and lower branches of larger trees) are willow-like, relatively small, have short petioles, and narrow (linear-oblong or narrowly lanceolate) leaf blades.

These riverine trees are often host to large clusters of *Psittacanthus calyculatus* (Loranthaceae), a mistletoe bearing showy, bright orange flowers. Tent caterpillars (*Malacosoma*) often defoliate the trees, especially in March.

Since construction of the large upriver dams, the annual or biannual Río Yaqui floods have generally diminished or ceased, and the river reduced to trickles and dusty depressions during dry seasons. By the 1980s, Yaqui cottonwoods were still quite common, although the gigantic trees were far fewer and large willows had also become rare. People in Pótam at that time said there were fewer cottonwoods than in the past, and although many of the trees were large, they were not huge as in earlier decades. Seedlings were sometimes abundant along the lower Río Yaqui in the late 1980s, but saplings were scarce. Young cottonwoods can be eliminated by cattle. By 2005, many of the larger riverine trees were dead, but smaller trees thrived along irrigation ditches, margins of irrigated farmland, and occasionally in roadside ditches in the Río Yaqui valley. In some places, however, such as along the Río Yaqui at the

Salicaceae, *Populus mexicana*, Río Mayo, Navojoa, 10 Dec 2021 (SC).

Salicaceae, Populus mexicana, Rio Mayo, Navajoa. 10 Dec 2021 (SC)

north edge of Vahkom (Bácum), there were still large, healthy cottonwoods, although in November 2006 the riverbed was dry.

The trees are appreciated and often planted or protected around homes and along the larger irrigation ditches and drainage canals that run through the Yoem Bwiara agricultural lands. However, there is also fear from lighting strikes, people are cautioned to stay away from the trees during storms, and many people do not like to have their homes near cottonwood trees. They say that cottonwood trees are struck by lightning more than any other trees.

Utilitarian: *Avaso* wood has been used for wagon wheels in Sonora. Lucas Chavez related, "there was only one wagon in the whole of Tórim in those days [early 1880s] and it had solid wheels cut from cottonwood trees" (Spicer 1988:110). In 2019 we saw a large, homemade wheelbarrow in Tórim with solid wheels made from cross-sections of avaso wood trunk.

Sonora: Tosai Bwia, near Kompuertam (Compuertas) on banks of Río Yaqui, 5 km SE of Vícam Pueblo, 11 Mar 1989 *Felger 89-70 & Molina* (ARIZ, MEXU, USON). Yaqui River, crossing at Corral, 2 Mar 1933, *Shreve 6135*.

Salix gooddingii

Common names: WATA; *sauce*, *sauz*; Goodding willow

Arizona and Sonora. Large shrubs to large trees, the trunk well-formed. Bark on trunk and limbs gray-brown and fissured, the twigs often yellow-brown. The leafy stems and leaves are generally drooping. The leaves are green on both surfaces, mostly 6 to 12 cm long, narrow, and have very short leaf stalks (petioles). Male flowering catkins are yellow. Peak flowering occurs in early spring as the new leaves unfold and some flowering occurs sporadically through the summer.

These trees and shrubs are common in wetlands and are often planted around homes. Willows are common along irrigation canals in agricultural areas of the Yoem Bwiara. Large willow trees were especially prominent with cottonwoods in the riverine forests of the lower Yaqui and Mayo rivers as well as the Sinaloan rivers. Large willow trees began to disappear after construction of river dams in the mid-twentieth century. A number of people told us that *wata* trees along the Río Yaqui were more beautiful and larger before the dams. The trunks were big and there were many very large mesquites. One man said these trees were more beautiful before the dams because there was more water, more water bringing nutrients. By the 1970s and 1980s, large willow trees had become scarce along most of the lower Río Yaqui. In the 1980s and 1990s, willows were still fairly common along irrigation and drainage ditches in Yoeme agricultural lands but were mostly shrubs and small trees.

Goodding willow was a common riverine and wetland tree in southern Arizona and is still common where there is sufficient water. It is the most widespread tree in Sonora, extending to each corner of the state in wetland habitats.

The term *wata* appears in the name of a number of places in Sonora, for example, the Yoreme town of Huatabampo (in the waters of the willow). Basilio (1890 [1634]:190) cited the term *huata* for "*sauz, árbol*."

Basketry: The *wata waari* (willow basket) was made with young, slender branches from trees found along rivers, such as the Santa Cruz River near Tucson. These baskets were made in Arizona until the 1930s and 1940s. Some of these baskets were large and made for sale, some serving as laundry hampers. Mariano Tapia was a well-known basket-maker in Old Pascua. Wata waarim

Salicaceae, *Salix gooddingii*, Santa Cruz County, Arizona. 29 Sep 2019 (SC).

were also made in the Yoem Bwiara from trees on the Río Yaqui.

Cultivation and Fencing: Willows are commonly cultivated in Yoem Bwiara, but are less often grown in Arizona because they require large amounts of water. Like the cottonwoods, they can easily be propagated from cuttings, best made during late December and early January. Fence posts made from live wood can form roots and develop into a new tree.

Household: Some furniture is made from the flexible, slender stems of willow saplings.

Masks: Yoeme artists in Tucson sometimes have used willow wood to carve pahko'ola masks.

Medicine: Willows are used to relieve high fever (Antonia Amarillas-Flores, Pótam, Dec 1988). Willow is also used in combination with other medicinal plants such as *batamote* (seepwillow, *Baccharis salicifolia*, Asteraceae).

Ritual: House crosses, the Holy Cross, *Santa Kuus*, hung on the outside wall of a house, are made from leafy willow branches and decorated with paper flowers. Several willow crosses are sometimes hung in a mesquite tree that might hang over a house; the crosses are put there to protect the house. The old, last year's Santa Kuusim are burned on the vesper of Holy Cross, May 2, around 8 pm. They are set in piles around the house with other perishable holy regalia including old palm crosses and burned. The remaining ashes are spread by the winds in all directions to keep away evil or negative energy. People say if you don't burn your willow holy crosses on May 2, the Gila monster will come to remind you to do so.

In Arizona, fresh, leafy willow branches are sometimes used for the sides and roofing of the traditional *pahko heka* (ceremonial ramada), especially if vaaka (cane or *carrizo*, *Arundo donax*, Poaceae) is not available.

Arizona: Tucson, Santa Cruz [River] bottoms, *Thornber 30 Mar 1913*.

Sonora: Potam Viejo, *Felger 84-1419*. Cuesta Alta, *Felger 85-1374*. Río Yaqui & Highway 15, *Gentry 24 Oct 1961*.

SANTALACEAE – SANDALWOOD FAMILY (includes Viscaceae)

The common mistletoe species in the genus *Phoradendron* have traditionally been treated with the family Viscaceae, however recent studies suggest that these plants should be allied with Santalaceae. Two other mistletoe species in the Yoem Bwiara, *Psittacanthus calyculatus* and *Struthanthus palmeri*, belong to the Loranthaceae.

Phoradendron californicum

Common names: CHICHIHAM; *toji*; desert mistletoe

Arizona and Sonora. A widespread and common parasite on various desert trees and shrubs, especially woody legumes (Fabaceae). The stems soon become drooping and can form large, dense masses. The dark green stems appear leafless with only scale leaves. Male and female flowers are on separate plants. The small yellow flowers are fragrant. The berries are globose, pinkish-orange, and fleshy.

The host plant is indicated in the name, for instance when it is parasitic on brea palo verde (*Parkinsonia praecox*) it is called *cho'i chichiham* (palo verde *chichiham*), and on mesquite (*Prosopis*) it is called *hu'upa chichiham* (mesquite *chichiham*).

Medicine: This mistletoe was sometimes used to treat head sores.

Arizona: Saguaro National Park, Tucson Mountains, *Rondeau 89-1*.

Sonora: West of Estero Soldado, *Felger 84-538*. 3.5 mi by road northeast of Mex Hwy 15, road junction to Tórim, 14 Dec 1988, *Felger*, observation.

SAPINDACEAE – SOAPBERRY FAMILY

Cardiospermum corindum

Common names: TOO VICHOM (bull testicles), TOORA; *farolitos, huevo de toro, tronadór*; balloon vine

Sonora. Perennial, scandent shrubs or vines with tendrils. The leaves are highly divided, the flowers are white or pink and in small clusters. The fruits are 3-lobed and curiously inflated like a miniature balloon or paper lantern.

Balloon vine is abundant throughout the Yoem Bwiara in washes and arroyos, plains, and rocky slopes.

Music: The seeds are put into gourds as rattles (see *Lagenaria*, Cucurbitaceae).

Sonora: 3 mi east of Mex Hwy 15, east of Las Guásimas, 17 Dec 1988, *Felger & Molina*, observation. Guaymas, *Jones 22329* (UCR).

Sapindaceae, *Cardiospermum corindum*, San Carlos. 27 Jan 2018 (SC).

Sapindus saponaria

Common names: TUPCHE; *amolillo, chirrión, jaboncillo;* soapberry

Sonora. Trees mostly 5 to 8 m tall, or to 10+ m in the larger riparian canyons. Trunk smooth and well developed. Leaves 20 to 30 cm long and mostly evergreen, pinnate, with 5 to 10 large leaflets. Flowers in dense terminal inflorescences, small, white, and visited by hordes of flies, honeybees, native bees, and wasps. Fruits 1- or 2-lobed (rarely 3-lobed); each lobe 1-seeded, rounded, leathery, and yellow-brown; seeds hard and rounded.

Basilio (1890 [1634]:227) recorded *tupche* as the name for "*amolillo (árbol)*."

Riparian habitats, including larger washes, arroyos, and canyon bottoms. In western Sonora it does not range north of the Guaymas region but inland it extends along the east side of the desert about as far north as the extent of tropical thornscrub near Onavas.

A related species, *Sapindus drummondii*, occurs in similar situations in southern Arizona and northern Sonora. It has smaller leaves, with 10 to 19 leaflets, and is not frost-sensitive like *S. saponaria*.

Household use: The berries have been used for soap. The generic name, *Sapindus*, derives from "soap of the Indies."

Sonora: San Carlos, *Barr 67-17*. Cañón Bacatete along stream, 17 Dec 1988 & Molina, *Felger*, observation. 3 mi northeast of Ortíz, *Turner 69-47*.

Sapindaceae, *Sapindus saponaria*, San Carlos. 30 Nov 2014 (SC).

SAPOTACEAE – SAPOTE FAMILY

Sideroxylon occidentale [*Bumelia occidentalis*]

Common names: VAPSA; *bebelama;* bumelia

Sonora. Spiny shrubs or trees to about 8 m tall with a well-developed trunk, rough bark, thorn-tipped twigs, and a dense crown of small, grayish-green leaves. The fruits are about 1.5 cm in diameter, bluish-black, fleshy, and 1-seeded.

Seldom common in the Yoem Bwiara, usually found in arroyos and canyons, occasionally on rocky slopes.

Food: The fruit is edible "like a *ciruela*" (see *Spondias purpurea*, Anacardiaceae).

Medicine: *Vapsa* has been used to treat diabetes. Chunks of bark are boiled (*en pura agua*) and the liquid consumed. Sr. Saavedra collects the bark from trees in the Sierra El Aguaje and sells it in Empalme.

Sonora: San Carlos, *Barr 61-142*. 8 mi by road northeast of Vikam Suichi, canyon bottom, *Felger 88-117*. Northeast of Ortíz, *Turner 69-45*.

SAURURACEAE – LIZARD-TAIL FAMILY

Anemopsis californica

Common names: VAVIS; *hierba de manso*

Arizona and Sonora. Perennial herbs with thick, creeping, aromatic rootstocks and also forming long, above-ground stolons. Leaves simple, with long stalks (petioles), the leaf blades ovate or oval, often 5 to 15 cm long. Flowers minute, borne on a cone-shaped inflorescence. White, petal-like bracts at the base of the inflorescence give the appearance of a single large flower resembling a *Ranunculus* or *Anemone* flower (*Anemopsis* is Greek for anemone-like). The plants may flower throughout the warmer seasons and are dormant during the coldest months.

This is a wetland plant, growing in widely-scattered, moist, swampy areas, and at *aguajes* (water holes) in the mountains. It was probably more widespread before the loss of so many wetland places in the twentieth century. For the most part it is encountered in home gardens or purchased from herbal stores.

Hierba de manso is one of the most highly esteemed medicinal plants in western North America and continues to be widely used in Sonora and elsewhere (Felger and Moser 1985; Felger 2007; Moerman 2010, 2020). Members of the Saururaceae are characterized by having ethereal oil cells in the parenchyma tissue and tannin, which may account for the long and varied medicinal use of hierba de manso. The term *manso* may be translated as tame, meek, gentle, mild, quiet, soft, gentle, or lamb-like.

Gardens: *Vavis* is one of most commonly grown medicinal plants in home gardens, especially in Sonora, not only by Yoemem but by many people in the region.

Medicine: The leaves and roots are highly esteemed as a useful remedy for most afflictions, including body aches and pains, sunburn, heat exhaustion, and especially for fevers. Cruz Matus and others in Guaymas (March 1985) said, "*Es remedio casi por todo. Se machaca y pone en un trapo, encima el cuerpo, entonces la dolencia del cuerpo se va. Se toma* [as tea] *también. Las plantas vienen de la sierra donde hay agaujes.*" (It is a remedy for almost everything. It is chopped and wrapped in a cloth, then placed on the body, then the illness goes away. It is also drunk as a tea. The plants come from the mountains where there are waterholes.) At Kompuertam in March 1989, José María Jaimez gave us almost the same information, also telling us, "*Es remedio casi por todo.*"

Vavis is considered a "cool" medicine. Dry rootstocks and stems are generally kept on hand or the plant is grown in the home garden. A common method of preparation is to chop it, place it on the afflicted part of the body and hold it in place with a cloth, and this will alleviate the body ache or pain. It is commonplace to prepare vavis as tea.

Vavis is used as a remedy for rattlesnake bite (also see *Euphorbia*, Euphorbiaceae). It is sometimes used medicinally in conjunction with other herbs, including *brasil* (*Haematoxylum brasiletto*) and mesquite (*Prosopis*, Fabaceae), and elderberry (*Sambucus cerulea*, Viburnaceae).

Wagner (1936:83) reported, "for cuts and bruises...The *yerba del monzo* [sic] is beaten to a pulp. Rosemary, *yerba colorado*, and *alucema* seed, all beaten to a pulp, are added, making a paste. This is applied to the cut or bruise." Rosemary is *Rosmarinus officinalis*, and *alucema* is a common name in Mexico for lavender (*Lavandula*), both in the mint family, Lamiaceae. Here, *yerba colorado* may be *Rumex hymenosepalus* or *R. inconspicuus* (Polygonaceae).

Arizona: Silver Lake [Tucson, along the Santa Cruz River near present-day 29th street], 9 May 1901, *Griffiths 2712*.

Sonora: Kompuertam, in home garden, 13 Mar 1989, *Felger & Molina*, observation.

Saururaceae, *Anemopsis californica*, Cienega Creek, east of Tucson. 25 Apr 2017 (SC).

SIMMONDSIACEAE – JOJOBA FAMILY

Simmondsia chinensis

Common names: HOHOOVA; *jojoba*; jojoba

Arizona and Sonora. Densely-branched shrubs with hard wood, and simple, leathery leaves. Male and female flowers are on separate plants. The nut-like, leathery capsules are 1.5 to 2 cm long and hold a single seed that contains about 50 percent liquid wax or oil.

Simmondsiaceae, *Simmondsia chinensis*, Arroyo La Pirinola, Sierra El Aguaje. 11 Feb 2017 (SC).

Jojoba is common in southern Arizona and western Sonora southward to the Guaymas region in arroyos, bajadas, desert plains, and hillslopes. It is highly palatable to cattle, which can extirpate or seriously reduce it in a local area.

Medicine: Jojoba is said to be "good for swellings." Wax from the seeds is used as a skin treatment.

Shampoo: Oil from the crushed seeds is used to condition the scalp and make the hair lustrous.

Arizona: Tucson Mountains, Tumamoc Hill, 25 Feb 1905, *Thornber 2576*.

Sonora: Miramar, *Felger 85-528*. Cañón Nacapule, *Felger 85-549*.

SOLANACEAE – NIGHTSHADE FAMILY

***Capsicum annuum** (various cultivars)

Common names: KO'OKO'IM; *chiles*; chilies

These are the large, common, commercial chilies, which also are grown in home gardens in Sonora and Arizona. The ancestral domesticated chilies originated in Mesoamerica, from southern Mexico southward, and may have arrived in Sonora around the time of the early Jesuits.

Ko'oko'im (Cócorit) is also the name for one of the eight traditional Yoeme towns in Sonora.

Capsicum annuum var. **glabriusculum** [*Capsicum annuum* var. *aviculare*]

Common names: HUYA KO'OKO'I (wilderness chile); *chile pequín*, *chiltepín*, *chiltipiquín*; chiltepin, wild chile

Arizona and Sonora. Shrubs or subshrubs with slender, brittle branches, and small, smooth, bright green leaves. The small, rounded fruits, called chiltepins, turn bright red when ripe and dry to form a capsule with several seeds inside.

Basilio (1890 [1634]:207) cited *cócori* as the word for "*chile, chiltepín.*"

Huya ko'oko'i is native in the Yoem Bwiara, including areas along the Río Yaqui and in the larger mountains. This wild chile ranges from a few localities in southern Arizona to South America.

Beliefs: Dried, crushed chiltepins are sprinkled at house entrances to keep out evil forces, especially against a *choni* that might be sent by an enemy to be harmful. Some say a choni can kill you by choking you with its hair. The choni is described as a little, evil hairy being, and can be both good or evil depending on its mood.

Food and harvesting: The fruits are the well-known spicy condiment. Dried, crushed chiltepins are sprinkled on beans, soups, and stews, and cooked with bean tamales.

Huya ko'oko'i is wild-harvested, grown in gardens, or purchased commercially. People in Sonora have picked chiltepins for their own use and for sale. In the 1940s, Rosalio Moisés and his wife harvested wild foods "out in the brush" and augmented their income from huya ko'oko'i. "*Chiltipiquín* grew wild all through the brush, and we would pick a sackful and sell it to the Mexican storekeeper in Vicam; he paid five pesos a kilo. We could pick two or three kilos of *chiltipiquíns* a week" (Moisés et al. 1971:199).

Nowadays some Sonoran Yoemem would go to the Yoreme lands to harvest chiltepins in places

Solanaceae, *Capsicum annuum var. glabriusculum*. A. Nacapule, 28 Oct 2015 (SC). B & C. Cultivated, from Santa Cruz County, Arizona. 25 Aug 2015 (JV).

such as near the villages of Saneal and Vatacosa. We were told, "There are places in the foothills where the plants are abundant, and people like to travel and visit other places, and meet new people and make friends; it is a nice leisure time to go out gathering in the [huya ania] *campo*."

Medicine: Painter (1986:55) reported "chilipiquin...mixed with tobacco, is used for pain" and to treat earache (see *Nicotiana rustica*).

Arizona: East side of Tumacacori Mountains at bottom of Rock Corral Canyon, 7 Jan 1977, *Kaiser 993.*

Sonora: Cañón Nacapule, *Felger 84-581, 85-251.* Palo Parado, Río Yaqui, *Felger 85-1403.*

Datura discolor

Common names: TEBWI; *toloache*; desert thorn-apple, poisonous nightshade

Arizona and Sonora. Annuals, highly variable in size, with large leaves, large, white, funnel-shaped flowers, rounded, spiny fruits, and numerous black seeds. The flowers open in the evening and close in the heat of the day. All parts of the plant are dangerously toxic.

Widespread, common in disturbed habitats, on sandy soils of desert plains and bajadas, gravelly washes, and as roadside and agricultural weeds. This species appears to be the only *Datura* native to the Sonoran Desert.

Datura is poisonous, causing illness and even death if eaten (Shutler 1967:36; also see *Jatropha macrorhiza*, Euphorbiaceae). Contrary to some fictitious writings, the Yoemem did not traditionally use *Datura* or any other plants for hallucinogenic purposes.

Medicine: The leaves are used in the treatment of sores and the healing of blisters and infections, such as one caused by a thorn. The leaf is heated in a pan over an open fire or on a stove. To treat a sore, the afflicted area is bathed with alcohol and a *tebwi* leaf is tied on the sore with a white cloth such as a bandana or handkerchief. This treatment is said to draw or "suck" out the infection. "A leaf of the toloache, put on a splinter, softens it until it can be pulled out" (Painter 1986:55).

Arizona: Tucson, west of Tumamoc Hill, 13 Sep 1983, *Bowers 2734.*

Sonora: Playa del Sol, *Felger 85-1102.* Río Yaqui at Mex Hwy 15, *Van Devender 94-465.*

Solanaceae, *Datura discolor*, San Carlos. 3 Mar 2015 (SC).

Lycium andersonii

Common names: KUH KUTA, KUS KUTA (cross wood), ROIRA (*ro'i*, crippled or lame), ROIYA SIKRO'OPO; *salicieso*; desert wolfberry

Arizona and Sonora. Common desert shrubs, 1 to 2 m tall, with hard wood, light brown bark, thorn-tipped twigs, and small, narrow leaves. The flowers are small, tubular, and violet and white. The shrubs may be seasonally laden with small, juicy, red-orange berries with small seeds.

Food: The fruits are sometimes eaten in the Yoem Bwiara. Some people, however, said the fruits are not edible and that eating it will make you crippled. In other regions, however, the fruits are eaten and we had no ill effects from eating them.

Ritual: Rosary beads are made from the wood. For this purpose, long straight stems, about as thick as a pencil are cut and the twigs are trimmed away. The stem is sliced into small sections and these are fried in oil or lard until black and hard. When we went to the Vakatetteve Kawim (Sierra Bacatete) in March 1989 with José María Jaimez and María Valenzuela, they cut several straight *kus kuta* stems each about 1 m long, trimming away the small, thorny twigs. They took the stems back to Kompuertam for making rosary beads. Also see *Phaulothamnus spinescens* (Achatocarpaceae).

José Guadalupe ("Lupe") Flores, who lived in Tucson, had relatives bring him specific wood, *kus kuta*, to carve the rosaries he sold to community members during lent. He carved the beads with a pocket-knife and then boiled them in lard to darken them.

Arizona: Marana, rocky soil in wash, 23 Jul 1953, *Parker 8300*.

Sonora: 3 km north of Totoitakuse'epo (Hill of the Rooster), 13 Mar 1989, *Felger & Molina*, observation. South of Peón, *Turner 61-44*.

Lycium californicum

Common names: SIKRO'OPO, SIKROPO'I; California wolfberry

Sonora. Low, spreading shrubs with very hard wood and rigid stems, and sharp, thorn-tipped twigs. Common in low-lying and fine-textured soils near the coast, especially from near Empalme and Las Guásimas southward.

Tools: The wood is used to make the pitaya fruit-picking stick (see *Lycium brevipes*).

Sonora: Las Guásimas, *Felger 85-379*. Cocorit, Yaqui Valley, 17 Nov 1933, *Gentry 887-M*. Guaymas, *Palmer 178 in 1887* (UC).

Lycium fremontii

Common names: HUTU'UKI (Arizona), JUTUQUI (Sonora); *salicieso*; Frémont wolfberry

Arizona and Sonora. Desert shrubs, often spinescent, with small, simple leaves. The lavender flowers and juicy, orange fruits appear in spring. Male and female flowers are on separate plants.

Mostly in arroyos and fine-textured alluvial soils of the coastal plain in Sonora, and also in southern Arizona mostly in bottomlands.

Food: The fruits have a tart flavor and are eaten cooked or sometimes fresh.

Arizona: Cortaro Road, a quarter mile west of Casa Grande Hwy, 15 Jan 1953, *Caldwell 155*.

Sonora: Las Guásimas, *Felger 85-379*. Cardonal, near Empalme, *Felger 85-635*.

Lycium brevipes [*Lycium richii*]

Common names: SIKRO'OPO, SIKROPO'I; *salicieso*; desert wolfberry

Sonora. Hardwood shrubs, often 2+ m tall, with small fleshy leaves, finely hairy twigs with tan or brown bark, small lilac flowers, and small red fruits.

Widely distributed in the Yoem Bwiara along the coastal plains and coastal strands, desert margins of mangroves and other salt scrub vegetation, in washes and canyon bottoms, and on rocky slopes. This is one of the most abundant and widespread lyciums in coastal habitats of the Gulf of California.

The fruits are edible (e.g., Felger and Moser 1985), although we do not have direct evidence of the fruits used for food by Yoemem.

Solanaceae, *Lycium brevipes*, San Carlos. 30 Oct 2015 (SC).

Tools: The wood of this and several other lyciums is used to make the pitaya fruit-picking stick (see *Stenocereus thurberi*, Cactaceae). These hardwoods are used because they do not impart a bitter taste to the fruit (also see *Cordia parvifolia*, Cordiaceae).

Sonora: Las Guásimas, *Felger 85-269*. Strand at Estero Soldado, *Felger 85-811B*. Yaqui River, *Palmer in 1869* (GH).

Unidentified Lycium

Common name: VA'AKO

This name was given in Sonora for an unidentified *Lycium* or similar shrub. This name is also used for *Phaulothamnus spinescens* (Achatocarpaceae).

*Nicotiana glauca var. glauca

Common names: RON HUAN; *don juan, san juan, san juanito, tabacón*; tree tobacco

Arizona and Sonora. Spindly shrubs to small trees 2 to 4 m tall. Leaves 5 to 20 cm long, bluish green, smooth, ovate, and somewhat thick. Flowers tubular and yellow, 3 to 4 cm long, and frequented by hummingbirds. Urban and agricultural weeds, and along roadsides and other disturbed habitats. Native to South America and now widely naturalized. Tree tobacco is highly poisonous if ingested or smoked.

Medicine: The leaves are used to treat *heka'uk* (air or wind, "*cuando le pega viento*"; see *Condea albida*, Lamiaceae). Leaves are placed on the forehead and temples and tied on with a scarf. This is said to cause sweating. Leaves stored in alcohol can be use in the same manner. The leaves also can be tied on the forehead in a similar manner to alleviate burning, irritated eyes; this treatment "takes out the heat." To alleviate sunstroke the leaves are put in a container with alcohol and shaken, and the fluid is then swabbed on the body.

Toys: Rosalio Moisés told of children using "a branch of the Don Juan plant" and throwing "mesquite thorns at it, like darts" (Moisés et al. 1971:22; also see *Prosopis*, Fabaceae).

Arizona: Santa Cruz River, San Xavier, *Maguire 17 Jun 1935*.

Sonora: Cerro El Vigía, *Felger 85-443*. Northeast of Pitahaya, *Felger 89-124A & Molina*.

Nicotiana obtusifolia [*Nicotiana palmeri, N. trigonophylla*]

Common names: WO'I VIVA (coyote tobacco); *tabaquillo de coyote*; coyote tobacco, desert tobacco

Arizona and Sonora. This wild tobacco is a perennial herb with sticky (glandular pubescent) foliage and tubular, white flowers. Widespread and common, often on rocky slopes, and less common along washes, desert plains, and bajadas.

We have no direct information that this wild tobacco was smoked by the Yoemem, although it was used by various other indigenous people in Arizona and Sonora, especially if superior tobacco was not available.

Arizona: Tucson, West Branch of the Santa Cruz River, 10 Sep 2001, *Mauz 21-133*.

Sonora: Cerro El Vigía, *Felger 85-466*. Kopas, *Felger 88-579*. Guaymas, Apr 1897, *Maltby 180* (NMC).

Solanaceae, *Nicotiana obtusifolia*, San Carlos. 9 Jan 2016 (SC).

Nicotiana rustica

Common names: HIAK VIVA (Yoeme tobacco); *tabaco*; tobacco

Arizona and Sonora. *Hiak viva* is the indigenous cultivated tobacco. These annual plants have large, sticky-haired leaves. *Viiva* is the term for cigarette. Homegrown tobacco was used for viivam, the hand-rolled cigarettes for the pahkom. Cigarettes were formerly rolled in corn husks or *saa tooro* bark (*Bursera fagaroides*,

Burseraceae). Since about the mid-twentieth century, commercial cigarettes have been substituted for the homegrown hiak viva.

Felipe reminds us that *hiak viva* is the only Yoeme noki name for cultivated tobacco, both in Arizona and the Yoem Bwiara. However, a number of references give *macuchi*, *macucho*, or *macuchus* as names for "Yaqui tobacco" (Escudero 1849:138; Sobarzo 1966; Moisés et al. 1971; Erickson 2008:170). These are not Yoeme terms and have different connotations elsewhere in Sonora and other places in Mexico (Robert A. Villa, personal communication, 2020). It is informative that Sobarzo (1966) reported *macuchi* to be a tobacco of inferior quality that is cultivated along the Río Yaqui, and Rosalio Moisés reported "native tobacco called *hiacbibam* or *machuco*" (Moisés et al. 1971:94).

Agriculture and gardens: *Hiak viva* was widely grown in household gardens and fields in the Yoem Bwiara from pre-contact to modern times. The seeds were planted during the corn-planting season and ready to harvest in early summer. *Hiak viva* was also grown in household gardens in Arizona, apparently until sometime after mid-twentieth century. Shutler (1967: 55) wrote, "The Pascuans today do not cultivate *hiyak vivam*, tobacco."

Beliefs, medicine, and narratives: There is a narrative about a girl named *Masa'asai* who was turned into a tobacco plant by God (see *Antigonon leptopus*, Polygonaceae; Giddings 1945). Spicer (1980:313) reminded the reader that, "A Yaqui knows that tobacco is a woman [and the pahkolam use this knowledge] for endless word plays."

Hiak viva could be used for a flying cigarette (Moisés et al. 1971; Painter 1986). It is said that some powerful people will use this cigarette as an information-gathering source or as an evil force to harm people. What happens is that the person smokes hiak viva and talks to it, and requests that it visits a friend or families far away. The smoker releases the cigarette and it takes off, flies away and goes to its destination and gathers information. And when the flying cigarette returns to the sender, the person reads the information and then he knows everything is o.k. with his friends or relatives. On the negative side, some people use the flying cigarette to harm people. Francisca Martinez, told Felipe that her cousin was hit with a hiak viva, and fell ill, had a high fever, and almost died. An *hitevi* (healer), worked hard on him and helped him recover. Felipe relates:

> I witnessed a flying cigarette at Ignacio ('Nacho') Sombras' house in Potam in 1987. The family thought it is was coming to harm the family members, so they dunked burlap sacks in water and tried to knock it down. But the cigarette was too fast for the men folk. This happened about 2:00 a.m. The flying cigarette went into one of the rooms and came out into the yard, landed on a mesquite branch and finally flew south to the neighbor's house. It managed to out-maneuver the family menfolk. It was quite a spectacle. *Viva nenne'eme* (cigarette that flies) would be the description but people don't use that phrase only, hiak viva.

A cigarette is said to be a good ghost repellent. If a ghost is nearby and is scaring you, just light up a cigarette and the ghost should go away. A ghost indicator is a *chichial*, a bug or something that makes a strange noise; if you hear one, it means there is a ghost nearby. Chichial is a spiritual rather than actual animal.

Shutler (1977:189) provides additional information for tobacco from interviews she conducted in Pascua (she disguised the name as Navidad) in 1958 and 1959:

> The ash and smoke of any cigarette can be used in curing ordinary illness but *hiyakvivam* grown in Sonora and made into a corn husk wrapped cigarette has added power to cure, to strengthen the deer singers so they can sing all night, and to warn of and combat witchcraft. The presence of a witch, Carlos said, can be detected at night by placing a lighted cigarette near the ramada. It will be seen to glow redly [red] as the invisible witch, lurking in the dark, cannot resist smoking from it. Sometimes the cigarettes of the *hitevi* and the witch can be seen fighting in the air at night. While smoking a cigarette,

the hitevi can become clairvoyant. Pedro said that during the wars with the Mexicans, *hitevim* accompanied war parties and smoked to spy out the movements of the enemy. There does not seem to be any suggestion of a trance involved with this clairvoyance. I saw Carlos do this, and he simply smoked a rather strong homemade cigarette and coolly told me what he saw.

Hiak viva has been an been an important medicinal plant. For ear ache, a *hitevi* smokes the cigarette and blows the smoke into a patient's ear. Painter (1986:55–56) reported several medical uses: "For earache, the curer grinds the silk of green corn with chilipiquin and Yaqui tobacco, blows it into the ear, and plugs it in"; tobacco mixed with chiltepins is used to treat pain; "Yaqui tobacco may be smoked, pure or mixed with corn silk, beeswax, or chilipiquin, over the head, the ear, or the teeth for pain" (see *Capsicum annuum*); "beeswax ground with tobacco eases an aching tooth."

A vine cured with cigarettes, called *viivai a hittone* (cure with cigarettes), is tied around the chest as a remedy when someone is sick. This might be nearly any kind of vine (*wiroa*) cured with cigarette smoke.

Rosalio Moisés wrote, "At the time of my baptism in San José de Pima, I was unable to walk because I had been born with a 'dry' leg. A *curandera* made my leg well by applying a warm mixture of olive oil and tobacco" (Moisés et al. 1971:14).

Ritual: During a pahko the pahko'olam distribute tobacco to the participants and audience. Nowadays commercial cigarettes (from *Nicotiana tabacum*) are used.

When a family plans a pahko they choose a *moro* (the pahko manager). They provide him with a number of viivam (cigarettes), which he offers to the individual participants—the *maaso* group and the pahko'ola group. When accepted, the moro lights a cigarette for the participant and one for himself and they smoke together. This seals the agreement and cannot be broken unless there is serious sickness. More recently the moro usually calls the participants by phone and viivam are provided at the ceremony.

Pérez de Ribas (1645; Reff et al. 1999:90) wrote:

> During...celebrations there were also offerings of tobacco, which is commonly used by all of these barbarous peoples. When one nation invites another to forge an alliance for war they convey the invitation by sending a number of reed canes [*Arundo donax*, Poaceae] filled with tobacco. They light these, enjoying the same habit of smoking that originated with these people and has now spread throughout the world. To accept this offering was to accept an invitation to forge an alliance for war.

He added, "This was the subject of their barbarous sermons, which were accompanied by their ritual smoking of tobacco" (Pérez de Ribas 1645; Reff et al. 1999:332).

Rosalio Moisés told about ritual tobacco use by a man known as a great *sabio* (one who has great knowledge) in the early years of the 1900s: "Our house was a meeting place for Yaquis in Hermosillo...When he arrived, he stood as he smoked a cigarette held cupped in his right hand. First he blew smoke to the east, then to the north, west, and south" (Moisés et al. 1971:24).

***Nicotiana tabacum**

Common names: VIIVA; *tabaco*; tobacco

Sonora. Commercial tobacco is a different species from *hiak viva* (*Nicotiana rustica*).

Agriculture: Tobacco was sometimes grown for sale. Spicer (1988:110) told of Lucas Chavez helping his father, Loreto, in "taking loads of native tobacco" and other produce from Tórim to sell in Guaymas. This would have been in the early 1880s, and they transported the produce by burros.

During the time of the haciendas in Sonora, in the late nineteenth and early twentieth centuries, "new crops were introduced such as... tobacco" (Spicer 1980:259). The ensuing revolutionary times and warfare put an end to such commercial ventures.

Gardens: Ornamental varieties with fragrant, multi-colored flowers are sometimes grown in Yoeme gardens in Arizona and Sonora.

Smoking: Commercial cigarettes have supplanted the native *hiak viva* (*Nicotiana rustica*).

Physalis

Several wild ground-cherries are found in well-watered, disturbed and irrigated places in southern Arizona. These include *Physalis acutifolia*, *P. angulata*, and *P. hederifolia*, which are annuals growing with warm weather and summer rains.

The foliage of the various species is reported to be highly poisonous, and there is some information that unripe fruits of certain species might be toxic. Ripe fruits, however, are edible fresh or cooked and have a pleasant, tart flavor. The fruits are encased in a loose-fitting, baggy calyx or husk that enlarges as the fruit develops and becomes papery at maturity.

Food: Meregilda Ochoa said that her mother, Romana Sanchez and others used to eat the fruits of *tomatillos* that grew along the edges of cotton fields at Marana. The fruit was washed and eaten fresh as a snack or with other foods.

Physalis angulata: Arizona: Tucson, Santa Cruz Valley, common weed, *Thornber 27 Nov 1902. Physalis hederifolia*: Arizona: Northeast of Marana, base of Tortolita Mountain bajada, 10 Sep 1984, *Reichhardt 84-13-a* (ASU).

Physalis acutifolia [*Physalis wrightii*]

Arizona and Sonora. Warm-season annuals with a well-developed main axis. Leaves relatively thin. Flowers (corollas) white with a yellow center; anthers blue-gray, the pollen yellow. Low-lying, intermittently marshy savanna of the coastal plain of the Yoem Bwiara, and weedy in well-watered agricultural and urban habitats in Arizona and Sonora.

Food: In 1877, Edward Palmer reported, "Fruit fleshy and edible, at Guaymas (175)" (Watson 1889:64).

Arizona: Upper part of Avra Valley, 30 mi west of Tucson, moist rich irrigated soil between cotton rows and along irrigation ditches, 14 Jul 1953, *Parker 8283*.

Sonora: Northwest of Pitahaya (Belem), *Felger 85-1283*. Guaymas, *Palmer 175 in 1887*.

Physalis crassifolia [*Physalis crassifolia* var. *versicolor*, *P. versicolor*]

Common names: KAU TOMA'ARISI (mountain *tomatillo*); *tomatillo del Desierto*; desert ground-cherry

Arizona and Sonora. Short-lived perennials with spreading branches and sticky-viscid foliage. The flowers are yellow, and the berries are enclosed in a papery husk. Flowering and fruiting at various times of the year. Widespread and common in many habitats including rocky slopes and gravelly arroyo margins.

Food: The small fruits are eaten fresh or prepared like the *tomatillo* (*Physalis philadelphica*).

Arizona: Tucson, Santa Cruz Valley, 15 Nov 1901, *Thornber 4429*.

Sonora: Kuubwa'e Kawi, lava rock hill, 1 km west of Kompuertam, *Felger 89-83 & Molina*. Guaymas, 2 Nov 1926, *Jones 22594* (RSA).

***Physalis philadelphica** [*Physalis ixocarpa*]

Common names: TOMA'ARISI, TOMATILLO; *tomate*; husk-tomato

Cultivated in household gardens, mostly in Sonora, and purchased in markets. A cultivated plant originally native to southern Mexico; not native in Sonora.

Food: The popular green salsa, *salsa verde*, is made from the ripe fruits. Homemade green salsa is prepared from *tomatillo* fruit mixed with onions and spices.

Sonora: Sirebampo vicinity, 9.5 km south on Mexico 15 from Las Bocas Road turnoff, 11.5 km S (by air) San José de Masiaca, flowers yellow with purple center, *tomatillo*, *tomate*, 18 Jan 1995, *Friedman 094-95* (ASU).

***Physalis pubescens** [*Physalis pubescens* var. *integrifolia*]

Common names: TOMA'ARISI; *tomatillo*; hairy ground-cherry

Sonora and limited occurrences in Arizona. Annuals with low, slender, spreading branches and swollen nodes, and pale yellowish flowers with a maroon spot at the base of each petal.

Wet soil, often along streams in riparian canyons, and along the lower Río Yaqui. Probably native to eastern United States, now weedy in warm regions worldwide.

Food: The fruits are sometimes eaten.

Sonora: Las Guásimas, *Felger 85-291*. Cuesta Alta, Río Yaqui, *Felger 85-1373*. Cañón Bacatete (Bacateve), *Felger 89-117 & Molina*.

***Solanum americanum** [*Solanum nodiflorum*]
Common names: MAMYA, MAMYAM; *chichiquelite*; black nightshade

Arizona and Sonora. Herbaceous plants that grow and flower during warm weather, and become dormant during colder weather, or in Arizona may freeze in winter. Distinguished by flowers 0.5 to 1 cm wide, white and often tinged with lavender, anthers 1.2 to 1.7 mm long, and styles barely longer than the anthers. The fruits are rounded berries, at first green, becoming dark purplish or blackish when ripe and 5 to 8 mm wide.

Mamyam are sometimes found in moist places and as urban and agricultural weeds, and are common weeds in orchards, canal banks, and hedgerows in the Yoem Bwiara. It is widely cultivated or protected in dooryard gardens in Arizona and Sonora. It is found nearly worldwide and probably native to South America.

Food: Mamyam are harvested as greens from wild or garden plants. Young leafy stem tips and leaves are boiled and then lightly fried, and often eaten with corn or flour tortillas: "*Quelite, muy buena con tortillas de maíz o harina*" (Moisés et al. 1971:134). The purplish-black berries are like sweet tomatoes, edible fresh or cooked. When people cook and eat mamya then they may say *mamyam kia* (mamyam are delicious).

Solanaceae, *Solanum americanum*, San Carlos. 10 Mar 2015 (SC).

In 1887, Edward Palmer recorded the common name as "*yerba mora*" in Guaymas and that "the young leaves and tops are much used by the Indians in cooking" (Watson 1889:64).

The plants can be toxic, although cooking the leaves and shoots removes the toxin; upon fully ripening the berries are edible or the toxicity is greatly reduced (Edmonds and Chweya 1997). Proper identification of the black nightshades is important—there have been cases of severe poisoning by eating similar-appearing species of wild nightshades in the Sonoran Desert region.

Gardens: Mamya is widely cultivated in Arizona and Sonora. The seeds are often harvested and saved from year to year.

Medicine: Mamya greens (cooked) are said to be "good for hangovers."

Arizona: [Marana], Santa Cruz River just west of Sanders Road crossing, sandy banks along river channel, 1969 ft, *Makings 4677* (ASU).

Sonora: Potam, orchard weed, *Felger 88-588*. Las Guásimas, *Felger 88-614*. Stream at Rancho Bacatetito, *Felger 89-170 & Molina*. Guaymas, *Palmer 9 in 1887* (GH).

Solanum erianthum
Common names: SOOSA
TOOKO HUYA, TOROKO HUYA (gray or light blue plant); *cornetón del monte*; tree nightshade

Sonora. Small trees or large shrubs with an open, spreading crown. The leaves are large and densely pubescent with star-shaped hairs. The flowers are white and star-shaped, with large yellow anthers. The fruits are rounded, dark-colored when ripe, with many seeds.

Found as an understory shrub along the banks of the Río Yaqui as it courses across the coastal plain. Native to the tropical dry forest in mountains east and south of the Yoem Bwiara. Occasionally grown in gardens. The name *toroko huya* is used for a number of other plants, and is a descriptive rather than a specific name.

Food: The fruit (*taaka*) is said to be edible fresh, and to have a greasy (*mantecosa*) taste with a slight orange flavor. (This information should be further verified. While some plants in this family have edible fruits, many others are toxic.) The fruits are used to curdle milk.

Medicine: In 1988, Antonia Amarillas, of Pótam, said that *soosa* is a good medicine to alleviate diabetes, rheumatism, and high and low blood pressure.

Tools: The wood has been used for axe handles.

Sonora: Río Yaqui, Palo Parado, riverbank, *Felger 85-1395*. Potam, canal bank, *Felger 88-554*.

***Solanum lycopersicum** [*Lycopersicon esculentum*]
Common names: VAKOT MUTEKA (snake pillow); *tomate, tomatillo*; tomato

Sonora. Semi-vining, wild tomato plants, often 2 to 3 m tall growing through shrubs. Flowers yellow. The plants produce large clusters of small red tomatoes each about 1 cm wide. Naturalized in the Río Yaqui region, and often growing in hedgerows and brushy areas along irrigation ditches. Native to South America.

Commercial tomatoes are called *tomaate*.

Food: The fresh fruit is sweet and juicy, often eaten as a snack food while walking along tree-shaded lanes in the Yoem Bwiara. The fruit is also harvested and brought home, and made into salsa.

Sonora: Río Yaqui, Cuesta Alta, *Felger 85-1363*. Southeast of Potam, *Felger 88-592*.

***Solanum tuberosum**
Common names: PAAPA, PAAPAM; *papa, papas*; potato, potatoes

Medicine: Potatoes can be used as a substitute for the tuberous roots of the *noono* cactus (*Peniocereus*, Cactaceae). To relieve headache and fever, fresh, thin potato slices are pricked with a fork or knife to make holes, and these slices are tied on the forehead and temples with a white cloth. The slices are replaced as often as needed, or when the potato turns gray. After a while the potato slices are said to "heat up" and turn gray, and soak up or draw out the fever.

STEGNOSPERMATACEAE – STEGNOSPERMA FAMILY

Stegnosperma halimifolium [*Stegnosperma watsonii*]
Common names: WOKKOI AAKI (mourning-dove pitaya); *ojo de zanate*;

Sonora. Large, sprawling, spineless shrubs with nearly evergreen foliage. Leaves semi-fleshy. Flowers white, about 1 cm wide, star-shaped and fragrant. Fruits red and berry-like.

Common near the shores and often ranging inland through washes and arroyos.

Medicine: The leaves are mashed and applied to soothe "burning feet and painful legs." Leafy stems were burned (singed or toasted), mixed with *seve'e choa* (*Cylindropuntia fulgida*, Cactaceae), and applied to the body to treat measles.

Sonora: Sierra Bacatete, 1 km north of ruins of the cuartel, *Felger 88-622*. Kuubwa'e Kawi, a lava rock hill 1 km west of Kompuertam, 12 Mar 1989, *Felger & Molina*, observation.

STERCULIACEAE, see MALVACEAE

TAMARICACEAE – TAMARISK FAMILY

***Tamarix aphylla**
Common names: PIINO; *pino salado*; athel pine, saltcedar, tree tamarisk

Arizona and Sonora. Large, fast-growing trees. The scale leaves on the slender twigs or branchlets are evergreen and beset with minute, salt-excreting glands. (The older people said that the salt from this tree ruins the soil.) The flowers are white, very small, and inconspicuous.

Tamarisk trees are planted in many communities in Arizona and Sonora. This species is native to North Africa and the Eastern Mediterranean.

Construction: In both Arizona and Sonora, the long, straight branches are used as beams for ramadas and houses. The wood is not considered a good material for fence posts because in the ground it rots quickly.

Cultivation: Athel trees are planted for shade in Arizona and Sonora, mostly near homes and also near the *Hiak Vatwe* (Río Yaqui). It is readily propagated by cuttings. Once established, it thrives without additional care. The first ones in

Marana were planted in the 1930s. Photos of Yoem Pueblo in 1941 show large tamarisk trees. The large tamarisk trees were removed in 1996 primarily because the roots were damaging the sewer system.

Fuel: If mesquite is not available, tamarisk is burned, but the people say, "the smoke will make you sick, and this wood does not make coals." Felipe tells us:

> In 1976, during a rare snowstorm during Holy Week, the Pharisees society from New Pascua came to my house to ask for the large tamarisk branches that I had cut the previous fall, to burn at the ceremonies because there was no time to gather mesquite in the desert as it was raining and dark. So they took all my stacked tamarisk wood, about six feet high. I told them to be careful, that the smoke will make you sick, but they already knew that because they were older men.

Medicine: Wagner (1936:84) reported, "Abortions are sometimes induced by drinking a tea made by boiling corcho (a cork-like pine) in water. A lump of sugar is put into a cupful and drunk once a day for three days. It makes the woman deathly sick. She has spasms and occasionally becomes perfectly rigid. Another kind of tea for the same purpose is made from the root of the immortal plant." *Corcho* is cork, although "a cork-like pine" indicates *Tamarix aphylla*, which somewhat resembles a pine and has cork-like bark.

Tamaricaceae, *Tamarix aphylla*, San Carlos. 30 Nov 2015 (SC).

Utilitarian: The smaller pieces of the traditional rope-twister could be made from a piece of salt cedar wood (see *Carnegiea gigantea*, Cactaceae).

Arizona: Tucson, West Branch of Santa Cruz River, 18 Aug 2001, *Mauz 21-115*.

Sonora: Potam Viejo, *Felger 85-1429*. 5 km east of Empalme, *Reina-G. 2002-1054*.

*Tamarix chinensis

Common names: PIINO, PIINO MORAOM SESEWAME (*piino* + purple (plural) + the-one-that-flowers; purple flowering pine); *pino salado*; saltcedar

Arizona and Sonora. Shrubs or small trees. The leaves are small and scale-like, and the minute flowers white to pink or lavender, borne in showy arrays.

These plants are common along floodplains, arroyos, fields and ditches, and other low-lying, disturbed places. Native to the Old World, it has become weedy and invasive in drainages across much of the Sonoran Desert region since about the mid-twentieth century.

Tamarix chinensis and *T. ramosissima* are difficult to distinguish and both or their hybrids may be present in Arizona and Sonora.

Arizona: Tucson, University of Arizona campus, Botanical plant introduction gardens, 9 Apr 1913, *Thornber 10043*.

Sonora: West of Estero Soldado, *Felger 84-528*. Vícam, railroad track, 6 Mar 2016, *Reina-G.*, observation (SEINet 2020).

THEOPHRASTACEAE (*Bonellia*, *Jacquinia*), see PRIMULACEAE

TYPHACEAE – CATTAIL FAMILY

Typha domingensis

Common names: TUULI; *tule*; cattail

Arizona and Sonora. Robust, perennial herbs with thick, starchy rhizomes. The leaves grow erect, 2 to 3+ m tall, and are rather thick, spongy, and flat on one side and convex on the other side.

The flowering stalks are as tall or taller than the leaves. Flowers unisexual, extremely numerous and minute, densely packed on tall cylindrical spike-like inflorescences, male portion of inflorescence above and separated from the female portion by a gap (a section without flowers). Fruits minute, with fine hairs aiding in wind-dispersal.

Wetland areas, including stream channels, arroyo beds, canyon bottoms, temporary ponds, roadside depressions, savannas, wet places at leaky pipes, irrigation ditches and drainage canals, brackish water of estero margins and wetland habitats at edges of bays.

Construction: Cattails have been used for house roofing in Sonora. In 1989 we saw cattails incorporated into roofs at Kompuertam.

A dense and expansive stand of cattails grew in the wetland at the inland side of the *estero*, or lagoon, at Miramar, northwest of Guaymas; known as *Tular*, it was where Yoemem in Guaymas went to harvest *tules*. In 1985, Cruz Matus, in Guaymas said that before he was born it was a common practice to use *tules* for roofing houses. Tules were laid across the roof and then covered with earth. There were still tule-roofed houses when he was a boy. Sr. Matus (1985) said:

> *para techar casas, para techos de tierra. Se pone tules y la tierra por arriba. Antes, aquí en Guaymas, hicierón techos con tule...En un lugar que llama Tular en Miramar. Había mucho en Tular. Era común.*
> ("for roofing houses, for earthen roofs. The cattails are laid down and covered with earth. In the past, here in Guaymas they made roofs with cattails...in a place called Tular in Miramar. There was a lot [of cattail] in Tular. It was common").

In 2020, cattails were still common in the El Tular wetland at Miramar, although the estero had been severely altered by development for a marina that was subsequently abandoned.

Arizona: [Tucson], Tumamoc Hill, 30 Aug 1983, *Bowers 2725*.

Sonora: Abundant on both sides of Calzada (causeway) Bacochibampo, 25 Jan 2018, *Carnahan*, photos. Estero de Miramar, SE side of estero at Miramar, abundant, S and E margin of the lagoon, in near pure stands, plants 2 to 3 m tall, leaves 17 to 22 mm wide, *Felger 85-507* (ARIZ, USON). Kompuertam, 13 Mar 1989, *Felger & Molina*, observation.

Typhaceae, *Typha domingensis*, El Tular, Miramar. 10 Apr 2019 (SC).

ULMACEAE (*Celtis*), see CANNABACEAE

VERBENACEAE – VERBENA FAMILY

Citharexylum flabellifolium

Common name: KUH KUTA, KUS KUTA (cross wood)

Sonora. Woody shrubs 1.5 to 2+ m tall. Branches rigid, the twigs thorn-tipped and square in cross-section. Leaves drought deciduous. During dry seasons it is an unpleasant, cloth-ripping shrub, but following rains it is covered with attractive lavender flowers and green leaves which contrast with the dark-colored bark. Common on desert plains and bajadas.

Sonora: Mountain ravines about Guaymas, *Palmer 237 in 1887* (GH). South of Peón, *Turner 61-41*. Mex Hwy 15 north of Las Guásimas, *Warren 16 Aug 1975*.

Lippia palmeri

Common names: OREGANO; *orégano*

Sonora. This shrub has slender brittle stems. Leaves are 1 to 3 cm long, highly aromatic, rough-surfaced, with deeply incised veins. New foliage appears after each soaking rain and the leaves are drought-deciduous. The minute yellow flowers are borne in small, cone-shaped inflorescences. Flowering response is apparently non-seasonal, but dependent on soil moisture.

Basilio (1890 [1634]:178) cited *orehan* as a word for "*orégano de la tierra.*"

Common throughout the Yoem Bwiara, from near sea level to peak elevations, slopes, mesas, and canyons.

Food and harvesting: The dried, crushed leaves are used for seasoning food. The flavor is like that of commercial oregano, although generally considered to be superior. It is sometimes commercially harvested and sold in Sonoran markets and also exported, marketed in the U.S. as "Mexican oregano." It could be a valuable agricultural crop for the Yoem Bwiara region.

Medicine: The dried, crushed leaves are taken as a medicinal tea.

Sonora: Cerro el Bachoco, south of Cruz de Piedra, 26 Apr 1985, *Burgess 6942*. Guaymas, basaltic hill slopes, oregano, 23 Oct 1939, *Gentry 4685*. Guaymas, *Palmer 277 in 1887* (US).

Verbenaceae, *Lippia palmeri*, Cerro El Vigía. 3 Oct 2015 (SC).

VIBURNACEAE – VIBURNUM FAMILY

***Sambucus cerulea** [*Sambucus cerulea* var. *neomexicana*, *S. neomexicana*, *S. nigra* subsp. *cerulea*]
Common names: SAUKO; *sauco* (flowers), *tápiro* (the tree); blue elderberry, elderberry

Arizona and Sonora. Small trees with very soft wood and scaly bark. Leaves large, pinnately compound. Flowers numerous, small, whitish, and borne in rather large flat-topped clusters. The fruits are small, berrylike, juicy, and dark blue or blackish. Growing and flowering mostly during the cooler seasons, from fall to spring.

This tree is cultivated in Arizona and Sonora. Elderberry also occurs wild in mountains above the desert in Arizona and Sonora, and is widespread elsewhere in western North America.

The unripe fruits and fresh leaves are poisonous.

Food: Elderberry trees were commonly planted at ranches and towns in Arizona and Sonora, largely for the berries, which sometimes are eaten fresh but mostly made into preserves, jelly, and wine, and also used for medicinal purposes.

Medicine: Sauko is considered a "cool" plant. The berries are boiled and the tea consumed as a remedy for fevers. Also as a remedy for fevers as well as chills, one or two flower clusters are mixed with *negrita* (an herb sold in Mexican herbal stores, see globe mallow, *Sphaeralcea*, Malvaceae) and/or *vavis* (*Anemopsis californica*, Saururaceae) and boiled in water. One only has to drink a small amount of the tea to "get rid of a fever real fast."

Viburnaceae, *Sambucus cerulea*, Santa Cruz County, Arizona. 4 May 2021 (SR).

Tea made from the flowers is given to newborn babies to drink to "wash out" their intestines. Mateo González said, "It is a medicine for children—they drink tea made from the leaves for colic. For older people, if you have a churning stomach, drink some and it cleans the stomach." Tea made from mesquite resin (*Prosopis*, Fabaceae) and *sauco* is taken to alleviate a cold.

Arizona: End of pavement south of Tucson on Nogales road, 25 Mar 1926, *Kearney 1394*.

Sonora: Koasepe, 3 km southeast of Kompuertam, 14 Mar 1989, *Felger*, observation.

VISCACEAE, see SANTALACEAE

VITACEAE – GRAPE FAMILY

Cissus trifoliata

Common names: YUKU WIROA (rain vine); *tumbacasa*; Arizona grape-ivy, sorrel vine

Arizona and Sonora. This common, robust vine has thick stems and becomes leafless during dry seasons. Leaves semi-succulent, with 3 leaflets or 3 lobes, the margins coarsely-toothed.

Arroyos and low places on the coastal plain of the Yoem Bwiara. It is said that this plant can actually knock down a traditional Yoeme house because it can grow so thick and heavy on top of the roof.

Arizona: Coyote Mountains, *Harrison 8011.*

Sonora: Rincón del León, *Felger 85-914.* Playa de Sol, *Felger 85-1122B.*

Cissus verticillata [*Cissus sicyoides*]

Common names: YUKU WIROA (rain vine); *tumbacasa*

Sonora. Rank-growing, perennial vines. Leaves of the juvenile growth are colorful and very different from those of the adult growth. Leaves simple, ovate-cordate, semi-succulent, with edges (margins) finely bristle-toothed.

Common along the lower Río Yaqui and southward into Sinaloa.

Sonora: Cuesta Alta, Río Yaqui, *Felger 85-1410.* Potam, canal bank, *Felger 88-554.*

Vitaceae, *Cissus verticillata*, San Carlos. 10 Nov 2014 (SC).

Unidentified Cissus

Common names: YUKU WIROATA SAILA (rain vine's younger brother)

This may be a species of *Cissus* or another kind of vine.

***Vitis vinifera**

Common names: UUVA; *uva*; grape

Arizona and Sonora. The common, cultivated grape is sometimes grown in gardens, especially for shade on ramadas. *Paa'asim* are raisins.

ZOSTERACEAE – EELGRASS FAMILY

Zostera marina

Common names: MOOSENINO, VAA VASO (water grass); *trigo del mar*; eelgrass

Sonora. This seagrass grows fully submerged in shallow, protected seawater. It occurs in the Guaymas region and southward along the coast of the Yoem Bwiara. It has long, thin, ribbon-like green leaves and produces seeds about 3 mm long. It dies back during hot weather and grows during the cooler seasons. Substantial quantities of eelgrass may drift ashore in spring, such as at Las Guásimas and farther south to the vicinity of Altata, Sinaloa.

Eelgrass grain was a staple food of the Seris (Felger and Moser 1985), but Yoeme fishermen at Chiinim and Las Guásimas in the 1980s did not know of eelgrass being eaten by people.

Animal food: While we were looking at shore grass (*vaa vaso, Distichlis littoralis*, Poaceae) on the coast at Chiinim in December 1988, several men said that another "grass, known by the same name exists in the water [sea]; it has the same name but grows under water. It is eaten by *mosenim* [green sea turtles, *Chelonia mydas*]." Some men used the name "*moosenino*," which implies it is eaten by green sea turtles. Others, such as fishermen at Las Guásimas, simply called it *vaso* and said that cows eat it when it washes ashore in late April.

Food: Pérez de Ribas (1645; Reff et al. 1999:88) wrote about coastal people who "at the time of the maize harvest...go up [river] to trade fish in the pueblos of their friends who are agriculturalists, at other times they collect a small seed from a plant that grows below the water, which they use for bread." This plant can only be eelgrass and the location could be from the Guaymas region southward to northern Sinaloa, especially near the deltas of the Yaqui, Mayo, or Fuerte rivers. Translation of *pan* from original (1645) does not necessarily imply bread, but it does suggest it "was as important to the Indians as bread was to Europeans" (Felger and Moser 1985:379).

Sonora: San Carlos Bay, blown in by wind, 22 Mar 1973, *Brown B-362-a* (ASU). San Carlos, Estero Soldado, beach drift, *Felger 13 Feb 2000*.

ZYGOPHYLLACEAE – CALTROP FAMILY

Guaiacum coulteri

Common names: HUYA'AWO, HUYAWO; *guayacán*

Sonora. Large shrubs or small trees with extremely hard wood and smooth, gray bark. The leaves have 3 to 5 pairs of dark green leaflets. Spectacular masses of indigo-blue flowers are produced in May and June during the hottest, driest weather.

Basilio (1890 [1634]:214) recorded *huiahuonahua* as the term for *guayacán*.

Common across much of Sonora. Widespread thought most of the Yoem Bwiara. including the coastal plain and rocky slopes. *Guayacán* is sometime cultivated in southern Arizona although it is frost-sensitive.

Household: The bark can be used for soap. Basilio (1890 [1634]:138) recorded *huiahuonahua* as "*Arbol, cuya corteza suele servir de jabon a los pobres*," a tree whose bark is used as soap by the poor.

Medicine: The bark is peeled away, then toasted and ground, and applied to sores. The mashed bark and leaves are applied to burns, or the bark and leaves are put in water until the mixture starts to foam. This liquid is applied to a burned area until the wound "cools down," or one can just pour the liquid over the burn. Mateo González told us that *huya'awo* is used for medicine to ward off evil, and [also] applied to cuts and burns that do not heal. To treat infections, the leaves are ground and washed with alcohol, and applied to the afflicted area.

Narrative: In the story "Topol the Clever," Giddings (1959:57) included, "Topol...cut himself a stick of the wood of a plant which has blue flowers." The plant probably would be *Guaiacum coulteri* (see javelina, *Tayassu tajacu*, Tayassuidae, and jaguar, *Panthera onca*, Felidae).

Sonora: 4 km north-northeast of Las Guásimas, *Felger 88-611*. Guaymas, *Palmer 326 in 1892* (UC). Sierra Bacatete, Arroyo El Álamo, 6.8 km (línea recta) al NE de Vícam, 11 Sep 2008, *Sánchez-Escalante*, observation

Kallstroemia grandiflora

Common names: *baiborín*, *mal de ojo*; orange caltrop

Summer–fall annuals. Leaves with 5 to 9 pairs of leaflets. Flowers orange, large and showy. Widespread, sometimes carpeting the desert and spreading orange across the landscape.

Sonora: Guaymas, Isla Chaperona, *Suárez-Gracida 16 Sep 1999* (USON). Guaymas, *Palmer 177, 225 in 1887* (Watson 1889:43).

Zygophyllaceae, *Guaiacum coulteri*, east of Hermosillo. 31 May 2016 (SC).

Larrea tridentata [*Larrea divaricata* subsp. *tridentata*]

Common names: KOVANAO (from the Spanish, *gobernador*); *gobernadora* (governor for its dominance of a landscape), *hediondilla* (little stinky for its distinct aroma); creosotebush

Arizona and Sonora. Multiple-stem shrubs with very hard wood. The new leaves and bright yellow flowers develop in response to rainfall at almost any time of the year. The resinous foliage produces a strong aroma, which is especially prominent after a rain. This characteristic aroma is reflected in the Yoeme, Spanish, and English names.

Larrea is an abundant and characteristic shrub of the deserts in Arizona and Sonora, and ranges south to the northern margins of the Yoem Bwiara.

The branches are sometimes encrusted with a red-brown lac, called *goma de Sonora* in Mexico, from the excretions of a scale insect (*Tachardiella larreae*).

Medicine: Someone who is ill and told not to take an ordinary bath can bathe with creosotebush water. A leafy branch is boiled in a bucket of water. This liquid is then mixed with cold water in the summer or warm water in winter and the liquid is poured over the body.

As a remedy for rheumatism, small leafy branches are warmed in embers and applied to the skin of the afflicted area; the branch is tied on with a cloth. While using this remedy, it is said that one is not supposed to take a bath because it may cause the person to become crippled sometime in the future. While visiting in Marana in 1987, Teresa Amarillas saw the plant and said, "Take some small leafy twigs and bunch them into a ball, a fist-sized ball, and when you boil it the leaves will separate and come to the top of the water. If you use too much, it is too strong and makes your body swell up. Apply the leaves to an area of the body with rheumatism."

Zygophyllaceae, *Larrea tridentata*, San Carlos. 10 Apr 2019 (SC).

Pulverized leaves are used as a foot powder for smelly feet. The leaves are also used as a remedy for athlete's foot fungus.

Creosotebush branches also are used for purification. For this purpose, four fresh branches are used to purify oneself by fanning the branches all around the body. This is done on various occasions such as when one feels sick, when something is mentally or spiritually wrong, to purify oneself after visiting the deceased at a funeral, or before going into the mountains or desert for purification—just to be there for spiritual reasons or just for enjoyment.

Mateo González said you can boil *kovanao* and take a bath with the water when one has chills. People also used to drink the tea cool or warm.

Wagner (1936:82) reported, "for dysentery the medicine man [*hitevi*] takes three small sticks of *goma de Sonora* and boils them in water. Then he adds *cominos* [coriander seeds], cinnamon bark, and essence of mint. This mixture is strained and some alcohol added to preserve it. The patient is given the equivalent of a small whiskey glass three times a day." Three of the four ingredients are from non-native, commercial plants, and even the goma de Sonora was probably commercially purchased or traded. Goma de Sonora has been widely used for medicinal purposes in Mexico.

Kavanoa continues to be a favored medicinal plant. For example, in the Tucson area in 2016, one popular preparation was as follows:

> Put as much kavanoa as possible into a large, wide metal pot with minimal water, and boil it for 3 or 4 hours. Cook down the water to end up with a black concentrate, resulting in a black salve. Kovanoa salve can used directly or as applied as poultice as a remedy for infection and other ailments; and is presumed to have antibacterial properties.

Ritual: The lac was used to form the eyes of the deer head used in the deer dance (see *Odocoileus virginianus*, Artiodactyla: Cervidae).

Arizona: Tucson, 9 Aug 1901, *Thornber* 4096. Tucson, Tumamoc Hill, *Turner* 68-63.

Sonora: 19 mi east of Guaymas on Mex Hwy 15, *Goldberg 76-53*. Ejido Cruz de Piedra, 13 km east of Empalme on Mex Hwy 15, coastal thornscrub, *Van Devender 2003-1070*.

***Tribulus terrestris**

Common names: CHIVA KOVAM (goat heads), TOOROM (bulls), WICHA'APOI, WICHA'APOLI (*wicha*, thorn + *'apoli*, not translatable; the same name used for bur grasses, *Cenchrus*, Poaceae); *toboso*, *torito*; goathead, puncture vine

Arizona and Sonora. Weedy, warm-season annuals native to the Old World. The stems trail across the ground. The flowers are small and yellow, and the fruit is a bur that breaks into five sharp, tack-shaped segments.

Arizona: [Tucson], Tumamoc Hill, 24 Aug 1934, *Shreve 6603*.

Sonora: San Carlos, 10 Mar 2015, *Carnahan SC1006*. Guaymas, *López 20 Jul 1983* (USON).

UNIDENTIFIED PLANTS

ALUCEMA

Wagner (1936:83) mentioned "*alucema* seed" in a medicinal remedy (see *Anemopsis californica*, Saururaceae). Alucema is lavender, *Lavendula intermedia* (Lamiaceae), but it is doubtful that *alucema* in this case is lavender.

ASELLA

Perhaps from the *acelga*, for chard or greens. This is a wild green, almost like a pigweed (*Amaranthus*, Amaranthaceae). It is an important, edible green, found in summer. Felipe heard a lot about it when he was growing up:

> The way I remember the people say the word is 'ah sehl lah.' I remember one plant that would always grow in my grandmother's garden. It was about two feet tall and the leaves were slender. It had a distinct fragrance. My grandparents didn't eat it because they were happy that it grew there voluntarily. This plant always came out in the summer years of the late 1950s to the mid 1960s. After that I have not seen it anymore.

BUCSNINI

A plant with a large leaf, found southwest of Pótam (Guillermo Moroyoki, 19 November 1985).

BWIA SEEWAM (earth flowers)

In her discussion of Yoeme "amulets," Painter (1986:42) recorded, "The *buia sewam* (earth flowers), on which the deer likes to browse, give one luck in cards and love." Buia sewam might be flowers fallen to the ground, such as those of the *palo santo* tree (*Ipomoea arborescens*, Convolvulaceae), which Gentry (1942:213) says are eagerly eaten by deer. Painter's (1986) *buia sewam* is *bwia sewam* in our orthography.

BUABÁ'IRU

Johnson (1962 257) told of "*buabá'iru, especie del árbol lenguehe; raíz del arbol que crece cerce del río; cuando machacan la raiz da un olor muy agredable. Los cazadores bañan con esto para quitarse el olor de gente.*" The locality is Pótam.

ESEI

An unidentified plant. Antonia Flores, of Pótam said that they used to eat the seeds.

HOHOHNA

Perhaps a species of *Acacia*?

Immortal plant

Wagner (1936:82–83) reported, "for typhoid fever a preparation made from the immortal plant is made. The leaves and fine stem are boiled and made into a tea. This is drunk at intervals. It is very bitter and probably contains quinine."

KAU MANSANIA

Probably in the Asteraceae family, perhaps *Perityle emoryi* or another species. In March 1989 at Kompuertam we told it is "around most of the time, more so than yellow-flowered *mansaniata* [*Perityle californica*]."

KOKITOM

Sonora. Edible like acorns.

KOONI SAAWA

Sonora.

KUUCHA PUUSI (fish eye)

Not here anymore [Kompuertam], good food for goats, flat-growing plant with lilac flower.

LIA

Wagner (1936:82–83) reported, "when a patient is delirious he is given a warm tea made from the leaves and stems of the *lia* plant. He drinks a cupful every half hour." Like many of the plant names used by Wagner, *lia* is probably not a Yoeme name.

MAVEM

MOLANISCO

This is likely a Mexican rather than a Yoeme name. Wagner (1936:82) reported that a "medicine man" in Pótam treated "Diarrhea...with *molanisco*. The root is beaten to a pulp. This is steeped in cold water, and the liquid is taken several times a day."

NAUTO'ORI

Unidentified herb. Purchased at Mexican herbal stores in Arizona and Sonora.

Possible translation: *nau*, together + *to'oria*, laid object. An example of to'oria: *huiwa to'oria*, quiver (*huiwa*, arrow + *to'oria*, again laid; where the arrows are actually standing).

Medicine: Used for dizziness. Grind up and apply to the forehead. Felipe's grandmother always had it in the house in Marana. Also see *Aristolochia watsonii* (Aristolochiaceae).

SAALOCHI

SOMOHKO

Near Las Guásimas or Yasicuri.

Spanish dagger

Wagner (1936:83) reported, "After there has been a case of whooping-cough it is customary to fumigate the house by burning Spanish dagger in it." In southwestern United States, Spanish dagger is often applied to various species of *Yucca*, but there are no yuccas in the Yoem Bwiara. Could this be *Koeberlinia spinosa* (Koeberliniaceae), that the Seris burned to fumigate their homes and clothes against diseases after exposure in Hermosillo?

VEA HUYA (bark plant)

VOONE

Lamb's quarters (*Chenopodium*)? Wash it thoroughly, boil and add onions and salt. (Meregilda Ochoa, Marana, Jan 1994).

VOVO'E

This plant is said to be related to *chinita* (*Sonchus*, Asteraceae) and is poisonous (Meregilda Ochoa, Marana, January 1994).

WICHA MAMYAM (spine *mamyam*)

This is an herbaceous plant, described as a *quelite* that is "very good with corn or flour tortillas." Mamyam is *Solanum americanum* (Solanaceae). A spiny *Solanum* might be *S. elaeagnifolium*, which is not known to be an edible quelite.

PART 4
ANIMAL LIFE IN THE YOEME WORLD

INTRODUCTION

We include more than 604 species or taxa (any taxonomic category) of animal life, as follows: invertebrates, 91 (including 46 insects); fishes, 50; amphibians, 19; reptiles 66 (including 36 snakes), birds, 280; and mammals, 98.

We include many kinds of animal life known or expected to occur in the regions treated in this book (see Part 1: Yoem Bwiara). Even if Yoeme names and written records for these taxa are lacking; we consider this information useful and relevant to the bio-cultural heritage of the region. The distributions given are not intended to account for the entire range of the species in nature.

Some general categories which might be called ethno-taxa include:

aakame	rattlesnake, mostly large rattlesnakes
auli	clam
chikul	mouse, native genera and house mouse
hupa	skunk
huvahe	spider
kaureepa	large toad
kuchu	fish
malon	squirrel
mansom	domestic mammals
mo'el	wren
sochik	bat
taawe	hawk
tamekame	shark
tori	rat, native *Neotoma* and Old World *Rattus*
vaakot	snake
vaatosai	seagull
vaisevoli, vaisevo'i	butterflies or large moths
vatat	frog
voovok	toad
wiirum	vultures
wikichim	birds
wikui	lizard, ordinary lizards
wokovavase'ela	swallow
yoawam	terrestrial animals, especially mammals
yoeriam	insects

In the following accounts, we first list the invertebrates followed by the vertebrates. The arrangements within these sections are by major formal or informal groups, for some groups including taxonomic Order, then (usually) alphabetically by family, genus, and species. These scientific names and taxonomic arrangement are not, however, static and are subject to change with new information—science marches on. The accepted scientific names for genera and species (or other taxonomic groupings) are in **bold** font, and selected synonyms are in *italics* within brackets [—]. Yoeme names follow the scientific names and are shown in small capital letters (SMALL CAPS), and translations are provided when known (but not for core or unanalyzable names). Common names, when known or worthwhile, follow in Spanish (*italics*) and English (plain text). Yoeme names are not always available, and the same holds for English- and Spanish-language common names. An asterisk (*) indicates non-native status (introduced since European arrival).

INVERTEBRATES

"WORMS"

Bwichia is the general term for a worm (*gusano*) and worm-like creatures such as earthworms, maggots, and caterpillars. Basilio (1890 [1634]:163) recorded *buichia* as the term for *gusano*.

Chunkuriam is the term for human parasitic worms and is also used for midge larvae (see Diptera: Chironomidae).

CRUSTACEANS

DECAPODA – CRABS, LOBSTERS, AND SHRIMPS

Acha kaari (sideways house) is the general term for a crab. Basilio (1890 [1634]:148) recorded *achacari* as a term for *cangrejo*.

OCYPODIDAE – FIDDLER CRABS

Uca

Common names: ACHA KAARI; *cangrejo violinista*; fiddler crab

These little crabs are seen scurrying sideways over tidal flats and in mangrove areas at low tide. The male has one oversized and one tiny claw. Richard Brusca tells us that eight species of *Uca* live in the Yoeme coastal region.

PORTUNIDAE – SWIMMING CRABS

Callinectes bellicosus

Common names: ACHA KAARI, HAIVA; *jaiba azul, jaiba guerrera*; blue crab

This edible and rather large swimming crab is commercially harvested. It is an important marine resource at Las Guásimas and elsewhere along the Sonora coast (Hernández-Moreno 2000; Nevárez-Martínez et al. 2003). Johnson (1962:270) listed "*kóchimai*" as "*jaiba*," although we found *kochimai* also to be the name for lobster and shrimp (see Palinuridae).

PALINURIDAE – SPINY LOBSTERS

Panulirus

Common names: LONWOSTA; *langosta*; lobster, spiny lobster

Three species of spiny lobsters occur in the region. While they are commercially harvested in the Gulf of California, lobster harvesting is not an older, traditional endeavor and the Spanish *langosta* is the usual name.

Panulirus gracilis

Common names: *langosta barbona*; green spiny lobster

Panulirus inflatus

Common names: *langosta azul*; blue lobster

The blue lobster is the more common of the three *Panulirus* species throughout the Gulf of California (Brusca 2004).

Panulirus interruptus

Common names: *langosta mexicana;* Mexican spiny lobster, red lobster, California spiny lobster

PENAEIDAE – PENAEID SHRIMPS

Common names: KAMARON, KAMAROONIM (plural), KOCHIMAI; *camarón*; shrimp

There is some confusion concerning the Yoeme terms for shrimp and lobster, perhaps because commercial harvesting is not a traditional endeavor. *Vaa koochi* was used for shrimp in Las Guásimas, although we were told that it is a Yoreme term. Andrew Semotiuk (personal communication, 2019) tells us *ba'a kochim* is the Yoreme name for shrimp. Basilio (1890 [1634]:147) recorded *cochi* as the word for *camarón*.

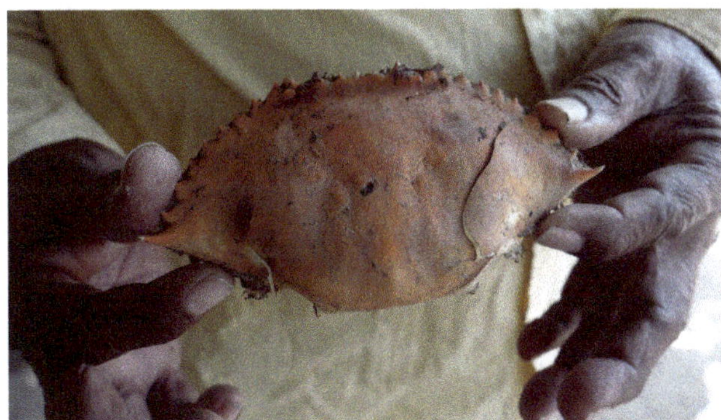

Callinectes bellicosus, Las Guásimas. 31 May 2019 (PB).

From the mid- to late twentieth century, wild-caught penaeid shrimp were an important economic resource in the Yoeme region, especially at Lobos (*Loovos*) and Las Guásimas. During this time, there was a thriving Yoeme shrimp-harvesting co-op. The artisanal harvest during the 1980s was described as follows:

> From the middle of August to Christmastide, Guásimas and Lobos on the coast become centers of intense fishing. The year-long residents of Guásimas and the seasonal occupants of the Lobos camp set out soon after daybreak in motor-powered, fiberglass boats of about eighteen feet, to net-cast for shrimp. Returning to shore around noon, the crews of two or three decapitate the catch, weigh it, and pack it in refrigerated trucks for transport to Guaymas, then on to processors across the border in San Diego. Once the catch is packed and shipped, the fishermen return to their houses—mostly tarpaper shacks at Lobos, cinderblock houses up the coast at Guásimas—and relax, eat, gossip, and

mend their *atarayas* [cast nets]. (McGuire 1986:123)

The artisanal Yoeme shrimp harvests were in the large esteros, or lagoons, such as at Las Guásimas and Bahía de Lobos. These harvests involved small crews operating from pangas powered by outboard motors. Commercial shrimpers operated from large, diesel-powered vessels with mechanical trawling equipment.

McGuire (1986) listed the Pacific brown shrimp (*Penaeus californiensis*) and the blue shrimp (*Penaeus stylirostris*) as the two commercial species along the Yoeme coast and the ones landed in Guaymas. However, a number of others might have been caught by Yoeme fishermen working out of Las Guásimas and Lobos.

If you were on a Mexican commercial shrimp boat at nighttime from the 1960s to the 1980s you would have seen myriad creatures from the sea floor dumped out of huge trawl nets onto slippery decks: thousands of wriggling fishes and invertebrates of all descriptions, once in a while a sea turtle, and here and there large, turgid shrimp. Sitting on low wooden stools, the crew picked out the shrimp and put them in big buckets. The dead and dying by-catch was shoveled overboard amid frantic seabirds. Crews, especially from Guaymas, were often Yoeme men.

Intensive commercial shrimp trawling escalated after the mid-twentieth century and peaked around the 1970s. High takes occurred from 1979 to 1981, during which time up to 285 shrimp trawlers and a dozen or so Japanese boats worked the Gulf. Between 1983 and 1987, the shrimp fishery collapsed due to overharvesting, destruction of the seafloor habitat from excessive trawling, and reduced freshwater from rivers entering the Gulf. Largely as a result of the demise of wild populations and increased market demand, shrimp farms have been established along the Sonora coast. These farms have contributed to destruction of coastal ecosystems and pollution of ocean waters.

Penaeus brevirostris [*Farfantepenaeus brevirostris*]
Common names: *camarón cristal*, *camarón cristalino*, *camarón rojo*; crystal shrimp, pink shrimp

This is the least common commercial shrimp taken in the Gulf.

Penaeus californiensis [*Farfantepenaeus californiensis*]
Common names: *camarón café*; Pacific brown shrimp

This shrimp is brown to reddish brown and has 2 or 3 ventral teeth on the rostrum. This was the most common offshore shrimp taken by Yoeme fishermen (McGuire 1986).

Penaeus stylirostris [*Litopenaeus stylirostris*]
Common names: *camarón azul*; blue shrimp

This species is whitish to pale blue and has 3 to 8 ventral teeth on the rostrum. This is a shrimp of the littoral zone, common in esteros, feeding on microorganisms in estero mouths and along currents (McGuire 1986).

Penaeus vannamei [*Litopenaeus vannamei*]
Common names: *camarón blanco del pacífico*; Pacific white shrimp

This species is documented from Las Guásimas, and is prevalent southward to coastal Sinaloa.

SICYONIIDAE – ROCK SHRIMP
Sicyonia
Common names: *camarón cacahuate*; target shrimp

There are 11 species of *Sicyonia* in the Gulf, 10 of which occur at depths of less than 20 m (65 ft) and would thus be accessible with nets from pangas. *Sicyonia penicillata* [*Penaeus penicillata*] might be the most common, but probably all the others are taken by fishermen. Like *S. penicillata*, most species have the distinctive spot or "bull's eye" marking on the side of the body.

MYRIOPODA – CENTIPEDES AND MILLIPEDES
Orthoperus ornatus
Common names: EYE'EKOE (*eeye*, ant + *koe*, a core word); *milpiés*; giant desert millipede

Arizona and Sonora. These harmless desert creatures can be 10 to 13 cm (4 to 5 in) long and the diameter of a pencil. They are often seen at night during the summer rainy season, moving along like miniature trains. Their sun-bleached carcasses are often found on the desert.

Scolopendra heros
Common names: MASIWE; *cienpiés, cienpiés gigante del Desierto;* giant desert centipede

Arizona and Sonora. This large centipede, 15 to 20 cm (6 to 8 in) long, is fast-moving and aggressive. Although it is not considered deadly, and incidents of centipede bites are rare, it can deliver an excruciating and toxic bite lasting several hours to days. The "bite" is actually a pinch delivered from paired pincer-like appendages (called gnathosomes or gnathopods) in front of the legs.

It is said that the *masiwe* can attach to you with their claw-like feet, and the only way to get them off is with fire. Although we do not consider this view to be accurate, masiwem can be dangerous.

Masks: Centipede images are sometimes painted on the side of a pahko'ola mask, traditionally to appeal to Huya Ania, the spirit of the Wilderness World.

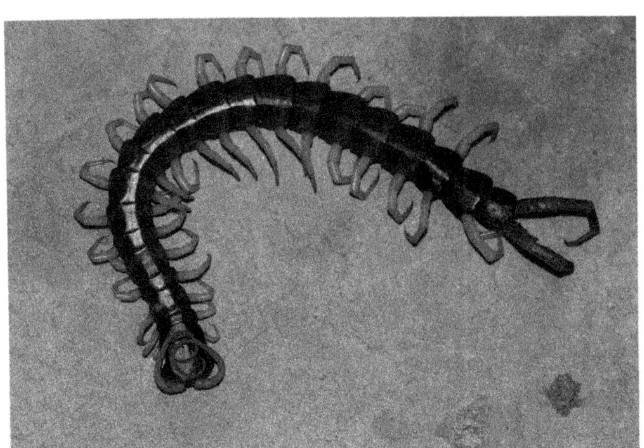

Scolopendra heros, Tucson Mountains, 28 July 2007, (AR).

CHELICERATA – SCORPIONS, SPIDERS, AND THEIR KIN

ACARI – CHIGGERS AND TICKS
Common names: TEMA'I; *garrapata;* tick

Before about the 1980s, domestic animals with ticks were washed with a decoction of the green twigs of *vachomo* (*Baccharis sarothroides*, Asteraceae).

Basilio (1890 [1634]:162) cited *temai* as the word for *garrapata*.

Common names: CHINCHIM; *baiburín;* chiggers

Sonora and parts of Arizona above the desert. These tiny mites can cause annoying and often painful skin irritations. They are prevalent during hot, humid weather on grasses and undergrowth such as along irrigation ditches at Vícam and Bácum, the riparian forests all along the Río Yaqui, and in non-desert foothill and low-mountain vegetation. The term *chinchim* is also used for scabies and bedbugs (*Cimex lectularius* and *Triatoma*, Hemiptera: Cimicidae and Reduviidae).

ARANAEA –SPIDERS
Common names: HUVAHE; *araña;* spider

Arizona and Sonora. *Huvahe* is the general term for spider. Basilio (1890 [1634]:138) cited *hubare* for *araña*, and *hubare iore huhame* (spider itchy) as the term for "*araña ponzoñosa*" (poisonous spider). He also reported *huitosaa* for *telaraña* (spiderweb) and *turus* as an "*especie de araña*" (a kind of spider) (Basilio 1890 [1634]:192, 227).

Medicine: Wagner (1936) reported that the cobweb of a special spider, when spread over a wound, will heal it fast.

Aphonopelma including **Aphonopelma chalcodes**
Common names: MAISOOKA; *tarantula, tarántula desértica;* desert tarantula

Arizona and Sonora. These large, hairy spiders might be feared but are essentially harmless. It is almost always the adult males that are seen traversing open ground and roadways during the summer-fall monsoon season as they search for females in their burrows. Males will wander after a rainstorm when it is cool and cloudy, especially at night. Adult females seldom leave their burrows. The females feed by using hairs on their legs to detect vibrations caused by wandering prey, then pounce on the prey near the burrow entrance.

Narrative: Giddings (1959:49) provided a story about the tarantula and tarantula hawk (see *Pepsis* and *Hemipepsis*, Hymenoptera: Pompilidae). She correctly identified "*masioka*" as the tarantula, but also confused tarantula and trapdoor spiders (see *Ummidia*). The tarantula in the story is said to be a male: "He lives in the ground, making a house in the ground." Consistent with tarantula behavior, this would be a female tarantula.

Aphonopelma, tarantula, Las Cocinas, Sierra El Aguaje. 2 Aug 2018 (SC).

Latrodectus
Common names: CHUKUI HUVAHE (black spider); *viuda negra*; black widow

Arizona and Sonora. These venomous and potentially dangerous spiders (including *Latrodectus hesperus*, western black widow, and *L. mactans*, southern black widow) are common, but bites are rare. Medical treatment should be sought in the case of a bite (also see *Euphorbia*, Euphorbiaceae).

Loxosceles reclusa
Common names: HUSAI HUVAHE (brown spider); *araña reclusa, araña reclusa parda, araña violín*; brown recluse spider, fiddleback spider, violin spider

Arizona and Sonora. These small, rather inconspicuous spiders are known for their venomous, necrotic (tissue-destroying) venom. A recluse spider bite produces a wound that can be slow and difficult to heal. Luckily the incidents of such bites are rare.

Ummidia
Common names: *araña terafosa*; trapdoor spider

Arizona and Sonora.

Narrative: In the story of the tarantula and spider wasp, Giddings (1959:49) mentioned, "He puts a top on his hole, which is the door to his house." This would be a trapdoor spider, which is also preyed on by spider wasps (*Pepsis* and *Hemipepsis*, Hymenoptera: Pompilidae).

SCORPIONES – SCORPIONS
Common names: MAACHIL; *alacrán*; scorpio

Arizona and Sonora. *Maachil* is the name for all scorpions. There are a number of species in the region. Scorpions often hide in firewood, and one should be careful when bringing firewood from the desert.

Masks: Scorpion images are sometimes painted on the side of a pahko'ola mask as a sign of respect for animals.

Centruroides sculpturatus
Common name: Arizona bark scorpion

This species is often found hiding in firewood. It is also found in houses. Its sting can be dangerous and sometimes life-threatening. The Arizona and Sonora populations were formerly classified as *Centruroides exilicauda*, a species now considered to be endemic to the Baja California Peninsula.

Centruroides sculpturatus, bark scorpion, Tucson Mountains, 22 Aug 2008 (AR).

Hadrurus arizonensis
Common name: desert hairy scorpion

This, and perhaps related, species are seldom found in houses. Their sting is painful but generally not dangerous.

SOLIFUGAE – SUN SPIDERS

Eremobates

Common names: KOVATARAU (*kova*, head + *tarau*, a description of big and awkward); *araña panzona*, *matavenado*; camel spider, sun spider, wind spider

Arizona and Sonora. These spider-like creatures are seen in summertime at night. *Kovataraum* sometimes occur in houses and can run quite fast. They are harmless, although the enormous pair of "jaws" (chelicerae) might inflict a pinch if one were fast and foolish enough to catch a sun spider. People in Marana used to be scared of them because they feared they could sting like a scorpion and are rather fierce looking.

INSECTS

Yoeriam is the general term for insects.

COLEOPTERA – BEETLES

BRUCHIDAE – BRUCHID BEETLES and CURCULIONIDAE – WEEVILS

Pérez de Ribas (1645; Reff et al. 1999:87) wrote, "To keep weevils from getting into the food, they stored it on a ramada covered with thorny branches." He was probably referring to bruchid beetles or possibly weevils, which typically infest mesquite pods and other legume crops such as beans. It is interesting to know that the people in the seventeenth century used some method of control for these pests, but thorny branches probably would not have been effective.

CERAMBYCIDAE – LONGHORN BEETLES

Derobrachus hovorei [*Derobrachus geminatus*]

Common names: YUKU; *escarabajo de palo verde*, *torito*; palo verde beetle, palo verde root-borer

Arizona and Sonora. The term *yuku* means 'rain,' and in this case signifies 'rainbug.' This blackish-brown beetle, often more than 75 mm (3 in) long, is seen at night during the summer rainy season. They can bite if handled but are otherwise harmless. The large larvae feed on roots of palo verdes, especially *Parkinsonia aculeata* (Fabaceae), and make obvious exit holes in the soil beneath the trees.

Derobrachus geminatus, palo verde root-borer, Tucson Mountains. 15 Jul 2008 (AR).

LAMPYRIDAE – FIREFLIES

Common names: KUUPIS, KUPISIM (pural); *copéchi*, *luciérnaga*; firefly

Sonora. *Kupisim* are likely to be fireflies (Lampyridae), which tend to occur where soils are moist. Click beetles (fire beetles) in the genus *Pyrophorus* (Elateridae) also luminesce (light up and blink at night) and are often confused with true fireflies. *Kuupis* can be seen at night around canals and shrubbery near Pótam.

SCARABAEIDAE – SCARAB BEETLES

Canthon

Common names: BWITA MAIVAL (*bwita*, excrement/shit + *maival*, beetle); *escarabajo pelotero*; dung beetle

These humble beetles make small dung balls

Canthon, dung beetles, Wilcox, Arizona. 14 Jul 2014 (AR).

and roll them to suitable locations where they bury them as food for their larvae. Elders in Marana told children not to play with these beetles because they are dirty, that they come from animal feces.

Cotinus mutabilis

Common names: MAIVAL; *mayate, mayate verde*; green fig beetle

Arizona and Sonora. These large, bright, iridescent green and tan beetles are seen during the summer rains, often around houses and gardens. They feed on figs and other fruits.

Play: Children in Marana would tie a little thread around them, in front of the wings, and the beetle flies around like a little airplane.

TENEBRIONIDAE – DARKLING BEETLES

Eleodes

Common names: HUVACHINAI (sticking up stink); *pinacate*; darkling beetle, pinacate beetle, stink beetle

Arizona and Sonora. These are large, shiny black beetles. The Yoeme name refers to the defense mechanism of this beetle—when annoyed, it sticks up its rear end, puts its head down, and emits a foul-smelling substance. El Pinacate, the great volcanic field near the head of the Gulf of California, is named for this beetle.

Medicine: Wagner (1936:83–84) reported "Medicine for smallpox is got from a small black insect called *pinocate* [sic]." Also see *Tabebuia impetiginosa* (Bignoniaceae) for another smallpox remedy reported by Wagner.

DIPTERA – FLIES

CERATOPOGONIDAE – BITING MIDGES

Culicoides

and/or

Dasyhelea

Common names: TEPU; *jején, jejénes* (plural); biting midges

Sonora. These obnoxious, tiny flies produce painful, irritating bites. They are encountered in summertime along the coast of Sonora and are common among mangroves.

Basilio (1890 [1634]:174) cited *curu* as a "*mosquito, que llaman jején.*"

CHAMAEMYIIDAE – ACALYPTERATE FLIES

Paraleucopis mexicana

Common names: WETEPO'I ; *bobito, bobo*; eye gnat

Arizona and Sonora. During hot weather these harmless but annoying little flies crawl in and out of eyes, nose, and ears seeking moisture and salt, but do not bite. They are called *bobos* in Sonora. Basilio (1890 [1634]:215) recorded *huotepoli* as "*bobito, especie de mosquito que se mete en los ojos.*"

CHIRONOMIDAE – MIDGES

Common names: CHUNKURIAM (*chun* could mean anus + *kuriam*, from *kuria*, stirring or screwing); midge larvae

Arizona and Sonora. These small, reddish brown, fresh water "worms" are probably larvae of midge flies.

They are seen at Marana in ditches, or ponds, anchored in mud, "standing up" and wiggling like tiny spaghetti. If children are fidgeting, the adults might say, "you act like you have *chunkuriam.*"

CULICIDAE – MOSQUITOS

Common names: WOO'O; *mosquito, zancudo*; mosquito

Arizona and Sonora. Places with lush vegetation or standing water, as well as irrigated agricultural lands, are known for mosquito infestations. There are many species and genera, both native and introduced, in the region. Some mosquitoes in Sonora are known to carry malaria and dengue, and West Nile disease arrived in Arizona at the end of the twentieth century.

Basilio (1890 [1634]:175) recorded *huo* as "*mosquito otro, que hace ruido,*" and *sebehuoo* as "*mosquito otro.*"

In Marana, people used to gather dried cow manure and burn it to smoke out mosquitoes around the houses. Especially when sitting outdoors telling stories in the summertime at night, people would burn cow dung fires to control the hordes of mosquitoes. This method of control was widespread across the Sonoran Desert and elsewhere, especially among river-dwelling people, for example in the delta region of the Río Colorado (Felger 2007). Dung burns low, smoky, and for a long time.

MUSCIDAE – HOUSEFLIES

Musca domestica

Common names: SEVO'I; *mosca*; housefly

Arizona and Sonora. This is the common, ubiquitous housefly.

Basilio (1890 [1634]:174) gave *seebori* as one word for *mosca*. Seebori is more like a Yoreme than a Yoeme word. The term is similar to the word for tadpole, *sivo'oli*, and perhaps related to a larval stage.

SYRPHIDAE – HOVERFLIES

Common names: NAHI SEVO'I (nahi fly); hoverfly

Arizona and Sonora. The same Yoeme name was used for a leafcutter bee, however the description is clearly that of a hoverfly: *Nahi sevo'i* is described as a kind of fly less than an inch long, smaller than a honeybee. They are black and white (no yellow), with fluttering wings. The body (abdomen) is the most obvious part, it is banded black and white with a white *chovek* ('rear end' or abdomen), the head is scarcely or not visible; you always see the rear end, the wings are transparent (you can see through them). They are seen in summertime and are like a fly or little bee. These insects go really fast from one area to another and then stop and hover, then dart quickly to another place and do the same.

Celebration: The nahi dances have been performed at pahkom, mostly by boys and girls. They dance to the drum and the songs.

Taub (1950) reported that the nahi sevo'i dance and songs had not been performed for decades or more. He gave a description of the Nahi Dance and a Nahi Song:

> The *nahi*, according to Arizona informants, is a small, common fly which moves its wings so rapidly it can remain hovering in one place in mid-air, much as the hornet [sic] does. The dancer, not a member of any special society, imitates the fly by stretching out his arms with bundles of corn-leaves in each hand, which he quivered as the fly's wings. He wore a cloth head-covering and danced alone to a drum accompaniment.
>
> Mr. Ambrosio Castro describes the insect as a large, green fly....He describes the dancer as wearing a red headband with feathers and mother-of-pearl shell ornaments, carrying an ear of corn in his right hand plus a bow in his left, and wearing strings of beads and dried pods around his neck. (Taub 1950:120–122)

In 1981, Porfirio Yokiwa of Marana taught Felipe and some boys and girls the nahi dance and songs and demonstrated the regalia worn for the dance. The dancer has two white scarves or cloths that are tied around the head and hang loose over the ears, and a red sash is worn around the waist. Dry cornhusks are held in both hands as wings and moved to follow the drumbeat and the song, while a rustling noise is made to represent the insect. The dancer has to be agile and move around real fast like the insect itself, jumping from one place to another while dancing, imitating the insect. Porfirio told Felipe, "That is how the dancers have to be, moving like the bee [sic], dancing in one place and then quickly jumping, moving to the next place."

Painter (1986:119) suggested that "Traditional dances mentioned but seldom seen are the Nahi Dance, said to be from Opata Indians of Mexico." The Nahi Dance is also discussed by Johnson (1962:277).

TABANIDAE – HORSEFLIES

probably **Tabanus**

Common names: TEEKA SEVO'I (heaven fly); *tábano*; horsefly

Arizona and Sonora. These large flies can inflict a painful bite to the legs and arms if you are not paying attention.

Basilio (1890 [1634]:174) cited *tecasebori* as one word for *mosca*.

Beliefs: This fly is said to be a harbinger of death. In the Deer Song book (Evers and Molina 1987:183), Felipe writes, "This kind of fly is considered a warning of somebody's death by Yaquis...We believe that these flies usually visit somebody at home or when he is out in the wilderness. When the fly lands close by, it means that somebody known to the person will meet with death."

HEMIPTERA – TRUE BUGS

CICADIDAE – CICADAS

Diceroprocta apache

Common names: MATE; *chicharra, cigarra*; citrus cicada, Sonoran Desert cicada

Arizona and Sonora. This is the common lowland species of cicada in the Sonoran Desert. These rather large, harmless insects are often found on mesquite trees. They are heard in summertime making their constant high-pitched sounds.

Basilio (1890 [1634]:149) cited *matem* as the word for *cigarra*.

Diceroprocta apache, Apache cicada, Huachuca Canyon, Arizona. 4 Aug 2015 (AR).

Play: Children in Marana liked to catch them and play with them, hold them and listen to their sound, then let them go.

CIMICIDAE – BED BUGS

***Cimex lectularius**

Common names: CHINCHI; *chinche*; bedbug

Arizona and Sonora. These bloodsucking household pests are of Old World origin and quickly spread in the New World after European contact.

REDUVIIDAE – ASSASSIN BUGS

Triatoma

Common names: CHINCHI; *chinche besucona, chinche picuda*; cone-nosed bug, kissing bug

Arizona and Sonora. Packrats are the usual hosts for these blood-sucking bugs. Especially during the dispersal season around May and June, the *chinchim* are flying at night and are attracted to lights and warmth. This is when the most bites are sustained. The bite can produce a painful, itchy swelling. Some people can have an anaphylactic allergic reaction to the bite of these insects. Mosquito netting is recommended for susceptible people and children in vulnerable areas. *Triatoma* can be vectors for Chagas disease.

HYMENOPTERA – ANTS, BEES, AND WASPS

APIDAE – BEES

Basilio (1890 [1634]:224) used *sitori* for "*miel, colmena*" (honey, hive) but we do not know if he was referring to the honeybee or native wild bees or even the honey wasp (*Brachygastra mellifica*). He also gave descriptive phrases for seven kinds of bees (Basilio 1890 [1634]:127–128).

***Apis mellifera**

Common names: MUUMU; *abeja*; honeybee

Arizona and Sonora. Honeybees are native to Europe. They were introduced into New Spain in early colonial times "and eventually went wild in the Sierra Madre Occidental and Sierra Madre Oriental, where today [in 1970] there is the greatest concentration of wild small black European bees" (Brand 1988).

The word *muumum* is used for honey as well as the bees. A beehive is *kolmeena* from the Spanish *colmena*. Other terms for a beehive are *mumu hoara* (bee home) or *paneelam* (from the Spanish *panela* for sugar cane).

Food: In addition to keeping beehives, some of the farmers in the Yoem Bwiara find wild beehives and bring the honeycombs home for the family.

Medicine: Beeswax combined with tobacco has been used medicinally (see *Nicotiana rustica*, Solanaceae).

Bombus sonorus

Common names: VIIKO; *abejorro*; Sonoran bumblebee

Arizona and Sonora. These large, fuzzy, black and yellow bees are seen at Marana and are widespread in the Sonoran Desert region.

Melipona
and/or
Trigona

Common names: *abejas que no pican, abejas sin aguijón*; stingless bees

Sonora. These tiny, honey-producing, stingless bees are widespread in the American tropics and range as far north as Sonora (Bennett 1964), where they are common near Álamos in the southeastern part of the state. These bees occur southeast of the Yoem Bwiara and might be found at the inland margins of the Yoeme region. Oral history of native, wild honey may relate to stingless bees, but more likely to the Mexican honey wasp (see *Brachygastra mellifica*, Vespidae).

Food and wine: We were told about *maikom* (beehives) that have white honey, almost like that of the honeybee. Phil Jenkins (personal communication, 2012) told us about these bees in the Yoreme region of southeastern Sonora:

> I saw stingless bees in shoeboxes or some such on porches. First time was on a hike in the Sierra de Álamos...came on a house on a trail, and we didn't know what the box was, so we asked. They showed us and then again around Los Llanos, on the way to Santa Barbara, I saw a box and was told, "*Si, abejas que no pican* (Yes, bees that do not sting)." So, passing homes in the Sierra, I have seen a number of such boxes hanging on porches, in containers that, I have imagined, were chosen such that *los murcielagos* [bats] can't get them. People in Álamos said the bees were rare, somehow the Sierra folks often had them.

Xylocopa

Common names: KUKUSAKA; *abeja carpintera*; carpenter bee

Arizona and Sonora. These are large, black bees. The females are black and excavate tunnels and chambers in various dry wood, or stems of plants such as agave stalks and chinaberry trees (*Melia azedarach*, Meliaceae, in Arizona). Males are yellow with green eyes. The females have a painful sting, but have to be provoked (such as capturing them) before doing so. Bernaldo Valencia said *kukusaka* live in holes in *vaaka* stems (cane or *carrizo*, *Arundo donax*, Poaceae). See *Euphorbia* (Euphorbiaceae) for treatment of the sting.

Xylocopa, carpenter bee, Tucson Mountains. 9 Sep 2008 (AR).

FORMICIDAE – ANTS

Common names: EEYE; *hormiga*; ant

Arizona and Sonora. The ant fauna of the region is diverse (Wheeler and Rissing 2015). Ants feature prominently in Yoeme oral and written narratives and culture, including the Talking Tree. Several major kinds of ants are named, including eeye, the general term for large ants. *Eeye masakame* (ant with wings) are winged ants, or alates, the reproductives that are the founders of new nests. Swarms of winged ants of both sexes are seen during the summer rainy season.

Basilio (1890 [1634]:165) recorded *eie* as the term for *hormiga*, and *eietecoa* as the term for *hormiguero*, or anthill.

Beliefs and narratives: Some narratives tell that when the Surem decided to go underground, some became ants. This was the time of the prophecies from *Huya Nokame*, the Talking Tree. The Surem had the power of transformation; they could change themselves to a different life form. In some versions there were no ants before the arrival of the Spaniards. In other versions the Surem chose to exist as an enchanted people and whenever it was necessary, they transformed themselves into animals including ants, or into plants.

It is said that we human beings come from another spiritual place, and when we come to this earth we get our human bodies. When a person is born, the umbilical cord is saved and put in a pouch to be later given to the red ants—because they were here before us. We are just here for a short while and the introduction of the umbilical cord to the ants means that we will respect life here on earth and be recognized by the *Huya Ania* (Wilderness World) as being "one of us." And in turn the animals and insects will respect us. This information is from the literature as well as from Felipe's mother, Juanita Paula Valle and his maternal grandmother, Anselma Tonopuame'a-Castillo. Offering the umbilical cord to the red ants was practiced widely by the women in Marana in the 1950s and 1960s. (These red ants probably are harvester ants, *Pogonomyrmex*.)

When the female relatives take the umbilical cord of a newborn infant to the ant hill, the narrative relates that the ants are a little bit hesitant, like they are saying, "should we take it in, or should we not?" That's one of the big discussions that the ants have, and it is said that they take longer with a girl's umbilical cord. Some children's umbilical cords are not offered to the red ants, and it said that those children are usually the ones who get stung by ants.

Felipe learned, "If ants are in your garden or at home, you talk to them to leave the area because we don't want them to bite or sting the children. Just talk to ants when they are bothering the family. So, the elders would offer food to the ant nest (ant hole); best would be grain like wheat, barley, or rice. That would keep the ants busy and away from the home. But nowadays people use ant poison."

Painter (1986:10) asked a woman, "Do ants speak Yaqui?" and was told, "Yes, my father told them to move out of the kitchen because he did not want them to bite a little boy, and they did."

Atta mexicana

Common names: MOCHOMO; *hormiga arriera*, *mochomo*; leafcutter ant

Sonora. These ants are common in Sonora, but in Arizona are known only from Organ Pipe Cactus National Monument and adjacent Tohono 'O'odham lands. These ants can strip wild and cultivated plants of their leaves, which they use to cultivate fungus gardens in underground nests.

Basilio (1890 [1634]:165) recorded *mocho* as a term for "*hormiga arriera*" (mule or driving ant). Santamaria (2000) reported that *mochomo* is a Mexican term for desert ant, from the Yaqui and Mayo languages.

A different leafcutter ant, *Acromyrmex versicolor*, is a garden pest in Arizona and Sonora, but it is generally not referred to as mochomo.

Narrative: Giddings (1959:19) related a mochomo story involving "a chieftain of the ants who was driving a mule train of little mochomos."

Pogonomyrmex

Common names: SIKI EEYE (red ant); *hormiga colorada*; harvester ant

Arizona and Sonora. These are large, red, seed-harvesting ants. They form wide, low craters in cleared areas around the nest entrance, with refuse piles of discarded chaff from seed- and grain-harvesting. They also deposit small gravel-sized pebbles around the nest from their extensive underground excavations.

Arizona and Sonora each have about 12 species of "pogos." *Pogonomyrmex* can inflict a painful sting, which can cause a severe allergic reaction. The venom is the most toxic of any known insect (Schmidt 2016).

Ceremonial regalia: Small pebbles from red anthills are put into *tenevoim* (cocoons) to produce a rattling sound (see *Eupackardia* and *Rothschildia*, Lepidoptera: Saturniidae).

Harvester ants, Dragoon Mountains. 5 Sep 2011 (AR).

Unidentified ants

CHUKUI EEYE (black ant); *hormiga negra*; black ant

Arizona and Sonora. *Chukui eeye* is the general term for large black ants.

EESUKI

Arizona and Sonora. These are "little tiny" red-orange ants.

HOOVO'E

Arizona and Sonora. These are orange-colored, stinging ants, perhaps similar to *siki eeye* (*Pogonomyrmex*). *Hoovo'e* is said to be more slender than siki eeye.

MEGACHILIDAE – LEAFCUTTER BEES

Megachile?

Common names: NAHI SEVO'I (also the name for hoverfly, Syrphidae, and perhaps not correct here); *abeja cortadora de hojas*; leafcutter bee

Arizona and Sonora. Female leafcutter bees form barrel-shaped brood cells made from nipped, semi-circular leaf fragments. They stuff several brood cells single file, end to end into holes, crevices and cracks wherever they can find them. The brood cells are packed mainly with pollen as provisions for the larvae. There are several species of leafcutter bees in Arizona and Sonora, but these might belong to the genus *Megachile* (subgenus *Chalicodoma*). In Marana, Felipe learned that the *nahi sevo'i* likes to make its home in unfired adobe walls, or in the walls of the traditional wattle and daub houses, and packs its nest with leaves. These descriptions are consistent with leafcutter bee behavior.

Play: Children in Marana liked to find the nests and take them apart to see how they are put together.

MUTILLIDAE – VELVET ANTS

Dasymutilla

Common names: SOOTO'OLI; *hormiga aterciopelada*; velvet ant

Arizona and Sonora. These solitary wasps, resembling a large, furry ant, can be combinations of brown, brownish orange, black and white, yellow, orange, or red. The females are wingless and are seen especially in summertime running over the ground. They can produce a potent and painful sting.

Narrative: Rosario Vakame'eri-Castillo told Felipe in 1976 that in Marana the elders would tell the children to respect the *sooto'oli* and not to kill or step on one intentionally because, "When a person kills a sooto'oli, a small hairy bug, it will rain hard with strong winds."

POMPILIDAE – WASPS

Brachygastra mellifica

Common names: *avispa de miel mexicana*; Mexican honey wasp

Sonora. The Mexican honey-producing wasp is widespread in Mexico (Sugden and McAllen 1994). These wasps build their nests on tree branches.

Food and wine: We were told about *maikom* (beehives) that have white honey, almost like that of the honeybee. These maikom apparently are from the Mexican honey wasps.

Pérez de Ribas (1645; Reff et al. 1999:94) gave a detailed and accurate description of a honey wasp nest and the harvesting of the honey. The locality might have been in present-day southern Sonora or northern Sinaloa:

> Another hunt...is for the honeycombs from wild beehives...[in] wooded areas and montes. Although their bees do not produce a wax, being no larger than flies, they do produce a very fine honey. In softness, sweetness, and aroma it is better than the best honey from Castile. The hive is round and measures two-thirds of a vara [about 56 cm, or 22 in] in length or, if full of honey, a full vara [about 84 cm, or 33 in]. The material covering the hive where the combs of honey are stored is similar to a leaf, like wasps' nests in Castile, and there is an entrance that is just large enough for the bee's tiny body...they construct them high up on a branch that has some kind of a hook to hold the hive so the wind cannot blow it down. Because the honey is made from sweet-smelling flowers, it, too, smells good.

> In the spring...the Indian who is looking for honey searches for a pool of water, often in areas of the monte where the streams have overflown. There he waits for the bees to arrive to drink the dew they need to make honey. When a bee flies off he runs after it, keeping its flight in his sight until he locates the hive. When he finds it, he cuts down the branch from which the hive is hanging and carries it to his house. There he savors its contents, not only the honey but also the bees' larvae, which are still shapeless, fragile little worms in their tiny houses in the comb. He places these on the coals and, when roasted, they are eaten as a delicacy...The Indian who searches out the hives must possess good eyesight in order to be able to see the little bee in the wind. For the same reason, one should not try to find them on a cloudy day.

Additionally, Pérez de Ribas (1645; Reff et al. 1999:90) observed, "Among all the wines they made, the most popular and flavorful was made from honey-combs, which are harvested at a particular time of the year."

Alberto Búrquez (personal communication, 2016) provided us with this firsthand account of his experience with honey wasps in the Guaymas region, in former Yoeme territory:

> When I was a teenager, in the late 1960s, we used to wander along the mountains and canyons of Sierra El Vigía that overlooks Guaymas Bay, Punta Colorada and Cabo Haro (the mountains that border Guaymas Bay to the southeast). There, I learned about these strange desert squirrels that chirped in the canyons, *coas* [*Trogon elegans*, Trogoniformes] singing loud near the *aguajes*, and paper wasps that yielded sweet honey. My recollections of the paper wasps are of great fear of being stung (although I never was stung), and the reward of a little honey after we destroyed the nest among clouds of smoke and enraged wasps. The silver-grayish nests were not uncommon, but widely spaced. They were nested in large shrubs or small trees supported by a large branch that seemed to go across the nest. An orifice at the bottom of the nest was the entrance to this papery mansion. Their size ranged from small nests of about 15 centimeters diameter to large, somewhat deformed ellipsoids up 50 centimeters diameter. The small wasps buzzed around the colony and patrolled all over the surface of the nest.
>
> Twice in my life we raided a colony (I probably was 12 to 14 years old) by making a smoky fire under the tree, and later, with the aid of poles ripping apart the nest. They consisted of papery layers arranged from top to bottom that enclosed chambers with wasps in different stages of development and some chambers with a clear, somewhat fluid honey in some cells. For a teenager, the flavor was great, sugary and wild with a tinge of some strange animal bouquet. The amount of honey was rather small, and we only got a taste of what I think was less than a thimble for each of the four raiders of the nest. Today, as a biologist, I guess the nests are those of a species that, as many other taxa, reach their northernmost range of distribution in Sonora.

Hemipepsis and **Pepsis**, and related genera
Common names: HIMA'AWIKIA (*hima'a*, burying + *wikia*, string; throwing string); *avispa cazadora de arañas*; spider or tarantula wasp, tarantula hawk

Arizona and Sonora. These large wasps may be all dark blue-black or blue-black with orange wings. The female spider wasp paralyzes the tarantula with her sting, drags it into a burrow, and lays a single egg on the spider. The larva feeds on the paralyzed tarantula, leaving the vital organs for last. These wasps also prey on other kinds of large spiders (see tarantula, *maisooka*, *Aphonopelma*, and trapdoor spider, *Ummidia*).

Narrative: A story told to us by Ignacio Amarillas relates how this insect and the tarantula will fight and that the little insect will win and drag the tarantula into its hole. Most people would

Pepsis, spider wasps, on *Asclepias albicans* (Apocynaceae) near Loreto, Baja California Sur. 22 Feb 2018 (SC).

incorrectly think that the large tarantula could overpower such a relatively small creature.

The story of *Maisoka* and *Hima'awikia*, told by Giddings is given here in abridged form. It is essentially the same as one related to us by Ignacio Amarillas:

> Maisoka was in his house one day and about the time that he stuck his head out of his door, Hima'awikia was walking about. Maisoka said, 'Who walks on top the house of the house of the King?' He said that because he considered himself a King.
>
> Hima'awikia is an insect with wings. It is a little animal, a bit reddish and it flies. This is he who walked over the home of Maisoka making a noise with his wings, going thus: 'ronronron.'
>
> Maisoka heard the 'ronronron' and stuck his head out and said, 'Who is this imposter who walks on the roof of the house of the King?'
>
> 'Oh, pardon me, Sir,' said Hima'awikia, and Maisoka allowed him to enter his house.
>
> A few minutes later Hima'awikia came out carrying Maisoka between his teeth. It appears unbelievable that Hima'awikia, who is so small, always conquers Maisoka and eats him. Maisoka never escapes from the teeth of Hima'awikia. (Giddings (1959:49)

Polistes

Common names: VIICHA, VIICHAM (plural; may also apply to related wasps); *avispa*, *bitache*; paper wasp

Arizona and Sonora. These wasps, often yellow and black, can deliver a painful sting. They can be eliminated by smoking them out, as is done for honeybees. Bernaldo Valencia (in Tucson, 1994) said that wasps are especially numerous in Sonora at *tahsi'o* trees (*san juanico*, *Bonellia macrocarpa*, Primulaceae) when the fruit is ripe and falling.

Basilio (1890 [1634]:205) reported "*vitachi*, *abispa*" as *bitza*, which seems to be a Yoreme word.

ISOPTERA – TERMITES

Common names: POLIA; *comején*, *termite*; termite

Several genera of termites are common in Arizona and Sonora. Basilio (1890 [1634]:150, 174) cited *curu* as the term for "*comején*, *animal*," as well as a name for biting flies; see Diptera: Ceratopogonidae).

LEPIDOPTERA – BUTTERFLIES AND MOTHS

Common names: VAISEVO'I, VAISEVOLI; *mariposa*, *palomilla*; butterfly, (large) moth

Arizona and Sonora. *Vaisevoli* is a term for any large butterfly such as the monarch or swallowtail, and sometimes for large moths. The plural *mariposoom*, from the Spanish *mariposas*, is sometimes used instead of *vaisevolim*. The Spanish term *paloma* or *palomilla* is sometimes used for a moth. A caterpillar is called *bwiiwi* or *bwichia voala*.

Basilio (1890 [1634]:179, 184) recorded *buaruchim* for *polilla* (moth) and *baesebela* for *palomilla*.

Beliefs: A black butterfly (or perhaps a moth) is said to be a bad omen, a messenger bringing news of sickness or death from the spirit world.

EREBIDAE– TIGER MOTHS and their kin

Estigmene

Common names: BWICHIA VO'ALA (woolly worm); tiger moth caterpillar

Arizona and Sonora. We were told that these

caterpillars are mostly seen in September and October, often near cotton fields and homes in Marana. They are described as woolly, black and dark orange, and about 5 cm (2 in) long. These are probably tiger moth caterpillars, *Estigmene albida* and perhaps *E. acraea*. The caterpillars of these moths are abundant during the fall months in Arizona and are also found in Sonora.

SATURNIIDAE – SILK MOTHS
Eupackardia calleta
Common name: calleta silkmoth
and
Rothschildia cincta
Common names: *cuatro espejos* (referring to the four large, transparent wing spots for *Rothschildia* in Mexico); cincta silkmoth

Sonora and locally in Arizona. These are large, beautiful moths. Many people use the Spanish word *paloma* for a small or giant moth. These silkmoths are almost as large as a bat. When Yoemem see them, they may say *sochiktavena* (like a bat).

Cincta silkmoth (*Rothschildia*) caterpillars primarily feed on *Jatropha* (Euphorbiaceae) and *Hintonia* (Rubiaceae). Calleta silkmoth (*Eupackardia*) caterpillars feed on *Pleradenophora* (Euphorbiaceae), *Fouquieria* (Fouquieriaceae), perhaps *Vallesia* (Apocynaceae), and other plants. Cincta moths may have multiple broods during warm seasons. The last brood, in the fall, spends eight to nine months as cocoons. Calleta moths also spend the eight or nine months of the dry season as cocoons, but are capable of delaying emergence for years until favorable conditions return. During their brief adult life of a few days, these moths mate and the females lay eggs. The adults do not feed.

The cocoons (*tenevoi yoeria*, insect cocoon) are classified according to the kinds of plants on which they are found. Cincta moth cocoons found on the *sapo* shrub (*Jatropha cinerea*) are known as *sapo tenevoim*. Calleta moth cocoons found on the *sita'avao* shrub (*Vallesia glabra*) are called *sita'avao tenevoim*. Bernaldo Valencia told us that there are three kinds of tenevoim. In addition to the two common kinds, *vaawe tenevoim* (sea cocoons) are found on shrubs in mangrove (*paseo*) areas (*vaawe* refers to anything near or in the sea). Bernaldo said these cocoons are hard to obtain and nowadays seldom used. We have not seen this third kind of cocoon.

We were told that right after Lent (April) the cocoons are found hanging on the plants. May is said to be the best time to harvest them, "because if you pick them too soon they are soggy," but they also are gathered through the summer. Whole families of Yoeme or Yoreme people would go to the mountains for several days at a time to gather cocoons. Among the Yoremem, goat herder families often gather cocoons while they are out tending their animals. These cocoons have become scarce in the Yoeme lands since the late twentieth century, said to be the result of pollution from nearby cities. The cocoons have also become scarcer in the Yoreme lands. Since the late twentieth century the cocoons have become expensive. Some people in Arizona raise these moths in captivity for their cocoons.

In the 1970s, Rosario Vakame'eri-Castillo told Felipe that in the early 1900s he had seen these cocoons in the Tucson area, but they went away, but were still found on the 'O'odham reservation. He said they "went away because the city was growing and they don't like pollution."

Ceremony and tenevoim: Silkmoth cocoons are used to make the tenevoim leg rattles. Sapo tenevoim (*Rothschildia*) are usually regarded as the best kind for tenevoim because the cocoon walls are thinner and produce the most pleasing sound. Bernaldo Valencia said that the sound produced by vaawe tenevoim is not as strong as that of the sapo tenevoim. More than 90 percent of the tenevoim cocoons we have seen are from *Rothschildia*.

The cocoons are collected or purchased and kept in cloth bags for later use. One end of the cocoon is cut off with a knife, and whatever is inside is cleaned out. Five little pebbles are placed in each cocoon. If too many pebbles are used the strand will be too heavy. These little pebbles are often gathered from the gravel mounds surrounding red ant hills (see *Pogonomyrmex*, Hymenoptera: Formicidae). A hole is punched on each side and the cocoons are then tied together in

A. *Eupackardia calleta*, calleta silkmoth, Dragoon Mountains. 26 Jul 2009 (AR). B. *Eupackardia calleta* cocoon on ocotillo stem, Dragoon Mountains. 31 Dec 2017 (AR). C. *Rothschildia cincta*, cincta silkmoth, cocoon suspended from a *Jatropha* stem, Yoem Bwiara. 1988 (RF).

pairs with sturdy string; rawhide and sometimes *ixtle* (twisted agave fiber), and perhaps sinew, were used in former times. More recently the cocoons have been strung on monofilament (fishing line). At the ends of the strand are bundles of red cotton string representing *seewam* (flowers).

The method of assembling the cocoons was described in the 1940s:

> Around the ankles and extending up to the base of the rolled-up trouser legs, are strips of cocoons sewn on rawhide and wrapped around the legs. These are called teneboim. The cocoons have been opened, cleaned out and cured so that they resemble a soft white leather. In each cocoon are placed several pieces of gravel, and the entire cocoon is closed by sewing it onto the rawhide strip. (Wilder 1963:166)

Both the deer dancer and the pahko'ola dancers wrap their legs with tenevoim. While the maaso wears three coils of rattles, the pahko'olam may cover their legs with many coils up to the knee. The tenevoim strands, coiled around the leg, represent serpents to provide more spiritual power. The power goes up the leg, the body, and out from the topknot tied into the hair.

A toe loop at one end of the strand holds it in place as the dancer begins wrapping the tenevoim around the leg. Many of the older photos show pahko'olam dancers with tenevoim wrapped all the way to the knee. Because of increased scarcity and high prices, since the latter part of the twentieth century, one often sees pahko'olam wearing only 2 or 3 loops like the deer dancer wears. Many Arizona pahko'olam cut their tenevoim strands to share with other dancers because they are so expensive. The tenevoim in recent times are purchased in the Yoreme region in Sonora. The strung tenevoim should last a lifetime.

Tenevoim strands are sometimes dunked into a bucket of *kaal* (lime) solution to make them look brighter. Usually only the *chapeyekam* did this, not the deer dancer or pahko'ola dancers, because it would more be natural to leave them alone. When Felipe's uncle Juan Luis Garcia bought new tenevoim strands he would dunk them in lime.

Merced Maldonado, a pahko'ola and mask maker from *Waalupe* (Guadalupe), raised calleta silkmoths for cocoons used to make tenevoim (Peigler and Maldonado 2005). He said the cocoons were plentiful and easy to find in southern Arizona until the early 1990s, and then they disappeared.

In the ceremonies, the tenevoim are highly respected because they were once the homes of the moths.

MANTODEA – MANTISES

KAMPO MOOCHI; *campamocha*; praying mantis

Arizona and Sonora. A number of genera and species, both native and introduced, occur in

Arizona and Sonora. *Stagmomantis limbata* is a common, green praying mantis. These large, predatory insects are often seen in gardens.

ODONATA – DRAGONFLIES

Common names: VAIKUMAREEWI (*vai*, with water + *kumareewi*, a core word); *caballito del diablo*, *libélula*; dragonfly

There is a considerable diversity of dragonflies in Arizona and Sonora (Bailowitz et al. 2015). Dragonflies are often seen flying over water, the females repeatedly dipping their abdomen into the water to lay eggs, or some drop their eggs into the water bombardier style. Dragonflies are also seen over water capturing prey insects "on the wing." The larvae are aquatic and voracious carnivores.

Dragonfly, Aguaje Los Pilares, Sierra Bacatete, May 2008 (MB).

Song: *Vaikumareewi*, the dragonfly song, is sung in the evening during a pahko. Evers and Molina (1990) provided "Yoyo Vaikumarewi, Enchanted Dragon Fly" as a coyote song from the *Wiko'i Yau'ura* (Coyote or Bow Leaders' Society).

The song below describes the dragonfly and the place where it flies about, hovering over the water in a motion seeming to take on the form of a flower. This song was given to Felipe by his compadre Luis Valenzuela, Sr., in 1980. Luis recorded it from a deer singer in Pótam.

yoyo vaikumareewi empo sewa vam maneka'apo
sewa soisoiti koweme
empo yoyo vaikumareewi sewa vampo sewa
soisoiti koweme

yoyo vaikumareewi empo sewa vam maneka'apo
sewa soisoiti koweme
empo yoyo vaikumareewi sewa vampo sewa
soisoiti koweme

yoyo vaikumareewi empo sewa vam maneka'apo
sewa soisoiti koweme
empo yoyo vaikumareewi sewa vampo sewa
soisoiti koweme

hunamansu seyewailo saniloapo naisukuni
empo yo vam maneka'apo uvalika
sewa soisoiti koweme
yoyo vaikumareewi empo sewa vam maneka'apo
sewa soisoiti koweme

Enchanted enchanted dragonfly,
you sway and hover in a flower way in the place of the flower water.
You enchanted enchanted dragonfly,
in the flower water, you sway and hover flower-like.

Enchanted enchanted dragonfly,
you sway and hover in a flower way in the place of the flower water.
You enchanted enchanted dragonfly,
in the flower water, you sway and hover flower-like.

Enchanted enchanted dragonfly,
you sway and hover in a flower way in the place of the flower water.
You enchanted enchanted dragonfly,
in the flower water, you sway and hover flower-like.

Over there in the middle of the flower covered grove
you bathe and sway and hover in a flower way in the enchanted water place.

Enchanted enchanted dragonfly,
you sway and hover in a flower way, in the place of the flower water.

ORTHOPTERA – CRICKETS AND GRASSHOPPERS

ACRIDIDAE – SHORT-HORNED GRASSHOPPERS

Common names: WO'OCHI; *chapulín*; grasshopper

Arizona and Sonora. There are numerous genera and species of grasshoppers in the Sonoran Desert region.

Andrew Semotiuk (personal communication, 2018) tells us that for Yoremem, "I find that the same word wochi is *chapulín*. *Wochim* would be plural, *chapulines*."

Narrative: A number of stories or legends involve the grasshopper, such as the "Grasshopper and Cricket" (Giddings 1959:51). In this story, like a number of others, the little creatures and big ones like the mountain lion are adversaries, and the little ones end up winning (also see cricket, Gryllidae). A version of the "Grasshopper (*Wo'ochimea*, Magician Grasshopper) and the Serpent" is provided in the discussion for snakes.

GRYLLIDAE – TRUE CRICKETS

Common names: KIICHUL; *grillo*; cricket

Arizona and Sonora. Basilio (1890 [1634]:163) cited *quichul* as the term for *grillo*.

Narrative: Giddings (1959:50–51) provided narratives of "The Cricket and the Lion" and "Grasshopper and Cricket." In the first one, the large animals, the mountain lions, jaguars, bears, and wolves—"all the people of the claw"—combat the small creatures, such as crickets, bees and "all the insects that fly and sting, and also the scorpions and ants." The small creatures are victorious. The "Yaquis say there is no small enemy. Everyone can defend himself. The cricket continues to sing, 'chik chik chik.' He is not afraid."

Painter (1986:325) discussed the role of various small animals in rituals of the pahko'olam during non-Easter ceremonies. One man told her, "What the pascolas might say," for example, "Thou Holy Cricket, thou who is able to stay awake all night and sing for me to sleep. Let me not feel very hard about staying up all night tonight."

PHTHIRAPTERA – LICE

Pediculus humanus

Common names: ETE; *piojo*; louse

Etem are lice, and *natchi'ikam* are the nits (egg cases). Lice combs (or brushes) were made from the spiny fruits of the *echo* cactus (*Pachycereus pecten-aboriginum*, Cactaceae).

Basilio (1890 [1634]:183) cited *ete* for "*piojo de la cabeza*" and "*piojo de la ropa*."

SIPHONAPTERA – FLEAS

Ctenocephalides canis

Common names: CHUU ETEM (*chuu*, dog + *etem*, ticks); *pulga*; dog flea

Arizona and Sonora. It's a dog's life to be troubled with fleas. See desert broom (*Baccharis sarothroides*, Asteraceae) for treating fleas on domestic animals.

ZYGENTOMA – SILVERFISH

Lepisma saccharina

Common names: KULIICHI ; *pez plateado*; common silverfish

Arizona and Sonora. Although some silverfish live outdoors, the best-known ones are the ubiquitous household pest *Lepisma saccharina* that can damage books and other paper objects.

MOLLUSCA

The Gulf of California supports a rich mollusk fauna. Marlett's (2014) *Shells on a Desert Shore* documents the extensive knowledge and usage of mollusks by the Comcaac (Seris). Sea life would also have been prominent in the culture of the original coastal-inhabiting Yoemem, however we obtained very limited Yoeme information of marine life from the literature and Las Guásimas residents. Most of the Yoemem at Guásimas are from elsewhere, including Pótam, and tend to use Mexican or newer Yoem noki names for sea life. Only a few are mentioned here.

Commercially harvested mollusks include various clams, oysters, and pen shells (Bivalvia), gastropods (Gastropoda) such as the pink-mouth murex (*Hexaplex erythrostomus*), black murex (*Hexaplex nigritus*), conchs (*Lobatus galeatus*, *Persististrombus granulatus*, and *Strombus gracilior*), and octopi and squids (Cephalopoda). There is a general trend of declining populations due to habitat loss, invasive species, overharvesting, and pollution.

BIVALVIA – CLAMS AND THEIR KIN

Auli is the general term for clam. *Vea* is the term for shell or skin, such as an orange peel. The clam shell is *auli vea.* Mother-of-pearl, obtained from several common shells, has long been a popular ornament. The most likely sources in the Yoem Bwiara are several bivalves with a nacreous inner shell (mother-of-pearl), especially *Anomia peruviana*, *Pinctada mazatlanica*, and *Pteria sterna.*

Pérez de Ribas (1645; Reff et al. 1999:91) wrote, "Among some nations when a virgin bride was handed over to her husband they removed from her neck a carved shell that the young women wear as an emblem of their virginity." Basilio (1890 [1634]: lviii, xxxix) described a similar ritual involving a pendant made from a "*concha de nácar*," and mentioned the use of "*conchas de perla*" (mother-of-pearl). Pérez de Ribas (1645; Reff et al. 1999:91) reported, "Some principals who acted as captains used to go to war wearing a type of doublet with sleeves or a blue cotton cape adorned with shells or mother of pearl that shine brightly." In discussing the Cross, Pérez de Ribas (1645; Reff et al. 1999:104) wrote, "The sign Cabeza de Vaca and his companions gave them was the Cross...This healing sign made a deep impression on them, and they wear Crosses made of mother-of-pearl around their necks or on their foreheads."

Wilder (1963:166) observed, "It is usual for both pascola and maso to have a string of black and white beads with a mother-of-pearl cross hung around the neck."

Several genera of edible oysters, including *Magallana*, *Ostrea*, and *Saccostrea*, are represented along the coast of the Yoem Bwiara (Raith et al. 2016). *Saccostrea palmula* may be found attached to mangrove roots and others are found on rocks or hardened mud-rock. The oyster shell is *koo vea.* Oysters may be gathered in esteros at low tide.

Food: Pérez de Ribas (1645; Reff et al. 1999:85) reported that "in the estuaries, of which there are many along this coast...There one...finds oysters, clams, and other types of seafood that they eat." Basilio (1890 [1634]:208) cited *conasim* as the word for *ostiones.*

Spicer (1954:49) reported, "Most families in Potam eat oysters (*koyum*) at least during Lent and

Bivalves, Las Guásimas, *Chionista fluctifraga* (left), *Magallana gigas* (center), and *Chionopsis gnidia* (right). 31 May 2019 (PB).

Holy Week. In the winter season there when there is little agricultural work to do, small parties go regularly to the shore and gather oysters. The oysters are usually baked in the shell."

In May 2019, Yoemem at Las Guásimas said that nowadays, they do not harvest the oysters (*kooyom*) for food in the waters near Las Guásimas because they are contaminated. The sewage and waste from the city of Guaymas have ruined their consumption of oysters and their livelihood. Angel Flores said if people eat too many oysters they may get sick and have lifelong health issues. For that reason, he said the community has oyster farms nearby. He said the farmed oysters have thin shells rather than the thick shells of the real kooyom.

Anadara tuberculosa

Common names: MOA'IM, MUURA WOKIM (mule feet), TATTE'ERA (one who chokes); *pata de mula* (mule foot); black ark clam

These medium-sized bivalves are collected along the coast at low tide and sold locally. The Comcaac (Seris) give similar names to this and some other clams because they are said to cause a scratchy throat when eaten (Marlett 2014).

The black ark clam is part of a small-scale fishery in the Gulf of California. There is a trend of declining populations (Félix-Pico et al. 2011).

Food: Richard recalls:

> Soon after dawn on a crisp December day in 1988, at the home of Antonia and Nacho Amarillas in Pótam, a local boy comes by with a big bucket full of freshly collected clams that he calls *pata de mula*. He asks 300 pesos each for them, and I pay 10,000 pesos for a pile of them. [10,000 old pesos in 1988 was about $1.00 U.S.] It is suggested that the best way to eat them is as *ceviche*, but I say I like them cooked. Felipe, Bill Steen, and I get clam *caldo* (soup) for breakfast plus local-grown squash with ranch cheese, beans, salsa, tortillas, and strong coffee.

These clams are harvested (in season) November to February. They are commonly eaten grilled, or consumed as *sevichi arosimmake* (from Spanish, *ceviche* with rice).

Anomia peruviana

Common names: VEEKO; *cascabel peruano*, *papas fritas*, *papitas*; jingle shell, pearly monia

Veeko is described as a thin, fragile shell, lustrous and attractive on the inside. It is seen washed up on the shore in places such as Chiinim and Las Guásimas. Veeko is also applied to other Gulf of California shells yielding mother-of-pearl, as well as to imported abalone shell (*Haliotis*, Gastropoda).

Atrina maura

Common names: *hacha*; pen shell

This large pen as well as *Pinna rugosa* are commercially harvested for *mariscos* (seafood) called *callo de hacha*, and also sold as "scollops."

Chione and similar clams

Common names: AULIM; Venus clams

There are more than ten species of *Chione* and similar common clams along the shores of the Yoem Bwiara. They are found buried in the sand.

Food: These small clams are harvested on tidal flats, such as near Las Guásimas. We have had meals of these clams steamed and eaten with melted butter.

Chionista fluctifraga [*Chione fluctifraga*]

Common names: AROSERAM (the name refers to rice because this clam is eaten with *arroz*, hence *aros*); *almeja china*; smooth Venus clam

These small clams are commonly harvested, including in the Guásimas lagoon, both to be eaten locally and sold commercially.

Chionopsis gnidia [*Chione gnidia*]

Common names: TEEWI, TEEWIIM; *venus vistosa*; gnidia Venus clam

"Easily identified by its ornate, sharply ridged white shell, this clam was considered to be a good food" by the Seris (Marlett 2014:119). Found at Las Guásimas and harvested for local and commercial use.

Chionopsis gnidia, Las Guásimas. 31 May 2019 (PB).

***Magallana gigas** [*Crassostrea gigas*]

Common names: VATNAATAKA KOOYOM (in the beginning oysters, or oysters in the past); *ostión japonés*; Japanese oyster, Miyagi oyster, Pacific oyster

This large oyster is native to Asia and has become established on the Pacific Coast of North America, Europe, Australia, and New Zealand. It is sometimes considered an invasive species, replacing native oysters.

Food: It is cultivated in aquaculture including in the Estero Sargento lagoon near San Carlos.

Ostrea angelica [*Myrakeena angelica*]

WO'IM WOKIM (coyote feet)

This small, thick-shelled oyster sometimes occurs in clusters of shells cemented together. It is locally harvested and sold. The Yoem noki name was given to us in Las Guásimas in 2019.

Ostrea angelica, Las Guásimas. 31 May 2019 (PB).

Pinctada mazatlanica

Common names: *madreperla, ostra perla nacarada*; black-lipped pearl oyster, Panamic black-lipped oyster

and

Pteria sterna

Common names: *callo de árbol, concha nácar, ostra perlera viuda*; Eastern Pacific pearl oyster, rainbow-lipped mollusk, western winged pearl oyster

These two species produce pearls and mother-of-pearl. Since sixteenth century Spanish times, the Gulf of California was famous for the lustrous, dark pearls, called black pearls (Pérez de Ribas 1645; Reff et al. 1999). From the early pearling times Yoeme men were pearl divers. The lucrative pearl oyster beds were largely depleted by the late 1800s or earlier (Mosk 1931, 1934, 1939; Bowen 2000).

Pearling was a terrible venture for Yoemem and other Indians, who were pressed into service for pearl diving. Jesuits in the early seventeenth century viewed with some alarm the "small scale pearlers who had committed some of the worst outrages against the Indians" (Bowen 2000:73). "By the mid-eighteenth century, local divers [from the Baja California Peninsula] had largely been replaced by Yaqui Indians brought over from Sonora" (Mosk 1931:211).

In 1772, Johann Jakob Baegert wrote, "The divers are tied to a rope and lowered into the ocean. They pick up the shells and mother-of-pearl, or pry them loose from the bottom or the rocks, throw them into a sack, and when they can no longer hold their breath, they emerge and dump the trash or treasure brought up from limbo" (Brandenburg and Baumann 1952:45; Bowen 2000:75). The pearling was seasonal:

> May to October, when the waters of the Gulf were warm enough for naked diving. The *armadores* [ship owners, entrepreneurs] employed annually about 500 men, mostly Yaqui Indians from Sonora, who with their families were housed in temporary shore settlements. These reminded [José] Esteva [1865] of prison camps, so closely were the inhabitants watched and so bad were the living conditions in them.

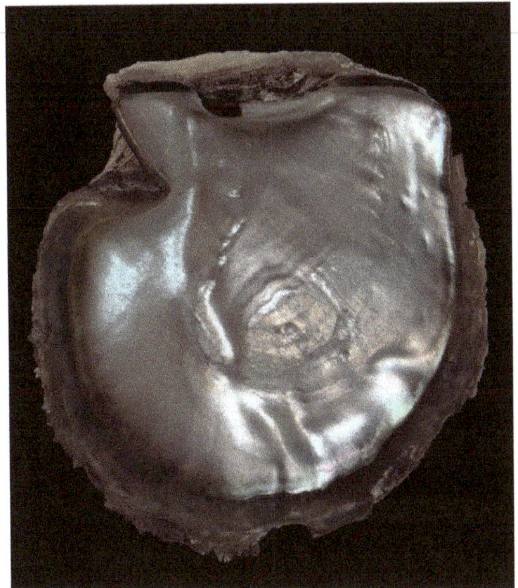

Pinctada mazatlanica, El Desemboque del Río San Ignacio. 2010 (CM).

Pteria sterna, El Desemboque del Río San Ignacio. 2010 (CM).

> This type of organization almost invariably led to peonage...The burden of debt and the tyrannical vigilance under which the divers worked could be shed only by running away, and Yaqui divers frequently escaped to the mainland in their employers' boats. (Mosk 1941:463–464; Bowen 2000:98)

Over two centuries of exploitation depleted the more accessible oyster beds (Bowen 2000:74). In 1939, the Mexican government imposed a *veda* (ban) on collecting pearl oysters. Many of the pearl oyster beds gradually recovered. In the latter part of the twentieth century, commercial aquaculture of *Pteria sterna* was successfully established at Bahía Bacochibampo (Miramar, near Guaymas). This pearl oyster farm produces the famed, richly colored Gulf of California pearls (Kiefert et al. 2004).

Pinna rugosa

Common names: *hacha*; hatchet shell, rugose pen shell

These large bivalves were seen at Las Guásimas and are commercially harvested. Also see *Atrina maura*.

Saccostrea palmula [*Crassostrea palmula*]

Common names: KOOYO; *ostión, ostra*; rock oyster

These oysters are found attached to mangrove roots and may also occur on rocks.

Food:They have been harvested for food, but have become scarce or extirpated in many localities.

Saccostrea palmula on mangrove roots (*Rhizophora mangle*), Bahía Concepción, Baja California Sur. 25 Feb 2016 (SC).

GASTROPODA – SNAILS AND THEIR KIN

***Haliotis**

Common names: VEEKO; *abalón, abulón*; abalone

Veeko is the name for abalone, as well as for *Anomia peruviana* (Bivalvia) and perhaps other Gulf of California mollusks yielding mother-of-pearl. Several North American abalone species occur along the Pacific Coast from Baja California Sur to Oregon, but do not occur in the Gulf of California. Abalone shells, however, have long been prized for jewelry and artwork among indigenous peoples across much of North America, and there was an extensive pre-contact trade in these shells across the continent. In modern times, abalone shells have been purchased by Yoeme craftsmen from specialty stores.

Hexaplex erythrostomus, Las Guásimas. 1 Jun 2019 (PB).

Adornment and ceremony: Abalone shell is incorporated into ceremonial regalia in order to ward off negative thoughts and energy. The main cross on the forehead of pahko'ola masks is sometimes decorated with pieces of abalone inserted into incised carvings. The cross on the mask is sometimes said to be the ancient sun symbol. The Coyote Society headdress features a large, disk-shaped shell ornament often made from abalone shell.

The deer dancer and pahko'ola dancers wear a specific necklace, called *hopo'oroosim* (*hopo* is the legume tree *Piscidea mollis*, which has leaves made up of several large, silvery leaflets; *oroosim* is not translatable). This necklace is made with glass beads and decorated with several crosses and triangles usually made from abalone shell, although brazilwood (*huchahko*, *Haematoxylum brasiletto*, Fabaceae) crosses have been used as well.

Yoeme craftsmen in Arizona purchase whole abalone shells from local stores. They cut the shell into square blanks and then into pattée-style crosses (like a St. Patrick's cross) to make the necklaces. They try to avoid breathing the dust produced from cutting the shell as it is considered dangerous to one's health. A dust mask is advised. One man kept the shell in a bowl of water and then dipped or sprinkled the shell blanks with water while cutting the cross shapes with a fine-toothed saw blade. Some Tucson mask makers use cut abalone pieces as decorative additions to pahko'ola masks. The designs could be small circles, triangles or tear-forms inlaid under the eyes (Kolaz 1985, 2007).

Hexaplex erythrostomus [*Chicoreus erythrostomus*; *Phyllonotus erythrostomus*]
Common names: *caracol chino*; pink-mouth murex

This large murex has been harvested in enormous, unsustainable quantities. In May 2019, extensive heaps of the discarded shells were seen along the shore at Las Guásimas. As with other mollusks, the meat is eaten as one of the common *mariscos*.

Hexaplex nigritus [*Muricanthus nigritus*]
Common names: *busano negro*, *caracol burro negro*; black murex

Lobatus galėatus [*Strombus galeatus*]
Common names: *caracol burro*; Eastern Pacific giant conch

This is one of the largest gastropods in the Gulf of California. It is occasionally harvested for food.

Melongena patula
Common names: *melongena coco*; Pacific crown conch

Persististrombus granulatus [*Strombus granulatus*]
Common names: granulated conch, knobby conch

Strombus gracilior
Common names: *cobo del Pacífico oriental*; Panama

fighting conch, smooth conch

CEPHALOPODA – OCTOPI AND SQUIDS

Dosidicus gigas
Common names: *calamar gigante*, *jibia gigante*; Humboldt squid, jumbo squid, jumbo flying squid and

Lolliguncula panamensis
Common names: KALAMAR; *calamar dedal*, *calamar dedal panameño*; Panama brief squid

The general term for squid is *kalamar*. There are two species of squid available for fishermen at Las Guásimas. *Lolliguncula panamensis* is a small species that occurs in warm shallow waters on the Pacific side of the Americas. It is fished when they go after shrimp; the squid is caught in gillnets, locally called *chinchorro de línea* or *red agallera*.

The second species, *Dosidicus gigas*, is seldom fished at Las Guásimas because it occurs in deep and far away areas. These large carnivores are caught with line and squid jigs or *poteras* (each being a device with a metal rod supporting several rings of hooks below and topped by a phosphorescent lure to attract and hook squid). The fishery for jumbo squid began in the Gulf of California as a small artisanal operation (using pangas and hand jigs) providing for local consumption. It was the basis of a newly developing fishery about 1974, and by 1978 it had become commercially significant. In 1978, part of the commercial shrimping fleet began jigging for this squid. After the collapse of the shrimp fishery between 1983 and 1987, and its slow build-up since the early 1990s, this squid was occasionally brought up by shrimp trawlers.

Octopus
Common names: VAA HUVAHE (sea spider); *pulpo*; octopus

Rosario Vakame'eri-Castillo said that *vaa huvahe* are strong in the water but weak out of water. He and other fishermen had numerous encounters with octopi in the early part of the twentieth century. Octopi are caught at Las Guásimas and elsewhere along the Sonora coast. The California two-spot octopus (*pulpo manchado*, *Octopus bimaculatus*) is the most common commercially harvested species in the Gulf of California, followed by Hubb's octopus (*Octopus hubbsorum*).

ECHINODERMATA – ECHINODERMS

Common names: ERISO; *erizo de mar*; sea urchin

At Las Guásimas we were told *erisom* are edible but are not eaten by Yoemem.

Common names: VAA CHOKI (water star); *estrella de mar*; starfish

Yoemem in Sonora sometimes hang various kinds of dried starfish in houses as ornaments, and also as a blessing from *Vawe Ania*, the Sea World.

Common species in the Las Guásimas region include the spiny sand star (*Astropecten armatus*) and the Panamic cushion star (*Pentaceraster cumingi*).

VERTEBRATES

FISH

Kuchu is the general term for fish. Various freshwater and ocean fishes, primarily in the estuaries or *esteros*, have provided significant food resources. Our brief and highly preliminary accounts of the fishes hints at a rich diversity. We extend special thanks to Lloyd Findley and Phil Hastings for sharing and reviewing information on fishes.

About a dozen or so species of freshwater fish are native to the lower Río Yaqui:

> Excluding marine taxa, most fishes of the lower reaches of the Río Yaqui systems are secondary, lowland species (*Poeciliopsis occidentalis*, *P. prolifera*, and other topminnows including all-female forms), some reaching the northern limits of their ranges (e.g., *Dorosoma smithi* and a cichlid, *'Cichlosoma' beani*). *Gila minacae*, *Catostomus bernardini*, and *Ictalurus pricei*, which occupy the main stem Río Yaqui, also live in smaller, upland habitats. Thus, there are no unique, big-river fishes like those of the lower Colorado, perhaps because of the

> lack of extensive lowland habitat over geologic time (Miller et al. 2005:32).

Fishes of the Río Yaqui watershed are greatly threatened and diminished since the introduction of exotic species (Hendrickson and Minkley 1980; Minkley and Marsh 2009). A considerable number of marine fish used to enter the brackish and even fresh water of the lower part of the river, and hundreds of species are found in the sea along the historic Yoeme lands (Campoy-Favela et al. 1989; Castro-Aguirre 1978; Hendrickson et al. 1981; Miller et al. 2005; Thomson et al. 2000). Basilio (1890 [1634]:136, 143, 182, 195) listed more than a dozen terms for different kinds of fish.

Padilla-Serrato et al. (2017) found 95 species of fish in 67 genera and 38 families in the Las Guásimas lagoon. Commercial fishermen, like those at Las Guásimas, generally use well-known Yori (non-Yoeme) names for fish, like most of the ones in the list below. In contrast, inland farmers or non-commercial fishers have tended to use traditional, Yoeme names for fish (Crumrine and Crumrine 1967). Hipolito Flores-Romero, at Las Guásimas in April 2007, provided local names of some of the best-known commercial fish:

botete
cabrilla
chivato
cochito
corvina
lenguado
lisa
manta
mojarra
pargo
pez dorado
roncacho
tiburón
totoaba ("*es prohibida, veda*")

Fishing with a net in the sea is called *kuchu sua* (fishing/killing in the sea with a net) and fishing with hook and line is called *vo'a*. There are coastal fishing camps and settlements, notably at Chiinim, Las Guásimas, and Lobos (*Loovos*) (Bartell 1963; Calderón and Campoy-Favela 1993). Spicer (1954:49) reported that during the 1930s and 1940s there were:

> ...three or four families in Potam territory who fish and gather shellfish regularly. Using nets and wooden plank boats they fish close to the shore on the Gulf of California. They use most of what they catch for themselves, but also sell or trade parts of large catches to others who dry and store them. Swordfish is the most commonly caught in large quantities [The swordfish might have been *pez espada*, *Xiphias gladius*.]

A Yoeme fishing cooperative was established in 1957 at Las Guásimas (McGuire 1986), where "For the first time in Yaqui history, an occupational specialty, namely, fishing became the basis for town formation" (Spicer 1980:272).

Pérez de Ribas (1645; Reff et al. 1999:85) reported, "Settlements near the mouths of the rivers... abound in several varieties of fish, especially mullet [*lisa*] and bass [*robalo*]...They fish with nets, sometimes out on the high sea and other times in the estuaries, of which there are many along this coast. Some fish with bow and arrow, especially in estuaries where the water is low." Basilio (1890 [1634]:187) recorded *hiteri* as the term for "*red para pescar*," a fishing net.

Fish traps were described in use in the Río Yaqui in the nineteenth century (Robertson 1968:101):

> When my father, Louis Robertson, explored the mouth of the Yaqui River in the year 1894, he found many fish traps still in use by the Yaqui Indians. These were in the form of brush fences similar to those used by the Indians of the Mayo and Fuerte Rivers. As did the Mayos and Zuaquis, the Yaquis barricaded with brush the outlets of little estuaries into the sea, placing the last barricade across the stream at high tide. As the tide receded, the fish were left stranded.
>
> Compadre Ignacio Amarillas likewise told us about fish traps: "Fish would swim into the river from the ocean. In small areas in the esteros the fishermen set up traps, weirs, and when the tide goes out, fish would get trapped."

Rosalio Moisés wrote about fishing in the Río Yaqui at Tórim in 1933 (Moisés et al. 1971:146):

> One day I found a small, sharp nail and made a fishhook. I started fishing in the water holes in the river. The first day I caught fourteen fish that weighed about two pounds each: twenty-eight pounds of meat. My sister-in-law was very happy. She said, 'this is the first time we are going to eat good fish steaks.' That night we had a big supper.
>
> Other Yaquis saw my fish, and the next morning several men and boys, with various kinds of homemade hooks, were trying to fish. Boys kept coming to me all day, asking what they should do.

RAYS AND SHARKS

Common names: AVATAAKA, HAVAATAKA; *manta, mantarraya*; manta ray

The term *manta* is used throughout Mexico to refer to many kinds of rays, just as the English 'ray' can refer to many different ones. Basilio (1890 [1634]:182) cited *habataca* as the term for *pescado raya*. *Manta* could indicate any of several species in five different families:

- Dasyatidae (*rayas látigo*, whiptail stingrays)
- Mobulidae (*mantas*, mantas, including the Giant Manta)
- Myliobatidae (*águilas marinas*, eagle rays)
- Rhinopteridae (*rayas gavilán*, cownose rays)
- Urotrygonidae (*rayas redondas*, round stingrays)

At Las Guásimas, Fernaldo Leyva-Flores told us of *muu'u* (owl) and two other kinds of mantas or rays: *manta cubana* and *manta gavilán*. The Giant Manta, or *manta gigante* (*Manta birostris*), was once common in the Gulf of California.

Common names: TAMEKAME (one with teeth); *tiburón*; shark

Tamekame is the general term for shark. In 1634, Basilio (1890:182) gave *tamecame* as the term for *pescado casón*. There are many species of shark in several different families in the Gulf of California. The three most commercialized are:

- Carcharhinidae, *tiburones gambuso*, requiem sharks; many species, including the Great White Shark.
- Sphyrnidae, *tiburones martillo*, hammerhead sharks; 2 or 3 possible species.
- Triakidae, *cazones*, hound sharks; 3 or 4 possible species.

Populations of large sharks in the Gulf of California have crashed because of overfishing (Bizzarro et al. 2009). They are long-lived and recruitment is low.

BONY FISH (OSTEICHTHYES)

ACHIRIDAE –*LENGUADOS REDONDOS*; AMERICAN SOLES

Common names: TAHKAI KUCHU (tortilla fish); *lenguados*; flatfishes

These are flatfishes, described as having both eyes on one side of the head (as do all the many flatfish and flounders). *Tahkai kuchu* or *lenguado* may apply to many different species of flatfishes in several different families, although the roundness alluded to by "tortilla fish" would tend to indicate Achiridae, the commonest species in the area being *Achirus mazatlanus* (*tepalcate*, Pacific Lined Sole) and *Paralichthys* (*lenguado*, flounder).

ARIIDAE – *BAGRES MARINOS*; SEA-CATFISH

Bagre pinnimaculatus

Common names: CHIWILI; *bagre, bagre barbón*; Long-barbed Sea-catfish, Red Sea-catfish

In the Las Guásimas region, sea-catfish are fished with hook and line and by gillnets in the estuaries and in the ocean. *Ictalurus* (Ictaluridae), the other well-known genus of catfish in the region, occurs only in fresh water.

CARANGIDAE – *JURELES, PÁMPANOS*; JACKS, POMPANOS

Oligoplites
and
Trachinotus

Common names: HO'OT WICHAKAME, PAMPANOM; *pámpano, zapatero*; Longjaw Leatherjack, Shortjaw Leatherjack

Juan Pablo Gallo-Reynoso identified *ho'ot wichakame* or pampanom as *Oligoplites*, however Fernaldo Leyva-Flores at Las Guásimas said that the spines of ho'ot wichakame will give a fever (*yee taiwettua*). Lloyd Findley (personal communication, 2016) tells us, "*Pámpano* indicates one of the species of jacks (*jureles, pámpanos*; in the family Carangidae) in the genus *Trachinotus*, and most likely for this coast is *T. kennedyi* (*pámpano gitano*, Blackblotch Pompano) or *T. rhodopus* (*pámpano fino*, Gafftopsail Pompano)." He also suggests that the mention of a venomous spine indicates a Stone Scorpionfish (see *Scorpaena mystes*, Scorpaenidae).

Basilio (1890 [1634]:182) recorded *cobuaera* for *pescado pámpano*.

CATOSTOMIDAE – SUCKER FAMILY

Catostomus bernardini

Common names: *matalote yaqui*; Yaqui Sucker

It was widely distributed in the Río Yaqui system. It is a bottom-feeder and occurs in Sonora and adjacent areas from southeastern Arizona to northern Sinaloa. It has been an important food fish.

Catostomus bernardini, Yaqui Sucker, Cajón Bonito, northeast Sonora. 2006 (WR).

CENTRARCHIDAE – *LOBINAS*; SUNFISHES

***Micropterus salmoides**

Common names: *lobina negra*; Largemouth Bass

This introduced fish is common in Presa Oviáchic, an artificial freshwater lake on the Río Yaqui. Like many introduced species, it has led to the decimation of the aboriginal fish species (Minkley and Marsh 2009).

CENTROPOMIDAE – *ROBALOS*; SNOOKS

These are predatory fishes typically found in mangrove-lined estuaries. They are excellent eating fishes. Basilio (1890 [1634]:182) gave *sateni* for *pescado robalo*.

Centropomus medius

Common names: *robalo aleta prieta*; Blackfin Snook

Ranging northward in coastal waters to Guaymas and sometimes entering rivers.

Centropomus robalito

Common names: *robalo aleta amarilla*; yellowfin Snook

Ranging northward to the Yoem Bwiara and entering the Río Yaqui.

CICHLIDAE – *GUAPOTES, MOJARRAS*; CICHLIDS

Nandopsis beani [*Cichlasoma beani*]

Common names: *mojarra, mojarra de Sinaloa*; Sinaloa Cichlid

Rivers and their tributaries from the lower Río Yaqui to Nayarit and Zacatecas. This is the northernmost species of native cichlid (found as far north as Texas). It was common in the Yoem Bwiara, but has become scarce because of the introduction of exotic fishes and loss of habitat. It reaches about 20 cm (8 in) in length.

Basilio (1890 [1634]:182) gave *rebera* for *pescado mojarra*.

***Oreochromis mossambicus**

Common names: *mojarra, tilapia*; Mozambique Tilapia, tilapia

By the late twentieth century, tilapia were well established in the Río Yaqui and had become staples of pond aquaculture. Tilapia are also found in canyons of the Sierra Bacatete and Sierra Santa Ursula (Bogan et al. 2014).

The common tilapia of the lower Río Yaqui is *Oreochromis mossambicus*, but there may be newer introductions in the area. These could include several species as well as hybrids, involving perhaps two or more genera (e.g., *Cichlosoma*, *Oreochromis*, and/or *Herichthys*).

There is a substantial commercial fishery for tilapia in Presa Oviáchic, exceeding catches of introduced catfish (see *Ictalurus*, Ictaluridae) and Largemouth Bass (*Micropterus salmoides*, Centrarchidae).

CLUPEIDAE – *SARDINAS*; HERRINGS AND SHADS

Harengula thrissina

Common names: *sardina plumilla*; Flatiron Herring

Coastal waters and entering lagoons and rivers.

Lile gracilis

Common names: *sardinita ague dulce*; Graceful Herring

Coastal lagoons, esteros, and river mouths.

Lile stolifera

Common names: *sardinita rayada*; Striped Herring

Esteros and lower parts of coastal rivers.

Unidentified sardine

Common names: PUSE'ELA (one with big eyes); *ojotón*; sardine

Fernaldo Leyva-Flores told us about *puse'ela*, also called *ojotón*, at Las Guásimas. There are several possible kinds of sardines in the Gulf of California.

CYPRINIDAE – *CARPAS, CARPITAS*; CARPS, MINNOWS

Agosia chrysogaster, Gila Longfin Dace, San Bernardino, Arizona. 2006 (WR).

Agosia

Common names: *pupo mexicana*; Mexican or Yaqui Longfin Dace

Streams and rivers from southeastern Arizona to northern Sinaloa including the lower Fuerte, Mayo, and Yaqui rivers. This is a distinct but yet-unnamed species or subspecies related to the more northern *Agosia chrysogaster*.

Codoma ornatus

Common name: Ornate Minnow

This small freshwater fish is the only member of its genus and is endemic to northwestern Mexico. It is documented for the Río Yaqui above 1,500 m but may have occurred at lower elevations. The males are brightly-colored in breeding season.

ELEOTRIDAE – *DORMILÓNES, GUAVINAS, PUGEQUES*; SLEEPERS

Eleotris picta

Common names: *guavina machada*; Spotted Sleeper

Gulf of California from the Río Colorado southward, entering the lower river systems.

ELOPIDAE – *MACHETES, SÁBALOS*; LADYFISHES, TARPONS

Elops affinis

Common names: *machete del Pacífico*; Machete

Coastal waters to lower parts of rivers.

ENGRAULIDAE – *ANCHOAS*; ANCHOVIES

Anchoa macrolepidota

Common names: *anchoa escamuda*; Bigscale Anchovy

Coastal waters in bays and esteros.

Anchoa walkeri

Common names: *anchoa persistente*; Persistent Anchovy

Near-shore waters with muddy or sandy bottoms and occasionally in rivers.

Cetengraulis mysticetus

Common names: *anchotón*, *anchoveta*, *anchoveta bocona*; Bigmouth Anchovy

At Las Guásimas, Fernaldo Leyva-Flores listed *anchetón* as a commercial fish. Lloyd Findley (personal communication, 2016) says: "*Anchotón*, the unofficial common name, obviously refers to a large species of anchovy, and the largest, most common one in the area is *C. mysticetus*, with *anchoveta bocona* being the official common name. The official common name in English is Anchoveta, but is sometimes referred to as the 'bigmouth anchovy.'"

GERREIDAE – *MOJARRAS*; MOJARRAS

Coastal waters and lagoons and entering fresh water. The "true" *mojarras* are marine or brackish-water fish in the family Gerreidae, including one of several species in the region in the genera *Diapterus*, *Eucinostomus*, *Eugerres*, or *Gerres*. The term *mojarra* is also applied to the cichlids (Cichlidae): *Nandopsis beani* and more recently to introduced tilapias (*Tilapia* and relatives).

GOBIIDAE – *GOBIOS*; GOBIES

Awaous banana

Common names: *gobio del río*; Banana Goby, River Goby

Coastal streams and rivers from Río Yaqui southward.

HAEMULIDAE – *BURROS, RONCAS*; GRUNTS

Haemulopsis leuciscus

Common names: *ronco ruco*; Racuous Grunt

Coastal waters and entering esteros and occasionally fresh water.

Pomadasys

Common names: *roncacho*; grunts

Coastal lagoons, esteros, and lower parts of rivers. Possibly up to three species may occur in the coastal waters of the Yoem Bwiara.

HEMIRAMPHIDAE – *PAJARITOS*; HALFBEAKS

Hyporhamphus rosae

Common names: *pajarito californiano*; California Halfbeak

Gulf of California coastal seawater and in fresh water in the Río Fuerte.

ICTALURIDAE – *BAGRES DE AGUA DULCE*; NORTH AMERICAN CATFISHES

***Ictalurus**

Common names: *bagre*; catfish

Since the latter part of the twentieth century, the *bagre* found in the large irrigation channels are mostly likely the introduced Channel Catfish (*bagre de canal*, *Ictalurus punctatus*). Some hybridization has occurred between this and the native species, *I. pricei*.

Ictalurus pricei

Common names: OMO'I; *bagre yaqui*; Yaqui Catfish

Lower parts of rivers including the Río Yaqui. Ranging historically to extreme southeastern Arizona, where it faces extirpation because of habitat loss and competition with introduced exotic species. Catfish are common in the Río Yaqui and large irrigation canals, where they are caught with hook and line and eaten: "We all enjoyed catfish from the river" (Moisés et al. 1971:193).

Basilio (1890 [1634]:143) gave *musu* for "*bagre, pescado*."

Ictalurus pricei, Yaqui Catfish, Cajón Bonito. 2006 (WR).

LUTJANIDAE – *HUACHINANGOS, PARGOS*; SNAPPERS

Lutjanus

Common names: SIKI KUCHU (red fish); *pargo*; snapper

Siki kuchu may be one of the several species of snapper (*huachinangos* and *pargos*), most likely the Pacific Red Snapper (*Lutjanus peru*).

The Yoeme (*siki kuchu*) and Spanish (*pargo*) names for these commercial ocean fishes were supplied by elders from Pótam at Chiinim in December 1988. Basilio (1890 [1634]:182) gave *pocato* for *pescado pargo*.

Lutjanus colorado

Common names: *pargo Colorado*; Colorado Snapper

Coastal waters and often entering tidal streams including the Río Yaqui.

Lutjanus novemfasciatus

Common names: *pargo prieto*; Pacific Dog Snapper

Coastal waters and sporadic occurrence in fresh water especially as juveniles, and documented in the Río Yaqui.

Lutjanus peru

Common names: *huachinango del Pacifica*; Pacific Red Snapper

MUGILIDAE – *LISAS*; MULLETS

Agonostomus monticola

Common names: *trucha de tierra caliente*; Mountain Mullet

Juveniles live in the lower parts of rivers and the adults live mostly in clear water of uppermost tributaries with strong currents, "where their troutlike behavior earned them the Spanish vernacular *trucha*" (Miller et al. 2005:181). Basilio (1890 [1634]:195) indicated "*pece huitao*" for *trucha*.

Chaenomugil proboscideus

Common names: *lisa hocicona*; Snouted Mullet

A marine fish occasionally found in fresh water, including "an irrigation canal near Ciudad Obregón" (Castro-Aguirre 1978:146).

Mugil cephalus

Common names: HUHUWO, LIISA; *lisa pardete, lisa rayada*; Flathead Gray Mullet, Striped Mullet

Abundant along the coastal waters of the Gulf of California, such as at Chiinim and Las Guásimas. They are common in esteros and move into rivers at least as far north as the Río Mayo, and probably enter the Río Yaqui. The names *huhuvo* or *huhuwo* for the common mullet were given to us by fishermen at Chiinim. Basilio (1890 [1634]:182) cited *huhuhuo* for *pescado lisa*. This species occurs in coastal waters nearly worldwide.

Food: Mullets, especially *Mugil cephalus*, are among the more common food fishes in the region and are considered delicious. Pérez de Ribas (1645; Reff et al. 1999:85) likewise mentioned mullet being an abundant food fish.

At Chiinim in December 1988, fishermen were casting nets from small *pangas* (boats) in the mangrove estero and catching mullet. We bought some and took them back to Pótam for dinner.

Mugil cephalus, Okinawa. (JER).

Mugil curema

Common names: TOSAI HUHUWO (white mullet); *lisa blanca*; White Mullet

Coastal waters and sometimes moving into rivers. This is the only other mullet in the area that would likely be caught and eaten, but the generally smaller size and less frequent occurrence in comparison to the striped mullet makes the latter the "mullet of choice" (Lloyd Findley, personal communication, 2016).

OPHICHTHYIDAE – SNAKE EELS

Common names: ANHILAM (fromSpanish), VAI VAKOCHIM (water snakes); *anguilas*; eels

Vai vakot is a kind of eel, told to us by Fernaldo Leyva-Flores at Las Guásimas. One of the many possibilities is *Myrophis*, which live buried in the mud and are eaten by egrets at low tide.

Basilio (1890 [1634]:136) gave "*bapu ilibacot*" (little snake in the water) for *anguila*.

POECILIDAE – *GUAYACONES, TOPOTES*; LIVEBEARERS

***Gambusia affinis**

Common names: *guayacón mosquito*; Western Mosquito-fish

Native to the Mississippi River and its tributaries and northeastern Mexico, it has been widely introduced for mosquito control. Although various small, native fishes (including *Poeciliopsis*) may be as effective or more in controlling mosquitoes (Miller et al. 2005), this little fish often outcompetes and replaces native fishes when introduced.

Poeciliopsis occidentalis

Common names: *guatopote de Sonora*; Gila Topminnow

This little livebearer was once quite common in rivers and streams of southern Arizona and Sonora, and was often the only fish to survive in pools along streams that are intermittent during dry seasons.

Poeciliopsis prolifica

Common names: *guatopote culiche*; Blackstripe Livebearer

West coast of Mexico from the Río Yaqui to Nayarit, in slow-moving fresh water to shallow water in mangrove esteros.

PRISTIDAE – *PECES SIERRA*; SAWFISHES

Two species of sawfish range through the Gulf of California and in Sinaloa and may enter estuaries. Sawfishes are globally threatened.

Pristis pectinata

Common names: *pez serrucho*, *pez sierra peine*; Small-tooth Sawfish

Pristis pristis

Common names: *pez serrucho*, *pez sierra común*; Large-tooth Sawfish

SCIAENIDAE – *RONCADORES*; CROAKERS AND DRUMS

Atractoscion

and

Cynoscion

Common names: TOSAI KUCHU (white fish); *corvina*; corvina

In December 1988, elders from Pótam told us about *tosai kuchu*. Corvinas are regionally important commercial fishes. "The term *corvina* is applied to several species in several genera of the family Sciaenidae, but usually is applied to those of the genera *Cynoscion* and *Atractoscion*, especially the former. It is likely that one of about four species of *Cynoscion* are caught in the area" (Lloyd Findley personal communication 2016).

Basilio (1890 [1634]:182) cited *siahuihuino* for *pescado curbina*.

Cheilotrema saturnum

Common names: HOROHTEME; *roncacho*; Black Croaker

Although *horohteme* might be *Cheilotrema saturnum*, Lloyd Findley (personal communication, 2016) has said that it could also be one of several species and genera in the families Sciaenidae or Haemulidae.

Totoaba macdonaldi

Common names: TOTOAVA; *totoaba*; totoaba

This enormous fish, once common in the Gulf of California, breeds at the Río Colorado delta. It is highly endangered and is protected by Mexican law. It probably once occurred along the Sonora coast as far south as the Yoeme lands. Pérez de Ribas (1645; Reff et al. 1999:84) found, "In the rivers...an abundance and variety of fish that enter from the sea, especially when they lay their eggs." He may have been referring to the totoaba, among others, which used to spawn in river deltas perhaps as far south as Sinaloa.

Fishermen at Las Guásimas are well aware of its endangered status and *veda* (prohibition).

SCORPAENIDAE – *ESCORPENAS*; SCORPIONFISHES

Scorpaena mystes

Common names: HO'OT WICHAKAME (one with spines on its back); *pez escorpión* roquero; Stone Scorpionfish

At Las Guásimas, Fernaldo Leyva-Flores mentioned *ho'ot wichakame* as a fish with spines and said it will give a fever (*yee taiwettua*). He told us that these fishes enter esteros and low salinity areas.

Lloyd Findley (personal communication, 2016) told us:

> The mention of a 'spine that will give a fever' pretty much leaves the pompanos out since their dorsal spines do not bear venom. More indicated for that description would be a scorpionfish (*escorpiones*, *rocotes*; family Scorpaenidae), which do possess such venomous dorsal spines. However, unlike the 'pelagic' pompanos, which are taken in gillnets, the benthic, generally rocky-bottom-occurring scorpionfishes would likely be taken by hook-and-line. The most

common, large species the Yaquis might take is *Scorpaena mystes* (*pez escorpión roquero*, Stone Scorpionfish).

The dorsal spines are connected to venom glands that can inflict a painful injury.

SERRANIDAE – *CABRILLAS, MEROS*; GROUPERS, SEA BASSES

Mero is a general term for sea basses and groupers (*cabrillas*, *meros*, sometimes also called *garropas*) of the family Serranidae, though is usually applied to the larger species of groupers (subfamily Epinepheliinae). *Mero* could be one of several commercial and sport fishing species and genera, though more likely in the genera *Mycteroperca* or *Epinephelus* (especially *E. quinquefasciatus*), or less likely *Paralabrax* (which are usually termed *cabrillas*). Basilio (1890 [1634]:182) gave *iabarau* for *pescado mero*.

Epinephelus quinquefasciatus and others
Common names: YAVARAI; *guasa*, *mero*, *mero guasa*; Goliath Grouper

Yavarai is a shape like a hard disk, or something uneven. A Yoeme song tells about a man who could not find a saddle, so he used the skin of the fish *yavarai*, which is saddle-shaped. The town of Puerto Yavaros in southwestern Sonora derives its name from this fish.

On July 27, 2017, a man from Huatabampo, Sonora, told Felipe that the yavarai are big fish and can weigh up to 90 kg (200 lbs). He is a Yoreme and he pronounced it *yavarau*.

Hyporthodus acanthistius
Common names: *baqueta*; Gulf Coney

This is a large marine fish caught by fishermen at Las Guásimas.

SYNGNATHIDAE – *CABALLITOS DE MAR*; SEAHORSES

Hippocampus ingens
Common names: KAVAYITO (little horse); *caballito de mar*; Giant Seahorse, Pacific Seahorse

Found in shallow waters of the eastern Pacific coastline where corals and seagrasses occur, including in the Gulf of California. It is protected from fishing in Mexico.

At Las Guásimas, they told us that you tie a dried seahorse with red ribbon and hang it in the house for good luck. The Comcaac (Seris) have a similar practice.

TRICHIURIDAE – *SABLES*; CUTLASS FISHES

Trichiurus nitens?
Common names: LIHTONIAM (from Spanish for ribbons), VELOHKO (shiny); *pez cinto*, *sable del Pacífico*
Beltfish, Largehead Hairtail Fish, Pacific Cutlassfish

At Las Guásimas, Fernaldo Leyva-Flores said it is a kind of *anguila* (eel) with a long dorsal fin. Lloyd Findley (personal communication, 2016) remarked: "There are dozens of eels, in several different families, in the Gulf of California. Even a list for the ones most likely occurring in the Yaqui area would be a long one. However, probably not many of them would be described as 'shiny' (in the sense of brilliance). There is a long-bodied, 'eel-like' scombroid that is silvery brilliant shiny that may be caught by the Yaquis, albeit in relatively deeper waters and usually as by-catch by shrimp boats, *Trichiurus nitens*."

AMPHIBIANS and REPTILES

Scientific names for amphibians and reptiles follow Lemos-Espinal et al. (2019). Organization of families follows Frost (2020) for amphibians, Pyron et al. (2013) for lizards and snakes, and Rhodin et al. (2017) for turtles. Standardized common names in Spanish and English generally follow Rorabaugh and Lemos-Espinal (2016), Crother (2017) for species in the United States, and Frost (2020) for amphibians. However some of these names are not in usage in Sonora and have been balanced with more practical common usage in Sonora. We extend special thanks to James C. Rorabaugh, Phillip C. Rosen, and Robert A. Villa for sharing and reviewing information on amphibians and reptiles.

AMPHIBIANS

CAUDATA – SALAMANDERS

***Ambystoma mavortium** [*Ambystoma tirgrinum* subsp. *mavortium*]
Common names: HIPUYESA'ALA (holding something, sits), POOWI; *ajolote*; *salamadra, salamandra tigre, ajolote tigre rayado*; waterdog, barred tiger salamander, western tiger salamander

Arizona. There are no native salamanders within the Sonoran Desert of Arizona or Sonora.

Hipuyesa'ala, the aquatic form of the barred tiger salamander, has external gills. It was seen in irrigation canals in Marana. Waterdogs are sold as fish bait and the ones in Marana were probably released into the canals or farm ponds and stock tanks. Aquatic hipuyesa'ala include larval salamanders and adults with gills. Larvae metamorphose into terrestrial salamanders with lungs or remain as aquatic, gilled adults that are often incapable of further metamorphosis to the terrestrial form. Hipuyesa'ala is also the name for house geckos (*Hemidactylus*, Lizards).

Poowi, the adult, terrestrial form, is described as slimy and similar to the hipuyesa'ala but does not have external gills, and also was seen in the water at Marana. Poowi is also the name for the desert iguana (*Dipsosaurus dorsalis*, Lizards).

ANURA – FROGS AND TOADS

Vatat is the general term for frog. Basilio (1890 [1634]:187) cited *batat* for *rana*. *Sivo'oli* is the term for tadpole, which may apply to the young of either frogs or toads. Frogs, such as species in the genus *Rana*, live in or near permanent water in the arid Sonoran Desert region.

Felipe heard that the way to get to the next world is to eat frogs or *kamam* (*Crocodylus acutus*, crocodile). Then you get passage to the next world. After you die. He heard that it is like a test, and God will look at you and say, "well, he did it, so he may come in." This is what Felipe heard in the 1960s and 1970s from his grandparents and other village elders in Yoem Pueblo.

Some long-standing scientific names for frogs and toads have changed (Rorabaugh and Lemos-Espinal 2016; Yuan et al. 2016), as follows:

BUFONIDAE
Bufo alvarius = **Incilius alvarius**
Bufo cognatus = **Anaxyrus cognatus**
Bufo kelloggi = **Anaxyrus kelloggi**
Bufo marinus horribilis = **Rhinella horribilis**
Bufo mazatlanensis = **Incilius mazatlanensis**
Bufo punctatus = **Anaxyrus punctatus**
Bufo retiformis = **Anaxyrus retiformis**

HYLIDAE
Pternohyla fodiens = **Smilisca fodiens**

PHYLLOMEDUSIDAE
Pachymedusa dacnicolor = **Agalychnis dacnicolor**

BUFONIDAE – TRUE TOADS
Common names: VOOVOK; *sapo, sapito*; toad

Voovok is the general name for toad. Basilio (1890 [1634]:190) gave *boboqui* as the word for *sapo*.

Toads have bumpy skin and a raised gland behind each eye (paratoid glands). Toads mostly emerge from subterranean retreats with the summer rains, but some species emerge in the spring. Many form large breeding aggregations and you hear the loud calls of males during the first nights of the summer rains.

Kaureepa is the term used for the largest toads, the Sonoran Desert toad (*Incilius alvarius*) and the cane toad (*Rhinella horribilis*). Kaureepa is also applied to the largest frog, the American bullfrog (*Rana catesbeiana*, Ranidae). *Voovo'okim* (toads) in traditional ceremonies and narratives would almost certainly be the Sonoran Desert toad.

Anaxyrus cognatus [*Bufo cognatus*]
Common names: *sapo de espuelas, sapo de las Grandes Planicias*; Great Plains toad

Arizona and Sonora. A medium-sized toad with dark green blotches. It is a prolific and widespread species. In the Yoem Bwiara it often breeds in canals and ditches.

Anaxyrus kelloggi [*Bufo kelloggi*]
Common names: *sapito mexicana*; little Mexican toad

Sonora. This is a regionally endemic species found in coastal thornscrub of southern Sonora north through the Plains of Sonora region of the Sonoran Desert. These toads are brown with a mottled, marbled pattern that resembles that of the Sonoran green toad (*Anaxyrus retiformis*).

Anaxyrus punctatus [*Bufo punctatus*]
Common names: *sapo de puntos rojas*, *sapo manchas roja*; red-spotted toad

Arizona and Sonora. A beige toad with red-orange spots. An opportunistic breeder, this little toad occurs at isolated waters and springs in the mountains, along the Río Yaqui and its tributaries, and it ranges north to Arizona in similar situations. It is one of the more common and frequently seen desert toads.

Anaxyrus retiformis [*Bufo retiformis*]
Common names: *sapito marmol verde*, *sapo sonorense*; Sonoran green toad

Arizona and Sonora. A small toad with a mottled green, yellow, and black pattern resembling malachite. This species is endemic to the Sonoran Desert from southern Arizona south to the vicinity of Guaymas.

Incilius alvarius [*Bufo alvarius*]
Common names: KAUREEPA; *sapo del desierto sonorense*, *sapo grande*; Colorado River toad, Sonoran Desert toad

Arizona and Sonora. Johnson (1962:271) listed *kuárepa* as *sapo grande*, which is probably this species. This large toad has relatively smooth, olive-green skin, often with orange dots, and a large gland behind each eye (paratoid gland). It ranges from Arizona and southwestern New Mexico southward to northern Sinaloa. This is the largest toad in Arizona, and in Sonora is second in size only to the cane toad (*Rhinella horribilis*).

This is one of the few toads emerging before the rains of summer, lending credence to the role of this species in rain ceremonies and stories. Some Sonorans have referred to this species as *rana toro* or *sapo toro*, perhaps confusing it with the large American bullfrog (Robert A. Villa, personal communication, 2020). *Kwareepa* is an alternative and apparently non-traditional spelling in Sonora.

Non-traditional outside influences have promoted smoking and inhaling paratoid gland secretions for psychedelic effects. There have been more than a few cases of cardiac arrest and death, although seldom reported. There is no evidence for this practice before the late twentieth century. Kaureepam have been locally extirpated at Loma de Bácum, the result of poaching for the psychedelic market (James C. Rorabaugh and Robert A. Villa, personal communication, 2019).

Incilius alvarius, Sierra Buenos Aires. 14 Aug 2016 (JR).

Beliefs and ceremony: Toads are seen as harbingers of rain. In Pótam, and elsewhere in the Yoem Bwiara, if there was a severe drought the people would have a procession with a toad (*kaureepa*) on a bier. This would take place in the dry season of early summer for the purpose of bringing rain. The toad is decorated with red ribbons and there would be songs and prayers. In the procession, following the toad, people carry a statue of San Isidro, the saint of farmers.

In Sonora, during the pahko for San Juan's Day, in the morning procession after the Vesper, toad sounds are made by the pahko'olam on their way to the Río Yaqui (*Hiak Vatwe*). The toad is connected to the rain and the people and the saints are blessed with water. This blessing can be with splashes of water, as the pahko'olam dramatize how voovok pulled rain to earth.

Narrative: There are different versions of how *Voovok* brought *Yuku*, the rain. Molina et al. (1999:129) tell that, "*Voovok* (Toad) brought the rain back from Yuku (Rain) by borrowing Bat's wings" (see Chiroptera, bats). Giddings based her Yuku story on Johnson's 1940 manuscript (published in Johnson 1962:140–146, as the "*yúku*" narrative), part of which is provided here:

> The man...spoke with the toad, saying, 'Toad, I want you to bring rain to us if you can.' 'I can bring it,' said the toad. He went up to the house of Yuku, sowing his children behind him on the road all the way. Then he said, 'Yuku, you know that man down there, don't you? He wants rain.' 'I know," said the God of Rain. 'You go on ahead of me and I'll come behind and catch up.'
>
> Toad began to return and in a short time the god caught up with him. Toad was singing along the road and before him on the road all of his sons also sang. Yuku went from one toad to the next and finally arrived where the corn was sown. He irrigated it and it grew. It formed ears, they ripened, and gave corn. Then men had corn. (Giddings (1959:18)

Incilius mazatlanensis [*Bufo mazatlanensis*]
Common names: *sapito de Mazatlán, sapo sinaloense*; Sinaloa toad

Sonora. This medium-sized toad is common throughout the Yoem Bwiara. It can be identified by darkened raised ridges on the head resembling spectacles or glasses (cranial crests).

Rhinella horribilis [*Bufo marinus horribilis*]
Common names: KAUREEPA; *sapo gigante, sapo grande*; cane toad, Mesoamerican cane toad, western cane toad

Sonora. This is the largest toad in Sonora and one of the few toad species emerging before the summer rains. It is brown with a giant raised gland behind each eye. The cane toad known from agricultural lands in southern Sonora and southward into the tropics. It spread into the Yoem Bwiara in the 1990s from more tropical regions in the south. It was found at Loma de Bácum in 2014 (James C. Rorabaugh and Robert A. Villa, personal communication, 2019). *Kwareepa* is an alternative and apparently non-traditional spelling in Sonora. "The skin secretions are quite toxic and have caused fatalities in humans and dogs" (Rorabaugh and Lemos-Espinal 2016:144).

Rhinella horribilis, Loma de Bácum. 14 Jul 2014 (JR).

HYLIDAE –TREEFROGS

Hyla arenicolor
Common names: *rana de cañón, ranita de las rocas*; canyon treefrog

Arizona and Sonora. These small, grayish to beige, mottled treefrogs are common in rocky, riparian canyons. They are often seen clinging to boulders and rock walls to which they camouflage well. Widespread in southern Arizona and eastern Sonora, and possibly in the Yoem Bwiara.

Smilisca baudinii
Common names: *rana arbolera, rana arborícola Mexicana*; Mexican treefrog

Sonora. This frog is variable in color and pattern. It has white lips and horizontal pupils aligning with a stripe that runs from snout to behind the tympanum, the ear structure behind the eye. Its distinctive call sometimes emanates from the forest canopy and can be heard from great distances. Found in thornscrub and tropical deciduous forest.

Smilisca fodiens [*Pternohyla fodiens*]
Common names: *rana chata, rana manchada arbolera*; lowland burrowing treefrog

Arizona and Sonora. This unique treefrog is uncharacteristically adapted to a fossorial (underground) existence during the dry season. It occurs in limited areas in south-central Arizona, southward through Sonora, and into the low tropical forests of southern Mexico.

Leptodactylus melanonotus
Common names: *rana del sabinal, rana del sabino, ranita manchada*; sabinal frog

Sonora. Small and secretive, the sabinal frog is a tropical species reaching its northern terminus in central Sonora. It is documented in ditches near Bácum and also at Rancho Bacatete in the Sierra Bacatete (Bogan et al. 2014). The adults are often heard but seldom seen. Egg masses are laid in foam nests next to or above streams and water. Rain washes the developing tadpoles into the water. Adults have a mottled to splotched pattern, and the males have two black spines on the thumb.

MICROHYLIDAE – NARROW-MOUTHED TOADS
Gastrophryne mazatlanensis [*Gastrophryne olivacea mazatlanensis*]
Common names: *ranita oliva, ranita olivo de Mazatlán*; Mazatlan narrow-mouthed toad, Sinaloan narrow-mouthed toad

Arizona and Sonora. This shy and diminutive frog is brown to gray with a narrow snout for eating ants and termites. It occurs in southern Arizona and throughout most of Sonora. Its distinctive call, heard during nighttime breeding aggregations with the early summer rains, is an intense buzz following a very brief, energetic squeak, resembling a high-pitched bleating of sheep.

PHYLLOMEDUSIDAE – MONKEY TREEFROGS
Agalychnis dacnicolor [*Pachymedusa dacnicolor*]
Common names: *rana verde arbolera, ranita verduzca*; Mexican leaf frog

Sonora. A large, green to brown, agile and arboreal treefrog, the Mexican leaf frog is found along the Río Yaqui and its tributaries. It deposits its eggs in shrubs and trees where branches overhang breeding ponds. The young emerge as tadpoles within seven days and complete the larval stage after dropping into the pond below.

RANIDAE – WATER FROGS
Rana [*Lithobates*]
Common names: *rana, ranas norteamericanas del agua, rana leopardo, rana manchada*; American water frogs, leopard frogs, North American water frogs

These frogs are typically characterized by their leopard-like markings. Some species in this genus may be difficult to distinguish without a trained eye and a careful examination of the distribution of the various species.

***Rana catesbeiana** [*Lithobates catesbeianus*]
Common names: KAUREEPA, KWAREEPA (alternative spelling in Sonora); *rana grande, rana mugidora, rana toro*; American bullfrog

Rana catesbeiana, Rancho El Aribabi, Cocóspera. 31 Aug 2014 (JR).

Arizona and Sonora. This is the largest frog in the region, distinguished by its mottled instead of leopard-like markings typical of other local ranid species. The bullfrog is a voracious predator and a major cause for the declines of native frogs and gartersnakes throughout the southwestern United States and northwestern Mexico: “It will eat just about anything that will fit into its mouth” (Grismer 2002:81). This frog is native to eastern North America and has become widespread across the Sonoran Desert region and many other regions worldwide.

The American bullfrog was already in the Tucson region by the 1960s. By the late 1980s, it was the only fully aquatic frog in Avra Valley, including Marana. Bullfrogs became established in agricultural areas and other wetlands in the Yoem Bwiara sometime before the 1980s.

Rana forreri [*Lithobates forreri*]
Common names: *rana leopardo, rana leopardo de Forrer, rana manchada*; Forrer’s leopard frog

Sonora. A robust, native leopard frog, it thrives in agricultural areas surrounding the lower Río Yaqui, where it is quite common.

Rana magnaocularis [*Lithobates magnaocularis*]
Common names: *rana leopardo, rana leopardo del noroeste de México, rana manchada*; northwest Mexico leopard frog

Sonora. This species is found along the Río Yaqui and its tributaries as well as in smaller streams and ponds in the foothills and mountains. Newly metamorphosed frogs of this species may be abundant in late summer. It can be difficult to discern this species from other leopard frogs.

Rana yavapaiensis [*Lithobates yavapaiensis*]
Common names: *rana leopardo, rana leopardo de Yavapai, rana manchada*; lowland leopard frog, Yavapai leopard frog

Arizona. Felipe describes a small, greenish, long-legged frog that was seen in the irrigation ditches at Marana, always in the water, at least during the 1960s.

In the 1960s this medium-sized frog was documented in the Tucson region at localities from around Camino del Cerro, Cortaro, and near Marana southeast of Trico and Silverbell Roads, in the area of irrigation sump-ponds. This frog was also elsewhere in Avra Valley and more recently has been replaced by the American bullfrog (*Rana catesbeiana*).

SCAPHIOPODIDAE – SPADEFOOT TOADS
Scaphiopus couchii
Common names: *cavador, sapo de espuela*; Couch's spadefoot toad

Arizona and Sonora. This little, desert-lowland specialist is distinguished by its large, iridescent, raised eyes with vertical pupils, and a black spade-like horn on the underside of each rear foot. This adaptation allows it to burrow deep into hard desert soils where it can remain until moisture is sufficient for it to emerge, feed, and reproduce.

REPTILES

CROCODILIA – CROCODILES
Crocodylus acutus
Common names: KAMA; *cocodrilo, cocodrilo amarillo, lagarto*; American crocodile

Sonora. This tropical animal occurs in the tropics on Atlantic and Pacific sides of the America. On the west coast it ranges northward to southern Sonora, including the deltas and lower reaches and wetlands of the Mayo and Yaqui rivers. Occasional strays are known as far north as Las Guásimas and Guaymas (Navarro 2003a, 2003b; Rorabaugh 2017).

There is no credible evidence that crocodiles occurred farther north than the Guaymas region. Let us look at four early accounts of regional crocodiles.

1550. As part of the massive Coronado Expedition of 1539–1542, Hernando Alarcón sailed up the Gulf of California. His ships reached the Colorado River delta in September 1540. Unlike the inland expedition headed by Coronado, Alarcón and his men were well received, and there, at the delta and lower reaches of the river, he made many inquiries of native people. From one high ranking individual, "I asked him whether he had information about a river called Totonteac. He replied to me no, but that he knew of well there was another great river where such large *lagartos* were found that they made round shields out of their skins" (Flint and Flint 2005:198).

The nearest possible "great river" with crocodiles would be the Río Yaqui. It is quite reasonable that he would know the Río Yaqui and such an exceptional animal as the crocodile. He and others had been to places as distant as Zuni (Cibola).

1634. Tomás Basilio (1890 [1634]:147) listed *camaa* as the word for *caimán*. *Kama* is also the term for squash, notably the native cushaw squash (see *Cucurbita argyrosperma* var. *callicarpa*, Cucurbitaceae).

1645. Andres Pérez de Ribas (1645; Reff et al. 1999:84–85) wrote, "Many caimans or crocodiles enter the mouths of the rivers in schools, fishing for food and even for men, whom they occasionally catch. Therefore, the Indians do not dare fish alone in the inlets where these beasts are found."

Large crocodiles could be dangerous, although the crocodiles of northwestern Mexico are not known to be man-eaters, at least not in modern times.

1772. Johann Baegert's "Observations in Lower California" (1952) includes a description that has been interpreted as claiming crocodiles at the Río Colorado (the Red River). A close look at the translation offers a different interpretation:

> The aforementioned Red River is, of all the waters which empty into the almost four-hundred-hour-long California ocean, the only one that deserves to be called 'river'...All others found on the maps are hardly more than rain-water courses, which during the greater part of the year have very little water....However, in all them abide alligators [sic] of considerable size, and since some of them are capable of devouring a full-grown man, it is necessary to be on guard while drawing water, bathing, or washing. I have seen several of these creatures. As everyone knows, they resemble a lizard, but are completely clad in armor like a turtle. Not many years ago it was discovered in America that the eyeteeth of alligators are a strong antidote. By applying to the wound or swallowing some of the powder scraped from such teeth, the lives of many who have been bitten by snakes have been saved. (Brandenburg and Baumann 1952:11)

Like other Jesuits sent back to Europe by Jesuit Expulsion of 1776, he did not write his book until a few years later. However, note. "All others...are hardly more than rain-water courses" and "in all of them abide alligators [sic]." This indicates he did not say crocodiles were at the Red River. Jesuits in the Baja Californian missions were in close communication with their brethren at the Río Yaqui missions. If his statement, "I have seen several of these creatures," is truthful, then the mostly likely places would be at the mainland Río Yaqui missions.

Narratives: When Felipe was growing up he learned from the older people that *kamam* live in the water and that they will eat people. His grandparents (living at that time in Marana) knew of *kamam* in the Yoem Bwiara, along the Río Yaqui (see frogs and toads, Anura).

The crocodile features in Yoeme stories, for example "Kaiman," related by Giddings (1959:35). A similar narrative is given by Painter (1986:52–53):

> Did you know that crocodiles turn themselves into human beings like us?...then there are fiestas, like pascolas and matachinis, they come out from the water or the swamps...and they roll themselves into the sand and take off their skins. And then they clothe themselves like us, and then they go to the fiestas.
>
> When they roll their self in the sand, then their skin comes off, and they are human like us. Then they dress like what we used to wear and then went to the fiesta. And they stood away from the wind, so the smell of the crocodile doesn't go to the people. Then when they get tired, and they go back again and get their skin that they have hidden and put it back on and go back into the water where they belong. My mother said that sometimes the boys would find the skins and throw them away so they would have to stay on land.
>
> The father of (a certain Yaqui) married a lady that used to be a crocodile. She didn't like the food we had, and she got skinny, and she passed away here...They don't harm anyone...they must all be females that do this, as they all wear women's clothes.

Shields and combat: Pérez de Ribas (1645; Reff et al. 1999:91) reported that "The most important leaders used small shields made from caiman [sic] or crocodile hides, which are very tough. These shields resist arrows as long as they are not shot at close range or by someone with a very powerful arm."

Knowledge crocodile hides used for shields apparently was widespread in pre-contact times (see Flint and Flint 2005, above).

SQUAMATA: SAURIA – LIZARDS

Wikui is the general Yoeme term for ordinary lizards, such as whiptail lizards (*Aspidoscelis*) and spiny lizards (*Sceloporus*). Spanish terms for lizards include *chachora*, *cachorón*, *lagartija*, and *lagarto*.

Masks: Lizards and other special animals are often painted on the sides of pahko'ola masks.

Narrative: Painter (1986:325) mentioned the role of the lizard and other small animals in rituals of the pahko'olam during non-Easter ceremonies. One man told her, "What the pascolas might say," for example, "Oh Saint Lizard...thou who can resist the heat of the summer and bury thyself in the soft dust of the desert, help me." Also see *Phrynosoma solare*, horned lizard.

CROTAPHYTIDAE – COLLARED AND LEOPARD LIZARDS

Crotaphytus nebrius

Common names: *cachorón*, *cachorón de las piedras de collár*, *cachorón de Sonora*; collared lizard, Sonoran collared lizard

Arizona and Sonora. Large colorful lizards with a prominent collar of bands around the neck. Specialized to living and perching on rocks to assert territory and scope prey, insects and small lizards. Arizona southward to the Guaymas and likely in the Yoem Bwiara.

Gambelia wislizeni

Common names: *cacharón leopard*; long-nosed leopard lizard

Arizona and Sonora. A medium-size lizard, more slender and less colorful that the collared lizard. Arizona southward to the Guaymas region and likely in the Yoem Bwiara.

EUBLEPHARIDAE – EYELID GECKOS

Coleonyx variegatus

Common names: WAITOPIT; *geco de bandas occidental*, *salamanquesa de bandas*; western banded gecko

Arizona and Sonora. These small, delicate, nocturnal lizards often hide in dark corners and under rocks and boards during the day. The skin is smooth with a speckled and banded pattern, and the large eyes have prominent eyelids. The banded gecko is harmless, but some people may mistake it for a baby Gila monster. Even a newly hatched Gila monster is many times larger than the largest banded gecko.

Coleonyx variegatus, north of Bahía Kino. 19 May 2019 (JR).

GEKKONIDAE – COMMON GECKOS

***Hemidactylus**

Common names: HIPUYESA'ALA (holding something, sits), WATTA'AKAILI; *besucona*, *salamanquesa*; house geckos

The two house geckos listed here are native to Europe and the Middle East. These geckos are similar to the native leaf-toed gecko *Phyllodactylus homolepidurus* (Phyllodactylidae), but lack leaf-like structures on the toes.

Hemidactylus commonly occur in and on houses and other buildings. Like many geckos, these harmless little lizards can scamper up and down walls and even walk on ceilings. They are often seen on warm nights at lights on walls waiting for insects. House geckos emit a variety of squeaks and chirps, sometimes quite loud. House gecko and leaf-toed geckos have relatively large eyes but lack eyelids (in contrast to the banded gecko).

Johnson (1962:265, 292) listed "*hípuyésa'alam*, *lagartijas de la casa*" and also "*wáta'aklim*, *lagartilla como salamanquesa*." Anselma Tonopuame'a-Castillo remembered *wata'akaili* as a *salamanquesa* from Guaymas.

The translation of *hipuyesa'ala* (holding something, sits) likely refers to geckos on walls. Hipuyesa'ala is also applied to the larval form of the tiger salamander (*Ambystoma mavortium*, Amphibians).

***Hemidactylus frenatus**
Common names: *besucona, salamanquesa*; common house gecko

Sonora. The dorsal side of this gecko has few or no tubercles. It is documented from Guaymas and Navojoa and undoubtedly occurs in the Yoem Bwiara. This gecko makes "loud chirps to the point of being annoying if you're trying to sleep, being much more vocal than *H. turcicus*" (James C. Rorabaugh, personal communication, 2019).

***Hemidactylus turcicus**
Common names: *besucona, geco del mediterráneo, salamanquesa*; Mediterranean house gecko

Arizona and Sonora. It is common in houses in southern Arizona and is known from western Sonora. It is likely to expand its range in Sonora. The dorsal side of this gecko is dotted with small tubercles, distinguishing it from the common house gecko.

Hemidactylus turcicus (juvenile), Puerto Peñasco. 15 Feb 2104 (JR).

HELODERMATIDAE – BEADED LIZARDS
Heloderma

The beaded lizards (four closely related species) and Gila monster are the only dangerously venomous lizards in the New World.

Basilio (1890 [1634]:158) reported *sacau* as the word for *escorpión*. He probably would have known the Gila monster and especially the beaded lizard.

Heloderma exasperatum [*Heloderma horridum exasperatum*]
Common names: *escorpión, escorpión del Río Fuerte, lagarto enchaquirado*; beaded lizard, Río Fuerte beaded lizard

Sonora. This very large lizard is known from tropical deciduous forest near the eastern margin of the Yoem Bwiara. This species occurs in southeastern Sonora, southwestern Chihuahua, and northern Sinaloa. This lizard is larger, has a proportionally longer tail, and is more yellow and darker in color than the Gila monster. Beaded lizard bites are rare, but the venom is dangerously potent, needing immediate medical treatment. Nearly all bites are to people foolish enough to handle one. A noxious liquid like rubbing alcohol may be applied to the snout to cause the animal to release.

Heloderma suspectum, Dragoon Mountains, Arizona. 1 Jun 2014 (AR).

Heloderma suspectum
Common names: SAKKAU; *escorpión, monstruo de Gila*; Gila monster

Arizona and Sonora. This venomous lizard is rather shy and retreating. The bite is generally not life-threatening but is painful and serious, requiring urgent medical treatment. Nearly all bites are to people foolish enough to handle a Gila monster.

Yoeme elders said these animals are known to bite and that they don't let go in order to inject their venom. To make one let go, "get a stick and poke their nose." Indeed, on the rare occasion of a Gila monster bite, they are known to hold on tenaciously and chew, which allows greater delivery of the venom in their saliva. (A noxious liquid like alcohol may be used to cause the animal to release.)

Beliefs: People say if you don't burn your willow holy crosses on May 2, that the Gila monster will come to remind you to do so (see *Salix gooddingii*, Salicaceae).

IGUANIDAE – IGUANAS

Ctenosaura macrolopha

Common names: KUTA WIKUI (wood lizard); *garrobo de Sonora, iguana, iguana cola espinosa*; Sonoran spiny-tailed iguana

Sonora. This large, heavy-bodied lizard is common in the desert and semi-tropical regions of central and southern Sonora. It has a prominent crest running down the back and a spiny tail. Basilio (1890 [1634]:166) reported *huicuim* as the term for *iguana.*

Ceremony: For a Lutu Pahko, bread is made in the shape of the iguana, and the pahko'ola dancers often jokingly pretend to play with the bread as if hunting an iguana.

Food: *Kuta wikuim* have been hunted since ancient times (Moisés et al. 1971:193). Pérez de Ribas (1645; Reff et al. 1999:93–94) reported:

> These Indians also hunt two other kinds of small animals that abound in the montes. Both are prized as food. The first they call *iguanas*, a reptile very similar to a lizard but uglier in appearance. It inhabits the hollows of trees and also the water, so it is both terrestrial and marine. For this reason its flesh can be eaten even on fast days. It is a healthy and delicious food source. The little stones these animals produce (like bezoar but whiter) are very medicinal and highly prized as a remedy to aid in the retention of urine. They are not, however, found in all iguanas. To catch these animals in the hollows of trees, the Indians first reach very carefully for them with their hand, breaking their jaws so they cannot bite down on the hand like they usually do. They bring back bunches of live ones with their jaws broken, and these can be kept alive for eight to fifteen days in this way, thrown into a corner until they are ready to be eaten.

The spiny-tailed iguana is a terrestrial animal and certainly not "marine," but Pérez de Ribas probably classified it as "marine" so it could be eaten on "meatless" Catholic days when meat other than fish and other marine life is taboo.

In Las Guásimas, Ángel Flores-Romero told us that if you eat the meat of an iguana, which is nutritious, you will have a good dosage of vitamins for a month or more.

Dipsosaurus dorsalis

Common names: POOWI; *cahorón güero, iguana del desierto*; desert iguana

Arizona and Sonora. This is a medium-large lizard, whitish or light gray overlain with a darker gray or reddish-gray netlike pattern or barring. It is widespread in the Sonoran Desert, mostly in creosotebush flats, and ranges southward along the coast of Sonora to northwestern Sinaloa.

Basilio (1890 [1634]:168) reported *porohui* as a kind of lizard. Johnson (1962:281) listed "*póowim*" as "*iguana chica y güera*" (small and light-colored iguanas). A Yoeme consultant from Sonora identified desert iguanas at the Desert Museum in Tucson as *porowi* (*porohui* in Spanish orthography).

Dipsosaurus dorsalis, near Álamos. Apr 2014 (JR).

Sauromalus ater

Common names: *cacharón de las piedras, cacharón de roca*; common chuckwalla.

Arizona and Sonora. These large, vegetarian lizards inhabit rocky desert areas from the Guaymas region northward into the Sonoran and Mojave desert regions of the United States.

PHRYNOSOMATIDAE – HORNED LIZARDS AND KIN

Callisaurus draconoides

Common names: CHUULI WIKUI (little-dog lizard); *cachora arenera, perrita*; zebra-tail lizard

Arizona and Sonora. These lizards have a short face with a shovel-like snout for burrowing and are fast runners. When alarmed they raise and wag their zebra-striped tail, dart ahead, stop and wag their tail again, earning them the name *perrita*, or little dog. They are frequently seen in areas with sandy soils, including beaches. *Callisaurus* has prominent external ear openings (compare with *Holbrookia elegans*).

Holbrookia elegans

Common names: *lagartija sorda elegante*; elegant earless lizard

Arizona and Sonora. This shovel-nosed lizard is distinguished from zebra-tail lizards by the absence of an external ear opening and the absence of banding on the underside of the tail.

Phrynosoma solare

Common names: MOTCHO'OKOLI ; *camaleón, camaleón real*; regal horned lizard

Arizona and Sonora. Basilio (1890 [1634]:147) recorded *mochócol* as the word for *camaleón*.

Horned lizards feed on ants. Elders in Marana told the children not to bother them, that they spit blood at you. Indeed, these harmless lizards can eject drops of blood from burst blood vessels at the corners of the eyes as a defense mechanism.

Phrynosoma solare, Cocóspera. 1 Sep 2018 (JR).

Beliefs: There are reports that the "horned toad" has supernatural significance, along with the "water serpent" (Spicer 1988:26).

Felipe tells of "santo mocho'okoli" invoked by pahko'olam (Evers and Molina (1987:83; see Amphibians and Reptiles). Painter (1986:325) discussed the role of the "horned toad" and other small animals in rituals of the pahko'olam. One man told her "what the pascolas might say," for example, "Saint Horned Toad (Santo Mochokol), thou who has a crown like the Virgin of Guadalupe. Help me."

Sceloporus

Common names: VEHO'ORI; *vejore, cachorra, cachorón, lagartijas espinosos;* spiny lizards

Basilio (1890 [1634]:205) recorded *behori* as the word for "*cachorra, lagarto pequeño.*"

Beliefs: Felipe tells of "santo vehori invoked by pahko'olam" (Evers and Molina (1987:83).

Sceloporus magister, Dragoon Mountains, Arizona. 7 Jun 2014 (JR).

Sceloporus clarkii

Common names: *vejore de Clark*; Clark's spiny lizard

Arizona and Sonora. Distinguished by its barred hind leg, the body may be brightly colored, and with a dark collar around the neck. These rather hefty lizards are about the same size as the desert spiny lizard and generally occur in mountains and rich Sonoran desertscrub.

Sceloporus magister

Common names: vejore del Desierto; desert spiny lizard

Arizona and Sonora. These spiny lizards are common in desert regions of southern Arizona south through the lowlands to Sinaloa. They were often seen in Marana on the bark of large tamarisk trees.

Urosaurus ornatus

Common names: VEHO'ORI; *roñito ornado*; ornate tree lizard

Arizona and Sonora. These small, slender, arboreal or rock-dwelling lizards are common and widespread. The males have bright blue or blue-green, metallic belly patches, or may be almost entirely blue beneath. The topside is dark brown, grayish and blackish, and appears roughened like the bark of a mesquite on which they are often found. The males have large bright blue or blue-green metallic belly patches, but their entire skin may become suffused with blue during breeding season.

Uta stansburiana

Common names: VEHO'ORI, WIKUL; *lagartija mancha lateral*; common side-blotch lizard.

Arizona and Sonora. This small lizard is common in western North America and the western deserts of Arizona and Sonora. It is marked by a black spot just behind its forelegs.

PHYLLODACTYLIDAE – LEAF-TOED GECKOS

Phyllodactylus homolepidurus

Common names: HIPUYESA'ALA (holding something, sits), WATTA'AKAILI; *besucona*, *salamanquesa*, *salamanquesa de Sonora*; Sonoran leaf-toed gecko

Sonora. This little, nocturnal, native gecko has flattened scales on each toe resembling leaves. It is secretive and mostly found in rocky places, but also seen around houses. This gecko is similar in appearance to the non-native house geckos (*Hemidactylus*).

TEIIDAE – WHIPTAILS AND KIN

Aspidoscelis [*Cnemidophorus*]

Common names: WIKUI; *huico*, *lagartijas corredoras*; whiptail lizards

Sonora and Arizona. Fast and agile, slender lizards with a narrow snout, and usually with "racing stripes" along the length of the body. Certain species are hybrids with three sets of chromosomes, and comprised entirely of females reproducing by parthenogenesis.

Beliefs: Felipe tells of "santo wikui" invoked by pahko'olam (Evers and Molina (1987:83).

Aspidoscelis burti [*Cnemidophorus burti*]

Common names: WIKUI; *huico*, *huico manchado de cañón*; canyon spotted whiptail

Sonora. Coastal areas of the Guaymas region.

Aspidoscelis costata [*Cnemidophorus costatus*]

Common names: WIKUI; *huico*, *huico llanero*; western Mexico whiptail

This species occurs near the mouth of the Río Yaqui, and ranges from east-central Sonora to southwestern Mexico.

Aspidoscelis costata, Álamos, 17 Jul 2014 (JR).

Aspidoscelis tigris [*Cnemidophorus tigris*]
Common names: WIKUI; *huico*, *huico tigre*; tiger whiptail

Arizona and Sonora. This is the whiptail of the North American deserts, ranging from thornscrub in Sinaloa and Sonora to the Great Basin in the United States.

SERPENTES – SNAKES

Common names: VAAKOT; *culebra, serpiente*; snake

Vaakot is the general term for snake. Basilio (1890 [1634]:153, 197) indicated *bacochim* (plural) or *bacot* (singular) as the term for "*culebra general*," and provided twelve additional descriptive phrases for different kinds of snake. Snakes feature prominently in Yoeme culture.

Beliefs and narrative: Snakes feature in a number of spiritual accounts, often similar but differing in detail according the narrator, place, and time. We offer some versions.

Traditional people say that snakes help one who is on a spiritual path. That person has many dreams about snakes. When the person has gained much spiritual power, he or she will dream about *teweli vaakot* (dark-blue snake), a translucent blue snake. It is a spiritual snake. When you see it, it smiles. It is seen at homes, in the desert, or in dreams. It is said that if by accident one sees such a snake out in the open, that some people become blind—the snake is trying to offer you spiritual powers and you should not be afraid of it. All snake dreams are spiritual empowerment. Some devout Christians, however, often think that snakes are bad, or evil, and believe that snakes are connected to the devil and are afraid of snake dreams.

It is said that if you kill a snake, don't let it turn over to show its belly because it is trying to reach flowers, *sewa'u*, before you do. Also, never keep a rattle from a rattlesnake because the lightning will strike your house. A pregnant woman can make a snake heavy. If a snake is in the area and a pregnant woman happens to be walking by the snake, it will become inactive and will try to move away but won't be able to.

Snakes also appear in ceremonial practices. "Closely associated with the *yo ania* are several animals including a large goat and a snake. Both of these appear in the power dreams through which a boy can learn that he is to become a *pascola*. Both figure in *pascola* symbolism. The ridgepole of the harp, the *tampaleo's* flute, and the *pascola's* sashes have all been likened to snakes" (Griffith and Molina 1980:29). The tenevoim strands, coiled around the leg, represent serpents, to provide more spiritual powers (see *Eupackardia* and *Rothschildia*, Lepidoptera, Saturniidae). Snakes painted on the side of a pahko'ola mask are said to have the purpose of teaching people to respect the snake, and the same goes for any animal painted on a mask. Such an animal can bite or sting you, but you still have to respect it.

Chupiarim are said to be serpientes grandes (giant snakes). There are various versions of narratives involving giant serpents that come out of the sea every seven years (e.g., Giddings 1959; Moisés et al. 1971:10–11). *Suawaka* (the one with intelligence) destroys them. When a big flash of light is seen coming down from the sky, the path of a meteorite, it is said to be Suawaka. In 1988, Alfonso Leyva told a version of the story to Felipe in Las Guásimas. Here is an abridged account of Alfonso's narrative:

> In ancient times *chupiari* came out of *Vakochevampo* (water of the snakes, or snake water), the place nowadays called bahías Bacochibampo (Miramar) and San Carlos, northwest of Guaymas. *Wo'ochime'a* (Killing Grasshopper or Killer Grasshopper; he was a Surem) slew the chupiari. Wo'ochime'a flew up and hit the snake's head with his green axe and that became *Tenhaawe* (Open Mouth Mountain). On the highway you see a sign saying *Boca Abierto*. (This is the lava peak northwest of Las Guásimas on the north side of Highway15). So then Wo'ochime'a jumped up again and hit the body, and the body flew up and landed at the place now called Chilikoote Kawim (Coral-bean Mountain; Cerro El Vigía; see *Erythrina flabelliformis*, Fabaceae). He jumped up again and hit the last part of the snake's body, and it flew up in the air and landed in the water, and this is now called *Yasikue* and if you are very observant, as you

> drive north on Highway 15 near Las Guásimas you can see the snake's head, body and tail in a chain of lava peaks. That was the magic that Wo'ochime'a performed long ago. There is a mountain named after him, *Wo'ochime'a Kawim* (Killer Grasshopper Mountain).

Giddings (1959:56) provided a similar narrative about Suawaka and "serpents with seven heads" that lived at Takalaim, known today as Cerro Tetas de Cabra at Bahía San Carlos. One part of the story tells that, "When these serpents come out, they make a terrific wind and floods. Suawaka is up above watching for them. He knows that they come out every seven years, and their middle head appears first. Then Suawaka throws a harpoon of fire. This is the shooting star we see at night." Beals (1945:201) wrote that "The horned water serpent has been identified by the Yaqui with the seven-headed serpent of Spanish folklore, and is called *bákot gobúsan kóbaka* (serpent seven head)."

Another narrative involves a giant snake and a cave at Takalaim (Cerro Tetas de Cabra at San Carlos). Rosario Vakame'eri-Castillo told Felipe:

> When you go into the cave there is a big, gigantic snake there, bigger than a person. Takalaim is the enchanted home of the Surem (ancestors). On a certain day when you are given the opportunity to enter the cave you will hear music at noon. That music will lure you to the entrance on the west side of the Takalaim. The door to the cave will open and you will go inside and see the Surem forming two lines and they all have something to offer you. You start picking up what you want and it could be an instrument, tool, or other objects to help you in life. So your arms are full of these spiritual materials, but they are not heavy even though you get more and more. You thank them and continue walking through the cave to the east and right there in the center of the cave will be a big, gigantic snake, menacing or trying to scare you. But you can't be afraid, it is just air and you walk right into the mouth of the snake. Then you make it out to the other side, the east side, and your friend who came with you will say, 'you were in there for two weeks,' but you will say, 'but I thought it was only one day.' It is common to go in pairs, and one sits outside and waits while you are inside.

Songs: According to Painter (1986:268), "snake songs were played at the Lenten season pahkos" in Arizona in the 1940s "long after dawn" (see *Boa sigma*). Since at least the 1970s, snake songs have been played only at weddings; these songs are played with violin and harp, and are non-verbal.

BOIDAE – BOAS

Boa sigma [*Boa constrictor sigma*]
Common names: KURUES; *corúa, limacoa, mazacoatl*; boa constrictor, western Mexico boa constrictor

Sonora. The boa constrictor is the largest and heaviest snake in Sonora. It is often found in the mountains, especially near springs. It is generally a docile snake. This species ranges from western Sonora through southwestern Mexico. It reaches about 2 m (6.5 ft) in length in Sonora but can be larger farther south. *Kurues* is also the Yoeme term for a rainbow, and as the name for boa constrictor, likely refers to its iridescent skin.

Basilio (1890 [1634]:153) cited *curuas* as the name for a "*culebra gorda y grande*."

Narrative: Rosario Vakame'eri-Castillo told Felipe that when he was with Pancho Villa's army he had prisoners from the mountains, and they were walking in the desert and saw a beam of light radiating from the mountains in rainbow colors. They got up to the light source and discovered a large animal, and it was a kurues lying in the water with all of the lights coming out of its back. They wanted to get water at the waterhole but decided to just continue on because they saw this animal, which the people were calling kurues.

Spicer (1980:64) wrote, there were "springs on the eastern margin of the Bacatetes, or the very heart of the mountains, where, for example, snakes with rainbows on their foreheads lived and swam in the water. These serpents possessed special forms of power of the yo aniya, and they could make the power available to human beings."

There are stories about a priest who had a relationship with a woman, so he was punished by

God and turned into a boa constrictor. It is said, "When you see a boa constrictor that has a cross on its forehead, that is the priest from long ago." Indeed, the Sonoran boa constrictor usually has a cross-like pattern on top of the head.

Song: Writing about events in Tucson in the 1940s, Painter (1986:268) reported that during a pahko, "Snake songs...can only be played long after dawn, just before the whole thing is over. It is prohibited because it is about the Kurues (the mythical snake with a cross on its forehead). There are many of them on the Río Yaqui." Kurues of course is the boa constrictor and not "mythical."

Boa sigma, Álamos; it probably recently shed its skin, hence the rainbow sheen. 27 Aug 2005 (JR).

Lichanura trivirgata [*Charina trivirgata*]
Common names: *boa del desierto, boa de tres lineas, boa rosada*; rosy boa, three-lined boa

Sonora and southwestern Arizona. Cream with chocolate stripes running down its body. The head and tail are similar in appearance, perhaps to confuse predators. This gentle snake is found in rocky areas and upper bajadas from the Guaymas area and Sierra El Aguaje northward in western Sonora to southwestern Arizona.

COLUBRIDAE – COLUBRIDS
Arizona elegans
Common names: *culebra brillante arenícola*; glossy snake

Arizona and Sonora. Medium sized snake with distinct blotches or squares along the length of its back.

Chilomeniscus stramineus
Common names: *arenera de modelo variable, culebrita arenera*; variable sandsnake

Arizona and Sonora. Small, shovel-nosed burrowing snakes with a banded pattern, found in sandy or gravelly habitats.

Drymarchon melanurus
Common names: *babatuco, corúa* (and variants, see *Boa sigma*), *culebra arroyeras*; indigo snake, Central American indigo snake, Mexican west coast cribo

Sonora. This rather slender, swift-moving, iridescent black snake reaches 1.8 to 2.9 m (6 to 9.5 ft) in length and is often found near water. It has iridescent skin somewhat like the boa constrictor.

Gyalopion quadrangulare
Common names: *culebrita de colores, naricilla del desierto*; thornscrub hook-nosed snake

Arizona and Sonora. A small, colorful snake, red with black bands and white squares on the back.

Hypsiglena chlorophaea
Common names: *culebra nocturna, culebra nocturna verde oscuro*; desert nightsnake

Arizona and Sonora. A small, nocturnal snake. Gray with yellow to light-colored ring behind a blackish head, and a bright orange underside exposed when threatened, presumably to deter predators.

Lampropeltis californiae
Common names: CHUKUI VAAKOT (black snake); *culebra negra anillada, culebra real de California*; California kingsnake

Arizona. Mostly black or with whitish bands.

Beliefs: Painter (1986:33) reported, "An older man says that there is a dream of a king snake that enables one to have skill in writing. The snake becomes a pencil.

Lampropeltis nigrita [*Lampropeltis getula negrita*]
Common names: CHUKUI VAAKOT (black snake); *culebra negra*; western black kingsnake

Arizona and Sonora. Black, sometimes with faint flecking or faint whitish bands. Both kingsnakes eat many things including other snakes, even rattlesnakes.

Leptodeira punctatus
Common names: *culebra de pintas, escombrera del occidente*; cat-eyed snake, western cat-eyed snake

Sonora. Potentially in the Sierra Bacatete. Golden brown snake with dark spots and a dark patch behind the head. Vertical pupils resembling those of a cat.

Masticophis [*Coluber*]
Common names: *chirrioneras* (*chirrión* is the soapberry tree, *Sapindus saponaria*, Sapindaceae), *látigos*; coachwhips, racers, whipsnakes

Arizona and Sonora. Long, slender, and fast-moving diurnal snakes.

Masticophis bilineatus
Common names: *chirrionera rayada, culebra latigo*; Sonoran whipsnake

Widespread in Sonora and Arizona. Grayish with a lateral stripe on each side of its body.

Masticophis flagellum [*Coluber flagellum*]
Common names: SIK TAVUT (red cottontail); *chirrionera, culebra latigo*; coachwhip

Widespread in Arizona and Sonora. Basilio (1890 [1634]:153) cited *sitabut* as the word for "*culebra colorada*" (red snake). This common snake can be highly variable in color, including reddish to pink. It is sometimes called a red racer. "Red cottontail" may relate to these sometimes large, red colored snakes that can eat cottontail rabbits.

Narrative: It is sometimes said that this snake "will whip you." In snake stories if you harm or injure a snake that is not bothering you, it will take you to a spiritual trial. The snake will say, "Why did you injure (and almost kill) this girl, or boy." All the while the snakes have transformed themselves into human beings, but the young man or boy on trial does not know this. Most of the time if they are not guilty, they will let the person go and tell him or her never to harm any animal, whether or not it is a coyote, snake, mountain lion, etc. If the person is guilty, and has done it before, then the snake authority will have him whipped by the red racer.

Masticophis flagellum, near Álamos. 12 Sep 2004 (JR).

Masticophis mentovarius [*Coluber mentovarius*]
Common names: *chirrionera sabanera*; Neotropical whipsnake

Sonora. This tropical whipsnake occurs in riparian areas and mountains along the eastern side of the Yoem Bwiara. It is a large, tan-colored, fast-moving snake.

Oxybelis aeneus
Common names: WIROA VAKOT (vine snake); *bejuquilla parda*; Neotropical vinesnake

Sonora. Long, very slender snakes sometimes seen in shrubbery in the Yoem Bwiara. The head, elongated with a pointed snout, is wider than the body. It is a rear-fanged snake and is mildly venomous. It should not be handled. There is a folktale (incorrect) that this snake can hang down and choke an animal.

Basilio (1890 [1634]:153) recorded *huirobacot* as the word for a "*culebra larga y delgada*."

Phyllorhynchus browni
Common names: *culebra ensillada, culebrita ensillada de naríz hoja*; saddled leaf-nosed snake

Arizona and Sonora. Small gray snake with dark brown to black blotches or "saddles" on the back. The distinctive nose scale facilitates digging up eggs.

Phyllorhynchus decurtatus
Common names: *culebra naríz moteada, culebrita pinta naríz de hoja*; spotted leaf-nosed snake

Arizona and Sonora. Small beige snake with dark brown mottling. The distinctive nose scale facilitates digging up eggs.

Pituophis catenifer
Common names: AAKAME NAKAPIT (deaf rattlesnake), ALA'AMAI; *alicante, cincuate del desierto, víbora sorda*; bullsnake, gophersnake, Sonoran gophersnake

Arizona and Sonora. In Marana these snakes sometimes would come in the house to hunt for mice. Many people were afraid of them because these snakes would try to act like a rattlesnake. Gophersnakes sometimes vibrate their tail to sound like a rattlesnake and also have a specialized hiss that sounds like a rattle.

Ala'amai is the usual Yoem noki name for gopher snake, although *aakame nakapit* has also been used. Rural people in different parts of Sonora call this snake "*víbora sorda* (deaf snake), looks like a rattler but has no rattle" (Robert Villa, personal communication, 2019).

Basilio (1890 [1634]:153) gave *alamai* as the word for a "*culebra muy pintada y grande.*"

Pituophis catenifer, Northern Jaguar Reserve. 12 Apr 2009 (JR).

Rhinocheilus lecontei
Common names: *culebra anillada, culebra nariz-larga*; long-nosed snake

Arizona and Sonora. Black snake with red to white squares on the back and an elongated snout.

Salvadora deserticola
Common names: *cabestrillo, culebra nariz de parche*; Big Bend patch-nosed snake

Arizona and Sonora. Mottled beige with solid beige stripe on the length of its back and a broad nose scale. Found in the lowlands of southeastern Arizona, and the Yoem Bwiara.

Senticolis triaspis
Common names: *culebra ratonera verde*; green rat snake

Arizona and Sonora. Olive green, iridescent snake with cream-colored lips.

Tantilla hobartsmithi
Common names: *culebrita cabeza negra*; Smith's black-headed snake

Arizona and Sonora. Small, brown snake with a dark head.

Tantila yaquia
Common names: *culebra cabeza negra*; Yaqui black-headed snake

Arizona and Sonora. Small, brown snake with a dark head and a whitish neck ring.

Thamnophis
Common names: *culebras de agua, jarreteras*;; gartersnakes, North American gartersnakes

Arizona and Sonora. Harmless snakes generally associated with waterways, and may stink or bite when captured.

Basilio (1890 [1634]:153) recorded "*bapo bacot*" for "*culebra de agua*" (water snake).

Thamnophis cyrtopsis
Common names: *culebra de agua rayada, jarretera cuello-negro*; black-necked gartersnake

Arizona and Sonora. Dark green snake with light bands, distinguished by a pair of black spots behind the head. This is the most common gartersnake in Yoem Bwiara and southern Arizona riparian zones.

Thamnophis eques

Common names: *culebrita del agua rayada, jarretera mexicana*; Mexican gartersnake

Arizona and Sonora. Likely found in the eastern portion of the Yoem Bwiara. Extirpated from most of Arizona.

Trimorphodon lambda

Common names: *falsa nauyaca, ilamacoa de Sonora*; Sonoran lyresnake

Arizona and Sonora. A nocturnal snake; gray with brown patches, a cream center stripe along its back, and prominent chevrons on the head resembling lyre. Venom is delivered from grooved fangs at the back of the jaw. The bite is mildly venomous to humans.

ELAPIDAE – COBRAS, CORAL SNAKES, AND KIN

Hydrophis platurus [*Pelamis platurus*]

Common names: *culebra del mar*; yellow-bellied sea snake

Sonora. This sea snake ranges throughout the Gulf of California. It is variable in color and pattern, but often black above with a bright yellow belly and sides. The body is latterly compressed and the tail paddle-like, making it an efficient swimmer. Sea snakes give live birth at sea and do not come ashore, but are not known to breed in the Gulf of California.

This highly venomous snake should never be handled; it is an elapid snake, the same group that includes coral snakes and cobras. "Although the fishermen of the Gulf of California know that *P. platurus* is extremely poisonous and can be quite aggressive when taken out of the water, they pay little regard to this danger. Sea snakes are often found in the haul of fishing nets; the fishermen simply remove them by hand and throw them back into the water" (Grismer 2002:318).

Micruroides euryxanthus

Common names: SIK KUCHA'A (red spoon); *coralillo, coralillo occidental*; Sonoran coral snake

Arizona and Sonora. This colorful snake is banded black, red, and cream. Its venom is a potentially dangerous neurotoxin, but a bite from this small snake would be unlikely unless one is foolish enough to handle one. People leave them alone, knowing they are venomous. Some people, however, kill them. There are several small, harmless snakes (e.g., *Chilomeniscus* and *Chionactis*) that resemble the coral snake. Do not handle a snake you do not know is venomous or not.

"Spoon" is unusual in a snake name. Robert Villa (personal communication, 2019) asks if "spoon" might refer to this coral snake raising and curling its tail when threatened, often making a small squeak with its cloaca.

There are two coral snakes in the Yoeme and Yoreme regions. Basilio (1890 [1634]:197) gave "*siccucha alaa*" as the name for "*víbora otra que llaman coralillo.*" He might have been referring to *Micruroides euryxanthus* and/or *Micrurus distans.*

Micruroides euryxanthus, Sierra Aconchi. 2 Sep 2004 (JR).

Micrurus distans

Common names: *coralillo, coralillo bandas claras*; West Mexican coral snake

Sonora. This snake, mostly red with black and yellow bands, can reach 1 m (3.3 ft) long; it is "large enough to deliver a dangerously venomous bite and should not be handled. This is a quick and nervous snake, making it all the more dangerous" (Rorabaugh and Lemos-Espinal 2016:543). It occurs in eastern Sonora and likely occurs in the Sierra Bacatete.

LEPTOTYPHLOPIDAE – BLINDSNAKES

Rena humilis [*Leptotyphlops humilis*]

Common names: *culebrilla ciega de occidente*; western blind snake, western threadsnake

Arizona and Sonora. A tiny, slender snake resembling an earthworm.

VIPERIDAE: CROTALINAE – PITVIPERS

Crotalus

Common names: AAKAME; *cascabel*, *víbora*, *víbora de cascabel*; rattlesnake

Arizona and Sonora. *Aakame* is the name for any large rattlesnake, especially the western diamondback (*Crotalus atrox*). Basilio (1890 [1634]:197) gave *aiacame* as the word for *víbora*, noting, "*que tiene cascabel, aia es cascabel.*"

Beliefs: Never keep a rattle from a rattlesnake because lightning will strike your house.

Songs: Songs about rattlesnakes are sung mostly around 3 a.m. during a pahko. The "Enchanted Rattlesnake" is a play song, sung by the *Woko'i Ya'ura* (Coyote or Bow Leaders' Society): "The dancers dance all the way out during the repetitions of the first stanza as usual but when the concluding stanza begins, 'when the drum calls them back,' they get down on the ground and slither like snakes" (Evers and Molina 1990). Sometimes the deer dancer will get down on the ground and slither around like a snake and rattle his *tenevoim* like a rattlesnake. This song is from Evers and Molina (1990):

yoyo aakame sevipo vo'oka	Enchanted enchanted rattlesnake in the cactus is lying
siirisiiriti hia siirisiiriti hia siirisiiriti hia	Siirisiiriti sounding siirisiiriti sounding, siirisiiriti sounding
hia hia hia hia	Sounding sounding sounding sounding
katikun taewalita sumeiyaka haivusu sevipo vo'oka	It is because afraid of the day already lying in the cactus
siirisiiriti hia siirisiiriti hia siirisiiriti hia	Siirisiiriti sounding siirisiiriti sounding siirisiiriti sounding
hia hia hia hia	Sounding sounding sounding sounding

Crotalus atrox

Common names: *cascabel de diamantes*; western diamondback rattlesnake

Arizona and Sonora. This common, widespread species reaches 1.6 m (5.2 ft) in length. This and the Mexican west coast rattlesnake (*Crotalus basiliscus*) are the largest rattlesnakes in the region.

Crotalus atrox, Cocóspera. 26 Jun 2012 (JR).

Crotalus basiliscus

Common names: *cascabel grande*, *saye*; Mexican west coast rattlesnake

Sonora. Southern Sonora and the Pacific Coast of Mexico. This is the largest rattlesnake in the region, reaching 2 m (6.5 ft) in length.

Crotalus cerastes

Common names: AWA'ALA; *víbora cornuda*, *viborita de cuernitos*; sidewinder rattlesnake

Arizona and Sonora, north of the Yoem Bwiara. A relatively small, beige to dark beige snake with darker blotches down the back. Distinguished by its sidewinding movements and horn-like scales above the eyes.

Deer songs: Sidewinder songs are sung anytime during a pahko. A sidewinder is implied, but not mentioned by name, in the Flower Badger deer song. Felipe tells:

> When I asked Don Lupe [Guadalupe Molina], about this one after he taught it to me, he told me it was about a badger and a sidewinder snake. The badger grabbed the sidewinder and killed it. Then the badger tossed the snake among the falling blossoms of the *masa'asai* [queen's wreath, *Antigonon leptopus*, Polygonaceae]...I guess that the sidewinder was coiled up, that's why the song says it was *temula*, rolled up in a ball. (Evers and Molina 1987:121)

Crotalus cerastes, north of Bahía Kino. May 2007 (JR).

Crotalus molossus

Common names: *cascabel de cola negra*; black-tailed rattlesnake

Arizona and Sonora, to the eastern and northern margin of the Yoem Bwiara. This colorful snake may be hued olive, yellow, brown, to relatively vibrant green.

Crotalus tigris

Common names: *cascabel tigre*; tiger rattlesnake

Arizona and Sonora. Named for its banded color pattern, which can be brown, gray, and orange. The head is small in proportion to the body

Unidentified rattlesnakes

AAKAME NAKAPIT (deaf rattlesnake). See *Pituophis catenifer*.

KUTKO SIARI AAKAME (dark green rattlesnake)

Arizona. Seen along the banks of the Santa Cruz River near Marana. It is likely to be the Mohave rattlesnake (*Crotalus scutulatus*) or the black-tailed rattlesnake (*C. molossus*).

SIARI AAKAME (green rattlesnake)

Arizona and Sonora. A light green rattlesnake, seen in the Silverbell Mountains. A similar snake is said to occur at Las Guásimas. Possibly *Crotalus molossus*.

TOSAI AAKAME (white rattlesnake)
Arizona.

Seen along the banks of the Santa Cruz River near Marana. It is also called *yoo vaakot* (it has grown, or lived, to its limit). Possibly a rattlesnake before shedding its skin when it becomes light colored, or covered in dusty silt.

TESTUDINES – TURTLES

CHELONIIDAE – SEA TURTLES

Five of the seven species of sea turtles in the world occur in the Gulf of California. Yoemem living along the coast or engaged in maritime economies would have known all five sea turtles. All sea turtles in Mexico are listed as vulnerable, endangered, or critically endangered by the International Union for Conservation of Nature (IUCN). Mexican federal law mandates a complete ban (*veda*) on the taking of sea turtles. Unfortunately, the veda is not universally respected and sea turtles continue to be captured and sold on the black market.

Felipe tells us that *moosen* is a general core word for all sea turtles.

Caretta caretta

Common names: *caguama amarilla, tortuga perica*; loggerhead sea turtle

The loggerhead has a large head and powerful jaws, which it uses to crush large mollusks that form a significant part of its diet. Loggerheads migrate from Mexico and elsewhere along the Pacific Coast of the Americas to Australia and the Orient to nest. These turtles have become scarce in the Gulf of California.

Chelonia mydas

Common names: MOOSEN; *caguama, caguama prieta*; green sea turtle

The green sea turtle, sometimes known locally as the "black turtle," is the most commonly encountered sea turtle in the region. It was the major commercial sea turtle and was once abundant in the Gulf of California.

Basilio (1890 [1634]:193) recorded *mosen* as a kind of *tortuga*. *Moosni* is the general term used by the Comcaac (Seris) for sea turtle, as well as the specific term for the green sea turtle (Felger and Moser 1985; Moser and Marlett 2010).

Food: Sea turtle meat was often made into a stew. Sea turtles were widely eaten in Sonora until the latter part of the twentieth century, by which time they had become rare.

Rosario Vakame'eri-Castillo told Felipe that they used to eat sea turtles at an island called *Sata Kawi*. (*Sata* is a red clay used as slips for making pottery; *Kawi* is mountain.) This is a small island at *Vakochevampo* (Bahía Bacochibampo). Rosario was stationed in the vicinity of Bacochibampo as part of Pancho Villa's army and said they kept eating moosen because it was the only thing they could get. At that time sea turtles were abundant in the Gulf of California and they hunted sea turtles from boats.

Medicine: Painter (1986:56) said that "Turtle soup is used for tuberculosis." She did not say what kind of turtle, although the green sea turtle is the one most likely to have been used for soup.

Chelonia mydas, Loreto, Baja California Sur. 1993 (JN).

Eretmochelys imbricata

Common names: *carey, tortuga carey*; hawksbill sea turtle

The lustrous carapace plates yield the tortoise shell of commerce. Hawksbills were often captured and stuffed to be sold as curios in Guaymas. Like most sea turtles, they are now rare.

Lepidochelys olivacea

Common names: *golfina, mestiza, tortuga golfina*; olive ridley

This sea turtle was once common in the Gulf of California and is the smallest sea turtle in the Gulf. It sometimes nests on the Sonora coast. Olive ridley populations have made a comeback as a result of vigorous conservation efforts, including protection of nesting sites on the Oaxaca coast. It occasionally nests on the Sonora coast.

DERMOCHELYIDAE – LEATHERBACK SEA TURTLE

Dermochelys coriacea

Common names: CHUKUI MOOSEN (black sea-turtle); *siete filos, tortuga laúd*; leatherback sea turtle

This is the largest turtle in the world. It is a pelagic (open ocean) animal and a powerful swimmer that is seen infrequently in the Gulf of California It is now globally highly endangered.

EMYDIDAE – POND, BOX, AND WATER TURTLES

Terrapene nelson [*Terrapene nelsoni klauberi*]

Common names: *caja de manchas, casquito pintada*; spotted box turtle

Sonora. A brown or mahogany turtle with yellow, light-colored, spots on the shell and a dark-colored belly. Likely in deeper more tropical drainages of the Sierra Bacatete and its eastern foothills

Trachemys yaquia [*Trachemys scripta yaquia*]

Common names: MAHAO, MAHAU; *jicotea del Yaqui, tortuga grande de río*; Yaqui slider

Sonora. These large freshwater turtles are endemic to Sonora, where they are found from the lower Río Sonora to the Río Mayo. They are common in irrigation canals in the Yoem Bwiara and are often seen sunning on steep slopes of cement-lined canals: "Coming along a back road out of Vicam that snakes in and out of fields and mesquite bosques until it opens up along one of the irrigation canals, we stop to look at all the turtles climbing up out of the water onto the concrete apron of the canal. They have a good spot to watch the sun go down, so we join them" (Evers and Molina 1987:19). José María Jaimez and others pointed out that they lay eggs in the ground and leave them, about 15 cm (6 in) deep, by the canals.

Basilio (1890 [1634]:193) cited *mahau* as the name of a *tortuga*.

Johnson (1962:261) listed "*habé'eko'i, tortuga de agua dulce*," a freshwater turtle, which likely refers to *Trachemys yaquia*. He also recorded "*sáwaimóchik, especie de tortuga amarilla*," which might refer to this species, which can be somewhat yellowish in coloration (see *Kinosternon*).

Food: This turtle is eaten. Although we lack specific information, we assume *mahaom* were significant food items because they are so common and large.

Trachemys yaquia, Río Mayo. 28 Nov 2004 (JR).

KINOSTERNIDAE – MUD TURTLES

Kinosternon

Common names: WA'IVIL; *casquito, casquito o tortuga de los charcos*; American mud turtles, mud turtles

Arizona and Sonora. Mud turtles are small and oblong. They are often found in wetlands including rivers, irrigation canals, and stock ponds.

Basilio (1890 [1634]:193) listed *huaibil* as the name of a *tortuga*. In 1939–1940, Johnson (1962:283) recorded "*sáwaimóchik, especie de tortuga amarilla*" This name might refer to *Kinosternon*, and in particular to the Arizona mud turtle.

Food: Guillermo Amarillas (Pótam, 16 December 1988) told us, "They are quite small, [found] in the water, and are eaten (*en agua, muy chico, se comen*)" and wrote the name as "*guaibil*."

Kinosternon alamosae

Common names: *casquito del agua, casquito de Álamos*; Alamos mud turtle

Sonora. Recorded from the Río Yaqui in central Sonora and southward to Sinaloa. It is difficult to discern from the Mexican mud turtle.

Kinosternon arizonense [*Kinosternon flavescens arizonense*]
Common names: *casquito del agua, casquito de Arizona*; Arizona mud turtle

Southern Arizona and Sonora southward to the Guaymas region. It is wide and flattened in appearance, yellowish brown in color, and the head and limbs are not mottled.

Kinosternon integrum
Common names: *casquito de fango mexicana*; Mexican mud turtle

Sonora. Known from central Sonora to Oaxaca. It is dark gray to blackish with a mottled neck and head. It is difficult to discern from Alamos mud turtle.

Kinosternon sonoriense
Common names: *casquito de Sonora*; Sonoran mud turtle

Arizona and Sonora. This turtle was found in the Tucson area into the 1950s and still occurs in streams in the Santa Catalina and Rincon mountains. It ranges from Arizona and New Mexico to central Sonora and western Chihuahua. It is dark gray to blackish or dark brown, with a mottled head and neck, and a pale stripe from eye into the neck.

Kinosternon integrum, Sierra Mazatán. 28 Apr 2014 (JR).

TESTUDINIDAE – TRUE TORTOISES
Gopherus evgoodei
Common names: KAU MOCHIK (mountain turtle), MOCHIK; *tortuga, tortuga del desierto, tortuga del monte*; desert tortoise, Goode's thornscrub tortoise, Sonoran Desert tortoise, thornscrub tortoise

Arizona and Sonora. The Sonoran Desert tortoise (*Gopherus morafkai*) was recognized in 2011 as distinct from Agassiz's desert tortoise (*G. agassizii*), which occurs only in the Mohave Desert. *Gopherus morafkai* occurs in Arizona and Sonora southward to the Guaymas region and northern portions of the Yoem Bwiara. A third desert tortoise species was named in 2016 as Goode's thornscrub tortoise (*Gopherus evgoodei*; Edwards et al. 2016). It occurs in southern Sonora, including the southern portions of the Yoem Bwiara, and southward to Sinaloa.

Basilio (1890 [1634]:193) cited *mochic* as the name of a *tortuga*.

Beliefs: Various consultants told Painter (1986:42), "If a pregnant woman takes the egg of a turtle or pulls the leg of one, her child will have *ute'a* and be strong and healthy." The concept of ute'a signifies spiritual power, strength, ability, and courage.

Painter (1986:325) also discussed the role of the "turtle" and other small animals in rituals of the pahko'olam during non-Easter ceremonies. One man told her "what the pascolas might say," for example, "Saint Turtle (Santo Mochik), thou who are known never to be afraid, even if you see our enemy coming to kill you, you always walk slow and never run away, help me."

Food: In common with other people in Sonora, tortoises are eaten. Guillermo Amarillas (Pótam, 13 December 1988) described the desert tortoise: "On land, it is eaten, very good, same as sea turtle—the

Gopherus evgoodei, Sierra de Álamos. 21 Jul 2015 (JR).

mountain turtle (*En la tierra, se comen, muy buena, mismo que cahuama—tortuga de la sierra*)."

Songs: Tortoise and turtle songs are sung during a *Luto Pahko* (death anniversary).

BIRDS

The avifauna of the Yoem Bwiara is rich and diverse, with over 280 species in 46 families and 21 orders. *Wikichim* is the general term for birds. Basilio (1890 [1634]) provided names of more than 60 birds, ranging from songbirds to waterfowl to birds of prey. Most of the better-known birds of the region are listed here, although relatively few have specific names in Yoem noki. We have relied on the distributional and other information from Russell and Monson's (1998) book on the birds of Sonora. Other significant works on bird life of the region include van Rossem (1945), Palacios and Mellink (1995), and Villaseñor-Gómez et al. (2010). We also consulted the Cornell Lab of Ornithology (birds.cornell.edu), Avibase–the World Bird Database (avibase.bsc-eoc.org), and the AOU (American Ornithologists Union) Check-list of North American Birds (Chesser et al. 2019). Susan Davis Carnahan, Walter Fertig, Aaron D. Flesch, Trica Oshant Hawkins, Osvel Hinojosa-Huerta, Juan Pablo Gallo-Reynoso, and Amadeo Rea have provided generous information and reviews. Information on birds in the Sierra Bacatete is mostly from Aaron D. Flesch, who contributed records of 87 species (Flesch 2003, 2008, and personal communication, 2019).

Natural history information for birdlife pertains to the Yoem Bwiara and nearby areas in southern Sonora unless otherwise stated. Information for residency, such as resident, migrant, transient, visitant, breeding (presumed or confirmed), is largely from Russell and Monson (1998).

Adornment: Feathers have been widely used for decoration and ritual (see accounts for the crow, *Corvus*, Passeriformes, and macaws and parrots, Psittaciformes).

Beliefs: It is said that one should never disturb a dead bird. Leave it alone. If you find it lying belly up on the ground, don't try to turn it over–this is for respect of the natural world.

Deer songs: In the traditional pahko, the night-long sequences of deer songs include bird songs (*wikit bwikam*), which are performed with the approach of dawn. These are "toward-the-morning-songs (*matchuo vicha bwikam*)" (Evers and Molina 1987:214–215).

Guano mining: Mining bird guano from certain Gulf of California islands was a regional industry, primarily during the nineteenth and early twentieth centuries (Bowen 2000; Velarde et al. 2005), and Yaquis were the usual workers. According to local narratives, conditions could be brutal, especially on Isla San Pedro Mártir where it is said there are many graves. The stinky guano was exported to Europe and elsewhere, primarily used as fertilizer and as a source of sulfuric acid (Cushman 2013). Nathaniel Goss mentions that the labor force mining guano on Isla San Pedro Mártir consisted of 135 Yaquis from Guaymas with their families (Goss 1888:240–241; Bowen 2000). Yaquis were also almost certainly the labor force on Isla Rasa at that time, since the same company mined guano on both islands in the late 1880s. But the record goes back to 1856 when Federico Craveri recorded 27 Indians on Isla Patos mining guano and loading it on ships (Craveri 2018:342). Craveri doesn't say Yaquis specifically, but the operation was run out of Guaymas, so the "Indians" almost certainly were Yaquis (Tomas Bowen, personal communication, 2019). Yaquis would have encountered numerous sea birds on the different guano islands including the following:

- Craveri's murrelet (*Synthliboramphus craveri*, Aiddae)
- Red-billed tropicbird (*Phaethon aethereus*, Fregatidae)
- Heermann's gull and yellow-footed gull (Laridae)
- Brandt's cormorant (*Phalacrocorax penicillatus*, Phalacrocoracidae)
- Blue-footed booby and brown booby (Sulidae).

Masks: Toward the latter part of the twentieth century, some Arizona and Sonora mask makers began producing pahko'ola masks representing birds including cardinals, hawks, parrots, and roosters. These masks were mostly made for sale

(Kolaz 1985, 2007).

Natural history information for birdlife pertains to the Yoem Bwiara and nearby areas in southern Sonora unless otherwise stated. Information for residency, such as resident, migrant, transient, visitant, breeding (presumed or confirmed), is largely from Russell and Monson (1998).

ACCIPITRIFORMES – EAGLES, HAWKS, KITES, and OSPREYS

ACCIPITRIDAE – EAGLES AND HAWKS

Taawe is a general Yoeme term for all hawks. Basilio (1890 [1634]:162) gave four names for kinds of hawk (*gavilán*): *tahue* "*gavilán por su especie pequeña*" (small hawks); *buasaca* "*gavilán por su especie grande*" (large hawks); plus *iochipai* and *tachi.*

Accipiter cooperi
Common names: *gavilán de Cooper*; Cooper's Hawk

Arizona and Sonora. A common winter visitor and transient; breeds in Arizona.

Accipiter striatus
Common names: *gavilán pajarero*; Sharp-shinned Hawk

Arizona and Sonora. This small, low-flying hawk is a winter visitor and transient in southern Sonora; documented in the Sierra Bacatete.

Aquila chrysaetos
Common names: *águila real*; Golden Eagle

Arizona and Sonora. The golden eagle occurs in Arizona and is reported in Sonora as far south as the Sierra El Aguaje but is unknown in the Yoem Bwiara. Like other large raptors, these huge birds can soar to great heights. Although proficient hunters, they will often feed on carrion.

Buteo albonotatus
Common names: *águila aura*; Zone-tailed Hawk

Arizona and Sonora. In the Sierra Bacatete and along the lower Río Yaqui; summer resident, breeding presumed.

Buteo jamaicensis
Common names: *aguililla colaroja*; Red-tailed Hawk

Arizona and Sonora. This is the most common large hawk in the region and is highly variable in coloration. The red-tailed hawk is "sometimes accused of killing poultry, but they seldom do. Their usual prey consists of small mammals such as cottontails and ground squirrels" (Russell and Monson 1998:72). Resident and winter migrant; breeding confirmed, including in Sierra Bacatete.

On November 1, 1940, Rosalio Moisés was looking for a friend about four miles from Tórim: "We sat under a big mesquite tree. Within a few minutes we heard a bird cry out. It was the deer guard bird, a big brown bird that looks like an eagle; this bird always cries when he sees a hunter to warn the deer. This deer guard bird thought Lucho was a deer hunter because he was carrying a Mauser" (Moisés et al. 1971:182). Perhaps Rosalio's "deer guard bird" is the red-tailed hawk. These large, mostly brown and white hawks will cry if an interloper approaches their nest.

Red-tailed hawk, Dragoon Mountains. 16 Dec 2009 (AR).

Buteo plagiatus
Common names: *aguililla gris*; Gray Hawk

Southern Arizona and widespread in Sonora, mostly in non-desert regions. It occurs along the Río Yaqui including the lower reaches of the river as well as in the Sierra Bacatete. Mostly breeding, summer resident.

Buteo swainsoni

Common names: *aguililla de Swainson*; Swainson's Hawk

Arizona and Sonora. Typically in agricultural areas and may be a frequent early spring migrate in both states.

Buteogallus anthracinus

Common names: *aguililla negra*; Common Black Hawk

Arizona and Sonora. This large hawk dwells in riparian forests of the Río Yaqui and mangrove esteros from Bahía Lobos to Las Guásimas, and is seen in San Carlos. Breeding range extends north to southern Arizona; resident and breeding in southern Sonora.

Circus hudsonius

Common names: *gavilán rastrero*; Northern Harrier

Arizona and Sonora. This common transient and winter visitant is often seen in Sonora flying low over farmlands and along coastal wetlands.

Elanus leucurus

Common names: *milano coliblanco*; White-tailed Kite

Sonora. Generally found in agricultural areas, it began to colonize Sonora from areas farther south in the late 1970s.

Geranospiza caerulescens

Common names: *azor zancón*; Crane Hawk

Sonora. This unusual raptor was observed in July 2006 in tall gallery forest of evergreen figs (*Ficus insipida*) in a Sierra Bacatete canyon (Flesch 2008). This locality is 115 km northwest of the northernmost record described by Russell and Monson (1998). This species may be expanding its range northward.

Haliaeetus leucocephalus

Common names: *águila calva*; Bald Eagle

Sonora. Bald eagles have been observed in the esteros at Las Guásimas before their migration north from March to May. Nests are known along the upper Río Yaqui.

Parabuteo unicinctus harrisi

Common names: *aguililla de Harris*; Harris's Hawk

Arizona and Sonora. Russell and Monson (1998:67) reported, "This dark-chocolate hawk with vivid white in the tail is a fairly common resident." It occurs in the Sierra Bacatete.

Unidentified hawks

BWASSA'AKA, a kind of hawk in Sonora.

HUCHACHI

Sonora. Johnson (1962:266) listed *huchachi* as a "*gavilán (?), que dice kíu*," a kind of hawk. *Huchachi* was also among the names of several birds listed by Basilio (1890 [1634]:142) as "*aves grandes*."

OORIS, a large hawk

Oris was also among the names of several birds listed by Basilio (1890 [1634]:142) for "*aves grandes*."

TAWE OVE'A (lazy hawk)

PANDIONIDAE – OSPREYS

Pandion haliaetus

Common names: SULUMAI (*sulu*, sliding + *mai*, in the sea); *águila pescadora*, *gavilán pescador*; Osprey

Resident in both Sonora and Arizona near bodies of water. *Sulumai* is described as a big fish-eating bird that "slides on the water," which would be when catching fish. These large birds, dark

Osprey, Parker Canyon Lake, Cochise County, Arizona. 6 Nov 2015 (AR).

brown above and white beneath, are seen along the Sonora coast where they build large stick-nests on columnar cacti and utility poles. They migrate through the interior in early spring in a broad range of upland environments in which they do not breed.

ANSERIFORMES – "WATERFOWL"

ANATIDAE – DUCKS, GEESE, AND SWANS
Common names: PAATO, VETA YEKA (flat nose); *pato*; duck (wild and domestic)

The Yoemem apparently do not distinguish most kinds of ducks. In fact, ducks appear to be under-differentiated (different kinds often not distinguished by name) in North American native taxonomies that have been carefully studied (Rea 2007). For example, Gila River Pimas say that ducks were probably not part of the aboriginal diet; they hunted ducks after obtaining shotguns (Rea 1983, 2007; Felger 2007). Perhaps the situation was similar for the Yoemem. Basilio (1890 [1634]:180) listed six names for different kinds of duck (*pato*): *bachaim "pato de tierra caliente colorado"*; *sibaro "pato ánsar"*; *ilibachau "pato chico"*; plus *totbio*, *bapo moatela*, and *tepciabiri*.

Anas acuta
Common names: *pato golondrino norteño*; Northern Pintail

Arizona and Sonora. Found along the coast and inland; winter transient.

Anas crecca
Common names: *cerceta aliverde*; Green-winged Teal

Arizona and Sonora. Common winter visitor throughout Sonora wherever there is water, from coastal wetlands to mountain streams.

Anas platyrhynchos
Common names: *pato de collar*; Mallard

Arizona and Sonora. Mallards are hunted by non-Yoeme sport hunters in Sonora and can be seen in wetlands in southern Arizona. Formerly common wintering along the lower Río Yaqui.

Anser albifrons
Common names: *ganso careto mayor*; Greater White-fronted Goose

Arizona and Sonora. Once abundant in the grain fields of the Río Yaqui valley, numbers have decreased substantially since the 1960s (Russell and Monson 1998).

Anser caerulescens [*Chen caerulescens*]
Common names: *ganso blanco*; Snow Goose

Arizona and Sonora. Flocks of migratory snow geese were once common in Sonora and are sometimes seen along the coast.

Aythya affinis
Common names: *pato-boludo menor*; Lesser Scaup

Arizona and Sonora. Winter transient along the coast in brackish and seawater.

Aythya americana
Common names: *pato cabecirrojo*; Redhead

Arizona and Sonora. Common migrant and wintering along the coast.

Aythya collaris
Common names: *pato piquianillado*; Ring-necked Duck

Arizona and Sonora. Rare transient and winter resident along the Río Yaqui.

Aythya valisineria
Common names: *pato coacoxtle*; Canvasback

Arizona and Sonora. Rare winter transient along the Río Yaqui and at San Carlos.

Branta bernicla nigricans [*Branta nigricans*]
Common names: *branta*, *ganso de collar*; Black Brant

Sonora. Brant feed mostly on eelgrass (*Zostera marina*, Zosteraceae). These small geese are present along the Sonora coast from fall to spring, and have been seen at Bahía Lobos, Presa Oviáchic, and occasionally at Estero Soldado near San Carlos (Sue Carnahan and Juan-Pablo Gallo-Reynoso, personal communication, 2019).

Bucephala albeola
Common names: *pato monja*; Bufflehead

Arizona and Sonora. Winter transient, on salt and fresh water.

Bucephala clangula
Common names: *ojodorado común*; Common Goldeneye
Sonora. Rare winter visitor in coastal wetlands.

Dendrocygna autumnalis
Common names: *pichichi*, *pichihuila*, *pijiji*; Black-bellied Whistling Duck
Sonora. Gregarious, cavity-nesting ducks; spring and summer resident. Basilio (1890 [1634]:226) recorded *tepciabiri* as the name for "*pato pichihuila.*"

Mareca americana [*Anas americana*]
Common names: *pato chalcuán*; American Wigeon
Arizona and Sonora. Winter transient; common along the coast during migration.

Mareca strepera [*Anas strepera*]
Common names: *pato pinto*; Gadwall
Arizona and Sonora. Transient and uncommon winter visitant, more common south of Sonora.

Mergus serrator
Common names: *mergo copetón*; Red-breasted Merganser
Arizona and Sonora. Coastal at rocky beaches; migrant and winter visitant.

Oxyura jamaicensis
Common names: *pato tepalcate*; Ruddy Duck
Arizona and Sonora. Transient and winter visitant; coastal and at freshwater reservoirs.

Spatula clypeata [*Anas clypeata*]
Common names: *pato cucharón norteño*; Northern Shoveler
Arizona and Sonora. Winter transient.

Spatula cyanoptera [*Anas cyanoptera*]
Common names: *cerceta castaña*; Cinnamon Teal
Arizona and Sonora. Mostly winter residents.

Spatula discors [*Anas discors*]
Common names: *cerceta aliazul*; Blue-winged Teal
Arizona and Sonora. Casual migrant and winter visitant.

Unidentified duck
VAKONI
Basilio (1890 [1634]:205) recorded *baconi* as the name for "*pato prieto*," a dark-colored duck.

APODIFORMES – HUMMINGBIRDS AND SWIFTS

APODIDAE – SWIFTS
Aeronautes saxatalis
Common names: *vencejo pecho blanco*; White-throated Swift
Arizona and Sonora. Common summer resident, including around large cliffs such in Sierra Bacatete where they nest.

Chaetura vauxi
Common names: *vencejo de Vaux*; Vaux's Swift
Arizona and Sonora. Spring and uncommon fall transient throughout the region.

TROCHILIDAE – HUMMINGBIRDS
Common names: SEMALULUKUT; *chuparosa*; hummingbird
Arizona and Sonora. Mateo González and others said that they make little nests, have little white eggs, and the hatchlings are bald.

Beliefs: Hummingbirds, or even their eggs, feature in varied and complex beliefs, as favorite fetishes, and often in dreams (Painter 1986:29). These fetishes have been used to make a loved one reciprocate and for curing a sick person. Dried hummingbirds are sold as fetishes all over Mexico, such as in the Mercado Sonora in Mexico City. These fetishes are related mainly with sexual abilities. When Felipe was little, he was told that if you caught one with your bare hands, it meant that you would receive a lot of spiritual power, adding "You can make a net from a coffee strainer to catch one."

Semalulukuchim are the first ones to announce Holy Saturday as they zoom up and down and make sharp humming sounds.

Deer songs: There are many deer songs telling about the hummingbird and what it does, like getting nectar from flowers (Evers and Molina 1990).

Seven species of hummingbirds are known from the Yoem Bwiara including the Sierra Bacatete.

Amazilia violiceps
Common names: *Colibri corona violeta*; Violet-crowned Hummingbird

Arizona and Sonora. Eastern Sonora and in the Sierra Bacatete especially in large canyons.

Archilochus alexandri
Common names: *colibrí barbinegro*; Black-chinned Hummingbird

Arizona and Sonora. Widespread in Sonora including the Río Yaqui.

Calypte costae
Common names: *colibrí de Costa*; Costa's Hummingbird

Arizona and Sonora. In the Yoem Bwiara including along the river, the coast, and the Sierra Bacatete.

Costa's Hummingbird, Tucson Mountains. 10 Jan 2008 (AR).

Cynanthus latirostris
Common names: *colibrí pico ancho*; Broad-billed Hummingbird

Arizona and Sonora. Widespread in Sonora including along the Río Yaqui. This is the most common hummingbird in Sonora.

Heliomaster constantii
Common names: *colibri picudo*; Plain-capped Starthroat

Sonora. Southernmost Sonora and the Río Yaqui above the Yoem Bwiara and locally in the Sierra Bacatete.

Selasphorus platycercus
Common names: *zumbador cola ancha*; Broad-tailed Hummingbird

Arizona and Sonora. Migrant across the region. This is one of the most common hummingbirds in Sonora.

Selasphorus rufus
Common names: *zumbador rufo*; Rufous Hummingbird

Arizona and Sonora. Widely scattered in Sonora.

CAPRIMULGIFORMES – NIGHTHAWKS AND ALLIES

Antrostomus ridgwayi [*Caprimulgis ridgwayi*]
Common names: *tapacamino tu-cuchillo*; Buff-collared Nightjar

Sonora. Eastern Sonora and recorded at the Río Yaqui north of Ciudad Obregón and in the Sierra Bacatete; breeding presumed.

Chordeiles acutipennis
Common names: KUTAPAPACHE'A; *chotacabras menor*, *tapacamino*; Lesser Nighthawk

Arizona and Sonora. In Tucson and Sonora, you can see these nighthawks flying around street lights at nighttime in summer. Migratory and some also apparent residents in the Yoem Bwiara including the Sierra Bacatete.

Nyctidromus albicollis
Common names: *chotacabras pauragu*e; Common Pauraque

Sonora. Typically occurring much farther south in western Mexico, "This bird was not discovered in Sonora until 26 May 1980, when it was considered common at Mexico Highway 15 north of Cd. Obregón" (Russell and Monson 1998:146).

Phalaenoptilus nuttallii
Common names: *pachacua norteña, tapacaminos tevii*; Common Poorwill

Arizona and Sonora. This poorwill is known to hibernate, a unique adaptation among birds. Richard reports, "My first trip to the Yoeme region and the tropical deciduous forest at Álamos was with my beloved high school biology teacher Nancy Thomas, her husband Peter Neeley, and graduate students studying the newly discovered poorwill hibernation with Professor Raymond Cowles, University of California, Los Angeles."

Common Poorwill. Dragoon Mountains. 30 Jul 2020 (AR).

CATHARTIFORMES –VULTURES AND CONDORS

Three kinds of vultures are recognized: wiiru (turkey vulture) and tekoe (black vulture).

Deer song: Evers and Molina (1987:156) provided a song and explanation involving a black vulture and a turkey vulture:

> The black vulture and the turkey vulture will meet where the white wood is standing. 'When we meet we will talk about the animal,' it says. They will talk together, the turkey vulture and the black vulture, about the deer. The black vulture wants to say that. The deer will dance with that. The turkey vulture and the black vulture want to talk together themselves where there is white wood standing.
>
> They come out here when they see something lying dead like cows or horses. They live somewhere here on top of us. They live on top of us. They want to come down here to eat. That turkey vulture, that black vulture. The song says that.
>
> The black vulture and the turkey vulture want to hunt, want to eat there. That's why they say this. Over there they say they will meet where the white wood is standing. Maybe it is a dead tree. Sitting together there, the sun hits them, warms them. They will talk about the animal, where they are going to overpower him. So these are the ones who are going to eat him. They are sitting somewhere out there.
>
> This is soon after dawn; two vultures are warming themselves in the sun, preparing to go aloft and to make their daily rounds in search of carrion. It is soon after dawn because this is one of the bird songs (*wikit bwikam*), which are performed with the approach of dawn (Evers and Molina 1987:214–215). These are toward-the-morning-songs (*matchuo vicha bwikam*).

According to Rea (2007:101), this pairing of the two species is fairly common. Black vultures most often watch and follow turkey vultures to a carcass. Turkey vultures have an excellent sense of smell (olfaction) but black vultures lack olfaction. "Both of course have astonishing visual powers, perhaps the best developed on any land animal" (Phillips et al. 1964).

Narrative: Giddings (1959:27) related a story of "The Man Who Became a Buzzard," meaning a vulture (also see *Spermophilus*, Sciuridae).

Cathartes aura
Common names: WIIRU; *aura, zopilote cabeza roja*; Turkey Vulture

Turkey Vulture, St. David, Arizona. 19 Sep 2016 (AR).

Arizona and Sonora. The turkey vulture is all black with a red head and a wingspan of up to 1.8 m (6 ft).

Basilio (1890 [1634]:199, 214) recorded *huiru* as "*aura, especie de buitre*" (a kind of vulture) and *huiru* as "*zopilote de cabeza colorada.*"

Hunting: In describing hunting, Pérez de Ribas (1645; Reff et al. 1999: 93) wrote:

> Nothing escapes their arrows. If an animal escapes with a fatal wound, they go find it the following day at the spot where they know for certain that it has fallen dead. Because they normally use poisoned arrows (even for hunting), it takes no longer than twenty-four hours for an animal to die...They find the place where the animal has dropped by looking up to the sky, where the *zopilotes*, or buzzards (a type of eagle that feeds on carrion, of which there are many in this land) are flying in circles.

Buzzards belong to an Old World family of hawks, but the term has been applied colloquially to vultures in the Americas. The original wording is "*los zopilotes, género de águilas que hay muchos en esta tierra, que se sustenan de carnes muertas*" (Pérez de Ribas 1645:13). The arrow poison in this account would be from *hierba de la flecha* (*Pleradenophora bilocularis*, Euphorbiaceae).

Coragyps atratus

Common names: TEKOE; *zopilote común*; Black Vulture

Arizona and especially common in Sonora. Basilio (1890 [1634]:198) recorded *tecoe* as a word for *zopilote* other than the red-headed *zopilote*.

It is intriguing to note that "van Rossem (1945) remarked that the black vulture must have extended its range during this century, as early explorers did not encounter it until they arrived at tropical localities. The species did not extend into Arizona until the 1920s" (Russell and Monson 1998:60; also see Rea 1983). The black vulture has a black head and a wingspan of about 1.5 m (5 ft).

Gymnogyps californianus?

Common names: KAU SATEMAI (mountain vulture); *cóndor Californiano*; California Condor

Deer song: A deer song tells of *kau satemai* in the mountains of the Yoem Bwiara, "*Kau satemai* is one kind that we call mountain buzzard. They say that mountain buzzards were very big, but that they haven't seen them for a long time. Perhaps they were condors" (Evers and Molina 1987:112–113).

There are no historical records of the California condor being in Sonora, however the birds were present at least prehistorically (and possibly as late as the 1930s) in the northern reaches of the Gulf of California (Tesky 1994; Croxen et al. 2007; Rea 2000). Condors are all black above, like the other two vultures in the region, but are distinguished by white on the underside of the wings and a much larger wingspan of 3 m (10 ft). These largest of North American birds search for food over wide areas in a variety of open habitats, soaring up to 150 miles in a day. They tend to nest on cliffs or in rocky areas of canyons and coastal mountain ranges.

The deer song clearly indicates that kau satemai is a kind of vulture. Another possible candidate is the king vulture (*zopilote rey*,

Sarcoramphus papa), a Neotropical species known to have once ranged northward in western Mexico as far as central Sinaloa (Graham 1987; Howell and Webb 1995; Clinton 2000; Holste et al. 2014). Nowadays the king vulture ranges from southern Mexico to South America. It is mostly white above and below, with black bordering the wings, although juveniles are mostly black; the birds can have a wingspan of about 1.5 m (5 ft).

Black Vulture, Rio Rico, Arizona. 4 Nov 2016 (AR).

Imte kuyutai	Here, we, where the mescal
kuyuliti weyekapo	like mescal agave stands,
naute yayaine	together we will meet.
Imte kuyutai	Here, we, where the mescal
kuyuliti weyekapo	like mescal agave stands,
naute yayaine	together we will meet.
Imte kuyutai	Here, we, where the mescal
kuyuliti weyekapo	like mescal agave stands,
naute yayaine	together we will meet.
Imte kuyutai	Here, we, where the mescal
kuyuliti weyekapo	like mescal agave stands,
naute yayaine	together we will meet.
Emposinto yoyo wilutakaine	And you are an enchanted, enchanted black vulture,
Emposinto yoyo tekoyelikaine	And you are an enchanted, enchanted turkey vulture,
Imte tosali kutala weyekapo	Here, we, where the white wood stands,
naute yahkai	together we meet.
inikate yowata naute etehone	together we will talk about this animal.
Imte kuyutai	Here, we, where the mescal agave,
kuyuliti weyekapo	like mescal agave stands,
ka naute yayaine	together we will not meet.

CHARADRIIFORMES – GULLS AND "SHOREBIRDS"

CHARADRIIDAE – LAPWINGS AND PLOVERS

Charadrius nivosus

Common names: *chorlito niveo*; Snowy Plover

Arizona and Sonora. Coastal; fall, winter, and spring migrant.

Charadrius semipalmatus

Common names: *chorlito semipalmeado*; Semipalmated Plover

Arizona and Sonora. Coastal; winter resident and transient.

Charadrius vociferus

Common names: TUIVIT; *chorlo tildío*; Killdeer

Arizona and Sonora. These active little birds are widespread in wetlands across Sonora and southern Arizona. Common resident, plus winter migrants from northern latitudes.

Charadrius wilsonia

Common names: *chorlito piquigrueso*; Wilson's Plover

Sonora. Common resident; beaches, tidal marshes, and estero margins.

Pluvialis squatarola

Common names: *chorlo gris*; Black-bellied Plover

Sonora. Common transient and winter visitant; beaches, tidal marshes, and estero margins.

HAEMATOPODIDAE – OYSTERCATCHERS

Haematopus palliatus

Common names: *ostrero americano*; American Oystercatcher

Sonora. These strange striking shore birds have a large red beak and white stripe through their black wings. Common resident along the Sonora coast, often seen feeding on tidal flats at low tide.

LARIDAE – GULLS, SKIMMERS, AND TERNS

Vaatosai (white water) is the general name for seagull. At least nine species of gulls occur along shores and waters of the Gulf of California and some are also seen inland in the Yoem Bwiara. Basilio (1890 [1634]:162) provided names of four kinds of gull (*gavilán*): *buasaca* "*gavilán por su especie grande*"; *tahue* "*gavilán por su especie pequeña*"; plus *iochipai* and *tachi*.

Chlidonias niger

Common names: *charrán negro, golondrina marina negra*; Black Tern

Arizona and Sonora. Uncommon spring and fall migrant; coastal and Río Yaqui.

Chroicocephalus philadelphia

Common names: *gaviota de Bonaparte*; Bonaparte's Gull

Arizona and Sonora. Transient and winter resident; mostly coastal.

Hydroprogne caspia [*Sterna caspia*]

Common names: *charrán cáspica, golondrina marina cáspica*; Caspian Tern

Sonora. Coastal and riverine spring and fall migrant.

Larus argentatus

Common names: *gaviota plateada*; Herring Gull

Sonora. Mostly coastal; transient and winter visitant.

Larus californicus

Common names: *gaviota californiana*; California Gull

Sonora. Coastal; fall migrant and winter visitant.

Larus delawarensis

Common names: *gaviota piquianillada*; Ring-billed Gull

Arizona and Sonora. Coastal and "the only common gull of interior Sonora, to be found chiefly on large reservoirs in late fall and winter" (Russell and Monson 1998:111).

Larus heermanni

Common names: *gaviota de Heermann*; Heermann's Gull

Sonora. Common coastal resident; breeding on Gulf islands.

Larus livens
Common names: *gaviota patamarilla*; Yellow-footed Gull
Sonora. Coastal resident, breeding on Gulf islands.

Larus occidentalis
Common names: *gaviota occidental*; Western Gull
Sonora. Coastal winter transient.

Leucophaeus atricilla [*Larus atricilla megalopterus*]
Common names: *gaviota reidora*; Laughing Gull
Sonora. Winter transient; coastal and inland.

Leucophaeus pipixcan [*Larus pipixcan*]
Common names: *gaviota de Franklin*; Franklin's Gull
Arizona and Sonora. Rare spring migrant; primarily encountered along the coast of the Yoem Bwiara, and a migrant in Arizona.

Rhynchops niger
Common names: *rayador americano*; Black Skimmer
Sonora. Rare transient and winter visitant at the coast.

Sterna forsteri
Common names: *charrán de Forster, golondrina marina de Forster*; Forster's Tern
Arizona and Sonora. Coastal transient and winter resident.

Sternula antillarum [*Sterna antillarum browni*]
Common names: *charrán mínimo, golondrina marina minima*; Least Tern
Sonora. Coastal; uncommon breeding, summer resident.

Thalasseus elegans [*Sterna elegans*]
Common names: *charrán elegante, golondrina marina elegante*; Elegant Tern
Sonora. Coastal, spring through fall and uncommon in winter, and more common on the Pacific Coast; breeding on Isla Raza.

Thalasseus maximus [*Sterna maxima*]
Common names: *charrán real, golondrina marina real*; Royal Tern
Sonora. Coastal winter transient and less often in summer; nesting on Gulf islands.

RECURVIROSTRIDAE – AVOCETS AND STILTS

Himantopus mexicanus
Common names: *candelero americano*; Black-necked Stilt
Arizona and Sonora. Common in coastal esteros and sometimes seen inland at wetlands.

Recurvirostra americana
Common names: *avoceta americana*; American Avocet
Arizona and Sonora. Like the stilt, this avocet is found in coastal esteros and also seen inland at wetlands.

SCOLOPACIDAE – PHALAROPES AND SANDPIPERS

About 20 species of these small wetland birds occur in the Yoem Bwiara, mostly along the shores of the Gulf of California, and some also at inland wetlands.

Actitis macularius
Common names: *playero alzacolita*; Spotted Sandpiper
Arizona and Sonora. Widespread coastal and inland wetlands; common transient and winter resident.

Arenaria interpres
Common names: *vuelvepiedras rojizo*; Ruddy Turnstone
Sonora. Coastal, winter transient.

Arenaria melanocephala
Common names: *vuelvepiedras negro*; Black Turnstone
Sonora. Rocky shores; transient and winter visitant.

Calidris
Common names: *vaa wo'i* (water coyote), *player*; Sandpiper
Sonora. Six species of these active little birds are seen along the coast.

Calidris alba
Common names: *playero blanco*; Sanderling
Arizona and Sonora. Transient and winter visitant.

Calidris alpina
Common names: *playero dorsirojo*; Dunlin
Sonora. Migrant and winter resident.

Calidris canutus
Common names: *playero gordo*; Red Knot
Sonora. Common transient and winter resident.

Calidris mauri
Common names: *playerito occidental*; Western Sandpiper
Arizona and Sonora. Common transient and winter resident.

Calidris minutilla
Common names: *playerito mínimo*; Least Sandpiper
Arizona and Sonora. Transient and winter resident.

Calidris virgata [*Aprisa virgata*]
Common names: *playero de marejada*; Surfbird
Sonora. Transient and winter resident.

Gallinago delicata
Common names: *agachona de Wilson*; Wilson's Snipe
Arizona and Sonora. Along the coast and at inland wetlands.

Limnodromus griseus
Common names: *costurero piquicorto*; Short-billed Dowitcher
Arizona and Sonora. Winter migrant, mostly along the Sonora coast.

Limnodromus scolopaceus
Common names: *costurero piquilargo*; Long-billed Dowitcher
Arizona and Sonora. Transient and winter visitant.

Limosa fedoa
Common names: *picopando canelo*; Marbled Godwit
Sonora. Common migrant and winter resident, and also in summer.

Numenius americanus
Common names: *zarapito picolargo*; Long-billed Curlew
Arizona and Sonora. Common transient and winter visitant.

Numenius phaeopus
Common names: *zarapito trinador*; Whimbrel
Sonora. Common migrant and winter resident.

Phalaropus lobatus
Common names: *falaropo cuellirrojo*; Red-necked Phalarope
Arizona and Sonora. Common transient over Gulf waters and rare along the shore.

Tringa melanoleuca
Common names: *patamarilla mayor*; Greater Yellowlegs
Arizona and Sonora. Transient and winter visitant.

Tringa semipalmata [*Catoptrophorus semipalmatus inornatus*]
Common names: *playero pihuiui*; Willet
Arizona and Sonora. Transient and winter visitant.

CICONIIFORMES – STORKS

Mycteria americana
Common names: *cigüeña americana*; Wood Stork
Sonora. "Van Rossem (1945:42) considered the Wood Stork 'common in summer...from Guaymas southward,' but we fear that its numbers have considerably decreased in recent years" (Russell and Monson 1998:41).

COLUMBIFORMES – DOVES AND PIGEONS

Basilio (1890 [1634]:179) listed five names for kinds of dove (*paloma*): *huocou* for "*paloma torcaz*" (a kind of pigeon); *meretau* for "*paloma parda*"; plus *batui*, *omócoli*, and *cucu*.

*Columba livia

Common names: PALOMA; *paloma*; Rock Pigeon

Arizona and Sonora. Some Yoemem raise pigeons but apparently not for food. Feral pigeons have become common from ranches and villages to cities throughout the region. The common domesticated pigeon is an Old World bird that has become ubiquitous nearly worldwide.

Columbina inca

Common names: OMMO'OKOLI; *tórtola colilarga*; Inca Dove

Arizona and Sonora. Widespread including the Sierra Bacatete breeding confirmed. Basilio (1890 [1634]:179) included *omócoli* as a word for *paloma*.

Columbina passerina

Common names: *tórtola común*, *tórtola coquita*; Common Ground-Dove

Arizona and Sonora. Widespread including the Sierra Bacatete; breeding confirmed.

Leptotila verreauxi

Common names: *paloma arroyera*; White-tipped Dove

Sonora. This dove is a denizen of eastern and southern Sonora, especially thornscrub and tropical deciduous forest, and also occurs in the vicinity of San Carlos and in the Sierra Bacatete.

Patagioenas flavirostris [*Columba flavirostris restricta*]

Common names: *paloma morada*; Red-billed Pigeon

Sonora. These pigeons breed in the tropical deciduous forest of southern Sonora and in the Yoem Bwiara are seen in the river delta.

*Streptopelia decaocto

Common names: *tórtola turca*; Eurasian Collared-Dove

Arizona and Sonora. This dove is native to Eurasia and has rapidly spread across North America, especially around agricultural and urban places. In July 2006 it was seen near the Sierra Bacatete and in Yoem Bwiara towns (Aaron Flesch, personal communication, 2019).

Zenaida asiatica

Common names: KUUKU; *paloma aliblanca*, *paloma pitayera*; White-winged Dove

Arizona and Sonora. Basilio (1890 [1634]:208) recorded *cucú* as the term for "*paloma llamada en México pichón*," which is probably this dove. These birds are especially conspicuous in the warmer months:

> The song of white-winged doves is the lusty sound of summer in the Sonoran Desert. With Egyptian blue eyeliner and iridescent breast feathers, males belt their sonorous 'who-cooks-for-you!' from the top of saguaros all through the summer...Although doves are dearly loved, in Arizona hordes of hunters kill them by the thousands in the fall...Their relationship with the mighty saguaro is tight but also ambiguous....They are saguaro mutualistic pollinators, but they are also parasitic seed predators. White-winged doves are wild desert denizens, but they can take advantage of human crops, and therefore the fate of their populations has been shaped by humans. (Martínez del Rio 2007:305)

As the name *paloma pitayera* indicates, it is also closely linked to the organpipe cactus (*Stenocereus thurberi*). This dove is common in the Yoem Bwiara including the Sierra Bacatete.

Hunting: Pérez de Ribas (1645; Reff et al. 1999:93) wrote, "Boys often hunt by themselves, particularly for doves and quail. These are very abundant, so they kill many." The doves may have mostly been white-winged doves.

This dove is hunted by Yorim and North Americans in southern Sonora and northern Sinaloa in large, unsustainable numbers.

Zenaida macroura

Common names: WOKKOI; *paloma huilota*; Mourning Dove

Arizona and Sonora. Widespread in both states. In the Yoem Bwiara including the Sierra Bacatete.

CORACIIFORMES – KINGFISHERS

Chloroceryle americana

Common names: *martin-pescador verde*; Green Kingfisher

Sonora. This small kingfisher is found along freshwater streams and esteros. It ranges through the eastern half of Sonora and approaches the eastern margin of the Yoem Bwiara. Also found at brackish esteros near San Carlos and Estero Soldado.

Megaceryle alcyon [*Ceryle alcyon*]

Common names: *martin-pescador norteño*; Belted Kingfisher

Arizona and Sonora. "This large blue kingfisher is a fairly common transient and winter visitant throughout Sonora at elevations below 1000 m [3280 feet], wherever bodies of water can provide its diet of small fish and crustaceans" (Russell and Monson 1998:165). It frequents mangroves and freshwater sites.

FALCONIFORMES – CARACARAS AND FALCONS

Caracara cheriway

Common names: CHOAWAE; *caracara, quelele*; Crested Caracara

Arizona and Sonora. These large, attractively-colored birds are mostly scavengers. They often frequent human habitation and are common in the lowlands and marsh areas of the Río Yaqui, as well as the Sierra Bacatete; breeding confirmed.

Falco mexicanus

Common names: *halcón de pradera*; Prairie Falcon

Arizona and Sonora. Sometimes seen in the area of Sierra El Aguaje and San Carlos.

Falco peregrinus

Common names: *halcón peregrino*; Peregrine Falcon

Arizona and Sonora. Islands in bahías San Francisco and Bacochibampo near Guaymas and Isla San Pedro Nolasco. Breeding locally on tall cliffs in the Sierra Bacatete.

Falco sparverius

Common names: *cernícalo americano*; American Kestrel

Arizona and Sonora. Common breeding resident and winter visitor; transient throughout the region including Sierra Bacatete.

GALLIFORMES – CHICKENS, QUAIL, TURKEYS, AND ALLIES

ODONTOPHORIDAE – NEW WORLD QUAIL

Common names: SUVA'I, SUVA'U; *cordoniz*; quail

Three kinds of quail are known from the Yoem Bwiara: the elegant quail, Gambel's quail, and masked bobwhite

Food: Presumably these three species were significant food resources.

Hunting: Quail were hunted with bow and arrow, in communal fire-drive hunts, and were taken in nets and snares. The quail snare is set up on the ground and baited. Pérez de Ribas (1645; Reff et al. 1999:93) wrote, "Boys often hunt by themselves, particularly for doves and quail. These are very abundant, so they kill many."

Callipepla douglasii

Common names: SIKILI SUVA'I (red quail); *cordoniz cresta dorada*; Elegant Quail

Sonora. The elegant quail, with its loud call, is common in the tropical thornscrub and tropical deciduous forest of eastern and southern Sonora, as well as the Sierra Bacatete. This quail is largely reddish brown in color, especially the males, corresponding to the Yoeme name. Two traditional songs indicate *Sikili Suva'i* is the Elegant Quail.

Songs: Evers and Molina (1987:111) provide a Red Quail song, *Sikili Suva'i*, which Felipe learned from Don Jesús Yoilo'i: "Red quail are called bobwhite quail in English. Don Jesús said that the red quail live up in the mountain, in the Vakatetteve Mountains." The mountain habitat suggests that *sikili suva'i* is the elegant quail rather than the Evers and Molina (1987:26–27) tells us that Frances Densmore (1952) "recorded Juan Ariwares's songs on phonographic cylinders which are still preserved at the Smithsonian Institution." From these recordings, Felipe was able to transcribe a red quail song and provide a new translation. The characteristic call (*kukupopoli*) suggests that "Sikili" is the elegant quail. masked bobwhite (see

Colinus virginianus ridgwayi). The green cane is *carrizo* (*Arundo donax*, Poaceae).

SIKILI SUVA'I

Siali vakata weyeka'apo ne su
siali vaka heka vetukun
ne kateka
Siali suvawi sikili suvawiii

Siali vakata weyeka'apo ne su
siali vaka heka vetukun
ne kateka
Siali suvawi sikili suvawiii

Siali vakata weyeka'apo ne su
siali vaka heka vetukun
ne kateka
Siali suvawi sikili suvawiii

Ayaman ne seyewailo

naiyoli huya aniwapo
chewa yolemem

Siali suvawi sikili suvawiii

RED QUAIL

Where the green cane stands,
under the green cane breeze,
I sit.
Red quail, red quail.

Where the green cane stands,
under the green cane breeze,
I sit.
Red quail, red quail.

Where the green cane stands,
under the green cane breeze,
I sit.
Red quail, red quail.

Over there, I, in the flower-covered,
cherished, enchanted wilderness world,
I am more human.
Under the green breeze,
I sit.
Red quail, red quail.

Gambel's Quail, Dragoon Mountains, Arizona. 15 Jul 2017 (AR).

Sikili...
kaita va vemu weamksu
hakun kukupopoti husakai

Sikili...
kaita va vemu weamksu
hakun kukupopoti hiusakai

Iyiminsu seyewailo
huya nainasukuni
kaita va vemu weamkasu
hakun kukupopoti husakai

Sikili...
kaita va vemu weamkasu
hakun kukupopoti hiusakai

Little red [quail].
walking afar whence there is no water.
where do they make the kukupopoli sound?

Little red [quail].
walking afar whence there is no water.
where do they make the kukupopoli sound?

Over here, in the center
of the flower-covered wilderness.
walking afar whence there is no water.
where do they make the kukupopoli sound?

Little red [quail].
walking afar whence there is no water.
where do they make the kukupopoli sound?

Callipepla gambelii
Common names: SUVA'I, SUVA'U; *cordoniz chiquiri*; Gambel's Quail

Arizona and Sonora. This quail is common and widespread in Arizona and Sonora.

Food. Gambel's quail was hunted for food at least during traditional times in the Yoem Bwiara. In the early twentieth century, Geraldo ("Lalo") Alvarez and his "family lived in the desert at Kingsley Ranch near Arivaca" in southern Arizona, where they sometimes ate "scrambled quail eggs" (Molina et al. 2003:4). Quail eggs were a common food resource for various regional indigenous people (Rea 2007).

Colinus virginianus ridgwayi
Common names: TESAKI SUVA'I (arroyo quail); *cordoniz mascarita*; Masked Bobwhite

Arizona and Sonora. This western subspecies of the northern bobwhite (bobwhite quail) once ranged from southern Arizona to south-central Sonora in lowland grassy, weedy, and farmland habitats (Brown et al. 2012). Native, wild populations of masked bobwhite decreased during the twentieth century and ultimately became extinct in the wild, largely due cattle grazing.

This quail was documented on the east side of the Sierra Bacatete, from 1905 to the 1930s. Brown et al. (2012:313) tell that:

> Moore (1932:74) took 2 females from a covey of 8 near Tecoripa ...'an extensive rolling area bounded on the south by the Sierra de Bacatete (Yaqui country); on the north by a low broken mountain range...The sides and gentle slopes at the foot of the mountains are covered with brush—largely mesquite. The rolling area is covered with grass (at least six species) with brush, mesquite, cat-claw, etc., along the washes. The height of the grass varies from knee to shoulder high, the latter height being predominant.'

Following fieldwork in 1937, Ligon (1952:48) wrote: "Although the birds were present in considerable numbers on the 'Llanos' segment of the wide Yaqui Valley...their doom was already foreshadowed by the upsurge in the cattle business." Ligon's sites included "Agua Caliente Valley" [Valle de Agua Caliente] in an "area of Yaqui Indian Reservation."

Yoeme consultants said that the arroyo quail (*tesaki suva'i*) is different from Gambel's Quail.

Hunting: Yoeme people would trap tesaki suva'im in nets and sell them in Guaymas. We were told this hunting resulted in making them scarce in the Yoem Bwiara. Don Jesús Yoilo'i from Pótam told Felipe and Larry Evers that the tesaki suva'i was a popular bird for sale in Guaymas. This trapping for sale apparently occurred in the decades before the 1950s.

PHASIANIDA – PHEASANT FAMILY

***Gallus gallus**
Common names: TOTOI, TOTOI HAMUCHIA (chicken female; hen), TOTOI O'OWIA (chicken male; rooster); *pollo*; Chicken

Arizona and Sonora. Chickens, originally from the Old World, have long been kept domestically in Sonora. Basilio (1890 [1634]:166, 184, 216, 227) recorded names for hen, rooster, chicks, chicken egg, rooster's cockscomb, and a chicken's cackle.

Beliefs and narrative: *Totoitakuse'epo* (where the rooster crows), Hill of the Rooster or Cerro de Gallo, is a sacred mountain east of Las Guásimas. The mountain peak is shaped like a rooster's beak. People make pilgrimages or vision quests to the mountain, and should purify their minds, souls, and bodies for several weeks before going to the mountain. In 1927 the Yoemem fought their last major battle at *Totoitakuse'epo*. *Hill of the Rooster* is the title of a 1956 novel by William Holden, based on his extensive knowledge of Yoeme culture and experience.

It is widely believed that it is a bad omen when a hen tries to crow like a rooster.

Medicine: Eggs are used for curing. The entire head is massaged with a whole egg. Best is a fertile egg and not a store-bought one. The curing spans four days, during which time the patient abstains from eating eggs. A new egg is used for each day's session. After each session, the egg is broken into a glass half full of water and set on an altar for about half an hour or left overnight. The person who performed the curing looks at the egg content in the glass to see if an image appears, and it is usually an image of someone's eyes. Then if the curer identifies the person whose eye appears, he will know that person is doing something bad to the patient, like envy or jealousy. Then the curer will pour crushed chiltepín into the glass and cover it with a palm cross and let it sit overnight. The next morning he will bury it in the yard to the east. Each egg is buried at the end of each day's session.

A different version is provided by Rosalio Moisés: The *curandero* "asked for a glass of water...Into this he broke an egg and put in a few drops of holy water. A few minutes later the entire figure of a woman dressed in a green and white striped skirt, pink blouse, and *rebozo* appeared in the glass. Her face was quite clear, 'That is a woman I know in Pótam,' I said" (Moisés et al. 1971:237–238).

Sport: Mounted chicken pulls were popular at celebrations in Sonora. Rosalio Moisés wrote of one in 1933:

> The big fiesta at Vikam is the Día de San Juan, beginning June 24...The fiesta began about 10 o'clock in the morning with all the Yaqui cowboys galloping by a live chicken buried in the sand, trying to pull it up by the head. Whoever gets the chicken wins twenty-five pesos. Two cowboys tore the live chicken in two by running away from each other on horseback. Then a chicken leg is thrown to the women. (Moisés et al. 1971:146)

Meleagris gallopavo

Common names: CHIIWI (turkey, wild), KO'OVO'E (turkey, domesticated); *guajolote*; Turkey

Arizona and Sonora. The Mexican or Gould's turkey (*Meleagris gallopavo mexicana*) is known from mountains in eastern and northern Sonora and wild turkeys also occur in mountains in Arizona.

Basilio (1890 [1634]: 162, 207) recorded *cobore* and *cóbori* as names for *gallina de la tierra* and *guajolote*, respectively. The "r" in the names indicate Yoreme rather than Yoeme terms.

The Yoemem prized their domesticated turkeys, which they had until the warfare and deportations of the late nineteenth and early twentieth centuries. Rosalio Moisés tells about a group of people who had fled to the Sierras and were captured by Mexican soldiers. The prisoners were marched to Guaymas where they were eventually released. They walked home to Tórim:

> Upon their return they found that Mexicans had stolen all their chickens, turkeys, and goats. They were especially sad about the turkeys, as these had come from the Sierra Madre of Chihuahua. Yaquis have always made special trips to get turkeys from the Tarahumaras of Chihuahua. (Moisés et al. 1971:10)

People in the Sierra Madre Occidental in southeastern Sonora and southwestern Chihuahua, including the Guarijíos and Tarahumaras, have a special race of an elegant, small domestic turkey. These turkeys, presumably derived from the local wild subspecies, are far sleeker and much more agile than modern, commercial strains of turkeys, and notable with their large white wing feathers against a dark body. Richard saw these turkeys at the Guarijío village of Bavícora, Sonora, in November 1997. These turkeys are seen around houses in villages and rancherías and at night they roost in trees. Mayo people at Aduana, near Álamos, also have these turkeys.

Wild turkey, Chiricahua Mountains, Arizona. 5 Apr 2016 (AR).

Ortalis wagleri

Common names: CHAPARA; *chachalaca*, *cuiche*; Rufous-bellied Chachalaca

Sonora. There is Yoeme name for this conspicuous bird, although it is not documented for the Yoem Bwiara. Chachalacas occur in southeastern Sonora east of the Yoem Bwiara, and southward along the Pacific Coast of Mexico.

***Pavo cristatus**
Common names: PAAVO; *pavo royal*, *pavón*; Indian Peafowl, Peacock

Arizona and Sonora. We have no history on peacocks in the Yoeme communities. It is said they are like living ornaments in some family's yards.

GAVIIFORMES – LOONS

Gavia immer
Common names: *colimbo mayor*, *somormujo mayor*; Common Loon

Sonora. Like the Pacific loon, this shore and offshore bird is more common north of Guaymas than southward. It is present only in winter.

Gavia pacifica
Common names: *colimbo*, *somormujo*; Pacific Loon

Sonora. This shore and offshore bird is common north of Guaymas and less so to the south. It is present only in winter.

GRUIFORMES – CRANES, RAILS, AND ALLIES

GRUIDAE – CRANES

Antigone canadensis
Common names: *grulla canadiense*; Sandhill Crane

Arizona and Sonora. These large cranes were once common in winter along the lower Río Mayo valley and may have once been common along the Río Yaqui (van Rossem 1945). Elsewhere in Sonora they are sometimes seen in irrigated farmland (Russell and Monson 1998). They were also seen historically in the irrigated valleys near Tucson.

Pérez de Ribas (1645; Reff et al. 1999:84) reported that "At times during the year there is also an abundance of cranes."

RALLIDAE – COOTS, RAILS, AND ALLIES

Fulica americana
Common names: *gallareta americana*; American Coot

Arizona and Sonora. In Sonora these birds are often common in areas of open water along the coast and all along the Río Yaqui.

Gallinula galeata
Common names: *gallineta común*; Common Gallinule

Arizona and Sonora. Gallinules in the Yoem Bwiara are found in wetlands, including irrigation ditches and along the Río Yaqui, especially among cattails. They require places to hide.

Porzana carolina
Common names: *polluela sora*; Sora

Arizona and Sonora. This small rail frequents coastal esteros and freshwater marshes, especially among cattails along the Río Yaqui. It is also seen in artificial ponds at San Carlos.

Rallus limicola
Common names: *rascón limícola*; Virginia Rail

Arizona and Sonora. "This rail is a rare transient and winter visitant in salt- and freshwater marshes throughout Sonora; most records are from the coast" (Russell and Monson 1998:85).

Rallus obsoletus
Common names: *rascón costero del Pacífico*; Ridgway's Rail

Sonora. These secretive birds live deep in the mangroves and are seldom seen, but they are often heard calling during the breeding season in May and June.

PASSERIFORMES – "PERCHING BIRDS"

ALAUDIDAE – LARKS

Eremophila alpestris
Common names: MAAVISA; *alondra cornuda*; Horned Lark

Arizona and Sonora north of the Sierra El Aguaje.

BOMBYCILLIDAE – WAXWINGS

Bombycilla cedrorum
Common names: *ampelis americano*; Cedar Waxwing

Arizona and Sonora. Winter transient.

CARDINALIDAE – BUNTINGS, CARDINALS, AND GROSBEAKS

Cardinalis cardinalis

Common names: WICHALAKAS; *cardenal norteño*; Northern Cardinal

Arizona and Sonora. Basilio (1890 [1634]:148) cited *huichalaca* for "*cardenal, pájaro*."

Cardinalis sinuatus

Common names: *cardenal desértico*; Pyrrhuloxia

Arizona and Sonora. Summer nesting and winter transient.

Passerina caerulea [*Guiraca caerulea*]

Common names: *picogrueso azul*; Blue Grosbeak

Arizona and Sonora. Summer resident and nesting, especially among cottonwoods, willows, and mesquites.

Passerina ciris

Common names: *colorín sietecolores*; Painted Bunting

Arizona and Sonora. Uncommon fall migrant and winter resident; extending into the eastern margin of the Yoem Bwiara.

Passerina versicolor

Common names: *colorín morado*; Varied Bunting

Arizona and Sonora. Summer resident, breeding presumed.

Pheucticus melanocephalus

Common names: *picogrueso tigrillo*; Black-headed Grosbeak

Arizona and Sonora. Winter transient.

Piranga ludoviciana

Common names: *tángara occidental*; Western Tanager

Arizona and Sonora. Transient.

Piranga rubra

Common names: *tángara roja*; Summer Tanager

Arizona and Sonora. Summer resident.

CORVIDAE – CROWS, JAYS, AND MAGPIES

Calocitta colliei [*Cyanocorax colliei*]

Common names: CHAROWE; *urraca, urraca-hermosa cara Negra*; Black-throated Magpie-Jay

Sonora. This elegant bird has a spectacular long tail, especially notable as it glides from tree to tree in tropical deciduous forest of southeastern Sonora. It occurs in foothill and mountain terrain to the east of the Yoem Bwiara and eastward into the Sierra Madre Occidental. It was, however, known to the Yoemem. Johnson (1962:259) identified "*charówe*" as "*urraca*" in his Yaqui dictionary. Basilio (1890 [1634]:210) cited *chóiau* as the word for *urraca*, although this might be from Yoremem or a neighboring group.

Corvus corax

Common names: SANKU'UKUCHI; *cuervo, cuervo común*; Common Raven

Arizona and Sonora. These large, raucous birds are especially common residents in the Yoem Bwiara.

Corvus sinaloae

Common names: KOONI; *cuervo sinaloense*; Sinaloa Crow

Sonora. This crow is a common resident in agricultural areas and towns and villages in the Yaqui and Mayo river basins. Their call is said to sound like that of a kitten.

Basilio (1890 [1634]:153) recorded *coni* as the word for *cuervo*.

Beliefs: If a bunch of crows are seen flying overhead, it means an enemy is approaching—move away from your path to avoid attack. Beals (1945:197) reported, "The only animal the Yaqui

The Crow Song

tau tau tau tiakasu kooni	Caw-caw-caw, the croaking crow,
tau tau tau tiaka kooni	Caw-caw-caw, the croaking crow,
cholakti chepteka niokita kooni	He's a fluttery, frisking song-thrush of a crow,
cholakti chepteka niokita kooni	He's a fluttery, frisking song-thrush of a crow

associate with the wizard is the crow (kóni), who is said to be a son of the devil and a sign of death. The crow is also said to inform one of expected visitors, flying from the direction they are coming, encircling the house three times, and departing."

Dance and song: Based on the work of Ruth Giddings and the words of Ambrosio Castro in Sonora, Taub (1950:94) reproduced the Crow Song (Free Translation)(see above).

The performance of the Crow Dance with the Crow Song was also explained:

> In keeping with the actions of the Deer and Coyote dancers, the performer of this dance imitates the crow. He would squat down close to the ground, then hop, and finally try to fly. A black cape worn over the shoulders added to the imitative effect. The dancer also had a tuft of feathers (presumably crow feathers) sticking up from the back of his head. He carried an ear of corn in his mouth. This would seem to be the only such Yaqui ceremonial in which the music and song were performed by separate individuals. The musician played a special cane 'harp' made of a yard-long length of carrizo cane, with a long peg on one end and a single, heavy string attached. The singer also was supposed to imitate the crow. (Taub 1950:119–120)

Evers and Molina (1990) provided a different Coyote Song featuring the crow, *Kooni Mahai* (Crow Is Afraid).

Sinaloa crows, near Álamos. Aug 2004 (JR).

CUCULIDAE – ANIS, CUCKOOS, AND ROADRUNNERS

Coccyzus americanus

Common names: *cuclillo pico Amarillo*; Yellow-billed Cuckoo

Arizona and Sonora. Seen in cottonwoods and willows along the Río Yaqui and in the Sierra Bacatete; breeding presumed.

Crotophaga sulcirostris

Common names: *garrapatero pijuy*; Groove-billed Ani

Sonora. Russell and Monson (1998:134) described this bird as a "strange all-black cuckoo...usually near streams and often in the company of cattle." It mainly occurs east of the Yoem Bwiara. It is recorded from near Guaymas, and seen at least from fall to spring in San Carlos.

Geococcyx californianus

Common names: TARUK; *churea, correcaminos*; Greater Roadrunner

Arizona and Sonora. Widespread in both states.

Beliefs: Roadrunners are said to be harbingers of bad news, and it is considered a bad omen to see one running around unnaturally at the home. He is trying to tell something, but the people know what it is at that moment—it is suspected that there will be an accident or death.

Food. Although generally not considered a food animal, in the early twentieth century, Geraldo "Lalo" Alvarez and his family lived "in the desert at Kingsley Ranch near Arivaca" and they "ate roadrunner with rice" (Molina et al. 2003:4).

EMBERIZIDAE, see PASSERELLIDAE

FRINGILLIDAE – FINCHES

Haemorhous mexicanus [*Carpodacus mexicanus*]

Common names: VASO MO'EL; *fringílido mexicana*; House Finch

Arizona and Sonora. Mateo González and Martha González identified house finches at the Arizona-Sonora Desert Museum as *vaso mo'el*.

Widespread including the Sierra Bacatete; breeding confirmed.

Spinus psaltria [*Carduelis psaltria*]
Common names: *dominico dorsioscuro*; Lesser Goldfinch
Arizona and Sonora. Along the Río Yaqui and in the Sierra Bacatete; breeding presumed.

HIRUNDINIDAE – MARTINS AND SWALLOWS

Wokovavase'ela is the general term for swallow; it is also the term for the water-clover fern (*Marsilea vestita*, Marsileaceae). Basilio (1890 [1634]:163) listed *huocobabalis* as the word for *golondrina*. There are at least seven swallow or swallow-like species in the Yoem Bwiara.

Hirundo rustica
Common names: *golondrina ranchera*; Barn Swallow
Arizona and Sonora. Migrant and breeding; their mud nests often seen on buildings.

Petrochelidon pyrrhonota
Common names: *golondrina risquera*; Cliff Swallow
Arizona and Sonora. Summer breeding migrant. Mud nest on buildings and cliffs.

Progne subis
Common names: *martín azul*; Purple Martin
Arizona and Sonora. Summer breeding migrant.

Riparia riparia
Common names: *golondrina ribereña*; Bank Swallow
Arizona and Sonora. Transient in the region.

Stelgidopteryx serripennis
Common names: *golondrina aliserrada norteña*; Northern Rough-winged Swallow
Arizona and Sonora. Spring and fall migrant and breeding.

Tachycineta albilinea
Common names: *golondrina manglera*; Mangrove Swallow
Sonora. Mangroves and inland along the Río Yaqui.

Tachycineta bicolor
Common names: *golondrina arbolera*; Tree Swallow
Arizona and Sonora. Winter transient, coastal and along the Río Yaqui.

Tachycineta thalassina
Common names: *golondrina cariblanca*; Violet-green Swallow
Arizona and Sonora. Transient and nesting in spring, along the coast as far south as Guaymas and the Río Yaqui nearly to the Yoem Bwiara.

ICTERIDAE – BLACKBIRDS AND ORIOLES

Agelaius phoeniceus
Common names: CHANA; *tordo sargento*; Red-winged Blackbird
Arizona and Sonora. This bird has a red patch on the shoulder and is otherwise all black. In the Marana region it arrives in spring, at Easter time. These birds like to roost in the cotton fields, sit on cotton plants and especially the *vaaka* (*carrizo*, *Arundo donax*, Poaceae). They sing beautifully but you don't hear or see them in summertime. Felipe says that since the 1990s there have been fewer every year in Marana, as agriculture has been converted to urban sprawl. This bird thrives in farmland areas, and in many regions of Sonora the populations have increased over the last half of the twentieth century with the growth of agriculture.

Euphagus cyanocephalus
Common names: *tordo de Brewer*; Brewer's Blackbird
Arizona and Sonora. Winter visitant; often around farmland.

Icterus bullockii
Common names: *bolsero de Bullock*, *bolsero calandria*; Bullock's Oriole
Arizona and Sonora. Migrant; widespread including the Sierra Bacatete.

Icterus cucullatus
Common names: TAKOCHAE; *bolsero cuculado*, *calandria encapuchada*; Hooded Oriole
Arizona and Sonora. Common summer resident. This bird makes pendulous nests suspended in cottonwoods and palms. Pérez de Ribas (1645; Reff et al. 1999:85) wrote:

> There are some birds that look like and are the size of thrushes...The nests...are shaped like a sack or a long mesh bag, and they hang from a branch or the knot of a tree limb that is usually very high. This net is narrow at the neck, which is the opening or door; at the base it widens to become round. From top to bottom it measures one-half or two-thirds of a vara [about 42 to 56 cm, or 16 to 22 in]. Even though this nest is hanging and exposed to breezes or winds there is no danger of it coming loose.

This may be the hooded oriole, one of the most common orioles in the Yoem Bwiara including the Sierra Bacatete.

Icterus parisorum
Common names: *bolsero tunero*, *calandria tunera*; Scott's Oriole

Arizona and Sonora. Summer resident and transient.

Icterus pustulatus
Common names: *bolsero dorso rayado*, *calandria*; Streak-backed Oriole

Sonora. Common resident.

Icterus spurius
Common names: *bolsero castaño*, *calandria castaña*; Orchard Oriole

Sonora. Summer resident among Río Yaqui cottonwoods.

Molothrus aeneus
Common names: *vaquero ojirojo*; Bronzed Cowbird

Arizona and Sonora. Summer resident and breeding.

Molothrus ater
Common names: POUTE'ELA; *tordo cabeza café*, *vaquero cabecicafé*; Brown-headed Cowbird

Arizona and Sonora. This species thrives with farming activity and is a common summer breeding resident in both states.

Deer Song: This bird sings in the morning before dawn, which is when the song of the *poute'ela* is sung by deer singers during a pahko.

Quiscalus mexicanus
Common names: CHANATE, HIAK CHANA (Río Yaqui chana, or chanate); *zanate mayo*, *zanate mexicano*; Great-tailed Grackle

Arizona and Sonora. They are described as smaller than a crow, with tail feathers that fan out, and are most abundant around agricultural fields and towns. They vocalize a lot and like to roost in large trees such as the saltcedar (*Tamarix aphylla*, Tamaricaceae). Breeding residents and abundant in southern Sonora.

Basilio (1890 [1634]:209) reported *chanat* as the word for "*zanate, tordo*," which may be this species, or other icterids (Icteridae).

Sturnella neglecta
Common names: VÁE VICHO'OLA; *pradero occidental*; Western Meadowlark

Arizona and Sonora. Transient and winter visitant, coastal and inland.

Xanthocephalus xanthocephalus
Common names: *tordo cabeciamarillo*; Yellow-headed Blackbird

Arizona and Sonora. Common transient and winter visitant; often seen in cultivated fields and in towns.

ICTERIIDAE – YELLOW-BREASTED CHAT
Icteria virens
Common names: *gritón pechiamarillo*; Yellow-breasted Chat

Arizona and Sonora. Summer resident and migrant. This is the only species in its family; formerly considered a kind of wood warbler (Parulidae).

LANIIDAE – SHRIKES
Lanius ludovicianus
Common names: *alcaudón verdugo*, *lanio americano*, *verduguillo*; Loggerhead Shrike

Arizona and Sonora. Widespread across Sonora including the Yoem Bwiara.

MIMIDAE – MOCKINGBIRDS, THRASHERS, AND ALLIES

Mimus polyglottos

Common names: NEO'OKAI; *centzontle norteño*, *chonte*; Northern Mockingbird

Arizona and Sonora. Winter migrants or year-round residents.

Oreoscoptes montanus

Common names: *cuitlacoche de artemisia*; Sage Thrasher

Arizona and Sonora. Uncommon transient.

Toxostoma

Common names: WIIVIS, WIIVISIM (plural); *cuitlacoche*; thrasher

Basilio (1890 [1634]:214) recorded *huíribis* as the name for "*huitacoche, pájaro*," a kind of bird. Johnson (1962:293) cited *wíibis* as the term for *cuitlacochi*. Anderton (1991:451) indicated that in southern California "*huitacoche*" refers to a dark bird that "sings pretty," interpreted as a California Thrasher (*Toxostoma redivivum*). Other thrashers in the Yoem Bwiara likewise can be highly vocal.

Toxostoma bendirei

Common names: *cuitlacoche de Bendire*; Bendire's Thrasher

Arizona and Sonora. Winter residents and breeding, along the Río Yaqui and coast.

Toxostoma crissale

Common names: *cuitlacoche crissal*; Crissal Thrasher

Arizona and Sonora. Apparently resident with a breeding record near Empalme (van Rossem 1931).

Toxostoma curvirostre

Common names: *cuitlacoche piquicurvo*; Curve-billed Thrasher

Arizona and Sonora. Common breeding resident, often seen around homes and urban areas.

MOTACILLIDAE – PIPITS AND WAGTAILS

Anthus rubescens

Common names: *bisbita americana*; American Pipit

Arizona and Sonora. Transient and winter resident.

PARULIDAE – WOOD WARBLERS

Cardellina pusilla [*Wilsonia pusilla*]

Common names: *chipe de Wilson*; Wilson's Warbler

Arizona and Sonora. This is among most abundant of the migrant warblers in Sonora; often in mangroves.

Geothlypis tolmiei [*Oporornis tolmiei*]

Common names: *chipe de Tolmie*; MacGillivray's Warbler

Arizona and Sonora. Migrant.

Geothlypis trichas

Common names: *mascarita común*; Common Yellowthroat

Arizona and Sonora. Breeding resident, mangroves and inland.

Icteria virens, see ICTERIIDAE

Leiothlypis celata [*Oreothlypis celata*, *Vermivora celata*]

Common names: *chipe corona-naranja*; Orange-crowned Warbler

Arizona and Sonora. Winter resident and during migration.

Leiothlypis luciae [*Oreothlypis luciae*, *Vermivora luciae*]

Common names: *chipe de Lucy*; Lucy's Warbler

Arizona and Sonora. Spring breeder in riparian areas.

Leiothlypis ruficapilla [*Oreothlypis ruficapilla*, *Vermivora ruficapilla*]

Common names: *chipe de Nashville*; Nashville Warbler

Arizona and Sonora. Transient.

Myioborus pictus

Common names: *chipe ala blanca*; Painted Redstart

Arizona and Sonora. Rare in winter along the Río Yaqui and also rare in winter in the Sierra El Aguaje; not known in the Sierra Bacatete.

Parkesia noveboracensis [*Seiurus noveboracensis*]
Common names: *chipe charquero*; Northern Waterthrush
Arizona and Sonora. Mangroves along the Sonora coast.

Setophaga coronata [*Dendroica coronata*]
Common names: *chipe rabadilla amarillo*; Yellow-rumped Warbler
Arizona and Sonora. Common transient and winter resident.

Setophaga nigrescens [*Dendroica nigrescens*]
Common names: *chipe negrigris*; Black-throated Gray Warbler
Arizona and Sonora. Winter resident and during migration; entering the eastern margin of the Yoem Bwiara and occasional along the coast.

Setophaga petechia [*Dendroica petechia*]
Common names: *chipe amarillo*; Yellow Warbler
Arizona and Sonora. Common transient in southern Sonora. Summer breeder in riparian zones.

Setophaga petechia rhizophorae [*Dendroica petechia rhizophorae*]
Common names: *reinita de mangler*; Mangrove Warbler
Sonora. This distinctive subspecies is a summer resident breeding among mangroves.

Setophaga ruticilla
Common names: *pavito migratorio*; American Redstart
Arizona and Sonora. Rare migrant.

Setophaga townsendi [*Dendroica townsendi*]
Common names: *chipe de Townsend*; Townsend's Warbler
Arizona and Sonora. Uncommon transient.

PASSERELLIDAE – LONGSPURS, SPARROWS, AND TOWHEES

Ammodramus savannarum
Common names: *gorrión chapulín*; Grasshopper Sparrow
Arizona and Sonora. Winter resident.

Amphispiza bilineata
Common names: *gorrión gorjinegro*; Black-throated Sparrow
Arizona and Sonora. Found all year, but probably not all residents, mostly in dense desert and thornscrub.

Amphispiza quinquestriata [*Aimophila quinquestriata*]
Common names: *zcactonero cinco rayas*; Five-striped Sparrow
Arizona and Sonora. Characteristic of rocky places, including Sierra Bacatete. Breeding in summer and probably resident.

Calamospiza melanocorys
Common names: *gorrión alipálido*; Lark Bunting
Arizona and Sonora. Winter transient.

Chondestes grammacus
Common names: *gorrión arlequín*; Lark Sparrow
Arizona and Sonora. Migrant and winter resident; widespread including Sierra Bacatete.

Junco hyemalis
Common names: *junco ojioscuro*; Dark-eyed Junco
Arizona and Sonora. Wintering transient.

Melozone fusca [*Kieneria fusca, Pipilo fuscus*]
Common names: *rascador arroyero*; Canyon Towhee
Arizona and Sonora. Common, widespread resident.

Melospiza georgiana
Common names: *gorrión pantanero*; Swamp Sparrow
Arizona and Sonora. Rare winter visitant, Russell and Monson (1998) show records at Guaymas and the Río Yaqui.

Melospiza lincolnii
Common names: *gorrión de Lincoln*; Lincoln's Sparrow
Arizona and Sonora. Winter resident and migrant.

Melospiza melodia
Common names: *gorrión cantor*; Song Sparrow
Arizona and Sonora. Wintering migrant. Possible breeder along the Río Yaqui.

Passerculus sandwichensis
Common names: *gorrión sabanero pico grande*; Large-billed Sparrow
Sonora. Esteros and bordering saltscrub; feeding on eelgrass and seaweeds (marine algae) at low tide (Russell and Monson 1998).

Peucaea carpalis [*Aimophila carpalis*]
Common names: *zacatonero ala rufa*; Rufous-winged Sparrow
Arizona and Sonora. Resident; coastal and farmland to mountains including Sierra Bacatete.

Peucaea cassinii [*Aimophila cassinii*]
Common names: *zacatonero de Cassin*; Cassin's Sparrow
Arizona and Sonora. Winter transient.

Pipilo chlorurus
Common names: *rascador coliverde*; Green-tailed Towhee
Arizona and Sonora. Common transient and winter resident.

Pooecetes gramineus
Common names: *gorrión coliblanco*; Vesper Sparrow
Arizona and Sonora. Winter transient.

Spizella breweri
Common names: *gorrión de Brewer*; Brewer's Sparrow
Arizona and Sonora. Wintering migrant.

Spizella pallida
Common names: *gorrión pálido*; Clay-colored Sparrow
Arizona and Sonora. Wintering transient.

Spizella passerina
Common names: *gorrión cejiblanco*; Chipping Sparrow
Arizona and Sonora. Wintering migrant.

Zonotrichia leucophrys
Common names: *gorrión coroniblanco*; White-crowned Sparrow
Arizona and Sonora. Wintering migrant.

PASSERIDAE – OLD WORLD SPARROWS
***Passer domesticus**
Common names: *gorrión doméstico*; House Sparrow
Arizona and Sonora. Common breeding resident throughout the region, especially around cities and towns. Native to the Old World.

POLIOPTILIDAE – GNATCATCHERS
Polioptila caerulea
Common names: *perlita grisilla*; Blue-gray Gnatcatcher
Arizona and Sonora. Winter transient.

Polioptila melanura
Common names: *perlita colinegra*; Black-tailed Gnatcatcher
Arizona and Sonora. Widespread residents.

Polioptila nigriceps
Common names: *perlita sinaloense*; Black-capped Gnatcatcher
Arizona and Sonora. Common resident.

PTILOGONATIDAE – SILKY FLYCATCHERS
Phainopepla nitens
Common names: WIIVIS; *capulinero negro*; Phainopepla
Arizona and Sonora. Resident, nesting in March to May, but moving with ripening of mistletoe fruits, one of its major foods.

Wiivis is described as a black bird with red eyes. Wiivisim Pueblo in the Yoem Bwiara is named after this bird. Felipe says, "My cousins in Pótam said *wiivisim* sing in the days before I arrive to visit in Pótam."

REGULIDAE – KINGLETS
Regulus calendula
Common names: *reyezuelo sencillo*; Ruby-crowned Kinglet
Arizona and Sonora. Common winter resident and migrant.

REMIZIDAE – VERDINS

Auriparus flaviceps

Common names: *baloncillo*; Verdin

Arizona and Sonora. Breeding resident.

SITTIDAE – NUTHATCHES

Sitta carolinensis

Common names: *saltapalo pechiblanco*; White-breasted Nuthatch

Arizona and Sonora. Winter transient.

STURNIDAE – STARLINGS

***Sturnus vulgaris**

Common names: *estornino europeo*; European Starling

Arizona and Sonora. Starlings are common breeding residents around urban and agricultural areas. Native to the Old World and deliberately introduced to the Americas in 1890; by the 1950s they were in Arizona and were recorded in Empalme in 1973. Starlings are well established in southern Arizona and Sonora and are displacing some native birds.

SYLVIIDAE, see POLIOPTILIDAE

THRAUPIDAE, see CARDINALIDAE

TITYRIDAE – BECARDS AND TITYRAS

Pachyramphus aglaiae

Common names: *mosquero-cabezón degollado*; Rose-throated Becard

Arizona and Sonora. Summer breeding resident; building large, hanging nests.

TROGLODYTIDAE – WRENS

Mo'el is the general term for wren. There are at least eight species in the Yoem Bwiara, and a similar diversity is found in southern Arizona.

Campylorhynchus brunneicapillus

Common names: VAE WAAKAS; *matraca desértica*; Cactus Wren

Arizona and Sonora. Common in both states. Widespread resident in the Yoem Bwiara including Sierra Bacatete.

Catherpes mexicanus

Common names: KAU ROAKTE'A (mountain tumbler); *saltapared barranquero*; Canyon Wren

Arizona and Sonora. Common in both states. Resident in rocky places in the Yoem Bwiara including Sierra Bacatete.

Cistothorus palustris

Common names: *saltapared pantanero*; Marsh Wren

Arizona and Sonora. Winter transient, usually in wetlands.

Pheugopedius felix [*Thryothorus felix*]

Common names: *chivirín feliz*; Happy Wren

Sonora. Southeastern Sonora and extending into the Yoem Bwiara along the Río Yaqui. Resident in dense vegetation.

Salpinctes obsoletus

Common names: *saltapared roquero*; Rock Wren

Arizona and Sonora. Wintering migrant, seen in rocky habitats.

Thryomanes bewickii

Common names: *saltapared de Bewick*; Bewick's Wren

Arizona and Sonora. Migrant and winter visitant.

Thryophilus Sinaloa [*Thryothorus sinaloa cinereus*]

Common names: *chivirín sinaloense*; Sinaloa Wren

Sonora. Eastern Sonora and bordering the eastern margin of the Yoem Bwiara. Resident in places of large trees.

Troglodytes aedon

Common names: *saltapared continental norteño*; House Wren

Arizona and Sonora. Transient and winter visitant.

TURDIDAE – THRUSHES

Catharus guttatus

Common names: *zorzalito colirrufo*; Hermit Thrush

Arizona and Sonora. Migrant, winter visitant.

Catharus ustulatus
Common names: *zorzalito de Swainson*; Swainson's Thrush
Arizona and Sonora. Spring and fall migrant.

Turdus migratorius
Common names: *mirlo primavera, zorzal petirrojo*; American Robin
Arizona and Sonora. Winter transient; resident in eastern Sonora.

Turdus rufopalliatus
Common names: *zorzal dorsirufo*; Rufous-backed Robin
Arizona and Sonora. Resident in riparian habitats, nesting with summer rains.

TYRANNIDAE – FLYCATCHERS

Camptostoma imberbe
Common names: *mosquero lampiño*; Northern Beardless-Tyrannulet
Arizona and Sonora. Resident in the Yoem Bwiara including Sierra Bacatete.

Contopus pertinax
Common names: *pibí tengofrío*; Greater Pewee
Arizona and Sonora. Winter range extends into riparian areas in the Yoem Bwiara.

Contopus sordidulus
Common names: *pibí occidental*; Western Wood-pewee
Arizona and Sonora. Transient and summer resident in cottonwoods and willows along the Río Yaqui.

Empidonax difficilis
Common names: *mosquero californiano*; Pacific-slope Flycatcher
Or
Empidonax occidentalis
Common names: *mosquero barranqueño*; Cordilleran Flycatcher
Arizona and Sonora. One of these flycatchers occurs in Sierra Bacatete; transient, winter resident.

Empidonax fulvifrons
Common names: *mosquero pecho leonado*; Buff-breasted Flycatcher
Arizona and Sonora. One of the smallest and most distinctive flycatchers. It ranges through eastern Sonora and winters in the lowlands including the eastern margin of the Yoem Bwiara along the Río Yaqui.

Empidonax oberholseri
Common names: *mosquero oscuro*; Dusky Flycatcher
Arizona and Sonora. Common migrant and winter visitant.

Empidonax traillii
Common names: *mosquero saucero, papamoscas saucero*; Willow Flycatcher
Arizona and Sonora. Uncommon transient, likely in mangroves.

Empidonax wrightii
Common names: *mosquero gris*; Gray Flycatcher
Arizona and Sonora. Common winter resident and migrant.

Myiarchus cinerascens
Common names: *copetón gorjicenizo*; Ash-throated Flycatcher
Arizona and Sonora. Migrant and breeding winter resident.

Myiarchus nuttingi
Common names: *papamoscas de Nutting*; Nutting's Flycatcher
Sonora. Resident, summer nesting.

Myiarchus tuberculifer
Common names: *papamoscas triste*; Dusky-capped Flycatcher
Arizona and Sonora. Sierra Bacatete, nesting in tall thornscrub, and also in riparian trees along the Río Yaqui.

Myiarchus tyrannulus
Common names: *papamoscas tirano*; Brown-crested Flycatcher

Arizona and Sonora. Migratory, breeding in late spring and early summer; Río Yaqui, Sierra Bacatete, and probably coastal.

Myiodynastes luteiventris
Common names: *papamosca atigrado*; Sulphur-bellied Flycatcher
Arizona and Sonora. Summer resident; locally in Sierra Bacatete.

Pitangus sulphuratus
Common names: *luis bienteveo*; Great Kiskadee
Sonora. Winter breeding transient; Río Yaqui cottonwoods and willows.

Pyrocephalus rubinus
Common names: *chapaturrín*, *mosquero cardenal*, *pájaro bule*; Vermilion Flycatcher
Arizona and Sonora. Bright red males; resident and winter migrant.

Tyrannus crassirostris
Common names: *tirano poco grueso*; Thickbilled Kingbird
Arizona and Sonora. Migrant, nesting in early summer.

Tyrannus melancholicus
Common names: *tirano tropical*; Tropical Kingbird
Arizona and Sonora. Summer breeding resident along the Río Yaqui and Sierra Bacatete.

Tyrannus verti
Common names: *tirano occidental*; Western Kingbird
Arizona and Sonora. Spring and fall migrant, along the Río Yaqui and Sierra Bacatete.

Tyrannus vociferans
Common names: *tirano de Cassin*; Cassin's Kingbird
Arizona and Sonora. Wintering and transient.

VIREONIDAE – VIREOS

Vireo bellii
Common names: *vireo de Bell*; Bell's Vireo
Arizona and Sonora. Widespread and probably resident including the Sierra Bacatete.

Vireo cassinii
Common names: *vireo de Cassin*; Cassin's Vireo
Arizona and Sonora. Found in winter and as migrants.

Vireo flavoviridis
Common names: *vireo verdeamarillo*; Yellow-green Vireo
Sonora. Summer breeding resident.

Vireo gilvus
Common names: *vireo gorgojeador*; Warbling Vireo
Arizona and Sonora. Common spring and fall migrant.

Vireo plumbeus
Common names: *vireo plomizo*; Plumbeous Vireo
Arizona and Sonora. Migrant and summer breeding resident; mountains including Sierra Bacatete.

Vireo vicinior
Common names: *vireo gris*; Gray Vireo
Arizona and Sonora. Transient and winter resident, coastal to mountains including Sierra Bacatete; feeding on *Bursera* (Burseraceae) fruits as well as insects.

PELECANIFORMES – EGRETS, HERONS, PELICANS, AND ALLIES

ARDEIDAE – BITTERNS AND HERONS

Ardea alba
Common names: *garza blanca*; Great Egret
Sonora. Large, white egret in mangrove esteros and reservoirs, mostly in winter. Breeding at Río Yaqui upriver reservoirs and the coast near the Río Mayo.

Ardea herodias
Common names: KOBWABWA'I; *garza gris*, *garza Morena*; Great Blue Heron
Arizona and Sonora. This is the largest heron in North America. Common and widespread resident in wetlands, often nesting in colonies in cottonwoods and mangroves. *Kobwabwa'im* is the name for the original village at Pótam.

***Bubulcus ibis**

Common names: *garza ganadera*; Cattle Egret

Arizona and Sonora. This Old World egret reached South America, probably from Africa, in the late nineteenth century and spread through warmer regions of the New World, arriving in Sonora probably in mid-twentieth century. This pure white egret has become common in cattle and farming areas, often riding on backs of cattle. At Pótam we were told this is a "new" bird.

Butorides virescens

Common names: *garceta verde*; Green Heron

Arizona and Sonora. Small heron in mangroves and along permanent streams; it may have been common along the Río Yaqui.

Egretta caerulea

Common names: *garceta azul*; Little Blue Heron

Sonora. Apparently rare in the region; nesting in mangroves (Palacios and Mellink 1995).

Egretta rufescens

Common names: *garceta rojiza*; Reddish Egret

Sonora. Mostly in mangroves, apparently scarce, and probably once more common.

Egretta thula

Common names: *garceta pie-dorado*; Snowy Egret

Sonora. Mostly near the coast and at reservoirs.

Egretta tricolor

Common names: *garceta tricolor*; Tricolored Heron

Sonora. Occasionally near mangroves; apparently rare, although perhaps formerly more numerous.

Ixobrychus exilis

Common names: *avetoro minimo*; Least Bittern

Sonora. Uncommon summer breeding resident, among aquatic vegetation including mangroves.

Nyctanassa violacea

Common names: *huaco corona amarilla, pedrete corona amarilla*; Yellow-crowned Night Heron

Sonora. "Nests in mangroves and in mesquites close to the water" (Russell and Monson 1998:39).

Nycticorax nycticorax

Common names: *huaco corona negra, pedrete corona negra*; Black-crowned Night Heron

Arizona and Sonora. Migrant, nesting in early summer. Feeding at night, often in mangroves and inland wetlands.

PELECANIDAE – PELICANS

Pelecanus erythrorhynchos

Common names: CHATCHAAKAM; *pelícano blanco americano*; American White Pelican

Sonora. At certain times of the year, white pelicans are seen migrating along the coast. There is a population that spends the winter in Estero Soldado, near San Carlos.

Beliefs: It is said that if you have had a heart attack and you see a white pelican you will die of a heart attack.

Pelecanus occidentalis

Common names: TENWE; *alcatraz, pelícano café*; Brown Pelican

Sonora. Common along the coast, nesting mostly on islands.

Food: People at Las Guásimas said that in the past, in times of hardship and famine they used to get young pelicans to eat. They say that they are as good as turkey.

Brown pelicans, Todos Santos, Baja California Sur. 23 Feb 2010 (AR).

THRESKIORNITHIDAE – IBISES AND SPOONBILLS

Eudocimus albus

Common names: *ibis blanco*; White Ibis

Sonora. Coastal, nesting in May in columnar cacti and mangroves at Las Guásimas.

PICIFORMES – FLICKERS, SAPSUCKERS, AND WOODPECKERS

Common names: CHOLLOI; *carpintero*; woodpecker

Basilio (1890 [1634]: 210) recorded *chólou* as the name for "*pajaro llamada carpintero.*"

Colaptes auratus

Common names: *carpintero collarejo*; Northern Flicker

Arizona and Sonora. Winter transient in the Yoem Bwiara.

Colaptes chrysoides

Common names: *carpintero collarejo desértico*; Gilded Flicker

Arizona and Sonora. Common resident, including the Sierra Bacatete; nesting in columnar cacti and cottonwoods.

Dryobates scalaris [*Picoides scalaris cactophilus*]

Common names: *carpintero listado*; Ladder-backed Woodpecker

Arizona and Sonora. Widespread resident, including Sierra Bacatete.

Melanerpes uropygialis

Common names: *carpintero de Gila, carpintero del desierto*; Gila Woodpecker

Arizona and Sonora. Common, widespread resident, including Sierra Bacatete.

Sphyrapicus nuchalis

Common names: *chupasavia nuquiroja*; Red-naped Sapsucker

Arizona and Sonora. Winter resident along the Río Yaqui, generally in riparian trees.

PODICIPEDIFORMES – GREBES

Podiceps nigricollis

Common names: *zambullidor orejudo*; Eared Grebe

Arizona and Sonora. Common winter visitant and migrant. Coastal and large ponds and reservoirs. Also at sea near Las Guásimas and bahías San Francisco and Bacochibampo.

Podilymbus podiceps

Common names: *zambullidor pico grueso*; Pied-billed Grebe

Arizona and Sonora. These water birds probably were once common on the Río Yaqui. At sea near Las Guásimas and bahías San Francisco and Bacochibampo.

PSITTACIFORMES – PARROTS

Basilio (1890 [1634]:169) gave the names for five types of "parrot": *baro* for "*loro grande*"; *haro* for "*loro otro de cabeza amarilla*"; *tabelo* for "*loro pequeño por su especie*"; *ilitabelo* for "*loro más pequeño*"; and *chaoe* for "*loro huacamaya.*"

Amazona albifrons

Common names: TAVELO, VAARO; *cotorra frente blanca, loro frente blanca*; White-fronted Parrot

Sonora. These medium-sized, green and yellow parrots are year-round residents in southern Sonora. Often seen northeast of Vícam, near the foothills of the Vakatetteve and were undoubtedly more widespread in the past. In the late 1980s, Richard saw parrots in the Cardonal near Empalme.

Basilio (1890 [1634]:169) included *tabelo* as the term for "*loro pequeño por su especie*" in a list of parrot species. He also recorded *baro* as a name for *perico* (Basilio 1890 [1634]:205), likely to be the white-fronted parrot. A town in the mountains east of Ciudad Obregón is called Baroyeca, or *Vaaro Varoyeka* in Yoem noki, which means parrot nose. *Tavelo* is also the modern name for this parrot.

Parrots were sometimes captured and sold. Lucas Chavez told of helping his father, Loreto, in the 1880s, "taking loads of native tobacco, cheese, corn, and parrots" from Tórim to sell in Guaymas (Spicer 1988:110). They transported the loads on the backs of burros. Nowadays, some Mexican families in Pótam and elsewhere keep these parrots as pets.

Pérez de Ribas (1645; Reff et al 1999:84) reported "a variety of parrots [*papagallos*]" in what is now southwestern Sonora and western Sinaloa.

Adornment and Coyote Society

Parrot feathers were used to decorate the headdress (*choomo*) of the *Wiko'i Yau'ura* (Bow Leaders' or Coyote Society) captain and soldiers, describe ed by Spicer (1980:182).

A Coyote Society song, "Sontao Ya'uchim, Soldier Leaders," was provided by Evers and Molina (1990) in *Wo'i Bwikam*, Coyote Songs. Another coyote song was recorded by Taub (1950:91) that tells of officers playing with parrot wing-feathers (Felipe has made minor corrections to the orthography):

Tavelo masata yeewa tuulisa yeewa
Nulanulata yeewa nulanulata yeewa
Nulanulata yeewa nulanulata yeewa

Hitasa amani ha'abweka
Tavelo masata yeewa
Katikun sontao yaa'uchim amani ha'abweka
Tavelo masata yeewa tuulisi yeewa
Nulanulata yeewa
Nulanulata yeewa
Nulanulata yeewa

Playing with the parrot wings, nicely playing
Back and forth playing with it, back and forth playing with it
Back and forth playing with it, back and forth playing with it

What is standing over there?
Playing with the parrot wings
Remember the officers standing over there
Playing with the parrot wings, nicely playing
Back and forth, playing with it
Back and forth, playing with it
Back and forth, playing with it

Ara militaris

Common names: CHAOE; *guacamaya verde*; Military Macaw

Sonora. These large parrots are unmistakable, "With resplendent plumage, long tail, a massive bill, great size, and strident, ear-splitting calls" (Russell and Monson 1998:127). They are often seen in the middle Río Yaqui basin not too far from the Yoem Bwiara. Macaws have been observed at Presa Álvaro Obregón "El Oviáchic" and the lower Río Yaqui (Juan Pablo Gallo-Reynoso, personal communication, 2016), and in the Sierra Bacatete (Flesch 2008).

Macaws sometimes fly rather long distances. In 2005 and 2006, two military macaws were living on Isla San Pedro Nolasco but failed to establish a colony (Gallo-Reynoso et al. 2012). Macaws are more common in southeastern Sonora and in Sinaloa.

Basilio (1890 [1634]:169, 208) recorded *chaoe* as a name for *perico* and *chaoe* for "*loro huacamaya*."

White-fronted parrot, Arizona-Sonora Desert Museum, Arizona. 5 Oct 2005 (JR).

Adornment: The feathers were traded far and wide in pre-contact and early historic times. Pérez de Ribas (1645; Reff et al. 1999:84) reported that "At times during the year there is also an abundance of cranes and a variety of parrots and macaws. The latter are colored like parrots but are much larger and have plumage that is very prized for ornamentation." Pérez de Ribas, however, may have been referring to Yoremem or another neighboring group.

Macaws are generally summer residents in southern Sonora, giving credence to the indication of seasonality. Macaws apparently were also kept domestically. Pérez de Ribas (1645; Reff et al. 1999:91) wrote, "To go to war…They decorate their heads and hair with bright feathers and crests from birds that they either raise or hunt in the montes."

STRIGIFORMES – OWLS

STRIGIDAE – TYPICAL OWLS

Athene cunicularia [*Speotyto cunicularia*]
Common names: BWIA MU'U (ground owl); *tecolote llanero*; Burrowing Owl

Arizona and Sonora. These little owls can be found by the Santa Cruz River at Marana, often standing by their burrows. They are also found along the Sonora coast, such as near Guaymas, on the islands in Bahía Bacochibampo, and on Isla San Pedro Nolasco.

Bubo virginianus
Common names: MUU'U; *búho cornudo*; Great Horned Owl

Arizona and Sonora. This is the largest owl in the region. Widespread and common resident in both states. In Sonora it is also called *tecolote*, the general term for owl. Basilio (1890 [1634]:146) recorded *muu* as the word for "*buho ó tecolote*."

Beliefs: It is said that the owl comes to the house and announces death in the family. Usually the owl tells the person's name by the sound it makes. When someone is going to lose a loved one, the owl will mention the person's name who will lose someone and say "*haisa siika*?" (did he/she go?). It is just this owl that does this, because the barn owl and others do not talk.

Some people in Marana in the 1950s and 1960s would want to kill the owl, but the elders would say, "Why would you want to kill them? It is their job and duty in life to announce someone's demise." Some people would still shoot them, but others would feel sorry for the poor owls.

Glaucidium brasilianum
Common names: *tecolote bajeño*; Ferruginous Pygmy-Owl

Arizona and Sonora. This little owl is a permanent resident in the remaining tall galleries of riparian trees along the Río Yaqui. It is also found in and around Sierra Bacatete with dense thornscrub and tall columnar cacti.

Megascops kennicottii [*Otus kennicottii*]
Common names: *tecolote occidental*; Western Screech-Owl

Arizona and Sonora. This is one of the most common and widespread owls in the Sonoran Desert region. It often nests in woodpecker holes in columnar cacti and other trees.

Micrathene whitneyi
Common names: WICHIK; *tecolotito enano*; Elf Owl

Arizona and Sonora. This is the smallest owl in the world, weighing about 40 g (about 1.4 oz). They are seen in summertime, from the desert to tropical deciduous forest regions. Elf owls nest in woodpecker holes in columnar cacti or other trees.

TYTONIDAE – BARN OWLS

Tyto alba
Common names: BWAWIS; *lechuza de campanario*; Barn Owl

Arizona and Sonora. Basilio (1890 [1634]:168) cited *bahuis* as the name for a "*lechuza, ave nocturna*." Resident, probable all across the Yoem Bwiara.

SULIFORMES – CORMORANTS AND FRIGATEBIRDS

FREGATIDAE – FRIGATEBIRDS

Fregata magnificens
Common names: *fragata magnifico*, *rabihorcado magnifico*, *tijereta de mar*; Magnificent Frigatebird

Sonora. Frigatebirds are seen by fishermen during warmer months over coastal waters or when migrating near the shore. They nest on the rocky hill at Yasikue (Cerro Yasicuri) near Las Guásimas and on Isla San Pedro Nolasco (Juan Pablo Gallo-Reynoso, personal communication, 2016).

PHALACROCORACIDAE – CORMORANTS

Phalacrocorax auritus
Common names: *cormorán orejudo*
Double-crested Cormorant

Sonora, common resident along the coast and nesting in dead trees along the Río Yaqui above the Yoem Bwiara; migratory in Arizona.

Phalacrocorax brasilianus

Common names: *cormorán oliváceo*
Neotropic Cormorant

Sonora. Coastal although mostly a freshwater bird, common and breeding along the lower Yaqui and Mayo rivers.

TROGONIFORMES – TROGONS

Trogon elegans

Common names: *trogón elegante, coa*; Elegant Trogon

Arizona and Sonora. In Sonora this colorful bird is known as "Coa, after its call" (Russell and Monson 1998:164). Is not known from the Yoem Bwiara although it has been recorded in some nearby places including Nacapule Canyon near San Carlos, mountains and canyons around Guaymas, and Cañón La Pintada in the Sierra

Elegant Trogon, Huachuca Mountains, Arizona. 24 Jul 2013 (AR).

Libre (Alberto Búrquez, and Susan Carnahan, personal communications, 2019). It occurs along the Río Yaqui above El Novillo, and in eastern Sonora and borderlands in southeastern Arizona.

Unidentified birds

CHILIK

Arizona. This bird is heard calling at night, making a call like its name, *chilik*. At Marana a pair made a nest on top of a power pole. Chilic was also among the names of 20 birds listed by Basilio (1890 [1634]:180) in Sonora as "pájaro generalmente."

HOOPOPOL

Sonora. Johnson (1962:265) listed this bird as a "*pájaro cazador.*" *Hopopol* was also among the names of 20 birds listed by Basilio (1890 [1634]:180) in Sonora as "*pájaro generalmente.*"

Song: The *hoopopol* song is played with violin and harp music during *pahko'ola* dances. The dancers imitate the sound of the bird, rapidly repeating *puu puu puu puu*.

PUCHI'ILAA

Sonora. Johnson (1962:281) listed "*puchi'ilaa*" as "*especie de pajarito que canta por la noche en el monte.*"

Song: This bird features in a song played by the tampaleo.

MAMMALS

We include more than 98 species (taxa) of mammals. *Yoawam* is the term that signifies most terrestrial animals, especially mammals. Pérez de Ribas (1645; Reff et al. 1999:84) wrote:

> In the dense thickets there are many peccary, deer, and rabbits. There are also some leopards, although they are not as big or as fierce as the African variety. There are very many powerful tigers, but they are not man-eaters because they rarely leave the hills, where they find their prey. There are also a number of bobcats, coyotes (an animal very similar to a fox).

(Leopards do not occur in the Americas; Pérez de Ribas was referring to spotted phase jaguars (*Panthera onca*, Felidae). Likewise, tigers do not occur in the New World; the translators used tiger for *tigre*, which in Mexico is the jaguar.)

Scientific and Spanish common names for mammals occurring in Mexico generally follow Ceballos and Oliva (2005), and Ceballos (2014), and Caire (2019), as well from Juan Pablo Gallo-Reynoso (personal communication 2012–2020).

Narratives: It is said that wolves and other animals come to you when you are depressed, lonely, or sad. You will forget your depression, loneliness, or sadness because the wolf will startle and scare you, but then the wolf will comfort you. Other animals that are said to do the same are javelina, bear, and mountain lion. Coyote, kit fox, and gray fox will come to you and just comfort and make you happy, and if you have anger these animals will take it away. You have to offer a prayer, a song, or something from your house for the animals because they helped you recover.

The *koludo* is a large, kangaroo-like spiritual animal (Molina et al. 1999). When Felipe was in high school and staying up late typing a term paper, perhaps at 2 a.m., his grandfather, Rosario Vakame'eri-Castillo would tell him "go to sleep or a koludo will come and scare you." Porfirio Yokiwa said that one night he saw a koludo on the frontage road along the highway between Red Rock and Marana. It was after midnight and the koludo jumped in front of the car and hopped away.

Masks: Older, traditional pahko'ola masks are generally based on the human face. Goat-faced masks are also traditional. In the 1960s and 1970s, mask makers started making animal-faced masks: monkeys, dogs, wolves, cats, roosters, bats, mice, badgers, pigs, horses, jaguars, etc. (Griffith 1972; Kolaz 1985, 1997, 2007). Many traditional people who grew up with the ceremonies did not like those animal-faced masks.

DOMESTIC MAMMALS

The term *mansom* derives from the Spanish *manso*, meaning tame or docile, and refers to domestic animals. Old World domestic animals, including cattle, were brought to the earliest Spanish settlements in Sinaloa in the 1530s. Additional domestic animals were brought into northwestern Mexico with the Coronado expedition of 1540. Some were herded along the route to serve as food on the hoof, while others served as beasts of burden: "With the party were some 1,500 horses and mules, large herds of sheep (probably in the thousands), and apparently a number of cattle. Pigs were also purchased for the trip, but it is unknown whether they went as salted-down pork or on the hoof, as had been true of the Diego Guzmán expedition seven years earlier" (Riley 1997:3).

European domestic animals became part of the Yoeme economy in the 1500s and in early Jesuit times. On the return the failed Coronado expedition:

> Some Spaniards... remained to settle the Villa and Province of Culiacán...The alférez, Don Pedro de Tovar, established herds of livestock at a location that seemed well suited along the banks of one of the rivers of Sinaloa so that in the future they could serve the settlers of this province. (Reff et al. 1999:105)

During the seventeenth century, "herds of goats, sheep, cattle, and horses were steadily built up, the western Yaqui River areas being the most productive. At Huirivis by the 1680s there were flocks and herds of 40,000 animals" (Spicer 1980:30). By this time the Yoemem were producing surplus livestock and wheat, which the Jesuits used to supply other missions, especially in the Baja California Peninsula and later in Pimería Alta.

***Bos taurus**

Common names: TOORO, *toro*, bull. VESEO, *becerro*, calf. VOES, *buey*, ox. WAAKAS, *vaca*, cow. WAKASRA, *ganado*, cattle.

Early cattle acquired by the Yoemem from Spanish colonial outposts in Sinaloa did not seem well-adapted to the hot climate. Pérez de Ribas (1645; Reff et al. 1999:84) wrote:

> It is usually very cold during the months of December and January. The rest of the time is usually excessively hot, so much so that even the animals suffer greatly. Many times livestock become so tired while walking that the heat melts the fat in their bodies and they drop dead; other times the heat makes them so stiff that they cannot be used for a long time, and then only if they are bled immediately.

Cattle, however, soon became highly successful: "the introduction of oxen along with ploughs led to increased productivity of the rich bottomlands...By 1700 cattle and wheat...were regularly being widely exported" (Spicer 1980:30).

In modern times, breeds adapted to the hot, arid conditions, such as *zebu*, or Brahma cattle, were successfully introduced. Cattle have been tribally owned and managed by the village governors.

***Camelus dromedarius**

Common names: KAMEEYO; *camello*; camel

The Old World dromedary camel, *kameeyo*, appears in Catholic-based rituals in conjunction with Three Kings' Day or Epiphany, because the "three kings" rode camels to Bethlehem.

Canis familiaris

Common names: CHUU'U, CHUU HAMAT (female dog, bitch); *perra*, *perro*; dog

Basilio (1890 [1634]:182) indicated "*chuo, chuuhamut*" as the words for "*perro ó perra*."

Arizona and Sonora. Most households, especially in Sonora, would have several dogs.

Domestic dogs have been part of the scene in North America since early prehistoric times. The chronicler of Guzmán's first battle in 1533 told of the Yoeme leader wearing "a black *sambenito* like a scapulary, which was sown [*sembrado*] with meticulously crafted conchs of pearls, [representing] many little dogs" (Folsom 2014:25).

Beliefs: At Kompuertam we saw a puppy with a red ribbon around its neck because it was St. Lazarus Day (*Día de San Lázaro*). Dogs are well treated on Friday of St. Lazarus Day. They are fed at the house cross (the mesquite cross in the patio), sometimes from a little table or bench set by the cross. Some people have done the same in Arizona.

Masks: Griffith and Molina (1980:26) wrote that in the 1970s "recently made ones [dog/coyote] with canine snouts and bared fangs" began to be worn at ceremonies. Dog masks soon became popular in Sonora and Arizona (Kolaz 1985, 1997). "Masks that depict dogs, [are] sometimes referred to by collectors and dealers as coyote faces...If a buyer asks if the mask is of a coyote, a carver will generally say yes. But if simply asked what the mask represents, a carver will answer, 'a dog'" (Kolaz 1997:16).

***Capra hircus**

Common names: CHIIVA, CHIVA'ATU (ram, billy goat); *chivo*; goat

Domestic goats, of Old World origin, have been raised in the Yoem Bwiara since early Spanish colonial times.

Masks: Some pahko'ola masks are carved to resemble goat faces; furthermore, it has been said that all pahko'ola masks resemble goats. Indeed,

Goats, Tórim. 1 Jun 2019 (PB).

the pahko'ola himself has been likened to a goat, and details of his costume are mentioned in this connection" (Griffith and Molina 1980:29). Spicer (1954) documented a goat-faced pahko'ola mask at Pótam in 1942.

***Equus asinus**

Common names: VUURU; *burro*; donkey

Burros were formerly seen in Arizona and nowadays mostly in Sonora. Since their introduction by the Spanish, the animals have been used extensively in Sonora to haul hay, firewood, and perform other tasks.

Basilio (1890 [1634]:145) cited *buru* as the term for *borrico* (the Spanish word for donkey).

In Marana there was a village called *Kampo Vuuru.*

In the 1930s there was talk about a ghost in a village, and the people would not go out at night because they were terrified of the ghost. Rosario Vakame'eri-Castillo came from *Kampo Wiilo* to investigate the ghost and stayed up all night to witness it. And sure enough, after midnight he heard rustling and a bit of banging, so he went outside and found out that it was a burro wagging his tail and hitting the side of corrugated tin walls, and told the people to get up and see their ghost.

***Equus asinus × E. caballus**

Common names: MUURA; *mula*; mule

Mules, the sterile hybrids of a male donkey and a female horse, have long been used as beasts of burden in Sonora.

Basilio (1890 [1634]:219) recorded *mura* as the term for *mula.* During Spanish colonial times, the Camino Real, "which traversed the whole west coast from Nayarit to the Upper Pimería ran through Ostimuri, skirting the Yaqui country. Mule trains of 100 or more animals with their skilled crews of mule drivers made the trip from Mexico City in six months" (Spicer 1980:120).

***Equus caballus**

Common names: KAVA'I; *caballo*; horse

The Yoemem encountered horses with the Spanish contacts of the 1530s. By the 1600s they were buying or trading for horses with Spaniards at nearby mining camps. When Captain Hurdaide's third expedition to the Río Yaqui ended in defeat, some horses were captured with the invaders' baggage train. In ensuing peace negotiations in 1610 in Culiacán, Sinaloa, Captain Hurdaide asked the Yoeme representatives to return their horses. Pérez de Ribas (1645; Reff et al. 1999:341) reported that the Yaquis "excused themselves for failing to return the remaining horses [saying] that they were so high-spirited from [pasturing on] the grass along the river that they could not catch them. However, they [promised to] hand them over when they had ropes or leather straps with which to tie them" The Yoemem probably would not have been able to catch free-ranging horses. Captain Hurdaide's demand that the Yoemem capture high spirited horses, at least some of which undoubtedly were stallions (Spanish military men preferred stallions), and then to bring them 400 km (250 mi) to Culiacán was naive or calculated to be impossible. During the peace negotiations, however, the captain "gave some of them horses" (Pérez de Ribas 1645; Reff et al. 1999:340).

Spicer (1980:30) tells that even before 1645, Yoeme men were working in Spanish mines "in order to buy horses." Pérez de Ribas (1645; Reff et al. 1999:375) reported, "With the fruits of their labor many Yaquis have purchased horses, using them for travel and to transport loads."

In his account of Pótam in the 1930s and early 1940s, Spicer (1954:108) tells that young Yaqui men "Sometimes ride horses, particularly on fiesta nights, rarely dismounting, swaying in their saddles and passing their bottles from rider to rider."

Horse races have been popular in the Yoem Bwiara, especially during celebrations. Rosalio Moisés tells of horse races, such as at "Vicam Station" in 1933 (Moisés et al. 1971:138).

During the procession at Lent, Pontius Pilot rides a horse draped with flowers and colorful ribbons. Spicer (1980:72) describes "The Horseman Society...called *kabayum*" [*kavayeom* or *kavayom*]. The term is the Yaqui form of the Spanish word *caballeros*, meaning 'cavalry' or 'horsemen.' They are an integral part of the Pascua or Easter ceremonies." Felipe tells that they are called *kavayeom* in Arizona.

Until the 1940s people in Arizona still used horse-drawn wagons but sold their horses and got cars. Horse-drawn wagons and carts continued to be used in Sonora into the late 1990s, and Yoeme cowboys ride horses into the mountains to check on the tribal cattle.

Food: Although horse meat was not normally eaten, it was resorted to during hard times such as in the 1930s. In early summer 1938, Rosalio Moisés was selling his watermelons and reported that "Yaquis who could not pay money for the melons gave me sugar, coffee, or dried meat. When they gave me horsemeat, I used it to feed the two big dogs that helped me guard my field" (Moisés et al. 1971:166).

*Felis catus

Common names: MIISI; *gato*; cat

The domestic cat is from the Old World. Basilio (1890 [1634]:162) recorded *misi* as the word for *gato*.

Beliefs: Since the introduction of Christianity, many people say that the cat is from the devil.

Estefana Garcia said people should not keep cats inside the house because they will cause respiratory problems, and when you get older, it will cause you to make sounds like a cat. Traditionally people did not keep cats in the house. Anselma Tonopuame'a-Castillo said if you have a cat in the house, put it out at night, because when you are sleeping your body travels, your *seatakaa* (flower body), will go out at night when you are sleeping and when the seatakaa returns a cat can see it and will try to capture it. If the cat catches it, then you will die. Some people say that cats and dogs have an extra sense and can see things that humans cannot see.

*Ovis aries

Common names: BWALA, BWALA ASO'OLA (sheep baby, lamb; *cordera, cordero, borrego, oveja*; sheep

Basilio (1890 [1634]:178) gave the name *sopoc* as the term for *oveja*.

Bwala voa yabwaika im weama (sheep wool fluffy here-is walking), means "there are too many rumors (too much gossip) in the village."

Weaving: The Spanish brought Old World domestic sheep to northwestern Mexico in the 16th and early 17th centuries. "It is recorded that even before 1645 some Yaquis had begun individually to raise sheep to produce wool for sale to Spaniards, presumably in the mining towns not far to the east in order to secure cash for dress goods, and men were doing the same in order to buy horses" (Spicer 1980:30). Yoeme women began weaving with wool in the early seventeenth century.

*Sus scrofa domesticus

Common names: KOOWI; *cochi, puerco, puerca* (sow); pig, sow

Yoemem probably acquired pigs from the Spanish even before the Jesuits arrived in the early seventeenth century. Basilio (1890 [1634]:185) cited *cohui* as the term for "*Puerco ó puerca*."

Ceremony: The deer-hoof belt rattle, *rihhu'utiam*, worn by the deer dancer and *chaapayekam* are nowadays made with pig hooves. Pig hooves tend to be yellowish while deer hooves are blackish.

Food: Pigs are raised in Sonora and many Yoeme families in southern Arizona kept pigs until the 1950s. Lard was a staple part of the diet in Sonora and in Arizona until the latter part of the twentieth century.

Masks: Pig-face pahko'ola masks appeared in the 1970s, about the same time as the dog face style. The primary carvers were Antonio Bacasewa from Vícam and Cheto Alvarez from Pótam. In the early 1980s, Frank Martinez from Old Pascua in Tucson produced many pig-face masks that he called either pigs or javelinas (Kolaz 1985).

WILD MAMMALS

Hunting: Pérez de Ribas (1645; Reff et al. 1999:93) reported:

> They hunted a great deal, in part because the montes and woodlands...abound in deer, peccary, hares, rabbits, and other odd creatures...Indeed, we could say that these peoples' dwelling places were the same as those of the deer and wild animals and that they all lived together. On their hunts they sometimes killed tigers [*tigres*], lions [*leones*], wolves [*lobos*], and foxes, although the latter they sought not so much for

> their meat as for their pelts.
>
> Some hunts are communal and bring together one or many pueblos and rancherías. On occasion the Indian individual hunts for his own diversion and benefit...The manner in which they conduct their communal hunts is to encircle a thicket. If it is the dry season, they set fires all around it and then circle it with the bows and arrows in hand. The fire forces all the fowl and wild animals, even the serpents and snakes, to flee the monte, and nothing escapes their arrows. If an animal escapes with a fatal wound, they go find it the following day at the spot where they know for certain that it has fallen dead. Because they normally use poisoned arrows (even for hunting), it takes no longer than twenty-four hours for an animal to die. It is remarkable that the meat is not poisoned when the animal is killed with these venomous arrows. They find the place where the animal has dropped by looking up to the sky, where the *zopilotes*, or buzzards (a type of eagle that feeds on carrion, of which there are many in this land) are flying in circles.

As Reff et al. (1999) mentioned in a footnote, *tigres* are jaguars (*Panthera onca*), not tigers, and *leones* are mountain lions (*Puma concolor*, Carnivora: Felidae). Wolves (*Canis lupus baileyi*, Carnivora: Canidae) at that time were probably widespread in the region. *Zopilotes* are vultures, either turkey vulture (*Cathartes aura*) or black vulture (*Coragyps atratus*, Acipitriformes: Cathartidae). Poisoned arrows were made with *hoyo kuta* (*Pleradenophora bilocularis*, Euphorbiaceae).

ARTIODACTYLA – DEER, PIGS, PRONGHORN, AND SHEEP

ANTILOCAPRIDAE –PRONGHORN

Antilocapra americana sonoriensis

Common names: *berrendo*; Sonoran pronghorn

Southwestern Arizona and northwestern Sonora is the current range for the Sonoran pronghorn. Historically it extended southward in Sonora to the region north of Guaymas, probably north of the Sierra El Aguaje (Medellín et al. 2005). Pérez de Ribas (1645; Reff et al. 1999) did not mention *berrendo* and we have no Yoeme knowledge of it. Some modern authors incorrectly report that pronghorn ("antelope") were hunted by Yoemem.

BOVIDAE – COWS, GOATS, AND SHEEP

Ovis canadensis nelsoni

Common names: OVE'ESO (ram), TETE'ESO (ewe); *borrego cimarron*; desert bighorn sheep

Arizona and northern Sonora. Bighorn sheep in Sonora are now restricted to western Sonora from the Seri lands northward (Wilder et al. 2014). It is thought that the animals once inhabited Sierra El Aguaje near Guaymas, but there are no records, only oral tradition in Guaymas (Juan Pablo Gallo-Reynoso, personal communication, 2017). There are, however, strong suggestions that bighorn formerly occurred in the Yoem Bwiara. Edward Spicer (1980:64, 68) wrote:

> There were also wild sheep who lived in the mountains, the Bacatetes and others farther away, always in the highest places, and they were the proprietors of power that emanated from the yo ania.
>
> There was, for example, a conception of a mountain sheep who gave dance power to a human Pascola dancer, but this was no particular mountain sheep whom one could find among the rocky cliffs of the Bacatete or other mountains. It existed in a visionary world only; perhaps it was the essence of all mountain sheep.

Basilio (1890 [1634]:145) provided *hobeso* as the word for *borrego*.

CERVIDAE – DEER

Basilio (1890 [1634]:195) indicated names for four kinds of "deer" (*venado*): *maso*, *bura*, *supuchi*, and *ilimaso* ("*venado pequeño*," possibly a young deer or else a kind of small deer).

Another deer-like animal known as *soute'ela* has been described as small, perhaps 2 feet tall, with straight, unbranched antlers. The description seems to resemble that of the diminutive, tropical brocket deer or *temazate*, in the genus *Mazama*, of which about 11 species range from southeastern Mexico, including Yucatán, to South America. Felipe tells us "that one of the elders from Pótam who later moved to Tucson told me that the soute'ela is a small deer with two antlers that don't have branches like the other deer. Another man from Vikam said that it could be a pronghorn (*Antilocapra*). I have seen pictures of tiny deer in the jungles of the Amazon and those images remind me of how the elders from Pótam described it and not the pronghorn."

Although a man from Pótam said he has seen one. Soute'ela is regarded as a magical deer from Yo Ania. Francisco Onamea told of his encounter with a soute'ela when hunting with a friend. They were walking in the desert toward Vakatetteve (Sierra Bacatete) and shot a small deer, but to their amazement, the bullets failed to kill it: "Francisco realized later that the soute'ela was from *yo ania*, and that it wanted to offer them some powers. Unknowingly, they had refused" (Evers and Molina 1987:45–46).

Odocoileus

Two species, the white-tailed deer and the mule deer, occur in the Sonoran Desert region (Heffelfinger 2006). The most noticeable differences between white-tailed and mule deer are the size of their ears, the color of their tails, and the configuration of their antlers. The mule deer has larger ears and is larger than the white-tailed deer. The mule deer's tail is black-tipped, whereas the white-tailed deer's is not. Mule deer antlers fork as they grow; white-tailed deer antlers branch from a single main stem.

Hunting: In Arizona "in the early days" (first part of the twentieth century) deer were seldom hunted because the people feared the ranchers and *rinchim* (rangers). There were, however, some exceptions. For example, in the early twentieth century, Geraldo ("Lalo") Alvarez and his "family lived in the desert at Kingsley Ranch near Arivaca" and they hunted deer. "When Lalo's family killed a deer, they shared it with others in the community" (Molina et al. 2003:4).

Odocoileus hemionus eremicus

Common names: SEVIS MAASO; *bura, venado bura*; desert mule deer

Arizona and Sonora. The desert mule deer occurs near the northern limits of the Yoem Bwiara and ranges northward to Arizona and southeastern California. The southern limit is the Sierra El Aguaje and it has been seen in the area northwest of the Sierra del Bacatete, around Palo Verde, La Misa, San Marcial, and Tecoripa (Juan Pablo Gallo-Reynoso, personal communication, 2018). The *Maso kova* has sometimes been made from a mule deer when a white-tailed deer head was not available.

Hunting: Mule deer were hunted in Sonora north of the Yoem Bwiara, such as near Hermosillo. It is said that the males are protective of their harem and hunters walking around in the wilderness could not get close to them.

White-tailed Deer, Chiricahua Mountains, Arizona. 6 Jan 2008 (GA).

Odocoileus virginianus

Common names: MAASO (adult), MAISO (an old deer), MALICHI, MALIT (fawn); *venado cola blanca*; white-tailed deer

Arizona and Sonora. Two subspecies of white-tailed deer occur in Arizona and Sonora: subsp. *couesi*, Coues white-tailed deer, occurs north of Guaymas into Arizona; subsp. *sinaloae*, Sinaloa white-tailed deer, occurs in western Mexico, and its northernmost reach occurs in the Sierra El Aguaje northwest of Guaymas. "Those two subspecies are currently recognized but no one could produce a hard set of characteristics to tell them apart" (James Heffelfinger, personal communication, 9 Aug 2019). White-tailed deer in the Yoem Bwiara are smaller overall, including their antlers, than those in Arizona.

The white-tailed deer is called *maaso*. The word maaso is a noun, for example, *saila maaso* (little brother deer). The word will drop one of the letters "a" when used as an adjective, for example: *maso aawam* (deer antlers), *maso bwikam* (deer songs), *maso ye'e* (deer dancing), *maso yi'ireo* (deer dancer).

A number of plant names indicate an association with the *maaso*, for example: *maso pipi* (*Funastrum heterophyllum*, Apocynaceae), *maso kuta* (*Bebbia juncea*, *Porophyllum pausodynum*, and *P. gracile*, Asteraceae), and *mahkoapa'i* (*Euphorbia*, Euphorbiaceae). Painter (1986) and Evers and Molina (1987) and provide a wealth of information for the maaso.

Beliefs: A medicinal plant, maso kuta (*Porophyllum gracile* and/or *P. pausodynum*, Asteraceae) is used "to alleviate deer sickness caused by the witchcraft of a deer" (Painter 1986:55–56).

Rosalio Moisés related that "Rabbits and deer can be dangerous to hunt; they are very powerful...A person can go blind hunting...There are lots of bad rabbits and bad deer...which will hurt you. Deer with red and black hooves are *brujos*" (Moisés et al. 1971:39).

Deer tails were hung up in houses for good luck or success in hunting (Shutler 1967:5; Painter 1986: 280).

Ceremony: *Maso kova* is the deer-head headdress worn by the deer dancer. Red ribbons tied to antlers of the *maso kova* represent flowers, *sewam*.

The maso kova is prepared as follows: After a deer hunt, the whole deer is brought into the house yard. The hide and meat are removed, and the head is saved.

The hide is removed from the head. The top of the skull is cut off with the antlers. The deer head skin and the *aawam* (top of the skull with the antlers) are suspended out of reach of dogs. A piece of wood, such as *avaso* (cottonwood root, *Populus*, Salicaceae), *hooso* (*Albizia sinaloensis*, Fabaceae), or soft driftwood, is carved in the shape of a deer head. The neck is put together separately and when finished is attached (often glued) to the carved wooden head. The neck is formed from strips of flexible wood, such as willow (*Salix gooddingii*, Salicaceae), using two strips, one at the top and the other at the bottom of the neck. The "wall" of the neck is formed with vertical strips of the same wood and tied with string to form a cylinder. The hide or skin from the head is fitted over the whole wooden framework. Traditionally, the eyes were formed with lac from creosotebush (*Larrea tridentata*, Zygophyllaceae), and sometimes from the large sea-bean seed, *maso puusim* (deer eyes, *Mucuna pruriens*, Fabaceae).

Clothing: Pérez de Ribas (1645; Reff et al. 1999:92) reported, "Those [women] who did not have...cotton mantas wore short skirts made of buckskin. This is a material that they know how to work well, and they drew designs on some of them with red ocher, especially the young women." He also reported, "One or two principals...wore a manta made from the hides of deer" (Pérez de Ribas1645; Reff et al. 1999:392).

Deer songs: The maaso is the most spiritually important animal of the Yoemem. In the deer songs, performed during a pahko, the deer has many names. *Maso bwikam* (deer songs) represent a rich poetic tradition in Yoeme culture. There are hundreds of deer songs (Evers and Molina 1987:7; see Way of Life, the pahko, deer songs, and the Flower World).

In deer songs performed during a pahko, the maaso is called *sewa malichi* (flower fawn) during the first part of the ceremonies, from about 4 to 8 p.m. (Hours can vary according to circumstances. In Arizona the ceremony starts later due to work hours.) From about 8 p.m. to 2 a.m., the maaso is called *sewa yoleme* (flower person), *yo yoleme* (enchanted person), *sewat chepteme* (one who walks on flowers; every time the deer walks it steps on a spiritual flower, and is trying to teach humans to do the same), or *sewata ye'eme* (one who dances the flower). Before dawn, from about 2 to 4 a.m., the maaso is called *maiso* (elder deer). When the maaso (deer dancer) leaves the ramada in the morning procession some singers will refer to him as sewa malichi or sewa yoleme.

We have chosen two deer songs from the Deer Song book by Evers and Molina (1987:106 and 176–77; a few minor updates have been made in the orthography).

Felipe explained: "When I look at the sky in the morning before dawn after a pahko, I think of this song. The growing flower is the light of the sun pushing back the darkness of the night until it disappears to the west. Don Lupe [Guadalupe Molina] taught me this song."

SEWA YO'OTUME	GROWING FLOWER
Sewa yo'otume sewa yo'otume	Growing flower, growing flower
sewa yo machi hekamake sika	flower, with the enchanted dawn wind, went.
Machi hekamake hekamake chasime	With the dawn wind's air, you are flying,
yo yo machi hekamake sika	with the enchanted, enchanted dawn wind you went.
Sewa yo'otume sewa yo'otume	Growing flower, growing flower
sewa yo machi hekamake sika	flower, with the enchanted dawn wind, went.
Machi hekamake hekamake chasime	With the dawn wind's air, you are flying,
yo yo machi hekamake sika	with the enchanted, enchanted dawn wind you went.
Sewa yo'otume sewa yo'otume	Growing flower, growing flower
sewa yo machi hekamake sika	flower, with the enchanted dawn wind, went.
Machi hekamake hekamake chasime	With the dawn wind's air, you are flying,
yo yo machi hekamake sika	with the enchanted, enchanted dawn wind you went.
Ayamansu seyewailo	Over there, alongside the flower-covered,
vetana yo aniwata vevepa	on top of the enchanted world,
taa'ata aman weche vetana	far, on the top, you are flying,
uhyoli kala lipapati ansime	on the side, where the sun falls,
Empo yo machi hekamake sikaaa	beautifully, endlessly, sparkling, you go.
	With the enchanted dawn wind, you went.

Felipe explained: "It is going to rain." This song will be going toward the time when it will rain. The Black Cloud says that it is moving: "I am moving under on the distant earth," it says. "When the rain drops, here and there, down," it says, "I am moving toward there." All the time we sing, the pahko'olam stick out their tongues, like they are the lightning. And they hit the tampaleo's drum on the post of the *ramá* [ramada] or the ground for the thunder.

CHUKULI NAAMU

Ne ka yo toloko namutakaine
kiane chukuli namuta hekau vetukun
vichane weesime
Ne ka yo toloko namutakaine
kiane chukuli namuta hekau vetukun
vichane weesime

Ne ka yo toloko namutakaine
kiane chukuli namuta hekau vetukun
vichane weesime
Ne ka yo toloko namutakaine
kiane chukuli namuta hekau vetukun
vichane weesime

Ne ka yo toloko namutakaine
kiane chukuli namuta hekau vetukun
vichane weesime
Ne ka yo toloko namutakaine
kiane chukuli namuta hekau vetukun
vichane weesime

Ne ka yo toloko namutakaine
kiane chukuli namuta hekau vetukun
vichane weesime
Ne ka yo toloko namutakaine
kiane chukuli namuta hekau vetukun
vichane weesime

Ayaman ne seyewailo
mekka bwiaponesu
toloko bwiaponesu
polopoloti komisu a yuyumao
Kiane chukuli namuta hekau vetukun
Vichane weesime

THE BLACK CLOUD

I am not the enchanted light blue cloud,
I am just the black cloud blowing,
under, I am moving,
I am not the enchanted light blue cloud,
I am just the black cloud blowing,
under, I am moving.

I am not the enchanted light blue cloud,
I am just the black cloud blowing,
under, I am moving,
I am not the enchanted light blue cloud,
I am just the black cloud blowing,
under, I am moving.

I am not the enchanted light blue cloud,
I am just the black cloud blowing,
under, I am moving,
I am not the enchanted light blue cloud,
I am just the black cloud blowing,
under, I am moving.

I am not the enchanted light blue cloud,
I am just the black cloud blowing,
under, I am moving,
I am not the enchanted light blue cloud,
I am just the black cloud blowing,
under, I am moving.

Over there, I, on the flower-covered
distant earth, I am,
on the light blue earth, I am,
as, here and there, down it reaches.
I am just the black cloud blowing,
under, I am moving.

Food: The white-tailed deer is now rather scarce in Sonora, but it used to be a significant food resource. It was prepared in a variety of ways, often as a stew. The meat was eaten freshly cooked and also dried for future use, from which it could be prepared as a stew (Moisés et al. 1971:166).

Hunting: Painter reported, "the hunter must have a legitimate need for hunting, and must never hunt for sport" (1986:275, also see page 278). Hunting methods were described as follows:

> In older times, it is said that Yaqui stalked the deer by wearing masks and a deer skin disguise. The hunters held a stick in one hand and bow in the other to simulate the forelegs of the deer as they imitated his movements. The hunters were able to come quite close to the deer and to overhear their speech. In this way the hunters learned the secrets of the deer and their language; and that deer language came to be translated into the deer songs of the hunters. During the hunt the hunter should not let his mind wander but should concentrate. He would be successful, though, only if he didn't 'think,' thereby leaving his mind open to communication with the deer. (Evers and Molina 1987:47)

Deer were also captured with a snare, called *hite'i*, which is also the term for a fishing net. The snare was set up between trees and placed on a deer path or trail, and was made from agave fiber or horse-tail rope.

Medicine: Painter (1986:56) reported:

The wild deer has a special function in curing. The deer hide is used to bind broken bones 'because it is the one that has the curing power'....Deer bones are burned into powder or chalk and used as a plaster on the affected part of the body. They also said that a curer can cure a man bewitched by the fawn if he hunts only for sport, but he will never be able to kill any more deer.

Shutler (1967:53–54) was told, "deer hides are used in Pascua to bind broken or sprained bones. Deer hide is extraordinarily effective because of the *seataka* [compassionate powers] of the deer" and "that the deer's tail brought success in hunting but not in curing."

Narrative: There are numerous narratives involving the maaso, for example, "The First Deer Hunter" (Giddings 1959).

TAYASSUIDAE – PECCARIES

Tayassu tajacu

Common names: HUA KOOWI (wild/wilderness pig), POCHO'OKU KOOWI (in the desert pig); *cochi jabalí, pecarí, pecarí de collar*; collared peccary, javelina

Arizona and Sonora. Pérez de Ribas (1645; Reff et al. 1999: 84) reported, "In the dense thickets there are many peccary [*jabalíes*]."

Hunting: *Hua koowi* was hunted as a food animal in Sonora. Some Yoemem in Arizona, including at Marana, also hunted javelinas.

Narrative: Javelinas appear in a number of stories, such as "Topol the Clever" (Giddings 1959:57). In this story a *kobanao* (leader or governor) offers a beautiful maiden "to the man who could bring in, alive, without having struck it or touched it, a javelina to the fiesta....Topol was clever. He cut himself a stick of the wood of a plant which has blue flowers." The javelina bit the stick held out by Topol and would not let go, and in this manner Topol dragged him all the way to the where the fiesta was taking place. The plant with blue flowers probably would be *huyawo* (see *Guaiacum coulteri*, Zygophyllaceae). Giddings interpreted topol as the jaguar (see *Panthera onca*, Felidae).

In the story of "The Big Bird," there is a mention of a boy carrying "his quiver of javelina hide" (Giddings 1959:36).

CARNIVORA – BEARS, CATS, COYOTES, FOXES, SEA LIONS, AND WOLVES

CANIDAE – COYOTES, FOXES, AND WOLVES

Canis latrans

Common names: WO'I; *coyote*; coyote

Arizona and Sonora. Coyotes are ubiquitous throughout the region. The *nama wo'i* (leading coyote) is reported to be larger than the regular coyote and to be only in Sonora. (*Nama* is the term for leading or herding animals, e.g., *chivam nama*, leading or herding goats.)

Coyotes figure in innumerable jokes, narratives, songs, and stories.

Coyote Society: The *Wiko'i Ya'ura* (Bow Leaders' Society, Coyote Society, or Coyote Warrior Society) has served Yoeme communities for centuries as a military society. Ya'ura comes from the word *ya'ut* (leader). The Bow Leaders' Society has remained active in many of the towns along the Río Yaqui (Evers and Molina 1990), and the dances were revived among the Arizona communities in the 1980s. Many Yoemem use *Wo'im* as the name for the coyote society.

Coyote songs: Taub (1950) and Evers and Molina (1990) provided a repertoire of *Wo'i Bwikam* (Coyote Songs). We reproduce coyote songs for the dragonfly (see Odonata), crow (see *Corvus sinaloae*, Passeriformes: Corvidae), parrot (see *Amazona albifrons*, Psittaciformes), rattlesnake (see *Crotalus*, Snakes), and skunk (see Carnivora: Mephitidae).

Masks: Some masks referred as coyote faces are generally regarded as dog faces by the mask maker (Kolaz 1997; see *Canis familiaris*).

Medicine: Painter (1986:56) reported, "Coyote fat is healing to rheumatism and broken bones."

Narrative: Coyote is smart and tricky, but more often is made out to be a fool. Giddings (1959) related stories in which various animals such as dogs, lambs, a rabbit, a turtle, and others outwit a foolish coyote, often ending up killing or eating the coyote. In her "Kaiman" story, however, the coyote ends up outwitting and eating the crocodile. A rather gruesome story is told about a wax monkey tricking a watermelon-thieving coyote. In the story of "Topol the Clever," Giddings (1959:57) related a narrative that includes the coyote dance held after Topol died (see javelina, *Tayassu tajacu*, Tayassuidae, and jaguar, *Panthera onca*, Felidae).

Canis lupus baileyi

Common names: LOOVO; *lobo, lobo gris mexicana*; Mexican gray wolf

The Mexican gray wolf occurred across much of Mexico and the southwestern United States. Wolves were extirpated in the wild from Arizona and Sonora during the twentieth century, although some have been re-introduced in Arizona and New Mexico.

Coyote, Tucson. 6 Mar 2014 (AR).

Narrative: Wolves feature prominently in oral tradition, including this story from Matilda Estrella:

> She shared a family story about a wolf man who came to *Bwe'u Hu'upa*. She said that her mother, María, and her grandmother, Manuela, were up early one morning and were starting to cook in the kitchen. They saw a long, hairy arm groping through the smoke hole in the roof. They screamed and ran out of the house for help. Some *kau homem* (mountain dwellers) were in *Bwe'u Hu'upa* at that time. They came to the rescue and talked loudly about how to destroy the wolf man. The *kau homem* said that they needed a silver bullet with a cross etched on it that would destroy the wolf man. Ever since that loud discussion, the wolf man never came back. Still the people of *Bwe'u Hu'upa* were afraid to stay out late at night. (Molina et al. 2003:63)

A. Ocelot, Rancho El Aribabi, Sierra Azul. 24 Sep 2016 (JR). B. Bobcat, Tucson Mountains. 26 Feb 2014 (AR).

Urocyon cinereoargenteus
Common names: KAWIS; *zorra gris*; gray fox
Arizona and Sonora.

Ceremonial: The headdress for the *Wiko'i Yau'ura* (Bow Leaders' Society) dance includes a fox pelt.

Hunting: Pérez de Ribas (1645; Reff et al. 1999:93) reported that although foxes were sometimes hunted, "the latter they sought not so much for their meat as for their pelts."

Vulpes macrotis
Common names: MATA'E; *zorra del Desierto*; kit fox
Arizona. In Sonora, the kit fox occurs north of Guaymas and is not known from farther south. At Marana, Juan Garcia said kit foxes like to be in sandy areas. Indeed, these desert animals are most common in open, sandy places.

FELIDAE – CATS

Leopardus pardalis [*Felis paradalis*]
Common names: TOPOL; *gato galavi*, *ocelote*, *tigrillo*; ocelot
Sonora. These shy, medium-sized cats occur in Sonora, primarily in the eastern part of the state. Ocelots are spotted like the spotted-phase jaguar (*Panthera onca*), although ocelots are very much smaller.

Coyote Society: Paraphernalia of soldiers in the *Wiko'i Yau'ura* (Bow Leaders' or Coyote Society) include "rifles, bows, ocelot skin quivers of arrows" (Spicer 1954:67; 1980:182). The arrow quiver is called *huiwa to'oria* (arrow to-lay, meaning to lay arrows, to lay down inside the quiver).

Lynx rufus
Common names: BWAHI LOVON, BWAI LOVON, HUYA MIISI (wilderness cat); *gato montés*; bobcat
Arizona and Sonora. Bobcats are widespread in Arizona and Sonora. Pérez de Ribas (1645; Reff et al 1999:84) reported, "There are also a number of bobcats [*gatos monteses*]."

Basilio (1890 [1634]:162) gave *buahi* as the term for *gato montés*. Molina et al. (1999:27) provide *bwahilovom* as the name for bobcat. They define *bwahim* as "diaper, loin cloth, breechclout (originally the Yoeme men's garb)," etc. Nowadays bwahim is also the word for underwear. At first, Felipe and Richard thought "underwear" must be mistaken as the name or part of the name for bobcat, but the original name(s) makes sense when realizing that bobcat pelts were once used in men's clothing. Johnson (1962:257) gives "bwáhilábom, gato montés, var. [variation] buáhi" and Estrada et al. (2004:71) provide "bwai lobam, gato montes."

Beliefs: People in Arizona say that when a bobcat comes to a home and makes its loud noises at night, that they are bringing bad news or a bad omen.

Panthera onca
Common names: YO'OKO (spotted), YOOKO OUSEI (spotted mountain-lion); *jaguar*, *tigre*; jaguar

In historical times jaguars were widespread in Arizona and Sonora. They were once in the Sierra El Aguaje and some were still in the Sierra Bacatete in 2002 (López-Gonzáles and Brown 2001, 2002). There are regional conservation efforts to foster jaguar recovery, especially in northeastern Sonora and southeastern Arizona.

There are two color phases, spotted jaguars and black jaguars; the black or melanistic phase does not occur north of southern Mexico. It should be noted that ocelots (*Leopardus pardalis*) and mountain lion cubs (*Puma concolor*) and are also spotted.

Basilio (1890 [1634]:216) recorded *iooco* as the name for *tigre*, the jaguar, and in his dictionary, Johnson (1962:297) interpreted *yo'oko* as *tigre*.

In some literature the term *tigre* is mistakenly translated as leopard or tiger, which are Old World animals. Although *tigre* is the Spanish term for tiger, in Mexico it applies to the jaguar. For example, although Reff et al. (1999) translated *tigres* as tigers, they noted that jaguar is correct: "There are also some leopards [*leopardos*], although they are not as big or as fierce as the African variety. There are very powerful tigers [*tigres*], but they are not man-eaters because they rarely leave the hills, where they find their prey" (Reff et al. 1999:84).

Ceremony: Pérez de Ribas (1645; Reff et al. 1999:329) reported that high-ranking officials wore jaguar capes: "One or two principals...wore a manta made from the hides of deer, lions [sic], or tigers [sic], or from cotton." Griffith and Molina (1980:35) reported, "In ancient times the *pahkolam* wore *topol* (jaguar) skins, but more recently blankets have replaced these."

Narrative: In the story of "Topol the Clever" (Giddings 1959:57), a coyote dance was held after Topol died:

> At his funeral great fiestas were celebrated. And it was here, for the first time, that the dance of the coyote was performed. It was sung thus:
>
> *yoli yoli yoli tamewuk yoli tamewuk*
> *kayoli tamewukayoli tamewuk*
> *kayoli tamewuk*
> *punkie ponki pok pok pok.*
>
> With this coyote song they buried the chief, Topol. And from then until today, when a soldier dies, or a Yaqui chief, in his funeral services, they sing and dance the coyote.

Hunting: Pérez de Ribas (1645; Reff et al. 1999:93) mentioned, "On their hunts they sometimes killed tigers [*tigres*]."

Jaguar; photo by motion-triggered camera, northern Sonora, 2018; courtesy Northern Jaguar Project.

Puma concolor [*Felis concolor*]
Common names: OUSEI; *león*, *puma*; mountain lion, puma

Arizona and Sonora. Once more common, mountain lions are now scarce in Sonora. The cubs have spotted coats when very young. The mountain lion is known as the number-one hunter of the white-tailed deer.

Clothing: Pérez de Ribas (1645; Reff et al. 1999:329) reported that high-ranking officials wore capes of mountain lion hides: "One or two *principals*...wore a manta made from the hides of...lions [mountain lions]."

Deer songs: *Ousei Bwikam* (mountain lion songs) may be performed during the *Maso Me'ewa*, killing the deer ceremony at a death anniversary pahko. Evers and Molina (1987:142–149) provide a set of four with detailed interpretations:

Ousei Nakapit (deaf mountain lion)
Ousei Hamuchia (female mountain lion)
Ouseli Omtekai (mountain lion is mad)
Ousei O'owia (male mountain lion

Another mountain lion song is given below. In the song language, a mountain lion cub is called *ouseli*; the "*l*" gives an affectionate meaning to the word.

Mountain lion, Patagonia Mountains, Arizona. 1971 (RF).

Ai ouseli,	Ai little mountain lion,
ouseli,	little mountain lion,
omteka.	you are angry.
Sewa huyapo	In the flower wilderness
omteka.	you are angry.
Ouseli,	Little mountain lion,
ouseli,	little mountain lion,
omteka.	you are angry.
Sewa huyapo,	In the flower wilderness,
omteka.	you are angry.
Ai ouseli,	Ai little mountain lion,
ouseli,	little mountain lion,
omteka.	you are angry.
Sewa huyapo	In the flower wilderness
omteka.	you are angry.
Ayamansu	Over there,
sewa huyapo naisukuni.	In the middle of the flower wilderness.
Kaita yumalika,	You didn't catch (seize) anything,
omteka.	you are angry.
Ouseli,	Little mountain lion,
ouseli,	little mountain lion,
omteka.	you are angry.
Sewa huyapo,	In the flower wilderness,
omteka.	you are angry.

MEPHITIDAE – SKUNKS

Common names: HUPA; *zorillo*; skunk

Hupa applies to all skunks. There are four species of skunks in the region.

Song: Evers and Molina (1990) provided a song adapted from Taub (1950) about the skunk. The skunk in the song has a white stripe down the back, which is probably the hooded skunk (*Mephitis macroura*), but also may be the striped skunk (*M. mephitis*). The song is a tongue-twister.

HUPA	SKUNK
hu'upa kutpo	on the mesquite wood is sitting
to'i to'eti hia	to'i to'i sounding
to'i to'eti hia	to'i to'i sounding
to'i to'eti hia	to'i to'i sounding
hia	Sounding
hia	sounding
hia	sounding
hia	sounding
katikun	Remember
yo hu'upapo	on the mesquite wood
katikun	is sitting
to'i to'eti hia	to'i to'i sounding
to'i to'eti hia	to'i to'i sounding
to'i to'eti hia	to'i to'i sounding
hia	Sounding
hia	sounding
hia	sounding
hia	sounding

Conepatus leuconotus

Common names: *zorillo*; American hog-nosed skunk

Arizona and Sonora.

Mephitis macroura

Common names: *zorro listado*; hooded skunk

Arizona and Sonora.

Mephitis mephitis

Common names: *zorrillo rayado*; striped skunk

Arizona and Sonora, probably east of the Yoem Bwiara.

Spilogale gracilis

Common names: YO'OKO HUPA (spotted skunk); *zorrillo machado*; western spotted skunk

Arizona and Sonora, mostly east of the Yoem Bwiara. In Arizona these little skunks like to get into houses and go through garbage.

MUSTELIDAE – BADGERS, OTTERS, WEASELS, AND ALLIES

Lontra longicaudis annectens

Common names: VAACHUU'U (water dog); *nutria de río, nutria neotropical, perro de agua*; Neotropical river otter

Sonora. River otters are still found in the Río Yaqui, as far downriver as the Presa Oviáchic above Ciudad Obregón, and once ranged farther downriver (Gallo-Reynoso 1996; Rangel-Aguilar and Gallo-Reynoso 2003). In 1994, when Bernaldo Valencia saw live river otters at the Arizona-Sonora Desert Museum he called them *nutria de río*. He remembered them from his youth in *Ko'obwabwa'im* (Old Pótam).

Basilio (1890 [1634]:205) recorded *bachuo* as the word for "*nutria, perro de agua*."

Pérez de Ribas (1645; Reff et al. 1999:91) wrote:

> We must also mention the defensive gear they wear on their left wrist, which is struck by the bowstring after an arrow is shot. In order for the wrist not to be bruised, they bind it (and wisely so) with the soft pelt of a martin, this absorbs the blow of the bowstring.

Reff et al. (91) point out the martin is the otter and that Pérez de Ribas' (1645:10) original words were "*un pellejo de marta blando*."

Taxidea taxus

Common names: HUURI; *tejón*; American badger

Arizona and Sonora. Felipe says, "One has to be careful with a badger, they have very sharp claws—don't tease the badgers. Never heard of anyone hunting a badger."

Basilio (1890 [1634]:205) recorded *huri* as the name for *tejón*.

Deer songs and mask: Richard asked Felipe if the badger features in Yoeme art, and Felipe replied: "Oh yes. He is in our deer songs, always traveling underground, even under the water, under the floor of a pond or lagoon (but not a river). Also would be with pahko'ola dancing, when the tampaleo plays a badger song, the pahko'olam have to act it out."

The Flower Badger deer song is described for the sidewinder rattlesnake (see *Crotalus cerastes*). Pahko'olam badger masks are sometimes seen. A pahko'ola dancer may act out a badger song, but the mask would not represent a badger.

Narrative: Giddings (1959:19) included a story called "When Badger Named the Sun."

OTARIIDAE –SEALS

Zalophus californianus

Common names: VA'APO LOOVO (in-the-water wolf); *lobo marino*; sea lion

Sea lions are common in winter and spring in the areas offshore from Las Guásimas. They come to the area to feed on sardines and other fishes. They sometimes gather in floating rafts with their flippers extended out in the air, with as many as twenty *lobos*, before the reproduction season of May to August (Juan Pablo Gallo-Reynoso, personal communication, 2016).

PROCYONIDAE – RACCOONS AND RINGTAILS

Bassariscus astutus

Common names: *cacomixtle*; ringtail

Arizona and Sonora. *Cacomixtles* are in the mountains east of Pótam, part of the Vakatetteve (Bacatete) Mountains. We did not get a Yoeme name for this elusive nocturnal animal. A ringtail was seen in Marana, at Yoem Pueblo in the 1970s, and the people said, "*changotavena*" (looks like a monkey).

Procyon lotor

Common names: CHOPARAO; *batepi*, *mapache*; raccoon

Arizona and Sonora. Raccoons were often regarded as pests in fields (Moisés et al. 1971:146).

Dance and song: Painter (1986:119) described "the Raccoon Dance. This was danced as late as 1928 by a pascola, who, according to custom, painted himself with black and white stripes. This is the only mention of body paint by any informants." During the dance the raccoons run out from time to time while dancing and grab spectators' asses.

The only time Rosario Vakame'eri-Castillo told Felipe about a raccoon dance, it was about one at Old Pascua when someone made a mistake and used oil-based paint to decorate the dancer with black and white. After the ceremony he was walking around a long time with paint until it wore off.

Taub (1950:122–123) provided a raccoon song and notes about the dance: "Little is known of the Raccoon (*choparau*) dance. It seems to be somewhat similar to the *Nahi* dance and Jean

Johnson reported that twisted corn leaves were placed over the dancer's hands."

RACCOON SONG

tuasu ne vau'uvarikau vau'uvari
tuasu ne vau'uvarikau vau'uvari
tuasu ne vau'uvarikau vau'uvari

hunamani vamamani kaapo
tuasu ne vau'uvarikau
vau'uvarikau vau'uvari

Yes, indeed! I'm enjoying the fishing –
I enjoy the mountain water,
Yes, indeed! I've been wanting to fish
in the fresh water –
Yes, indeed! I've always enjoyed fishing –
when the water is fresh

Isn't that he – over there, in the water?
Yes, indeed! – It's me!
Been enjoying the fishing,
Been enjoying scooping them in,
fresh from the water.

During a pahko the violin and harp players may play a raccoon song and the pahko'olam will act out raccoon antics and a hunt:

> Then one will be made into a dog. The dog, the one who is made into a dog, will walk around the others. Then the one who is going to be a raccoon will be sent outside. When they have fallen asleep again, the raccoon will enter and will dig around. Then he will grab the asses of the ones who are sleeping. 'The dog, the sleepy dog! He did not sense it,' they will say. Then they will look for the raccoon.
>
> Then they will skin the raccoon and sell the skin....They will just ask for some wine for it....You see, father, we have finished. It almost ate up all our corn. (Evers and Molina 1987:140–141)

URSIDAE – BEARS

Ursus americanus

Common names: HOOSO; *oso, oso negro americano*; black bear

Arizona and Sonora. Black bears in Sonora occur in the mountains in the northeastern part of the state, but once ranged across most of the higher mountains of the northern half of Mexico, including the eastern margin of Sonora (Medellín et al. 2005). Although documented records do not show bears occurring as far west as the Yoeme lands, there are intriguing narratives involving bears.

Narrative: The story of "The Two Bears" related by Giddings (1959:22) begins as follows: "Once in the land of the Yaquis there were two very amusing men....One day while walking about in the countryside they killed a bear." The men skinned the bear and carefully prepared the hide by filling it "with grass so that it appeared to be a live bear." One of them dressed in the hide and pretended to be a bear. They had various adventures until they were found "chatting and smoking," and then ran away after being threatened by the furious rich man they had deceived.

CETACEA – DOLPHINS, PORPOISES, AND WHALES

Huhteme is the general term for whale. At Las Guásimas, Fernaldo Leyva-Flores used the same name for porpoise (*tonina*). The gray whale, *ballena gris* (*Eschrichtius robustus*) is often seen in winter and early spring entering Bahía Las Guásimas and surrounding areas.

Delphinus capensis

Common names: *cochinito, delfín de hocico largo*; long-beaked common dolphin

Found offshore in the region. Both this species and porpoises are sometimes seen stranded on beaches along the Yoem Bwiara.

Tursiops truncatus

Common names: KOCHIITO; *cochi, delfin, tonina*; common bottlenose porpoise

Porpoises inhabit coastal areas and enter esteros. The fishermen do not go fishing in an estero if *toninas* are there, because they say the toninas eat all the mullet before they can fish for them.

CHIROPTERA – BATS

Common names: SOCHIK; *murciélago*; bat

Arizona and Sonora. *Sochik* is the general term

for any bat. Basilio (1890 [1634]:175) cited *sochic* as the term for "*murciélago, ave nocturna.*" There are several families and numerous species of bats represented in the region.

Sonoran ranchers call all bats *murciélagos*; sometimes they use the term *trompudo* (large mouth), or *orejón* (for a long-eared bat), and *murcielagote* for very big bats (Juan-Pablo Gallo-Reynoso, personal communication, January 2019).

Deer songs and narrative: "Bats are common in Yoeme myths and stories; in the story of how Voovok (Toad) brought the rain back from Yuku (Rain) by borrowing Bat's wings" (Molina et al. 1999:129). Deer songs featuring bats are sung around midnight during a pahko.

Some of the species of bats recorded from the Yoem Bwiara are listed here.

Artibeus hirsutus

Common names: *murciélago frugívoro peludo*; hairy fruit-eating bat

Sonora.

Desmodus rotundus

Common names: *murciélago vampiro común*; vampire bat

Sonora. There are a few documented records of the vampire bat in the Yoem Bwiara and at Guaymas (Loomis and Davis 1965). Sonoran ranchers know the vampire bat, and call it *vampiro* or *chupa sangre* (Juan Pablo Gallo-Reynoso, personal communication, January 2019).

Leptonycteris curasoae

Common names: *murciélago magueyero*; lesser long-nosed bat

Arizona and Sonora.

Myotis vivesi

Common names: *murciélago pescador*; fish-eating bat

Sonora. This curious bat catches little fishes and marine crustaceans in its sharp, sickle-like claws as it skims over the water. It generally roosts in caves and rock crevices along the coast and on islands of the Gulf of California.

Pteronotus davyi

Common names: *murciélago lomo pelón menor*; Davy's naked-backed bat

Sonora.

Tadarida brasiliensis

Common names: *murciélago guanero*; Brazilian free-tailed bat

Arizona and Sonora.

DIDELPHIMORPHIA – OPOSSUMS

Didelphis virginiana

Common names: *tlacuache*; Virginia opossum

Arizona and Sonora. Opossums in Sonora are *Didelphis virginiana californica*; the ones in southern Arizona belong to the subspecies *virginiana* and are likely introduced or escaped from captivity.

LAGOMORPHA – HARES AND RABBITS

Beliefs: "Rabbits and deer can be dangerous to hunt; they are very powerful. If a rabbit shakes his ears from side to side, you should not shoot it, or it will make you sick and maybe kill you....A person can go blind hunting. There are lots of bad rabbits and bad deer, which will hurt you....Rabbits with worms (*gusanos*) in their neck are *brujos*." (Moisés 1971:39).

Lepus

Common names: PAAROS; *liebre*; jackrabbit

There are two species of hares or jackrabbits in the region. They are often seen in the open desert even during the day.

Beliefs: It is said that when you go for a walk in the desert and take a nursing baby along, you should never let him or her touch the ground because if their little footprints are left there, the jackrabbit will come and nurse the baby.

Lepus alleni

Common names: *liebre antílope*; antelope jackrabbit

Arizona and Sonora. This jackrabbit is common the Yoem Bwiara.

Hunting: Antelope jackrabbits, as well as cottontails were hunted in the Yoem Bwiara in communal drives.

Lepus californicus
Common names: *liebre cola Negra*; black-tailed jackrabbit

Arizona and Sonora north of the Guaymas region.

Back-tailed jackrabbit, Dragoon Mountains, Arizona. 7 Jun 2010 (AR).

Sylvilagus audubonii
Common names: TAAVU; *conejo del Desierto*; desert cottontail

Arizona and Sonora. This common little rabbit is smaller than the jackrabbits and often more secretive during the day when they are hiding under bushes.

Basilio (1890 [1634]:151) sited *tabu* as the word for *conejo*.

Desert cottontail, Dragoon Mountains. 11 Aug 2018 (AR).

Hunting: Cottontails were caught in snares made with *palo blanco* (*Mariosousa heterophylla*, Fabaceae) and were, along with jackrabbits, hunted in the Yoem Bwiara in communal drives.

RODENTIA – RODENTS

The concept of *chikul* (*ratón*, mouse) may include three genera of native mice (*Chaetodipus*, *Perognathus*, *Peromyscus*), as well as the common introduced house mouse (*Mus musculus*). Similarly, the term *tori* (*rata*, rat) applies to New World packrats (*Neotoma*) as well as introduced Old World rats (*Rattus*). In addition to *chiculi* and *tori*, Basilio (1890 [1634]:187) listed *baaia*, *bisuetata*, *manol*, *naaca*, and *sopec*, as other kinds of *ratón*.

CRICETIDAE – NEW WORLD MICE, RATS, AND VOLES

Neotoma
Common names: TORI; *rata magueyera*; packrat, woodrat

Arizona and Sonora. Two species are prominent in the Yoem Bwiara and one of them also occurs in Arizona.

Neotoma albigula
Common names: TORI; *rata de campo*, *rata nopalera*, *rata magueyera*; white-throated packrat

Arizona and Sonora. The white-throated packrat extends from Arizona to Sinaloa. The nests are often seen on the ground among bushes and cacti, and in the shelter of large rocks. It apparently does not occur in the floodplain of the Río Yaqui.

Neotoma phenax
Common names: TORI; *rata de campo*, *rata magueyera*; Sonoran woodrat

Sonora. This woodrat occurs on the coastal plains of the major river valleys from southeast of Guaymas to the northern half of Sinaloa below about 150 m (500 ft) elevation (Ceballos and Oliva 2005, Ceballos 2014). The town of Torim is named for these animals. They build massive nests in trees, large shrubs, or large cacti, at least 1.2 to 1.5 m (4 to 5 ft) off the ground. These huge ball-like nests can look like a large raptor nest. This woodrat is unusual in having arboreal nests and they can sometimes be seen jumping around in trees, and from tree to tree. The habit of building arboreal nests is an adaptation to the river floods on the coastal plains.

Nest of Sonoran wood rat (*Neotoma phenax*), northeast of Sierra Bacatete. 5 Nov 2011 (TV).

Food: Torim have served as food animals (Johnson1962:290). Spicer (1980:10) mentioned "a large gray tree-dwelling rodent" as an important game animal. Jane Holden Kelley wrote (Moisés et al. 1971:xii):

> Game provided... an important secondary food source when crops failed. Hunting of less desirable game, such as wood rats, was regarded by most as a final resort to avert starvation. As Rosalio [Moisés] puts it, one could judge how hungry people of Torim were by how close the big wood rats could be found to the village. In periods of food shortages, the wood rat population was cleaned out in increasingly larger concentric circles around each village or ranchería. Posole (soup) was prepared with *torim* meat, or the whole tori was cooked. Rosalio Moisés related that at Torim, in December 1933, he was offered *pozole*...made with beans, hominy corn, and meat. She gave me a bowl. The meat looked like chicken.
>
> 'What kind of meat is this?' I asked.
>
> 'That is a big rat. It is good, like chicken.'
>
> I tried to eat it, but I could not. It smelled like burning feathers. I picked out all the meat and put it back in the cooking bowl, then I ate the beans and corn. My sister-in-law's family was glad I did not like the rat. They ate it all. (Moisés et al. 1971:138–139)

The archeological record and historic accounts demonstrate that *Neotoma* were widely used for food in the American Southwest. *Neotoma* continue to be sold at *mercados municipales* in the Altiplano of Mexico, in San Luis Potosí and Zacatecas. They are offered fresh, skinned and cleaned. Tom Van Devender says that the meat is white, delicate, and delicious (personal communication, 2015).

Medicine: Johnson (1962:290) reported that this animal is "used as a remedy for a cough."

Onychomys torridus

Common names: *ratón*; southern grasshopper mouse, scorpion mouse

Arizona and Sonora.

Oryzomys couesi

Common names: *rata arrocera*; Coues' rice rat

Sonora.

Peromyscus eremicus

Common names: *ratón del campo*; cactus mouse

Arizona and Sonora.

Peromyscus merriami

Common names: *ratón del campo*; Merriam's deermouse, mesquite mouse

Arizona and Sonora.

Reithrodontomys burti

Common names: *ratón*; Sonoran harvest mouse

Sonora.

Sigmodon arizonae

Common names: *rata algodonera*; Arizona cotton rat

Arizona and Sonora.

ERETHIZONTIDAE – PORCUPINES

Erethizon dorsatum

Common names: WICHAKAME (one with spines), *puerco espín;* North American porcupine

Arizona and Sonora.

GEOMYIDAE – POCKET GOPHERS

Thomomys bottae

Common names: TEVOS; *topo*, *tuza*; Botta's pocket gopher

Arizona and Sonora. *Potam* are the mounds formed by gophers from soil brought up to the surface, and the name of one the major towns in the Yoem Bwiara. Basilio (1890 [1634]:195) cited *tebos* as the word for "*tuza, animal.*"

HETEROMYIDAE – KANGAROO RATS AND POCKET MICE

Chaetodipus baileyi

Common names: *ratón de abazones*; Bailey's pocket mouse

Arizona and Sonora.

Chaetodipus goldmani

Common names: *ratón de abazones*; Goldman's pocket mouse

Sonora. This species occurs in southern Sonora and northern Sinaloa.

Chaetodipus penicillatus

Common names: *ratón de abazones*; desert pocket mouse

Arizona and Sonora.

Chaetodipus pernix

Common names: *ratón de abazones*; Sinaloan pocket mouse

Sonora.

Dipodomys

Common names: VACHA'I; *rata canguro*; kangaroo rat

Dipodomys deserti is found in Arizona and in Sonora north of Bahía Kino; *D. merriami* occurs in Arizona and Sonora including the Yoem Bwiara.

Perognathus flavus

Common names: *ratón de abazones*; silky pocket mouse

Arizona and Sonora.

MURIDAE – OLD WORLD MICE AND RATS

***Mus musculus**

Common names: CHIKUL; *ratón*; house mouse

Arizona and Sonora. The ubiquitous house mouse, native to the Old Word, probably arrived in Sonora in early Spanish colonial times.

***Rattus**

Common names: TORI; *rata*

Tori is the name for New World woodrats and packrats as well as for Old World rats.

***Rattus norvegicus**

Common name: Norway rat

Arizona and Sonora. The common rats of Old World origin likely arrived in Sonora in early Spanish colonial times.

***Rattus rattus**

Common names: black rat, roof rat

Arizona and Sonora. An Old World rat now found in all inhabited regions of the world. It has also become invasive on certain islands of the Gulf of California.

SCIURIDAE – SQUIRRELS

Malon is the general term for squirrel. Basilio (1890 [1634]:139) provided the names for three "squirrels": *ile tecu* for "*ardilla, por su especie pequeña*"; *bueru tecu* for "*ardilla, por su especie grande*"; and *bamaoatecu* for "*ardilla que suele bajar el río.*"

The squirrels in the Yoem Bwiara include members of four genera.

Ammospermophilus harrisii

Common names: MALON; *ardilla antilope*; Harris's antelope ground squirrel

Arizona and Sonora north of Guaymas.

Neotamias dorsalis sonoriensis [*Tamias dorsalis sonoriensis*]
Common names: MALON; *chichimoco de Guaymas*; cliff chipmunk

Sonora. This subspecies is endemic to coastal Sonora. The southern part of its range includes the Sierra El Aguaje, Sierra Libre, and Sierra Bacatete. It is about the size of the antelope ground squirrel with which it could be confused.

Otospermophilus variegatus [*Spermophilus variegatus*]
Common names: TEKU; *ardilla, ardillón*; rock squirrel

Arizona and Sonora.

Xerospermophilus tereticaudus [*Spermophilus tereticaudus*]
Common names: *juancito*; round-tailed ground squirrel

Arizona and Sonora.

SORICOMORPHA –SHREWS

Notiosorex crawfordi
Common names: *musaraña*; desert shrew

Arizona and Sonora. These tiny nocturnal animals are seldom seen.

PART 5
LITERATURE CITED

Acevedo A.A., M. Lampo, and R. Cipriani. 2016. The Cane or Marine Toad, *Rhinella marina* (Anura: Bufonidae): Two genetically and morphologically distinct species. *Zootaxa* 4103:574–586. http://dx.doi.org/10.11646/zootaxa.4103.6.7

Adorno, R. and P.C. Pautz. 1999. *Álvar Núñez Cabeza de Vaca: His Account, His Life, and the Expedition of Pánfilo de Narváez.* 3 vols. University of Nebraska Press, Lincoln.

Anderton, A.J. 1991. The Spanish of John P. Harrington's Kitanemuk notes. *Anthropological Linguistics* 33:448–457. https://www.jstor.org/stable/30028222

Bailowitz, R., D. Danforth, and S. Upson. 2015. *A Field Guide to the Damselflies and Dragonflies of Arizona and Sonora.* Nova Granada Publications, Tucson.

Balbás, M. 1927. *Recuerdos del Yaqui. Principales episodios durante la campaña de 1899 a 1901.* Sociedad de Edición y Librería Franco Americana, S.A., Mexico City.

Barber, C. 1973. Trilingualism in an Arizona Yaqui village. In: P.R. Turner (editor), *Bilingualism in the Southwest.* University of Arizona Press, Tucson, pp. 295–318.

Barco, M. del. 1980 [1772]. *The Natural History of Baja California.* Translated by Froylan Tiscareno. Baja California Travel Series 43. Dawson's Bookshop, Los Angeles.

Bartell, G.D. 1963. *Directed Culture Change among the Sonoran Yaquis.* Ph.D. dissertation. University of Arizona, Tucson.

Basilio, T. 1890 [1634]. *Arte de la Lengua Cahita, por un padre de la Compañia de Jesús (con una introducción, notas y un pequeño diccionario) e publicado de nuevo, por E. Buelna.* Imprenta del Gobierno Federal, Mexico City. http://hdl.handle.net/2027/hvd.32044086533940

Beals, R.L. 1932. The Comparative Ethnology of Northern Mexico before 1750. *Ibero-Americana* 2. University of California Press, Berkeley.

Beals, R.L. 1943. The Aboriginal Culture of the Cahita Indians. *Ibero-Americana* 19. University of California Press, Berkeley.

Beals, R.L. 1945. The Contemporary Culture of the Cahita Indians. *Bureau of American Ethnology Bulletin* 142. Smithsonian Institution, Washington, D.C.

Bennett, C.F. Jr. 1964. Stingless-bee keeping in western Mexico. Geographical Review 54:85–92. https://doi.org/10.2307/213031

Bizzarro, J.J., W.D. Smith, J.F. Márquez-Farías, J. Tyminski, and R.E. Hueter. 2009. Temporal variation in the artisanal elasmobranch fishery of Sonora, Mexico. *Fisheries Research* 97:103–117. https://doi.org/10.1016/j.fishres.2009.01.009

Bogan, M.T., N. Noriega-Felix, S.L. Vidal-Aguilar, L.T. Findley, D.A. Lytle, O.G. Gutiérrez-Ruacho, J.A. Alvarado-Castro, and A. Varela-Romero. 2014. Biogeography and conservation of aquatic fauna in spring-fed tropical canyons of the southern Sonoran Desert, Mexico. *Biodiversity and Conservation* 23:2705–2748. http://doi.org/10.1007/s10531-014-0745-z

Bowen, T. 2000. *Unknown Island: Seri Indians, Europeans, and San Esteban Island in the Gulf of California.* University of New Mexico Press, Albuquerque.

Bowen, T. (ed.) 2002. *Backcountry Pilot: Flying Adventures with Ike Russell.* University of Arizona Press, Southwest Center Series, Tucson.

Brand, D. 1988. The honey bee in New Spain and Mexico. *Journal of Cultural Geography* 9:71–82. https://doi.org/10.1080/08873638809478475

Brandenburg, M.M. and C.L. Baumann (editors and translators). 1952. *Observations in Lower California, by Johann Jakob Baegert.* University of California Press, Berkeley.

Brewer, S.A. 1976. *The Yaqui Indians of Arizona: Trilingualism and Cultural Change.* Ph.D. dissertation. University of Texas, Austin.

Brown, D. E., K.B. Clark, R.D. Babb, and G. Harris. 2012. An analysis of masked bobwhite collection locales and habitat characteristics. *Proceedings of the National Quail Symposium* 7:305–328. http://trace.tennessee.edu/nqsp/vol7/iss1/117

Brusca, R., E. Kimrey, and W. Moore (editors). 2004. *A Seashore Guide to the Northern Gulf of California.* Arizona-Sonora Desert Museum, Tucson.

Burckhalter, D. 1992. Photographs from the Río Yaqui, 1980s. *Journal of the Southwest* 34(1):129–138.

Búrquez, A., A. Martínez-Yrízar, R.S. Felger, and D. Yetman. 1999. Vegetation and habitat diversity at the southern edge of the Sonoran Desert. In: R.H. Robichaux (editor), *Ecology of Sonoran Desert Plants and Plant Communities*. University of Arizona Press, Tucson, pp. 36–67.

Caire, W. 2019. The distribution of the land mammals of Sonora, Mexico. *Journal of the Arizona-Nevada Academy of Science* 48(1–2):40–219. https://doi.org/10.2181/036.048.0203

Calderón Aguilera, L.E. and J. Campoy-Favela. 1993. Bahía de las Guásimas, Estero los Algodones y Bahía de Lobos, Sonora. In: S.I. Salazar-Vallejo and N.E. González (editors), *Biodiversidad Marina y Costera de México*. Comisión Nacional para el Conocimiento y Uso de la Biodiversidad y CIQRO, Mexico City, pp. 411–441.

Campoy-Favela, J., A. Varela-Romero, and L. Juárez-Romero. 1989. Observaciones sobre la ictiofauna nativa de la cuenca del Río Yaqui, Sonora, México. Ecologica 1:1–29.

Carter, A.M. 1974. The genus *Cercidium* (Leguminosae: Caesalpinioideae) in the Sonoran Desert of Mexico and the United States. *Proceedings of the California Academy of Sciences*, ser. 4, 40:17–57. https://www.biodiversitylibrary.org/page/15774265

Castro-Aguirre, J.L. 1978. *Catálogo Sistemático de los Peces Marinos que Penetran a las Aguas Continentales de México, con aspectos zoogeográficos y ecológicos*. Serie Científica 19. Departamento de Pesca, Dirección General del Instituto Nacional de la Pesca, México, Mexico City.

Ceballos, G. and G. Oliva (coordinators). 2005. *Los Mamíferos Silvestres de México*. Comisión Nacional para el Conocimiento y Uso de la Biodiversidad, Fondo de Cultura Económica, Mexico City.

Ceballos, G. (editor). 2014. *Mammals of Mexico*. Johns Hopkins University Press, Baltimore.

Chesser, R.T., K.J. Burns, C. Cicero, J.L. Dunn, A.W. Kratter, I J. Lovette, P.C. Rasmussen, J.V. Remsen, Jr., D.F. Stotz, and K. Winker. 2019. *Check-list of North American Birds (online)*. American Ornithological Soc. http://checklist.aou.org/taxa

Clinton Eitniear, J. 2000. Zopilote rey. In: G. Ceballos and L. Márques Valdelamar (editors), *Las Aves de México en Peligro de Extinción*. Instituto de Ecología, Universidad Nacional Autónoma de México, Mexico City, pp. 105–107.

Colunga-GarcíaMarín, P. 2003. The domestication of henequén (*Agave fourcroydes* Lem.). In: A. Gómez-Pompa, M.F. Allen, S.L. Fedick, and J.J. Jiménez-Osornio (editors), *The Lowland Maya Area: Three Millennia at the Human-Wildland Interface*. Food Products Press, Binghamton, N.Y, pp. 439–446.

Coville, F.W. and D.T. MacDougal. 1903. *Desert Botanical Laboratory of the Carnegie Institution*. Publication 6. Carnegie Institution of Washington, Washington, D.C. https://www.biodiversitylibrary.org/page/17336226

Craveri, F. 2018. Journal of a Voyage: Federico Craveri and the Gulf of California in 1856. Translated from the Italian by Beatrice D'Arpa and Beppe Cavatorta, edited and annotated by Thomas Bowen. *Journal of the Southwest* 60:299–483.

Crother, B.I. (editor). 2017. Scientific and Standard English Names of Amphibians and Reptiles of North America North of Mexico, with Comments Regarding Confidence in Our Understanding. *Society for the Study of Amphibians and Reptiles, Herpetological Circular* 43:1–102.

Croxen, F.W. III, C.A. Shaw, and D.R. Sussman. 2007. Pleistocene geology and paleontology of the Colorado River Delta at Golfo de Santa Clara, Sonora, Mexico (reprinted from: Wild, scenic & rapid—a trip down the Colorado River trough, the 2007 Desert Symposium field guide and abstracts from proceedings, pp. 84–89, California State University, April 2007). https://www.yumpu.com/en/document/view/29739745/el-golfo-paper-arizona-western-college

Crumrine, N.R and L.S. Crumrine. 1967. Ancient and modern Mayo fishing practices. *Kiva* 33:25–33. https://www.jstor.org/stable/30247051

Cushman, G.T. 2013. *Guano and the Opening of the Pacific World: A Global Ecological History*. Cambridge University Press, Cambridge, U.K.

Dabdoub, C. 1964. *Historia de El Valle del Yaqui*. Librería Manuel Porrúa, México, D.F.

Das, S. 2016.Taxonomy and phylogeny of grain amaranths. In: *Amaranthus: A Promising Crop of the Future*. Springer, Singapore, pp. 57–94. https://doi.org/10.1007/978-981-10-1469-7

Dedrick, J.M. and E.H. Casad. 1999. Sonora Yaqui Language Structures. University of Arizona Press, Tucson.

Densmore, F. 1952. Yuman and Yaqui Music. *Bureau of American Ethnology Bulletin* 110. Smithsonian Institution, Washington, D.C.

Edmonds, J.M. and J.A .Chweya. 1997. Black Nightshades: *Solanum nigrum* L. and related species. *Promoting the Conservation and Use of Neglected and Underutilized and Neglected Crops*, no. 15, Institute of Plant Genetic and Crop Plant Research, Gatersleben, and International Plants Genetic Resource Institute, Rome.

Edwards, T., A.E. Karl, M. Vaughn, P.C. Rosen, C.M. Torres, and R.W. Murphy. 2016. The desert tortoise trichotomy: Mexico hosts a third, new sister-species of tortoise in the *Gopherus morafkai–G. agassizii* group. *ZooKeys* 562:131–158. https://doi.org/10.3897/zookeys.562.6124

Erickson, D.L., B.D. Smith, A.C. Clarke, D.H. Sandweiss, and N. Tuross. 2005. An Asian origin for a 10,000-year-old domesticated plant in the Americas. *Biological Sciences* 102:18315–18320. https://doi.org/10.1073/pnas.0509279102

Erickson, K.C. 2008. *Yaqui Homeland and Homeplace: The Everyday Production of Ethnic Identity*. University of Arizona Press, Tucson.

Escudero, J.A. 1849. Noticias estadisticas de Sonora y Sinaloa. Tipografía de R. Rafael. Mexico City. https://archive.org/details/noticiasestadist00escu

Estrada Fernandez, Z., C. Buitimea Valenzuela, A.E. Gurrola Camacho, M. Castillo Celaya, and A. Carlón Flores. 2004. *Diccionario Yaqui–Español y Textos: Obra de Preservación Lingüística*. Plaza y Valdés, México, D.F., and Universidad de Sonora, Hermosillo.

Evans, S. 2007. *Bound in Twine: The History and Ecology of the Henequen-wheat Complex for Mexico and the American and Canadian Plains, 1880–1950*. Texas A&M University Press, College Station.

Evers, L. and F.S. Molina. 1987. *Yaqui Deer Songs/Maso Bwikam: A Native American Poetry*. University of Arizona Press, Tucson.

Evers, L. and F.S. Molina (editors). 1990. *Wo'i Bwikam/Coyote Songs: Songs from the Yaqui Bow Leaders' Society*. Chax Press, Tucson.

Evers, L. and F.S. Molina. 1992. Hiakim: Yaqui homeland. *Journal of the Southwest* 34:1–139. https://www.jstor.org/stable/40169842

Felger, R.S. 1977. Mesquite in Indian cultures of southwestern North America. In: B.B. Simpson (editor), *Mesquite: Its Biology in Two Desert Scrub Ecosystems*. Dowden, Hutchinson, and Ross, Stroudsburg, PA, pp. 150–176.

Felger, R.S. 2002. Sinaloa Shootout. In: T. Bowen (editor), *Backcountry Pilot: Flying Adventures with Ike Russell*. University of Arizona Press, Tucson, pp. Pages 2–14.

Felger, R.S. 2007. Living resources at the center of the Sonoran Desert: Native American plant and animal utilization. In: R.S. Felger and B. Broyles (editors), *Dry Borders: Great Natural Reserves of the Sonoran Desert*. University of Utah Press, Salt Lake City, pp. 147–192.

Felger, R.S. and D.F. Austin. 2005. *Ipomoea seaania*, a new species of Convolvulaceae from Sonora, Mexico. *Sida, Contributions to Botany* 21:1293–1304. https://www.biodiversitylibrary.org/page/9311535

Felger, R.S., S.D. Carnahan, and J.J. Sánchez-Escalante. 2023. Oasis at the Desert Edge: Flora of Cañón del Nacapule, Sonora, Mexico. BRITT Press, Fort Worth Botanic Garden/ Botanical Research Institute of Texas, U.S.A.

Felger, R.S., S.D. Carnahan, J.J. Sánchez-Escalante. 2020. *The Desert Edge: Flora of the Guaymas-Yaqui Region of Sonora, Mexico*. Special publication of the Desert Laboratory, University of Arizona, Tucson.

Felger, R.S., M.B. Johnson, and M.F. Wilson. 2001. *Trees of Sonora, Mexico*. Oxford University Press, New York.

Felger, R.S. and E. Joyal. 1999. The palms (Arecaceae) of Sonora, Mexico. *Aliso* 18:1–18. https://doi.org/10.5642/aliso.19991801.11

Felger, R.S. and C.H. Lowe. 1976. The Island and Coastal Vegetation and Flora of the Gulf of California, Mexico. *Natural History Museum of Los Angeles County Contributions in Science* 285:1–59.

Felger, R.S. and F. Molina. 1988. Plants and Animals in Yoeme culture: Yoeme (Yaqui) Pascola Masks and

Deer Songs, plant and animal resource materials. Report to The Smithsonian Institute Folklife Programs. Washington, D.C. 23 pages + 12 illustrations.

Felger, R.S. and F.S. Molina. 2019. Giant reed in the Yoeme world of Sonora and Arizona. *The Plant Press, Arizona Native Plant Society* winter 2019:11–15. https://aznps.com/wp-content/uploads/anps-19ii-webfinal.pdf

Felger, R.S. and M.B. Moser. 1985. *People of the Desert and Sea: Ethnobotany of the Seri Indians.* University of Arizona Press, Tucson.

Felger, R.S., W.J. Nichols, and J.A. Seminoff. 2005. Sea turtle conservation, diversity and desperation in northwestern Mexico. In: J.-L.E. Cartron, G. Ceballos, and R.S. Felger (editors), *Biodiversity, Ecosystems, and Conservation in Northern Mexico.* Oxford University Press, New York, pp. 405–424.

Felger, R.S., P.L. Warren, L.S. Anderson, and G.P. Nabhan. 1992. Vascular plants of a desert oasis: Flora and ethnobotany of Quitobaquito, Organ Pipe Cactus National Monument, Arizona. *Proceedings of the San Diego Society of Natural History* 8.1–39. https://arizona.pure.elsevier.com/en/publications/vascular-plants-of-a-desert-oasis-flora-and-ethnobotany-of-quitob

Felger R.S. and D. Yetman. 2000. Roasting the *Hechtia* out of it: The use of *Hechtia montana* (Bromeliaceae) as a food plant in Sonora, México. *Economic Botany* 54:229–233. https://www.jstor.org/stable/4256295

Félix-Pico, E.F., M. Ramírez-Rodríguez, and O. Holguín-Quiñones. 2011. Growth and fisheries of the Black Ark *Anadara tuberculosa*, a bivalve mollusk, in Bahía Magdalena, Baja California Sur, Mexico. *North American Journal of Fisheries Management* 29:231–236. https://doi.org/10.1577/M06-050.1

Fikes, J.C. 1993. *Carlos Castaneda: Academic Opportunism and the Psychedelic Sixties.* Millenia Press, Victoria.

Flesch, A.D. 2003. Distribution, abundance, and habitat of cactus ferruginous pygmy-owls in Sonora, Mexico. Master's thesis, University of Arizona, Tucson.

Flesch, A.D. 2008. Distribution and status of birds of conservation interest and identification of important bird areas in Sonora, Mexico. U.S. Fish and Wildlife Service, Sonoran Joint Venture, Tucson, AZ. Cooperative Agreement No. 201816J827. https://sonoranjv.org/the-university-of-arizona-distribution-and-abundance-of-birds-of-conservation-interest-and-identification-of-important-bird-areas-in-sonora-mexico/

Flesch, A.D. 2019. Patterns and drivers of long-term changes in breeding bird communities in a global biodiversity hotspot in Mexico. *Diversity and Distributions* 25:499–513. https://doi.org/ 10.1111/ddi.12862

Flint, R. and S.C. Flint. 2005. *Documents of the Coronado Expedition, 1539–1542.* Southern Methodist University Press, Dallas.

Folsom, R.B. 2014. *The Yaquis and the Empire: Violence, Spanish Imperial Power, and Native Resilience in Colonial Mexico.* Yale University Press, New Haven.

Friedman, S.L. 1996. Vegetation and flora of the coastal plains of the Río Mayo Region, southern Sonora, México. Master's thesis, Arizona State University, Tempe.

Frost, D.R. 2020. *Amphibian Species of the World: An Online Reference, Version 6.1.* American Museum of Natural History, New York. https://amphibiansoftheworld.amnh.org/index.php. doi.org/10.5531/db.vz.0001

Gallo-Reynoso, J.P. 1996. Distribution of the Neotropical river otter (*Lutra longicaudis annectens* Major, 1897) in the Río Yaqui, Sonora, Mexico. *IUCN Otter Specialist Group Bulletin* 13:27–31. https://www.iucnosgbull.org/Volume13/Gallo_1996.html

Gallo-Reynoso, J.P., R.S. Felger, and B.T. Wilder. 2012. Near colonization of a desert island by a tropical bird: Military Macaw (*Ara militaris*) at Isla San Pedro Nolasco, Gulf of California. *Southwestern Naturalist* 57:459–462. https://tumamoc.arizona.edu/sites/tumamoc/files/galloreynoso_et_al._2012_desert_island_by_a_tropical_bird.pdf

Gentry, H.S. 1942. *Rio Mayo Plants, a Study of the Flora and Vegetation of the Valley of the Rio Mayo, Sonora.* Carnegie Institution of Washington Publication 527. Carnegie Institution of Washington, Washington, D.C.

Gentry, H.S. 1963. The Warihio Indians of Sonora–Chihuahua: An ethnographic survey.

Anthropological Papers 65, Bureau of American Ethnology Bulletin 186. Smithsonian Institution, Washington, D.C.

Gentry, H.S. 1972. *The Agave Family in Sonora*. Agriculture Handbook No. 399, United States Agricultural Research Service, Washington, D.C.

Gentry, H.S. 1982. *Agaves of Continental North America*. University of Arizona Press, Tucson.

Giddings, R.W. 1945. Folk literature of the Yaqui Indians. Master's thesis, University of Arizona, Tucson.

Giddings, R.W. 1959. *Yaqui Myths and Legends*. University of Arizona Press, Tucson.

Goss, N.S. 1888. New and rare birds found breeding on the San Pedro Martin Isle. *The Auk* 5(3):240–244. https://doi.org/10.2307/4067309

Graham, F. 1987. Back from oblivion. *Audubon Magazine* (September):20–26.

Griffith, J.S. 1972. Cáhitan pascola masks. *Kiva* 37:185–198. https://www.jstor.org/stable/30247766

Griffith, J.S. and F.S. Molina. 1980. *Old Men of the Fiesta: An introduction to the Pascola arts*. The Heard Museum, Phoenix.

Grismer, L.L. 2002. *Amphibians and Reptiles of Baja California*. University of California Press, Berkeley.

Heffelfinger, J. 2006. *Deer of the Southwest: A Complete Guide to the Natural History, Biology, and Management of Southwestern Mule Deer and White-Tailed Deer*. Texas A&M University Press, College Station.

Hendrickson, D.A., W.L. Minkley, R.R. Miller, D.J. Siebert, and P.H. Minkley. 1980. Fishes of the Río Yaqui basin, México and United States. *Journal of the Arizona-Nevada Academy of Science* 15:65–106. https://www.jstor.org/stable/40025038

Hernández Moreno, L.G. 2000. Aspectos sobre ecología y biología de las jaibas Callinectes arcuatus y C. bellicosus (Crustacea: Protunidae) en la laguna costera Las Guásimas, Sonora, Mexico. Thesis, Centro de Investigaciones Biológicas del Noroeste, Ensenada.

Hodgson, W.C. 2001. *Food Plants of the Sonoran Desert*. University of Arizona Press, Tucson.

Holden, W.C. 1936a. La Fiesta de Gloria. In: W.C. Holden et al., *Studies of the Yaqui Indians of Sonora, Mexico*. Texas Technological College Bulletin, Vol. 12, No. 1. Texas Technological College, Lubbock, pp. 34–54.

Holden, W.C. 1936b.Yaqui funerals. In: W.C. Holden et al., *Studies of the Yaqui Indians of Sonora, Mexico*. Texas Technological College Bulletin, Vol. 12, No. 1. Texas Technological College, Lubbock, pp. 55–66.

Holden, W.C. 1936c. Household economy. In: W.C. Holden et al., *Studies of the Yaqui Indians of Sonora, Mexico*. Texas Technological College Bulletin, Vol. 12, No. 1. Texas Technological College, Lubbock, pp. 67–71.

Holden, W.C. 1965. *Hill of the Rooster*. Henry Holt and Co., New York.

Holden, W.C., C.C. Seltzer, R.A. Studhalter, C.J. Wagner, and W.G McMillan. 1936. *Studies of the Yaqui Indians of Sonora, Mexico*. Texas Technological College Bulletin, Vol. 12, No. 1. Texas Technological College, Lubbock.

Holste, M., J.M. Ruth, and J.C. Eitniear. 2014. King Vulture (*Sarcoramphus papa*). In: T.S. Schulenberg (editor), Neotropical Birds Online. Cornell Lab of Ornithology, Ithaca. https://doi.org/10.2173/nb.kinvul1.01

Hovens, P., W.J. Orr, and L.A. Hieb (editors and translators). 2004. *Travels and Researches in Native North America, 1882–1883, by Herman ten Kate*. University of New Mexico Press, Albuquerque.

Howell, S.N.G. and S. Webb. 1995. *A Guide to the Birds of Mexico and Northern Central America*. Oxford University Press, Oxford.

Hrdlička, A. 1904. Notes on the Indians of Sonora, Mexico. *American Anthropologist* 6:51–89. https://www.jstor.org/stable/659297

Hu-DeHart, E. 1981. *Missionaries, Miners, and Indians: Spanish Contact with the Yaqui Nation of Northwestern New Spain, 1533–1820*. University of Arizona Press, Tucson.

Hu-DeHart, E. 1984. Yaqui Resistance and Survival: The Struggle for Land and Autonomy, 1821–1910. University of Wisconsin Press, Madison.

Johnson, J.B. 1962. *El Idioma Yaqui*. Departmento de Investigaciones Antropológicas 10. Instituto Nacional de Antropología y Historia, Mexico City.

INALI. 2019. Jiak noki ji'ojtei yoojio, Norma de escritura de la lengua jiak noki (yaqui). Instituto Nacional de Lenguas Indigenas, Mexico City. https://site.inali.gob.mx/Micrositios/normas/yaqui.html

Kaczkurkin, M.V. 1977. *Yoeme: Lore of the Arizona Yaqui People*. Sun Tracks, Tucson.

Kurath, W. and E. Spicer. 1947. A brief introduction to Yaqui: A native language of Sonora. *University of Arizona Social Science Bulletin* 15. Tucson.

Kay, M.A. 1996. *Healing with Plants in the American and Mexican West*. University of Arizona Press, Tucson.

Kelley, J.H. 1978. *Yaqui Women: Contemporary Life Histories*. University of Nebraska Press, Lincoln.

Kiefert, L., D. McL. Moreno, E. Arizmendi, H.A. Hanni, and S. Elen. 2004. Cultured pearls from the Gulf of California, Mexico. *Gems & Gemology* 40:26–38. https://www.gia.edu/gems-gemology/spring-2004-cultured-pearls-gulf-california-mexico-kiefert

Kingsbury, J.M. 1964. *Poisonous Plants of the United States and Canada*, 3rd edition. Prentice-Hall, Edgewood Cliffs.

Kolaz, T.M. 1985. Yaqui pascola masks from the Tucson area. *American Indian Art Magazine* 11(1):38–45. http://www.yoemecarver.com/aiam07.pdf

Kolaz, T.M. 1997. Yoeme dog pascola masks. *Archaeology Southwest* 22(3):16.

Kolaz, T.M. 2007. Yoeme Pascola masks from the Tucson communities: A look back. *American Indian Art Magazine* 32(3):50–61, 106. http://www.yoemecarver.com/aiam07.pdf

Kolaz, T.M. 2015. Pahko'ola mask. In: P. Hovens and B. Bernstein (editors), *North American Indian Art: Masterpieces and Museum Collections in the Netherlands*. University of Oklahoma Press, Norman, pp. 148–149.

Lemos-Espinal, J. A., G.R. Smith, and J.C. Rorabaugh. 2019. A conservation checklist of the amphibians and reptiles of Sonora, Mexico, with updated species lists. *ZooKeys* 829:131–160. https://doi.org/10.3897/zookeys.829.32146

Ligon, J.S. 1952. The vanishing masked bobwhite. *Condor* 54:48–50. https://doi.org/10.2307/1364527

Loomis, R.B. and R.M. Davis. 1965. The vampire bat in Sonora, with notes on other bats from southern Sonora. *Journal of Mammalogy* 46: 497. https://doi.org/10.2307/1377641

López-Gonzáles, C.A. and D. E. Brown. 2001. *Borderland Jaguars*. University of Utah Press, Salt Lake City.

López González, C.A. and D.E. Brown. 2002. Status and distribution of the jaguar in Sonora, Mexico. In: R.A. Medellín, C.-L.B. Chetkiewicz, C. Equihua, and P.G. Crawshaw, *El Jaguar en el Nuevo Milenio*. Universidad Nacional Autónoma de México, Fondo de Cultura Económica, Mexico City, pp. 379–392.

Lutes, S. 1977. *Alcohol Use Among the Yaqui Indians of Potam, Sonora, Mexico*. Ph.D. dissertation, University of Kansas, Lawrence.

Lutes, S.V. 1983. The mask and magic of the Yaqui paskola clowns. In: N.R. Crumrine and M.M. Halpin (editors), *The Power of Symbols: Masks and Masquerade in the Americas*. University of British Columbia Press, Vancouver, pp. 81–92.

Marlett, C.M. 2014. *Shells on a Desert Shore: Mollusks in the Seri World*. University of Arizona Press, Tucson. (also: https://shellsonadesertshore.com)

Martin, P.S., D. Yetman, M. Fishbein, P. Jenkins, T.R. Van Devender, and R.K. Wilson. 1998. *Gentry's Río Mayo Plants: The Tropical Deciduous Forest and Environs of Northwest Mexico*. University of Arizona Press, Tucson.

Martínez, M. 1969. *Las Plantas Medicinales de México*. Ediciones Botas, Mexico City.

Martínez del Rio, C. 2007. Long-nosed bats and white-winged doves: Travels and tribulations of two migrant pollinators. In: R.S. Felger and B. Broyles (editors), *Dry Borders: Great Natural Reserves of the Sonoran Desert*. University of Utah Press, Salt Lake City, pp. 303–309.

Maaso, M., F.S. Molina, and L. Evers. 1993. The Elders' Truth: A Yaqui sermon. *Journal of the Southwest* 35:225–327. https://www.jstor.org/stable/40169889

McGuire, T.R. 1983. The political economy of shrimping in the Gulf of California. Human Organization 42:132–145. https://doi.org/10.17730/humo.42.2.ku2m3176u7706747

McGuire, T.R. 1986. Politics and Ethnicity on the Río Yaqui: Potam Revisited. University of Arizona Press, Tucson.

McMillan, W.G. 1936. Yaqui architecture. In: W.C. Holden et al., *Studies of the Yaqui Indians of Sonora, Mexico*. Texas Technological College Bulletin 12(1), Lubbock, pp. 72–78.

Medellín, R.A., C. Manterola, M. Valdéz, D.G. Hewitt, D. Doan-Crider, and T.E. Fulbright. 2005. History, ecology, and conservation of the pronghorn antelope, bighorn sheep, and black bear in Mexico. In: J.-L.E. Cartron, G. Ceballos, and R.S. Felger (editors), *Biodiversity, Ecosystems, and Conservation in Northern Mexico.* Oxford University Press, New York, pp. 387–404.

Merrill, W.L. 2012. The historical linguistics of Uto-Aztecan agriculture. *Anthropological Linguistics* 54:203–260. https://www.jstor.org/stable/23621084

Miller, R.R., W.L. Minckley, and S.M. Norris. 2005. *Freshwater Fishes of México.* University of Chicago Press, Chicago.

Minkley, C.O. and P.C. Marsh. 2009. *Inland Fishes of the Greater Southwest: Chronicle of a Vanishing Biota.* University of Arizona Press, Tucson.

Moctezuma Zamarrón, J.L., H. López Aceves, E. Merino González, A.P. Pintado Cortina, M.V. Morales Muños, M.de G. Fernández Ramos, and C.J. Harriss Clare. 2016. Ritualidad en los Valles y la Sierra del Noroeste de México: La Semana Santa entre Yaquis, Mayos Tarahumaras y Guarijíos. In: A. Oseguera Montiel and A. Reyes Valdez (editors), *Develando la Tradición: Procesos rituales en las comunidades indígenas de México*, vol. IV. Secretaría de Cultura, Instituto Nacional de Antropología e Historia, Mexico City, pp. 233–336.

Moerman, D. 2010. *Native American Ethnobotany.* Timber Press, Portland.

Moerman, D. 2020. *Native American Ethnobotany: A Database of Foods, Drugs, Dyes and Fibers of Native American Peoples, Derived from Plants.* Botanical Research Institute of Texas. naeb.brit.org

Moisés, R., J.H. Kelley, and W.C. Holden. 1971. *A Yaqui Life: The Personal Chronicle of a Yaqui Indian.* University of Nebraska Press, Lincoln.

Molina, F.S. and D.L. Shaul. 1993. *A Concise Yoeme and English Dictionary.* Tucson Unified School District, Tucson.

Molina, F.S., A. Olivas, R. Tapia, and H. Valenzuela. 2003. *The Ones Who Lived Here in the Beginning: Wame Vatnataka Im Hohhoasukame.* Pascua Yaqui Tribal Council, Tucson.

Molina, F.S., H. Valenzuela, and D.L. Shaul. 1999. *Hippocrene Standard Dictionary Yoeme-English, English-Yoeme: With a Comprehensive Grammar of Yoeme Language.* Hippocrene Books, New York.

Moore, M. 1990. *Los Remedios: Traditional Herbal Remedies of the Southwest.* Red Crane Books, Santa Fe.

Moser, M.B. and S. Marlett (compilers). 2010. *Comcaac quih yaza quih hant ihiip hac: Diccionario seri-español-inglés*, 2nd edition. Plaza y Valdés Editores, Mexico City. http://www.mexico.sil.org/resources/archives/42821

Mosk, S.A. 1931. *Spanish Voyages and Pearl Fisheries in the Gulf of California: A Study in Economic History.* Ph.D. dissertation, University of California, Berkeley.

Mosk, S.A. 1934. The Cardona Company and the pearl fisheries of Lower California. *Pacific Historical Review* 3:50–61. https://doi.org/10.2307/3633457

Mosk, S.A. 1939. Subsidized hemp production in Spanish California. *Agricultural History* 13:171–175. https://www.jstor.org/stable/3739684

Nabhan, G.P. and R.S. Felger. 1978. Teparies in southwestern North America. *Economic Botany* 32:2–19. https://doi.org/10.1007/BF02906725

Navarro, C. 2003a. *Crocodylus acutus* in Sonora, México. *Crocodile Specialist Group Newsletter* 22(1):21.

Navarro, C. 2003b. Abundance, habitat use, and conservation of the American crocodile in Sinaloa. *Crocodile Specialist Group Newsletter* 22(2):22–23.

Nentvig, J. 1977. *El Rudo Ensayo.* SEP Instituto Nacional de Antropología e Historia. Proyectos Especiales 58. Colección Científica, Etnología, Mexico City.

Nuñez Cabeça de Vaca, Á. 1542. *La Relación y Comentarios del Governador Alvar Núñez Cabeça de Vaca, de lo Acaescido en las Dos Jornadas que Hizo a las Indias.* Pedro Hernández.

Nuñez Cabeça de Vaca, Á. 1555. *La Relación y Comentarios del Gouernador Alvar Nuñez Cabeça de Vaca de lo Acaescido en los Doze Jornadas que Hizo a las Indias.* Francisco Fernandez de Cordoua, Valladolid.

Olavarría, M.E. 1992. *Símbolos del Desierto.* Universidad Autónoma Metropolitana, Mexico City.

Padilla Ramos, R. and J.L. Moctezuma Zamarrón. 2015. Mazatán: Yaquis itinerantes entre el campo de batalla, la sala de un museo y el sepulcro de honor. La tribu yaqui y la defense de sus derechos territoriales. *Diario De Campo* 8:41–47. https://www.revistas.inah.gob.mx/index.php/diariodecampo/article/view/7426

Padilla Ramos, R. and J.L. Moctezuma Zamarrón. 2017. The Yaquis, a historical struggle for water. *Water*

History 9:29–43. https://doi.org/10.1007/s12685-017-0194-1

Padilla-Serrato, J., J. López-Martínez, J. Rodríguez-Romero, A. Acevedo-Cervantes, F. Galván-Magaña, and D. Lluch-Cota. 2017. Changes in fish community structures in a coastal lagoon in the Gulf of California, México. *Revista de Biología Marina y Oceanografía* 52:567–579. https://doi.org/10.4067/S0718-19572017000300013

Painter, M.T. 1986. Edited by E.H. Spicer and W. Kaemlein. *With Good Heart: Yaqui Beliefs and Ceremonies in Pascua Village*. University of Arizona Press, Tucson.

Painter, M.T. and E.B. Sayles, with E.H. Spicer. 1962. *Faith, Flowers, and Fiestas: The Yaqui Indian Year*. University of Arizona Press, Tucson.

Palacios, E. and E. Mellink. 1995. Breeding birds of Esteros Tóbari and San José, southern Sonora. *Western Birds* 26:99–103. http://www.westernfieldornithologists.org/archive/V26/26(2)%20p0099-p0103.pdf

Paredes Aguilar, R., T.R. Van Devender, and R.S. Felger. 2000. *Las Cactáceas de Sonora: Su Diversidad, Uso y Conservación*. Arizona-Sonora Desert Museum Press, Tucson.

Peigler, R.S. and M. Maldonado. 2005. Uses of cocoons of Eupackardia calleta and Rothschildia cincta (Lepidoptera: Saturniidae) by Yaqui Indians in Arizona and Mexico. *Nachrichten des Entomologischen Vereins Apollo* 26(3):111–119. https://www.zobodat.at/pdf/NEVA_26_0111-0119.pdf

Perez, G. and E. Vance. 1993. Glafiro: A young Yoeme Deer Dancer. *Native Peoples, Arts and Lifeways* (winter):46–50.

Pérez de Ribas, A. 1645. *Historia de los Trivmphos de Nvestra Santa Fee entre Gentes las Mas Barbaras, y Fieras del Nuevo Orbe*. A. de Paredes, Madrid.

Phillips, A., J.T. Marshall, and G. Monson. 1964. *Birds of Arizona*. University of Arizona Press, Tucson.

Pyron, R.A., F.T. Burbrinkand, and J.J. Wiens. 2013. A phylogeny and revised classification of Squamata, including 4161 species of lizards and snakes. *BMC Evolutionary Biology* 13:93. https://doi.org/10.1186/1471-2148-13-93

Raith, M., D.C. Zacherl, E.M. Pilgrim, and D.J. Eernisse. 2016. Phylogeny and species diversity of Gulf of California oysters (Ostreidae) inferred from mitochondrial DNA. *American Malacological Bulletin* 33:263–283. http://dx.doi.org/10.4003%2F006.033.0206

Rangel-Aguilar, O. and J. P. Gallo-Reynoso. 2013. Hábitos alimentarios de la nutria neotropica (*Lontra longicaudis annectens*) en el Río Bavispe-Yaqui, Sonora, Mexico. *Therya* 4:297–309. https://www.revistas-conacyt.unam.mx/therya/index.php/THERYA/article/view/59/html_94

Rea, A.M. 1983. *Once a River: Bird Life and Habitat Changes of the Middle Gila*. University of Arizona Press, Tucson.

Rea, A.M. 1997. *At the Desert's Green Edge: Ethnobotany of Gila River Pima*. University of Arizona Press, Tucson.

Rea, A.M. 2000. Cóndor californiano. In: G. Ceballos and L. Márques Valdelamar (editors), *Las Aves de México en Peligro de Extinción*. Instituto de Ecología, Universidad Nacional Autónoma de México, Mexico City, pp.100–105.

Rea, A.M. 2007. *Wings in the Desert: A Folk Ornithology of the Northern Pimans*. University of Arizona Press, Tucson.

Real Academia Española. 2019. *Diccionario de la Lengua Española*, 23rd edition. [version 23.3] https://dle.rae.es

Reff, D.T. 1991. *Disease, Depopulation, and Culture Change in Northwestern New Spain, 1518–1764*. University of Utah Press, Salt Lake City.

Reff, D.T., M. Ahern, and R.K. Danford (editors and translators). 1999. *History of the Triumphs of Our Holy Faith amongst the Most Barbarous and Fierce Peoples of the New World, by Andrés Pérez de Ribas*. University of Arizona Press, Tucson.

Rhodin, A.G.J., J.B. Iverson, R. Bour, U. Fritz, A. Georges, H.B. Shaffer, and P.P. van Dijk. 2017. Turtles of the World: Annotated checklist and atlas of taxonomy, synonymy, distribution, and conservation status (8th edition). *Chelonian Research Monographs* 7:1–292. https://doi.org/10.3854/CRM.7.CHECKLIST.ATLAS.V8.2017

Riley, C.L. 1997. Introduction. In: R. Flint and S.C. Flint, *The Coronado Expedition to Tierra Nueva: The 1540–1542 Route Across the Southwest*. University Press of Colorado, Boulder, pp. 1–21.

Robertson, T.A. (translator, in condensed form). 1968. *My Life Among the Savage Nations of New Spain, by Andrés Pérez de Ribas.* The Ward Ritchie Press, Los Angeles.

Rorabaugh, J.C. 2017. Crocodiles spotted in southern Sonora waters for the first time since 1973. *Sonoran Herpetologist* 30(4):68–69.

Rorabaugh, J.C. and J.A. Lemos-Espinal. 2016. *A Field Guide to the Amphibians and Reptiles of Sonora, Mexico.* ECO Herpetological Publishing and Distribution, Rodeo, New Mexico.

Rose, J.N. 1899. Notes on useful plants of Mexico. *Contributions from the United States National Herbarium* 5:209–259. https://doi.org/10.5962/bhl.title.43253

Russell, S.M. and G. Monson. 1998. *Birds of Sonora.* University of Arizona Press, Tucson.

Santamaria, F.J. 2000. *Diccionario de Mejicanismos,* 6th edition. Editorial Porrúa, Mexico City.

Sauer, J.D. 1967. The grain amaranths and their relatives: A revised taxonomic and geographic survey. *Annuals of the Missouri Botanical Garden* 54:103–137. https://doi.org/10.2307/2394998

Schmidt, J.O. 2016. *Sting of the Wild.* John Hopkins University Press, Baltimore.

SEINet Portal Network. 2020. http://swbiodiversity.org/seinet/index.php

Semotiuk, A.J., P. Colunga-GarcíaMarín, D. Valenzuela Maldonado, and E. Ezcurra. 2017. Pillar of strength: columnar cactus as a key factor in Yoreme heritage and wildland preservation. *Ambio* 47:86–96. https://link.springer.com/article/10.1007/s13280-017-0940-8

Shaul, D.L. 2014. *A Prehistory of Western North America: The Impact of Uto-Aztecan Languages.* University of New Mexico Press, Albuquerque.

Sheridan, T.E. 1988. How to tell the story of a "people without history": narrative versus ethnohistory through time. *Journal of the Southwest* 30:172–186.

Sheridan, T.E. 1996. The Yoemem (Yaquis): An enduring people. *In:* T.E. Sheridan and N.J. Parezo (editors), *Paths of Life: American Indians of the Southwest and northern Mexico.* University of Arizona Press, Tucson, pp. 35–40.

Shorter, D.D. 2009. *We Will Dance Our Truth, Yaqui History in Yoeme Performances.* University of Nebraska Press, Lincoln.

Shreve, F. 1951. *Vegetation of the Sonoran Desert.* Carnegie Institution of Washington Publication 591. Carnegie Institution of Washington, Washington, D.C.

Shutler, M.E. 1967. *Persistence and Change in the Health Beliefs and Practices of an Arizona Yaqui Community.* Ph.D. dissertation, University of Arizona, Tucson.

Shutler, M.E. 1977. Disease and curing in a Yaqui community. In: E.H. Spicer (editor), *Ethnic Medicine in the Southwest.* University of Arizona Press, Tucson, pp. 169–237.

Silva Encinas, M.C., P. Álvarez, and C. Buitimea. 1998. *Jiák nokpo etéjoim: Platicas en lengua Yaqui.* Universidad de Sonora, Hermosillo. Editorial Unison.

Sobarzo, H. 1966. *Vocabulario Sonorense.* Editorial Porrúa, Mexico City.

Spicer, E.H. 1940. *Pascua: A Village in Arizona.* University of Chicago Press, Chicago.

Spicer, E.H. 1943. Linguistic aspects of Yaqui acculturation. *American Anthropologist* 45:410–426. https://doi.org/10.1525/aa.1943.45.3.02a00060

Spicer, E.H. 1954. Potam: A Yaqui village in Sonora. Memoir 77. American Anthropological Association, Menasha, Wisconsin.

Spicer, E.H. 1969. Review: The Teaching of Don Juan: A Yaqui Way of Knowledge. *American Anthropologist* 71:320–322. https://doi.org/10.1525/aa.1969.71.2.02a00250

Spicer, E.H. 1980. *The Yaquis: A Cultural History.* University of Arizona Press, Tucson.

Spicer, E.H. 1983. Yaqui. In: A. Ortiz (Ed), *Handbook of North American Indians, vol. 10: Southwest.* Smithsonian Institution, Washington, D.C., pp. 250–263.

Spicer, E.H. 1988. Edited by K.M. Sands and R.B. Spicer. *People of Pascua.* University of Arizona Press, Tucson.

Spicer, E.H. and G. Hill. 1947. Yaqui villages past and present. *Kiva* 13:2–12. https://doi.org/10.1080/00231940.1947.11757487

Spicer, R.B. and N.R. Crumrine (editors). 1997. *Performing the Renewal of Community: Indigenous Easter Rituals in North Mexico and Southwest United States.* University Press of America, Lanham, MD.

Standley, P.C. 1923. Trees and Shrubs of Mexico. *Contributions from the Unites States National Herbarium* 23 (part 3):517–848.

Stevens, B.J. 2019. Mexican Dance Masks. http://mexicandancemasks.com

Stevens, P.F. 2001 onwards. Angiosperm Phylogeny Website. Version 14, July 2017 [and more or less continuously updated since]. http://www.mobot.org/MOBOT/research/APweb

Studhalter, R.A. 1936. Yaqui agriculture. In: W.C. Holden et al., Studies of the Yaqui Indians of Sonora, Mexico. *Texas Technological College Bulletin* 12(1), Lubbock, pp. 114–125.

Sugden, E.A. and R.L. McAllen. 1994. Observations on foraging, population and nest biology of the Mexican honey wasp, *Brachygastra mellifica* (Say) in Texas (Vespidae: Polybiinae). *Journal of the Kansas Entomological Society* 67:141–155. https://www.jstor.org/stable/25085503

Taub, A. 1950. Traditional poetry of the Yaqui Indians. Master's thesis, University of Arizona, Tucson.

Tesky, J.L. 1994. *Gymnogyps californianus.* Fire Effects Information System Online. U.S. Department of Agriculture, Forest Service, Rocky Mountain Research Station, Fire Sciences Laboratory. https://www.fs.fed.us/database/feis/animals/bird/gyca/all.html

Thapa, R. and M.W. Blair. 2018. Morphological assessment of cultivated and wild amaranth species diversity. *Agronomy* 2018, 8, 272. https://doi.org/10.3390/agronomy8110272

Thiers, B. 2020 [continuously updated]. Index Herbariorum: A global directory of public herbaria and associated staff. New York Botanical Garden's Virtual Herbarium. http://sweetgum.nybg.org/science/ih

Thomas, C. and J.R. Swanton. 1911. Indian languages of Mexico and Central America and their geographical distribution. *Bureau of American Ethnology Bulletin* 44. Government Printing Office, Washington, D.C.

Thomson, D.A., L.T. Findley, and A.N. Kerstitch. 2000. *Reef Fishes of the Sea of Cortez.* University of Texas Press, Austin.

Troncoso, F.P. 1905. *Las Guerras con las Tribus Yaqui y Mayo del Estado de Sonora.* Tipografía del Departamento de Estado Mayor, Mexico City.

Tropicos.org. 2020. Missouri Botanical Garden. [continuously updated]. http://www.tropicos.org

Turner, J.K. 1910. *Barbarous Mexico.* C.H. Kerr & Co., Chicago.

Turner, R.M., T.L. Burgess, and J.E. Bowers. 1995. *Sonoran Desert Plants, an Ecological Atlas.* University of Arizona Press, Tucson.

Uphof, J.C.T. 1968. *Dictionary of Economic Plants.* J. Cramer, New York.

van Rossem, A.J. 1931. Report on a collection of land birds from Sonora, Mexico. *Transactions of the of the San Diego Society of Natural History* 6:237–304.

van Rossem, A.J. 1945. A distributional survey of the birds of Sonora, Mexico. *Occasional Papers of the Museum of Zoology, Louisiana State University* 21.

Velarde, E., J.-L. E. Cartron, H. Drummond, D.W. Anderson, F. Frebon Gallardo, E. Palacios, and C. Rodriguez. 2005. Nesting seabirds of the Gulf of California's offshore islands: Diversity, ecology, and conservation. In: J.-L. E. Cartron, G. Ceballos, and R.S. Felger (editors), *Biodiversity, Ecosystems, and Conservation in Northern Mexico.* Oxford University Press, New York, pp. 452–470.

Villaseñor-Gómez, J.F., O. Hinojosa-Huerta, E. Gómez-Limón, D. Krueper, and A.D. Flesch. 2010. Avifauna, Appendice I, Especies de aves registradas en el estado de Sonora. In: F.E. Molina-Freaner and T.R. Van Devender (editors), *Diversidad Biológica de Sonora.* UNAM, Mexico City, pp. 385–420.

Wagner, C.J. 1936. Medicinal practices of the Yaquis. In: W.C. Holden et al., *Studies of the Yaqui Indians of Sonora, Mexico.* Texas Technological College Bulletin 12(1), Lubbock, pp. 79–90.

Watson, S. 1889. Contributions to American Botany: Upon a collection of plants made by Dr. E. Palmer in 1887, about Guaymas, Mexico, at Muleje and Los Angeles Bay in Lower California, and on the island of San Pedro Martin in the Gulf of California. *Proceedings of the American Academy of Arts and Sciences* 24:36–82. https://doi.org/10.2307/20021550

Wheeler, D.E. and S.W. Rissing. 2000. Ants. In: S.J. Phillips and P.W. Comus, *A Natural History of the Sonoran Desert.* Arizona-Sonora Desert Museum, Tucson, pp 349–351.

Wiggins, I.L. 1964. Flora of the Sonoran Desert. In: F. Shreve and I.L. Wiggins, *Flora and Vegetation of the Sonoran Desert*, 2 vols. Stanford University Press, Stanford, pp. 189–1740.

Wilder, B.T., J.L. Betancourt, C.W. Epps, R.S. Crowhurst, J.I. Mead, and E. Ezcurra. 2014. Local extinction and unintentional rewilding of bighorn sheep (*Ovis canadensis*) on a desert island. *PLoS ONE* 9(3):e91358. https://doi.org/10.1371/journal.pone.0091358

Wilder, C.S. 1940. The Yaqui Deer Dance, a study in culture change. Master's thesis, University of Arizona, Tucson.

Wilder, C.S. 1963. The Yaqui Deer Dance: a study in culture change. *Anthropological Papers 66, Bureau of American Ethnology Bulletin* 186.

Yetman, D. 2006. *The Organ Pipe Cactus.* University of Arizona Press, Tucson.

Yetman, D. and R.S. Felger. 2002. Ethnoflora of the Guarijíos. In: D. Yetman, *Guarijíos of the Sierra Madre: The Hidden People of Northwestern Mexico.* University of New Mexico Press, Albuquerque, pp. 174–230.

Yetman, D. and T.R. Van Devender. 2002. *Mayo Ethnobotany: Land, History, and Traditional Knowledge in Northwestern Mexico.* University of California Press, Berkeley.

Yuan, Z.-Y., W.W. Zhou, X. Chen, N.A. Poyarkov Jr.and J. Che. 2016. Spatiotemporal diversification of the true frogs (genus *Rana*): A historical framework for a widely studied group of model organisms. *Systematic Biology* 65:824–842. https://doi.org/10.1093/sysbio/syw055

Zambrano, F. 1961–1977. *Diccionario Bio-bibliográfico de la Compañia de Jesús en México.* 16 vols. Editorial Jus, Mexico City.

Zavala Castro, P. 1989. *Apuntes Sobre el Dialecto Yaqui.* Gobierno del Estado de Sonora, Secretaría de Fomento Educativo y Cultura, Hermosillo.

Zatarain Tumbaga, A. 2018. *Yaqui Indigeneity, Epistemology, Diaspora, and the Construction of Yoeme Identity.* University of Arizona Press, Tucson. https://doi.org/10.2307/j.ctt1zxsmtg

PART 6
APPENDICES

APPENDIX 6A. Yoeme plant names with scientific, Spanish, and English equivalent names. Non-native plants are marked with an asterisk (*).

Equivalents of Yoeme Plant Names			
Yoem noki	**Scientific name**	**Spanish**	**English**
aaki	*Stenocereus thurberi*	*pitaya dulce*	organpipe cactus
aapio	**Apium graveolens*	*apio*	celery
aasam	*Descurainia pinnata*	*pamita*	tansy mustard
aasos	**Allium sativum*	*ajo*	garlic
aiya	*Guazuma ulmifolia*	*guásima*	picklenut
alfidillo	**Erodium cicutarium*	*alfilerillo*	filaree
alvaaka	**Ocimum basilicum*	*albahaca*	sweet basil
alverkooki	**Prunus armeniaca*	*chabacao*	apricot
arosim	**Oryza sativa*	*arroz*	rice
au'ori	*Proboscidea parviflora*	*aguaro, cuernitos, torito, uña de gato*	devil's-claw, unicorn-plant
awa'aro	*Proboscidea altheifolia*	*cuernitos, gato, torito, uña de gato*	golden devil's-claw, desert unicorn-plant
avaso	*Populus deltoides*	*álamo*	Frémont cottonwood
avaso	*Populus mexicana*	*álamo*	Yaqui cottonwood
aveena	**Avena sativa*	*avena*	oat
aya'awi	*Cucurbita argyrosperma*	*calabaza*	cushaw squash
ayal	**Crescentia alata*	*jícaro, tecomate*	gourd tree
baiburilla	*Dorstenia drakena*	*baiburilla*	---
bwaarom	*Portulaca oleracea*	*verdolaga*	purslane
bwaarom	*Sesuvium verrucosum*	*verdolaga de playa*	western sea-purslane
bwasu'ubwila	*Aristolochia watsonii*	*hierba del indio*	desert pipevine, Indian-root
bugambilia	*Bougainvillea spectabilis*	bugambilia	bougainvillea
champuusi	*Rhynchosia precatoria*	*ojo de chanate*	precatory bean, rosary bean

Equivalents of Yoeme Plant Names			
Yoem noki	**Scientific name**	**Spanish**	**English**
chichiham	*Phoradendron californicum*	*toji*	desert mistletoe
chichiham	*Psittacanthus calyculatus*	*toji*	---
chichiham	*Struthanthus palmeri*	*toji*	---
chichivo, chi'ichivo	*Ambrosia confertiflora*	*estafiate*	slimleaf ragweed
chiihu	*Indigofera suffruticosa*	*añil*	indigo plant
chiini	**Gossypium hirsutum*	*algodón*	cotton
chiini	**Talipariti tiliaceum*	*majagua del mar*	wild cotton tree, sea hibiscus
chiiya	*Salvia columbariae*	*chia*	chia
chiiya	**Salvia hispanica*	*chia*	chia
chikul aaki, chikul hu'i	*Mammillaria grahamii, M. yaquensis*	*cabeza de viejo*	fishhook or pincushion cactus
chiniita	**Sonchus oleraceus*	*chinita*	common sow-thistle
chirikoote	*Erythrina flabelliformis*	*chilicote*	coral bean
chiva kovam	**Tribulus terrestris*	*toboso, torito*	goathead, puncture vine
chiva'ato himsi, chiva'ato himsita saila	*Clematis drummondii*	*barba de chivata*	old man's beard, Texas virgin's bower
cho'oko vaso	Poaceae	*un zacate*	a grass (unidentified)
choa	*Cylindropuntia*	*cholla, choya*	cholla
choali	*Chenopodium neomexicanum*	*choali, chual*	lamb's quarters
choi, cho'i	*Parkinsonia praecox*	*brea, palo brea*	brea palo verde
chukui kuta	*Vachellia constricta*	*mezquitillo*	white-thorn acacia
chukui tooro	*Bursera laxiflora*	*copalquín, torote prieto*	red-bark elephant-tree
chuuhi	**Chenopodiastrum murale*	*choali, chual*	netleaf goosefoot
chuuna	**Ficus carica*	*higuera*	fig
echo	*Pachycereus pecten-aboriginum*	*etcho*	---
ee oona	*Euphorbia* spp.	*golondrinas*	desert spurges
ehea	*Olneya tesota*	*palo fierro*	ironwood

Equivalents of Yoeme Plant Names			
Yoem noki	**Scientific name**	**Spanish**	**English**
ekonia	*Hechtia montana*	*aguamita, mescalito*	---
epasoote	**Dysphania anthelmintica*	*epazote*	wormseed
hachi'ihtia	*Encelia farinosa*	*incienso, rama blanca*	brittlebush
hachi'ihtia	*Verbesina encelioides*	*mirasol*	golden crownbeard
hapa'awi	**Casimiroa edulis*	*chapote, zapote blanco*	white sapote
havelina	*Ruellia californica*	*chuparrosa, rama parda*	---
hawesowi	*Pachycereus pringlei*	*cardón, sahueso*	cardon
heeko, heko	*Baccharis sarothroides*	*escoba amargo, romerillo*	desert broom
heeko nawia	*Ambrosia salsola*	---	white burrobush
heoko kuta	*Sphaeralcea coulteri*	*mal de ojo*	annual globe-mallow
heoko kuta	*Sphaeralcea emoryi*	*mal de ojo*	globe mallow
hese'i	*Ipomoea arborescens*	*palo santo*	tree morning-glory
heseim	*Phaseolus acutifolius*	*tépari*	brown tepary
hiak viva	*Nicotiana rustica*	*tabaco*	Yoeme tobacco
híchuiquia	*Desmanthus bicornutus*	*dais*	bundle-flower
hiitepoa	*Desmanthus covillei*	*dais*	Coville's bundle-flower
hi'ito	*Forchhammeria watsonii*	*palo jito*	lollipop tree
hiowe	*Ambrosia ambrosioides*	*chicura*	canyon ragweed
hohoova	*Simmondsia chinensis*	*jojoba*	jojoba
ho'opo, hopo	*Piscidia mollis*	*palo blanco*	fish-poison tree
hooso	*Albizia sinaloensis*	*joso, palo joso*	---
hosoina	*Randia echinocarpa*	*papache, papache borracho*	---
hoyo kuta	*Pleradenophora bilocularis*	*hierba de la flecha*	arrow-poison plant
hua muuni, hua se'elaim	*Phaseolus acutifolius*	*tépari*	wild tepary
huchahko	*Haematoxylum brasiletto*	*brasil*	brazilwood
hupa chumi	*Ruellia californica*	*chuparrosa, rama parda*	---

Equivalents of Yoeme Plant Names			
Yoem noki	**Scientific name**	**Spanish**	**English**
hupsi	**Diospyros sonorae*	*guayparín*	Sonoran persimmon
hupsi	*Randia thurberi*	*papache borracho*	---
hutu'uki	*Sarcomphalus obtusifolius*	*abrojo, marcha*	graythorn
hutu'uki	*Lycium fremontii*	*salicieso*	Frémont wolfberry
huu'upa	*Prosopis articulata*	*mezquite amargo*	bitter mesquite
huu'upa	*Prosopis glandulosa* var. *torreyana*	*mezquite*	western honey mesquite
huu'upa	*Prosopis velutina*	*mezquite*	velvet mesquite
hu'upa keka'ala	*Condalia globosa*	*crucillo*	bitter condalia
hu'upa keka'ala	*Senegalia greggii*	*uña de gato*	catclaw acacia
huvahe	*Vitex mollis*	*uvalama*	---
huvak vena	*Atamisquea emarginata*	*palo hediondo, palo zorillo*	desert caper
huya aasam	*Descurainia pinnata*	*pamita*	tansy mustard
huya ko'oko'i	*Capsicum annuum* var. *glabriusculum*	*chiltepín, chiltipiquín*	chiltepin, wild chile
huya tono'oara	*Euphorbia cymosa*	*candelilla, jumete*	---
huya'awo, huyawo	*Guaiacum coulteri*	*guayacán*	---
kaape	**Coffea arabica*	*café*	coffee
kaavansa, kaavapsa	**Cicer arietinum*	*garbanzo*	chickpea, garbanzo bean
kafe	**Coffea arabica*	*café*	coffee
kakawaate	**Arachis hypogaea*	*cacaguate*	peanut
kama	Cucurbita argyrosperma	*calabaza*	cushaw squash
kame'eroi	**Xanthium strumarium*	*huicholi, mata de cadillo*	cocklebur
kamoote	**Ipomoea batatas*	*camote*	sweet potato
kandelia	*Euphorbia lomelii*	*candelilla*	slipper plant
kannao	**Punica granatum*	*granada*	pomegranate
kanteela	*Euphorbia lomelii*	*candelilla*	slipper plant
kanutio	*Ephedra*	*canutillo, tepopote*	joint-fir

Equivalents of Yoeme Plant Names			
Yoem noki	**Scientific name**	**Spanish**	**English**
kapa	*Chenopodium neomexicanum*	*choali, chual*	lamb's quarters
kapo, kapo seewa	*Nymphaea elegans*	*capomo*	tropical royal-blue water lily
kartamo	**Carthamus tinctorius*	*cártamo*	safflower
kau chaani	*Ibervillea sonorae*	*guarequi*	cow-pie plant
kau chuuna	*Ficus insipida*	*chalate*	---
kau ee oona	*Tidestromia lanuginosa*	*hierba ceniza, hierba lanuda*	honeysweet
kau heeko	*Ambrosia monogyra*	*hierba de pasmo, jécota*	slender burrobush
kau howo	*Pleradenophora bilocularis*	*hierba de la flecha*	arrow-poison plant
kau huvahe	*Vitex mollis*	*uvalama*	---
kau ohasen	*Senna covesii*	*daisillo, hojasen*	desert senna
kau sapo	*Jatropha cardiophylla*	*palo sangrón, sangrengado*	limberbush
kau sapo	*Jatropha cordata*	*papelio*	---
kau sikro'opo	*Malpighia watsonii*	*granadilla*	---
kau tavachin	**Erythrostemon gilliesii*	*tabachín amarillo*	yellow bird-of-paradise
kau toma'arisi	*Physalis crassifolia*	*tomatillo del desierto*	desert ground-cherry
kau vattai	*Rumex hymenosepalus*	*cañaigre, hierba colorado*	Arizona dock
kau vattai	*Rumex inconspicuus*	*hierba colorado*	---
kaupo hi'u	**Brassica tournefortii*	*mostasa cimarona, quelite cimarona*	Sahara mustard
kevenia	**Ricinus communis*	*higuerilla*	castor bean
koapa'im	*Euphorbia* spp.	*golondrina*	desert spurges
koni woki	**Sisymbrium irio*	*pamita*	London rocket
ko'oko'im	**Capsicum annuum*	*chiles*	chiles
kooni saaki	*Cyperus esculentus*	*cebollín, coquillo amarillo*	yellow nutgrass
koowi choali	**Chenopodiastrum murale*	*choali, chual*	netleaf goosefoot

Equivalents of Yoeme Plant Names			
Yoem noki	**Scientific name**	**Spanish**	**English**
koowi saawa	*Jatropha macrorhiza*	---	ragged jatropha
koowi tami	*Vachellia campechiana*	*güinolo*	boat-spine acacia
koowi veyootam	*Quercus emoryi*	*bellota*	Emory oak
kopal ouwo	*Encelia farinosa*	*incienso, rama blanca*	brittlebush
kopal ouwo	*Encelia halimifolia*	---	----
kopalkin	*Encelia farinosa*	*incienso, rama blanca*	brittlebush
kopalkin	*Hintonia latiflora*	*amargo, copalquín*	---
korai	**Sonchus oleraceus*	*chinita*	common sow-thistle
kosawe, kosawi	*Krameria*	*cósahui*	ratany
kovanao	*Larrea tridentata*	*gobernador, hediondilla*	creosotebush
kowi bwaarom	*Trianthema portulacastrum*	*verdolaga de cochi*	horse purslane
kuh kuta	*Citharexylum flabellifolium*	---	---
kuh kuta	*Lycium andersonii*	*salicieso*	desert wolfberry
kuh kuta	*Phaulothamnus spinescens*	*mal de ojo, putilla*	snake eyes
kuka	*Vachellia farnesiana*	*huisache, vinorama*	sweet acacia
kumaro	*Celtis reticulata*	*cúmero*	canyon hackberry
kume'a ouwo	*Erythrostemon palmeri*	*palo piojo*	---
kunwo	*Celtis pallida*	*cúmero, garambullo*	desert hackberry
kus kuta	*Citharexylum flabellifolium*	---	---
kus kuta	*Lycium andersonii*	*salicieso*	desert wolfberry
kus kuta	*Phaulothamnus spinescens*	*mal de ojo, putilla*	snake eyes
kusim	*Quercus emoryi*	*bellotas*	acorns
kuta kama	*Cucurbita argyrosperma*	*calabaza*	cushaw squash
kuu'u	*Agave*	*maguey*	agave, century plant
kuu'u	*Agave angustifolia*	*bacanora, lechugilla*	narrow-leaf agave
kuu'u	*Agave colorata*	*maguey*	banded agave
kuu'u	**Agave fourcroydes*	*henequén*	henequen
laatiko	**Phoenix dactylifera*	*datillo*	date palm
laureel	**Nerium oleander*	*laurel*	oleander

Equivalents of Yoeme Plant Names			
Yoem noki	**Scientific name**	**Spanish**	**English**
lechuuwa	**Lactuca sativa*	*lechuga*	lettuce
liima	**Citrus aurantiifolia*	*limón*	lime
liima	**Citrus ×limon*	*limón*	lemon
lirio	*Hymenocallis sonorensis*	*cebolla de coyote*	Sonoran spider lily
lorio	**Tecoma stans* var. *stans*	*flor de fortuna, lluvia de oro*	yellow bells
machao	*Lysiloma watsonii*	*tepeguaje*	feather tree
mahkoapa'i	*Euphorbia*	*golondrina*	desert spurges
mako'ochiini	**Pithecellobium dulce*	*guamúchil*	Manila tamarind, monkey pod
malva	**Malva parviflora*	*malva*	cheeseweed
mamya	**Solanum americanum*	*chichiquelite*	black nightshade
mango	**Mangifera indica*	*mango*	mango
mansaana	**Malus* hybrids	*manzana*	apple
mansania	**Matricaria chamomilla*	*manzanilla*	chamomile
mansaniata saila	*Perityle californica*	---	California rock daisy
mansaniata saila	*Perityle microglossa*	---	---
manto	**Ipomoea carnea*	*palo santo de Castilla*	bush morning-glory
masa'asai, masa'asai wiroa	*Antigonon leptopus*	san miguelito	queen's wreath
maso kuta	*Bebbia juncea*	*hierba ceniza*	sweetbush
maso kuta	*Porophyllum gracile*	*hierba del venado*	odora
maso kuta	*Porophyllum pausodynum*	---	---
maso pipi	*Funastrum heterophyllum*	*güirote*	vining milkweed
maso puusim	**Mucuna pruriens*	*ojo de venado*	sea bean
mastaoka	*Passiflora arida*	---	desert passion flower
memrio	**Cydonia oblonga*	*membrillo*	quince
minai	**Cucumis melo* var. *reticulatus*	*melón*	cantaloupe
mochi	*Boerhavia*	---	spiderling

Equivalents of Yoeme Plant Names			
Yoem noki	**Scientific name**	**Spanish**	**English**
mo'oko	*Suaeda nigra*	*chamiso*	sea blite, seepweed
mo'oko vaso	**Cynodon dactylon*	*zacate bermuda, zacate inglés*	Bermuda grass
mohtaasa	**Brassica tournefortii*	*mostasa cimarona*	Sahara mustard
mohtaasa	**Sinapis arvensis*	*mostasa*	charlock
moochi	*Boerhavia*	---	spiderling
moora	**Morus alba*	*mora*	white mulberry
moora	*Piscidia mollis*	*palo blanco*	fish-poison tree
moosenino	*Zostera marina*	*trigo del mar*	eelgrass
mumsa	*Plantago ovata*	*pastora*	woolly plantain, Indian-wheat
mureo	*Fouquieria diguetii*	*palo adán*	Adam's tree
mureo	*Fouquieria macdougalii*	*ocotillo macho, palo adán*	tree ocotillo
mureo	*Fouquieria splendens*	*ocotillo*	ocotillo
museo	*Lophocereus schottii*	*músaro, sinita*	senita
muuni	**Phaseolus vulgaris*	*frijol*	common bean, pinto bean
naaka	---	---	shelf fungus
na'aso (fruit), na'aso ouwo (tree)	**Citrus sinensis*	*naranja*	orange
naavo	*Opuntia*	*nopal* (plant), *tuna* (fruit)	prickly-pear
nakkaim	*Opuntia bravoana*	---	---
nakkaim	*Opuntia gosseliniana*	*duraznillo*	Sonoran purple prickly-pear
nakkaim	*Opuntia santa-rita*	*duraznillo*	Santa Rita prickly-pear
nahnawa'ara	*Rhizophora mangle*	*mangle colorado*	red mangrove
naka'apuli, nakapuri	*Ficus pertusa*	*nacapule, nacapuli*	Central American banyan
namu rokoa	Algae	*algas marinas, lama*	algae, seaweed
naoto'oria	*Dorstenia drakena*	*baiburilla*	---

Equivalents of Yoeme Plant Names			
Yoem noki	**Scientific name**	**Spanish**	**English**
naowo	*Chloracantha spinosa*	---	spiny aster
naranhio	*Sarcomphalus amole*	*amole, saituna*	---
nata'e	*Tragia jonesii*	*ortiga, rama quemadora*	noseburn
nawi'o	*Mariosousa heterophylla*	*palo blanco*	
noono	*Echinocereus leucanthus*	---	---
noono	*Peniocereus greggii*	*reina de la noche, sarramatraca*	desert night-blooming cereus
noono	*Peniocereus marianus*	*sacamatraca, sarramatraca*	Sonoran night-blooming cereus
noono	*Peniocereus striatus*	*cardoncillo, sacamatraca, sarramatraca*	---
ochoko kuta	*Sphaeralcea emoryi*	*mal de ojo*	globe mallow
ono'e	*Ferocactus*	*biznaga*	barrel cactus
oregano	*Lippia palmeri*	*orégano*	---
orholiinim	**Sesamum indicum*	*ajonjolí*	sesame
ovei	*Brahea brandegeei*	*babiso, palmilla*	hesper palm
paapa, paapam	**Solanum tuberosum*	*papa, papas*	potato, potatoes
pamiitam	*Descurainia pinnata*	*pamita*	tansy mustard
paseo	*Avicennia germinans*	*mangle blanco*	black mangrove
paseo	*Tricerma phyllanthoides*	*mangle dulce*	
pato puusi	*Sagittaria longiloba*	*flechas de agua*	arrowhead, broadleaf arrowhead
pawis	**Ligusticum porteri*	*chuchupate*	osha
peonasim	**Pisum sativum*	*chicharos*	peas
petootam	*Blitum nuttallianum*	---	poverty weed
piino	**Tamarix aphylla*	*pino salado*	athel pine, saltcedar
piino, piino moraom sesewame	**Tamarix chinensis*	*pino salado*	saltcedar
piisi	*Randia echinocarpa*	*papache, papache borracho*	

Equivalents of Yoeme Plant Names			
Yoem noki	**Scientific name**	**Spanish**	**English**
piisi	*Randia obcordata*	*papache*	---
piisi	*Randia thurberi*	*papache borracho*	---
pinya	**Ananas comosus*	*piña*	pineapple
platano	**Musa ×paradisiaca*	*platano*	banana
pochoote	*Ceiba aesculifolia*	*pochote*	kapok tree, silk-cotton tree
pomahe	*Cordia sonorae*	*palo de asta*	---
ravano	**Raphanus sativus*	*rabano*	radish
repooyo	**Brassica oleracea*	*berza*	cabbage
riptia	*Abutilon palmeri*	---	---
riptia	*Sphaeralcea* spp.	*mal de ojo*	globe mallow
roira, roiya	*Lycium andersonii*	*salicieso*	desert wolfberry
ron huan	**Nicotiana glauca*	*don juan, san juanito, tabacón*	tree tobacco
roosam	**Rosa* hybrids	*rosas*	roses
rurahno	**Prunus persica*	*durazno*	peach
ruura	**Ruta graveolens*	*ruda*	rue
saa tooro	*Bursera fagaroides*	*torote amarillo, torote de venado*	fragrant elephant-tree
saa tooro	*Bursera microphylla*	*torote*	elephant-tree
saamo	*Coursetia glandulosa*	*sámota*	samota
saavila	**Aloe vera*	*sávila*	aloe vera
saawa	*Amoreuxia palmatifida*	*saiya*	---
sakovai, sakvai	**Citrullus lanatus*	*sandia*	watermelon
samo	*Coursetia glandulosa*	*sámota*	samota
sana	**Saccharum officinarum*	*caña de azúcar*	sugarcane
saneal	*Prosopis articulata*	*mezquite amargo*	bitter mesquite
sanooria	**Daucus carota*	*zanahoria*	carrot
santa puusim	*Rhynchosia precatoria*	*ojo de chanate*	precatory bean, rosary bean

Equivalents of Yoeme Plant Names			
Yoem noki	**Scientific name**	**Spanish**	**English**
sapo	*Jatropha cinerea*	*palo sangrón, sangrengado*	ashy limberbush
sapo	*Jatropha cuneata*	*matacora, sangrengado*	limberbush
sauko	**Sambucus cerulea*	*tápiro* (tree), *sauco* (flowers)	blue elderberry
sauwo	*Carnegiea gigantea*	*saguaro*	saguaro
se'elai	*Phaseolus acutifolius*	*tépari*	white tepary
seeva, seevam	**Hordeum vulgare*	*sevada*	barley
semalulukut kuta	*Tabebuia impetiginosa*	*amapa, amapa rosa*	
seve'e choa	*Cylindropuntia fulgida*	*choya, velas de coyote*	chain-fruit cholla, jumping cholla
sevii	*Cylindropuntia arbuscula*	*tasajo*	pencil cholla
sevii	*Cylindropuntia thurberi*	*siviri*	staghorn cholla
sevoa'a	*Cryptantha* spp., *Johnstonella* spp.	---	---
sevoa'ara	*Sphaeralcea coulteri*	*mal de ojo*	annual globe-mallow
sevora	**Allium cepa*	*cebolla*	onion
sewalka	*Cucurbita* sp.	---	---
sewalulukut	*Justicia californica*	*chuparrosa*	desert hummingbird-bush
si'iya	**Matricaria chamomilla*	*manzanilla*	chamomile
siari kuta	*Parkinsonia* spp.	*palo verde*	palo verde
siari paseo	*Laguncularia racemosa*	*mangle blanco*	white mangrove
siari viiva	**Cannabis sativa*	*marijuana*	marijuana
sikro'opo	*Lycium andersonii*	*salicieso*	desert wolfberry
sikro'opo, sikropo'i	*Lycium brevipes*	*salicieso*	desert wolfberry
sikro'opo, sikropo'i	*Lycium californicum*	---	California wolfberry
silaantro	**Coriandrum sativum*	*cilantro*	cilantro, coriander
silweela	**Prunus domestica*	*ciruela*	plum
sina	*Stenocereus alamosensis*	*sina*	snake cactus

Equivalents of Yoeme Plant Names			
Yoem noki	**Scientific name**	**Spanish**	**English**
sita'avao	*Vallesia glabra*	*citavaro, huevito*	
site'epoa	*Desmanthus covillei*	*dais*	Coville's bundle-flower
soosa	*Solanum erianthum*	*cornetón del monte*	tree nightshade
sooya	**Glycine max*	*soya*	soybean
sutu'ura	*Pisonia capitata*	*garabata, vainoro*	---
suva'u muunim	*Phaseolus acutifolius*	*tépari*	wild tepary
taa'ata vitchu	**Helianthus annuus*	*mirasol*	sunflower
taa'ata vitchu	*Verbesina encelioides*	*mirasol*	golden crownbeard
tabwikoseewa	*Caesalpinia pulcherrima*	*tavachín*	red bird-of-paradise
tabwikoseewa	**Erythrostemon gilliesii*	*tabachín amarillo*	yellow bird-of-paradise
tahiwechia naavo, taiwechia naavo	*Opuntia bravoana*	*nopal*	---
tahiwechia naavo, taiwechia naavo	*Opuntia engelmannii*	*nopal*	desert prickly-pear
tahkali	*Juniperus* spp.	*taskale, taskate,*	juniper
tahsi'o	*Bonellia macrocarpa*	*san juanico*	---
tai pusi	*Croton texensis*	*tortolita*	Texas croton
tako	*Sabal uresana*	*palma del taco*	Sonoran palmetto
tamarindo	**Tamarindus indica*	*tamarindo*	tamarind
tamko'okochi	**Martynia annua*	*aguaro, uña de gato*	cat's claw
tahsi'o, tasi'o, tassio	*Bonellia macrocarpa*	*san juanico*	---
tatchi'ina	*Argemone gracilenta*	*cardó*	prickly poppy
tava'i	*Sporobolus airoides*	*zacatón alcalino*	alkali sacaton
tavachin	**Delonix regia*	*flamboyán*	flame tree, royal poinciana
tavachin	**Leucaena leucocephala*	*guaje*	white popinac
tebwi	*Datura discolor*	*toloache*	desert thorn-apple, poisonous nightshade

Equivalents of Yoeme Plant Names			
Yoem noki	**Scientific name**	**Spanish**	**English**
teevo	**Melilotus albus*	*trébol agrio*	white-flowered sweet clover
teevo	*Melilotus indicus*	*trébol agrio*	yellow sour clover
te'owe	*Parkinsonia florida*	*palo verde, palo verde azul*	blue palo verde
teso	*Senegalia occidentalis*	*teso*	Mexican catclaw acacia
teta'ahao	*Cucurbita digitata*	*calabacilla*	coyote gourd
tiiko	**Triticum aestivum*	*trigo*	wheat
toh ouwo	*Atriplex polycarpa*	*chamiso cenizo*	desert saltbush
toma'arisi	*Physalis pubescens*	*tomatillo*	hairy ground-cherry
tomatillo	**Physalis philadelphica*	*tomate*	husk-tomato
too vichom	*Cardiospermum corindum*	*farolitos, huevo de toro, tronadór*	balloon vine
tooko huya, toroko huya	*Abutilon incanum*	---	---
tooko huya, toroko huya	*Abutilon palmeri*	---	---
tooko huya, toroko huya	*Atriplex elegans*	*chamiso cenizo*	wheel-scale orache
tooko huya, toroko huya	*Encelia farinosa*	*incienso, rama blanca*	brittlebush
tooko huya, toroko huya	*Krameria* spp.	*cósahui*	ratany
tooko huya, toroko huya	*Solanum erianthum*	*cornetón del monte*	tree nightshade
tooko huya, toroko huya	*Sphaeralcea emoryi*	*mal de ojo*	globe mallow
toora	*Cardiospermum corindum*	*farolitos, huevos de toro, tronadór*	balloon vine
tooro	*Bursera microphylla*	*torote*	elephant-tree
toorom	**Tribulus terrestris*	*toboso, torito*	goathead, puncture vine
toowo	*Tabebuia impetiginosa*	*amapa, amapa rosa*	---

Equivalents of Yoeme Plant Names			
Yoem noki	**Scientific name**	**Spanish**	**English**
tosai naavo	**Opuntia ficus-indica*	*nopal*	Indian-fig prickly-pear
tovei	*Brahea brandegeei*	*babiso, palmilla*	hesper palm
tronpio	**Ipomoea tricolor*	*trompillo*	morning-glory
tupche	*Sapindus saponaria*	*amolillo, chirrión, jaboncillo*	soapberry
tuuli	*Typha domingensis*	*tule*	cattail
uuva	**Vitis vinifera*	*uva*	grape
vaaka	**Arundo donax*	*baca, carrizo*	cane, giant reed
va'ako	*Phaulothamnus spinescens*	*mal de ojo, putilla*	snake eyes
vaa ko'oko'i	*Heliotropium angiospermum*	---	---
vaa minai	**Cucumis melo* var. *dudaim*	*melón de coyote, meloncillo de coyote*	coyote melon, dudaim melon, stink melon
vaa va'ako	**Cyperus rotundus*	*cebollin, coquillo morado*	purple nutgrass, purple nutsedge
vaa vaso	*Eleocharis macrostachya*	*zacate del agua*	mountain spikerush
vaa vaso	*Zostera marina*	*trigo del mar*	eelgrass
vaa vikam	*Sagittaria longiloba*	*flechas de agua*	arrowhead, broadleaf arrowhead
vachi	*Zea mays*	*maíz*	corn, maize
vachomo	*Baccharis salicifolia*	*batamote*	seepwillow
vaeka'a	*Rumex inconspicuus*	*hierba colorado*	---
vaekio	*Sesbania herbacea*	*bequilla*	river-hemp
vahewo	*Pisonia capitata*	*garabata, vainoro*	---
vai mamya	*Atriplex barclayana*	*saladillo*	coast saltbush
vai mansania	*Palafoxia linearis*	---	---
vai mo'oko	*Allenrolfea occidentalis*	*chamiso*	iodine bush
vai mo'oko	*Arthrocnemum subterminale*	---	pickleweed
vai muuni	*Batis maritima*	---	saltwort
vai muuni	*Sesuvium portulacastrum*	*verdolago de playa*	sea purslane

Equivalents of Yoeme Plant Names			
Yoem noki	**Scientific name**	**Spanish**	**English**
vai tava'i	*Sporobolus cryptandrus*	---	sand dropseed
vaka'apo, vaka'apoa, vaka'aporo	**Parkinsonia aculeata*	*bagota, retama*	Mexican palo verde
vakalaume, vakau	**Phyllostachys aurea*	*bambú*	Chinese bamboo, golden bamboo
vakau	*Otatea acuminata*	*otate*	weeping bamboo
vakot muteka	**Solanum lycopersicum*	*tomatillo, tomate*	tomato
vamyo	*Lysiloma divaricatum*	*mauto*	---
vapsa	*Sideroxylon occidentale*	*bebelama*	bumelia
vaso	Poaceae	*zacate*	grass
vattai	*Rumex inconspicuus*	*hierba colorado*	---
vauwo	*Ceiba aesculifolia*	*pochote*	kapok tree, silk-cotton tree
vavis	*Anemopsis californica*	*hierba de manso*	
veak vena	*Tricerma phyllanthoides*	*mangle dulce*	---
vetaveel	**Beta vulgaris*	*betabel, remolacha azucarena*	sugar beet
viiva	**Nicotiana tabacum*	*tabaco*	tobacco
visa'e	**Lagenaria siceraria*	*bule, guaje*	bottle gourd
vivino	*Condea albida*	*salvia*	desert lavender
voak	"molds"	*molde de alimentos*	food mold
vuru sisi, vuu sisi	*Battarrea* spp., *Podaxis pistillaris*	---	stalked puffballs (fungi)
vuu naka	*Sphaeralcea emoryi*	*mal de ojo*	globe mallow
waevas, waivas	**Psidium guajava*	*guayava*	guava
waharom	**Luffa aegyptiaca*	*estropajo*	loofah, sponge gourd
waharom	*Luffa quinquefida*	*estropajo de coyote*	coyote loofah
wata	*Salix gooddingii*	*sauce, sauz*	Goodding willow
wee'e	*Amaranthus fimbriatus*	*bledo, quelitillo*	fringed amaranth
wee'e	*Amaranthus palmeri, Amaranthus watsonii*	*bledo, quelite*	careless weed, pigweed

Equivalents of Yoeme Plant Names			
Yoem noki	**Scientific name**	**Spanish**	**English**
wicha'apoi, wicha'apoli	*Cenchrus palmeri*	*guachapori, huizapori, toboso*	giant sandbur
wicha'apoi, wicha'apoli	**Cenchrus spinifex*	*guachapori, huizapori, toboso*	common sandbur, field sandbur
wicha'apoi, wicha'apoli	**Tribulus terrestris*	*toboso, torito*	goathead, puncture vine
wicho'e, wicho'e kauwa'ara	*Funastrum clausum*	*huiroa*	---
wikit woki	**Sisymbrium irio*	*pamita*	London rocket
wivis kuu'u	*Tillandsia exserta*	*quiqui*	---
wivis kuu'u	*Tillandsia recurvata*	*quiqui*	ballmoss
wo'i ko'oko'i	*Rivina humilis*	*chile de coyote*	bloodberry, rouge plant
wo'i minai	*Cucumis melo* var. *dudaim*	*melón de coyote, meloncillo de coyote*	coyote melon, dudaim melon, stink melon
wo'i si'iya	*Pectis* spp.	*manzanilla de coyote*	cinchweed, desert chinchweed
wo'i si'iya	*Thymophylla concinna*	*manzanilla de coyote*	dogweed
wo'i vaaka	*Phragmites australis*	*carrizo*	common reed, reedgrass
wo'i va'am	*Parkinsonia microphylla*	*palo verde*	foothill palo verde
wo'i viva	*Nicotiana obtusifolia*	*tabaquillo de coyote*	coyote tobacco, desert tobacco
wo'i voa'am	*Parkinsonia microphylla*	*palo verde*	foothill palo verde
wokkoi aaki	*Stegnosperma halimifolium*	*ojo de zanate*	---
woko	**Pinus* spp.	*pino*	pine
wokohna	*Havardia sonorae*	*jócono*	Sonoran ebony
wokovavase'ela	*Marsilea vestita*	*trébol de agua*	hairy water-clover
wotovo	*Cordia parvifolia*	*vara prieta*	desert cordia, littleleaf cordia
yerba pahmo	*Baccharis pteronioides*	*hierba de pasmo*	hierba de pasmo
yerba pahmo	*Isocoma tenuisecta*	*hierba de pasmo*	burroweed, golden burrowed

Equivalents of Yoeme Plant Names			
Yoem noki	**Scientific name**	**Spanish**	**English**
yerba pahmo	*Xylothamia diffusa*	*hierba de pasmo*	---
yerva bwena, yerva wena	**Mentha* sp.	*hierba buena*	common mint
yoi muunim, yori muunim	**Vigna unguiculata* subsp. *unguiculata*	*guisante negro de ojos*	black-eyed pea
yoi sana	**Saccharum officinarum*	*caña de azúcar*	sugarcane
yonso vaso	**Sorghum halepense*	*zacate Johnson*	Johnson grass
yotui kova	*Mammillaria bocensis, Mammillaria johnstonii*	*biznaguita*	---
yo vakau	*Otatea acuminata*	*otate*	weeping bamboo
yo vakau	**Phyllostachys aurea*	*bambú*	Chinese bamboo, golden bamboo
yukateeko	**Ficus benjamina*	*yucateco*	Benjamin fig
yuku wiroa	*Cissus trifoliata*	*tumba casa*	Arizona grape-ivy, sorrel vine
yuku wiroa	*Cissus verticillata*	*tumba casa*	---
yuku wiroata saila	*Cissus* sp.?	---	---

APPENDIX 6B. Yoeme animal life names with corresponding scientific, Spanish, and English names. Non-native animals are marked with an asterisk (*).

Equivalents of Yoeme Animal Names			
Yoem noki	**Scientific name**	**Spanish**	**English**
	CRUSTACEANS	***Crustáceos***	**Crustaceans**
	DECAPODA	*camarones, cangrejos, langostas*	crabs, lobsters, shrimp
acha kaari	*Callinectes bellicosus*	*jaiba azul*	blue crab
acha kaari	Brachyura	*cangrejo*	crab
acha kaari	*Uca* spp.	*cangrejo*	fiddler crab
haiva	*Callinectes bellicosus*	*jaiba azul, jaiba guerrera*	blue crab
kamaron, kamaroonim	Penaeidae	*camarón*	shrimp
lonwosta	*Panulirus* spp.	*langosta*	lobster, spiny lobster
kochimai	Penaeidae	*camarón*	shrimp
	MYRIOPODA	***cienpiéses, milpiéses***	**centipedes, millipedes**
eye'ekoe	*Orthopterus ornatus*	*milpiés*	giant desert millipede
masiwe	*Scolopendra heros*	*cienpiés gigante del desierto*	giant desert centipede
	CHELICERATA	***alacránes, arañas***	**scorpions, spiders**
chinchim	Trombiculidae	*barbourin*	chigger
chukui huvahe	*Latrodectus* spp.	*viuda negra*	black widow
husai huvahe	*Loxosceles reclusa*	*araña reclusa parda*	brown recluse spider
huvahe	Araneae	*araña*	spider
kovatarau	*Eremobates* spp.	*araña panzona, matavenado*	sun spider, wind scorpion
maachil	Scorpiones	*alacrán*	scorpion
maisooka	*Aphonopelma* spp.	*tarántula desértica*	desert tarantula
tema'i	*Dermacentor* sp.	*garrapata*	dog tick

Equivalents of Yoeme Animal Names			
Yoem noki	**Scientific name**	**Spanish**	**English**
yoeriam	INSECTA	***insectos***	insects
bwichia vo'ala	*Estigmene* sp.	*oruga* (*de Estigmene*)	tiger moth caterpillar
bwita maival	*Canthon* spp.	*escarabajo pelotero*	dung beetle
chinchi	**Cimex lectularius*	*chinche*	bedbug
chinchi	*Triatoma* spp.	*chinche besucona, chinche picuda*	cone-nosed bug, kissing bug
chukui eeye	Formicidae	*hormiga negra*	black ant
chunkuriam	Chironomidae	*larvas de mosquito*	midge larvae
chuu etem	*Ctenocephalides canis*	*pulga*	dog flea
eesuki	Formicidae	*hormiga*	ant
eeye	Formicidae	*hormiga*	ant
ete	*Pediculus humanus*	*piojo*	louse
hima'awikia	*Hemipepsis, Pepsis,*	*avispa cazador de arañas*	spider or tarantula wasp
hoovo'e	Formicidae	*hormigas naranjas*	orange-colored ants
huvachinai	*Eleodes* spp.	*pinacate*	darkling beetle, pinacate beetle
kampo moochi	Mantodea	*campamocha*	praying mantis
kiichul	Gryllidae	*grillo*	cricket
kukusaka	*Xylocopa* spp.	*abeja carpintera*	carpenter bee
kuliichi	*Lepisma saccharina*	*pez plateado*	silverfish
kuupis	Lampyridae	*luciérnaga*	firefly
maival	*Cotinus mutabilis*	*mayate, mayate verde*	green fig beetle
mate	*Diceroprocta apache*	*chicharra, cigarra*	citrus cicada, Sonoran Desert cicada
mochomo	*Atta mexicana*	*hormiga arriera, mochomo*	leafcutter ant
muumu	**Apis mellifera*	*abeja*	honeybee
nahi sevo'i	*Megachile* sp.?	*abeja cortadora de hojas*	leafcutter bee
nahi sevo'i	Syrphidae	*sírfidos*	hoverfly
polia	Isoptera	*comején, termita*	termite

Equivalents of Yoeme Animal Names			
Yoem noki	**Scientific name**	**Spanish**	**English**
sevo'i	*Musca domestica*	*mosca*	housefly
siki eeye	*Pogonomyrmex* spp.	*hormiga colorada*	harvester ant
sooto'oli	*Dasymutilla* spp.	*hormiga aterciopelada*	velvet ant
teeka sevo'i	*Tabanus* spp.	*tábano*	horsefly
tepu	*Culicoides* spp., *Dasyhelea* spp.	*jején*	biting midge
vaikumareewi	Odonata	*caballito del diablo, libélula*	dragonfly
vaisevo'i, vaisevoli	Lepidoptera	*mariposa, palomilla*	butterfly, large moth
viicha	*Polistes* spp.	*avispa, bitache*	paper wasp
viiko	*Bombus sonorus*	*abejorro*	Sonoran bumblebee
wetepo'i	*Paraleucopis mexicana*	*bobito, bobo*	eye gnat
wo'ochi	Acrididae	*chapulín*	grasshopper
woo'o	Culicidae	*mosquito, zancudo*	mosquito
yuku	*Derobrachus hovorei*	*escarabajo de palo verde*	palo verde beetle
	MOLLUSCA	***moluscos***	**mollusks**
aroseram	*Chionista fluctifraga*	*almeja china*	smooth Venus-clam
auli	Bivalvia	*almeja*	clam
kalamar	Cephalopoda	*calamar*	squid
kooyo	*Saccostrea palmula*	*ostión, ostra*	rock oyster
moa'im, muura wokim	*Anadara tuberculosa*	*pata de mula*	black ark
tatte'era	*Anadara tuberculosa*	*pata de mula*	black ark
teewi	*Chionopsis gnidia*	*venus vistosa*	gnidia Venus-clam
vaa huvahe	*Octopus* spp.	*pulpo*	octopus
vatnaataka kooyom	**Magallana gigas*	*ostión japonés*	Japanese oyster, Pacific oyster
veeko	*Anomia peruviana*	*cascabel peruano, papas fritas*	jingle shell, pearly monia,

Equivalents of Yoeme Animal Names			
Yoem noki	**Scientific name**	**Spanish**	**English**
veeko	**Haliotis* spp.	*abulón*	abalone
wo'im wokim	*Ostrea angelica*	*un ostión*	an oyster
	ECHINODERMATA	***erizos, estrellas de mar***	**echinoderms**
eriso	Echinoidea	*erizo de mar*	sea urchin
vaa choki	---	*estrella de mar*	starfish
	FISH	*peces*	fish
anhilam	---	*anguilas*	eels
avataaka	---	*manta, mantarraya*	manta ray
chiwili	*Bagre pinnimaculatus*	*bagre, bagre barbón*	Long-barbed Sea-catfish, Red Sea-catfish
havataaka	---	*manta, mantarraya*	manta ray
ho'ot wichakame	*Oligoplites* spp., *Trachinotus* spp.	*pómpano, zapatero*	Longjaw Leatherjack, Shortjaw Leatherjack
ho'ot wichakame	*Scorpaena mystes*	*pez escorpión roquero*	Stone Scorpionfish
horohteme	*Cheilotrema saturnum*	*roncacho*	Black Croaker
huhuwo	*Mugil cephalus*	*lisa pardete, lisa rayada*	Flathead Gray Mullet, Striped Mullet
kavayito	*Hippocampus ingens*	*caballito de mar*	Giant Seahorse
lihtoniam	*Trichiurus nitens*	*sable del Pacífico*	Pacific Cutlassfish
liisa	*Mugil cephalus*	*lisa pardete, lisa rayada*	Flathead Gray Mullet, Striped Mullet
omo'i	*Ictalurus pricei*	*bagre yaqui*	Yaqui Catfish
pampanom	*Oligoplites* spp., *Trachinotus* spp.	*pómpano, zapatero*	Longjaw Leatherjack, Shortjaw Leatherjack
puse'ela	Clupeidae	*ojotón*	sardine
siki kuchu	*Lutjanus* spp.	*pargo*	snapper
tahkai kuchu	Achiridae	*lenguados*	flatfishes
tamekame	---	*tiburón*	shark
tosai huhuwo	*Mugil curema*	*lisa blanca*	White Mullet
tosai kuchu	*Atractoscion* spp.,	*corvina*	corvina

Equivalents of Yoeme Animal Names			
Yoem noki	**Scientific name**	**Spanish**	**English**
	Cynoscion spp.		
totoava	*Totoaba macdonaldi*	*totoaba*	Totoaba
vai vakochim	---	*anguilas*	eels
velohko	*Trichiurus nitens*	*sable del Pacífico*	Pacific Cutlassfish
yavarai	*Epinephelus quinquefasciatus*, etc.	*guasa, mero, mero guasa*	Goliath Grouper
	AMPHIBIA	***anfibios***	**amphibians**
hipuyesa'ala	**Ambystoma mavortium*	*ajolotel*	waterdog [tiger salamander larva]
kaureepa (kuareepa)	*Incilius alvarius*	*sapo grande, sapo del desierto sonorense*	Sonoran Desert toad
kaureepa (kuareepa)	**Rana catesbeiana*	*rana toro, rana mugidora*	bullfrog
kaureepa (kuareepa)	*Rhinella horribilis*	*sapo gigante*	western cane toad
poowi	**Ambystoma mavortium*	*salamandra tigre*	tiger salamander
vatat	Anura	*rana*	frog
voovok	Anura	*sapo*	toad
	REPTILIA	***reptiles***	**reptiles**
aakame	*Crotalus* spp.	*cascabel, víbora de cascabel*	rattlesnake
aakame nakapit, ala'amai	*Pituophis catenifer*	*cincuate, víbora sorda*	bullsnake, gophersnake
awa'ala	*Crotalus cerastes*	*víbora cornuda*	sidewinder
chukui moosen	*Dermochelys coriacea*	*siete filos, tortuga laúd*	leatherback sea turtle
chukui vaakot	*Lampropeltis californiae*	*culebra negra anillada*	California kingsnake
chukui vaakot	*Lampropeltis nigrita*	*culebra negra*	western black kingsnake
chupiarim	---	*serpientes grandes*	giant snakes
chuuli wikui	*Callisaurus draconoides*	*cachora arenera, perrita*	zebra-tail lizard
have'eko'i	*Trachemys yaquia*	*tortuga de agua dulce*	Yaqui slider
hipuyesa'ala	**Hemidactylus* spp.	*besucona, salamanquesa*	house gecko

Equivalents of Yoeme Animal Names			
Yoem noki	**Scientific name**	**Spanish**	**English**
kama	*Crocodylus acutus*	*cocodrilo, lagarto*	crocodile
kau mochik	*Gopherus* spp.	*tortuga del desierto, tortuga del monte*	desert tortoise
kukko siari aakame	*Crotalus scutulatus? Crotalus molossus?*	*cascabel*	rattlesnake
kurues	*Boa sigma*	*corúa*	boa constrictor
kuta wikui	*Ctenosaura macrolopha*	*iguana de cola espinosa*	Sonoran spiny-tailed iguana
mahao, mahau	*Trachemys yaquia*	*tortuga grande de río*	Yaqui slider
mochik	*Gopherus* spp.	*tortuga del desierto, tortuga del monte*	desert tortoise
moosen	*Chelonia mydas*	*caguama*	green sea turtle
motcho'okoli	*Phrynosoma solare*	*camaleón, camaleón real*	regal horned lizard
poowi	*Dipsosaurus dorsalis*	*iguana del desierto*	desert iguana
sakkau	*Heloderma suspectum*	*escorpión*	Gila monster
siari aakame	*Crotalus molossus?, Crotalus scutulatus?*	*cascabel*	rattlesnake
sik kucha'a	*Micruroides euryxanthus*	*coralillo*	Sonoran coral snake
sik kucha'a	*Micrurus distans?*	*coralillo*	west Mexican coral snake
sik tavut	*Masticophis flagellum*	*chirrionera*	coachwhip
tosai aakame	*Crotalus* sp.	*cascabel*	rattlesnake
veho'ori	*Sceloporus clarkii*	*cachorra, vejore de Clark*	Clark's spiny lizard
veho'ori	*Sceloporus magister*	*cachorón, vejore del desierto*	desert spiny lizard
veho'ori	*Urosaurus ornatus*	*largartija, roñito ornado*	ornate tree lizard
veho'ori	*Uta stansburiana*	*lagartija de mancha lateral*	side-blotch lizard
wa'ivil	*Kinosternon* spp.	*casquito del agua, tortuga de los charcos*	mud turtle
waitopit	*Coleonyx variegatus*	*salamanquesa de*	western banded gecko

Equivalents of Yoeme Animal Names			
Yoem noki	**Scientific name**	**Spanish**	**English**
		bandas	
watta'akaili	**Hemidactylus* spp.	*besucona, salamanquesa*	house gecko
wikui	*Aspidoscelis* spp.	*huicos*	whiptail lizards
wiroa vakot	*Oxybelis aeneus*	*bejuquilla*	Neotropical vinesnake
	AVES	***aves***	**Birds**
bwassa'aka	Accipitridae	*un halcón*	a hawk
bwawis	*Tyto alba*	*lechuza de campanario*	Barn Owl
bwia mu'u	*Athene cunicularia*	*tecolote llanero*	Burrowing Owl
chana	*Agelaius phoeniceus*	*tordo sargento*	Red-winged Blackbird
chanate	*Quiscalus mexicanus*	*zanate mayo, zanate mexicano*	Great-tailed Grackle
chaoe	*Ara militaris*	*guacamaya verde*	Military Macaw
chapara	*Ortalis wagleri*	*chachalaca, cuiche*	Rufous-bellied Chachalaca
charowe	*Cyanocorax colliei*	*urraca*	Black-throated Magpie-Jay
chatchaakam	*Pelecanus erythrorhynchos*	*pelícano blanco americano*	American White Pelican
chiiwi	*Meleagris gallopavo*	*guajolote*	Turkey (wild)
chilik	---	*un especie de ave*	unidentified bird
choawe	*Caracara cheriway*	*caracara, quelele*	Crested Caracara
cholloi	Piciformes	*carpintero*	woodpecker
coa	*Trogon elegans*	*trogón elegante*	Elegant Trogon
hiak chana	*Quiscalus mexicanus*	*zanate mayo, zanate mexicano*	Great-tailed Grackle
hoopopol	---	*un especie de ave*	unidentified bird
huchachi	Accipitridae	*un especie de halcón*	unidentified hawk
kau roakte'a	*Catherpes mexicanus*	*saltapared barranquero*	Canyon Wren
kau satemai	Cathartidae	---	unidentified vulture
ko'ovo'e	*Meleagris gallopavo*	*guajolote*	Turkey (domestic)
kobwabwa'i	*Ardea herodias*	*garza gris, garza Morena*	Great Blue Heron

Equivalents of Yoeme Animal Names			
Yoem noki	**Scientific name**	**Spanish**	**English**
kooni	*Corvus sinaloae*	*cuervo sinaloense*	Sinaloa Crow
kutapapache'a	*Chordeiles acutipennis*	*chotacabras menor, tapacaminos*	Lesser Nighthawk
kuuku	*Zenaida asiatica*	*paloma aliblanca*	White-winged Dove
maavisa	*Eremophila alpestris*	*alondra cornuda*	Horned Lark
mo'el	Troglodytidae	*chivirín*	wren
muu'u	*Bubo virginianus*	*búho cornudo*	Great Horned Owl
neo'okai	*Mimus polyglottos*	*centzontle norteño, chonte*	Northern Mockingbird
ommo'okoli	*Columbina inca*	*tórtola colilarga*	Inca Dove
ooris	Accipitridae	*un especie de halcón grande*	unidentified large hawk
paavo	**Pavo cristatus*	*pavo royal, pavón*	Peacock
paato	Anatidae	*pato*	duck
paloma	**Columba livia*	*paloma*	Rock Pigeon
poute'ela	*Molothrus ater*	*tordo cabeza café*	Brown-headed Cowbird
puchi'ilaa	---	*un especie de ave*	unidentified bird
sanku'ukuchi	*Corvus corax*	*cuervo común*	Common Raven
semalulukut	Trochilidae	*chuparosa*	hummingbird
sikili suva'i	*Callipepla douglasii*	*cordoniz cresta dorada*	Elegant Quail
sulumai	*Pandion haliaetus*	*águila pescadora*	Osprey
suva'i, suva'u	*Callipepla gambelii*	*cordoniz chiquiri*	Gambel's Quail
taawe	Accipitridae	*gavilán*	hawk
takochae	*Icterus cucullatus*	*bolsero cuculado, calandria encapuchada*	Hooded Oriole
taruk	*Geococcyx californianus*	*churea, correcaminos*	Greater Roadrunner
tavelo	*Amazona albifrons*	*loro frente blanca*	White-fronted Parrot
tawe huchachi	Accipitridae	*un especie de halcón*	unidentified hawk
tawe ove'a	Accipitridae	*un especie de halcón*	unidentified hawk
tekoe	*Coragyps atratus*	*zopilote común*	Black Vulture

Equivalents of Yoeme Animal Names			
Yoem noki	**Scientific name**	**Spanish**	**English**
tenwe	*Pelecanus occidentalis*	*alcatraz, pelícano café*	Brown Pelican
tesaki suva'i	*Colinus virginianus ridgwayi*	*cordorniz mascarita*	Masked Bobwhite
totoi	**Gallus gallus*	*pollo*	chicken
tuivit	*Charadrius vociferous*	*chorlo tildío*	Killdeer
vaa wo'i	*Calidris* spp.	*playero*	Sandpiper
vaaro	*Amazona albifrons*	*cotorra frente blanca, loro frente blanca*	White-fronted Parrot
vaatosai	Laridae	*gaviota*	seagull
vae vicho'ola	*Sturnella neglecta*	*pradero occidental*	Western Meadowlark
vae waakas	*Campylorhynchus brunneicapillus*	*matraca desértica*	Cactus Wren
vakoni	Anatidae	*un especie de pato*	unidentified duck
vaso mo'el	*Haemorhous mexicanus*	*fringílido mexicano*	House Finch
veta yeka	Anatidae	*pato*	duck
wichalakas	*Cardinalis cardinalis*	*cardenal norteño*	Northern Cardinal
wichik	*Micrathene whitneyi*	*tecolotito enano*	Elf Owl
wiiru	*Cathartes aura*	*aura, zopilote cabeza roja*	Turkey Vulture
wiivis	*Phainopepla nitens*	*capulinero negro*	Phainopepla
wiivis	*Toxostoma* spp.	*cuitlacoches*	thrashers
wiirum	Cathartidae	*zopilotes*	vultures
wokkoi	*Zenaida macroura*	*paloma huilota*	Mourning Dove
wokovavase'ela	Hirundinidae	*golondrina*	swallow
	MAMMALIA	***mamíferos***	**mammals**
bwahi lovon, bwai lovon	*Lynx rufus*	*gato montés*	bobcat
bwala	**Ovis aries*	*borrego, oveja*	sheep
chikul	*Chaetodipus, Perognathus, Peromyscus*	*ratón*	New World mice
chikul	**Mus musculus*	*ratón*	house mouse

Equivalents of Yoeme Animal Names			
Yoem noki	**Scientific name**	**Spanish**	**English**
choparao	*Procyon lotor*	*batepi, mapache*	raccoon
chuu'u	*Canis familiaris*	*perra, perro*	dog
hooso	*Ursus americanus*	*oso negro americano*	black bear
hua koowi	*Tayassu tajacu*	*cochi jabalí, pecarí de collar*	collared peccary, javelina
huhteme	Cetacea	*ballena*	whale
hupa	Mephitidae	*zorrillo*	skunk
huuri	*Taxidea taxus*	*tejón*	badger
kameeyo	**Camelus dromedarius*	*camello*	camel
kava'i	**Equus caballus*	*caballo*	horse
kawis	*Urocyon cinereoargenteus*	*zorra gris*	gray fox
kochiito	*Tursiops truncatus*	*cochi, delfin, tonina*	bottlenose porpoise
koowi	**Sus scrofa domesticus*	*cochi*	pig
loovo	*Canis lupus baileyi*	*lobo, lobo gris mexicano*	Mexican gray wolf
maaso	*Odocoileus virginianus*	*venado cola blanca*	white-tailed deer
malon	Sciuridae	*Juancito, juansito*	ground squirrel
mata'e	*Vulpes macrotis*	*zorra del desierto*	kit fox
miisi	**Felis catus*	*gato*	cat
muura	**Equus asinus× Equus caballus*	*mula*	mule
ousei	*Puma concolor*	*león*	mountain lion, puma
ove'eso	*Ovis canadensis nelsoni*	*borrego cimarrón*	desert bighorn sheep (ram)
paaros	*Lepus alleni*	*liebre antílope*	antelope jackrabbit
paaros	*Lepus californicus*	*liebre cola negra*	black-tailed jackrabbit
pocho'oku koowi	*Tayassu tajacu*	*pecarí, pecarí de collar*	collared peccary, javelina
sevis maaso	*Odocoileus hemionus eremicus*	*bura, venado bura*	desert mule deer
sochik	Chiroptera	*murciélago*	bat
taavu	*Sylvilagus audubonii*	*conejo del desierto*	desert cottontail

Equivalents of Yoeme Animal Names			
Yoem noki	**Scientific name**	**Spanish**	**English**
teku	*Ammospermophilus harrisii*	*ardilla antilope*	Harris's antelope ground squirrel
teku	*Otospermophilus variegatus*	*ardilla, ardillón*	rock squirrel
tete'eso	*Ovis canadensis nelsoni*	*borrego cimarrón*	desert bighorn sheep (ewe)
tevos	*Thomomys bottae*	*topo, tuza*	Botta's pocket gopher
topol	*Leopardus pardalis*	*gato galavi, ocelote, tigrillo*	ocelot
tori	*Neotoma* spp.	*rata magueyera*	packrat, woodrat
tori	*Neotoma albigula*	*rata de campo, rata nopalera*	white-throated packrat
tori	*Neotoma phenax*	*rata magueyera, rata de campo*	Sonoran woodrat
tori	**Rattus* spp.	*rata*	Old World rats
vaachuu'u	*Lontra longicaudis annectens*	*nutria de río, nutria neotropical, perro de agua*	Neotropical river otter
va'apo loovo	*Zalophus californianus*	*lobo marino*	sea lion
vacha'i	*Dipodomys* spp.	*rata canguro*	kangaroo rat
vuuru	**Equus asinus*	*burro*	donkey
wakasra	**Bos taurus*	*ganado*	cattle
wichakame	*Erethizon dorsatum*	*puercoespín*	porcupine
wo'i	*Canis latrans*	*coyote*	coyote
yo'oko, yo'oko ousei	*Panthera onca*	*jaguar, tigre*	jaguar

PART 7
INDEX TO PLANT AND ANIMAL NAMES

Names for taxonomic orders and families, as well as plant family common names, are all uppercase. Accepted scientifc names for genera, species, infraspecific taxa (subspecies or varieties, and hybrids) are in bold font. Latin taxonomic synonyms (scientific names) and Spanish common names are in italics. Yoeme names, authors of scientific names for plants, as well as English common names, are in regular font.

www.ingramcontent.com/pod-product-compliance
Lightning Source LLC
Chambersburg PA
CBHW041730100726
47973CB00010B/167
9798218371111